Modern Approaches in Solid Earth Sciences

Volume 17

Series Editors

Yildirim Dilek, Department of Geology and Environmental Earth Sciences,
Miami University, Oxford, OH, U.S.A.
Franco Pirajno, The University of Western Australia, Perth, Australia
Brian Windley, Department of Geology, The University of Leicester, UK

More information about this series at http://www.springer.com/series/7377

Bronislav Gongalsky • Nadezhda Krivolutskaya

World-Class Mineral Deposits of Northeastern Transbaikalia, Siberia, Russia

Bronislav Gongalsky
Institute for Geology of Ore Deposits
Petrography, Mineralogy and Geochemistry
Russian Academy of Sciences
Moscow, Russia

Nadezhda Krivolutskaya
Vernadsky Institute of Geochemistry and
Analytical Chemistry
Russian Academy of Sciences
Moscow, Russia

Responsible Series Editor: F. Pirajno

Copy editor: Alexander Yakubchuk

The Work was first published in Russian in 2015 with the following title: B. Gongalsky "Deposits of the Unique Metallogenic Province of Northern Transbaikalia" Moscow, VIMS, 248 p.
ISBN 978-5-9906776-4-7

English translation by Eugeny Kordukov and Nadezhda Krivolutskaya

ISSN 1876-1682   ISSN 1876-1690   (electronic)
Modern Approaches in Solid Earth Sciences
ISBN 978-3-030-03558-7   ISBN 978-3-030-03559-4   (eBook)
https://doi.org/10.1007/978-3-030-03559-4

Library of Congress Control Number: 2018960387

This Springer imprint is published by the registered company Springer Nature Switzerland AG.
The registered company address is: Gewerbestrasse 11, 6330 Cham, Switzerland

Foreword

This monograph is a new detailed study of large and unique ferrous, precious, base metal and rare metal deposits in Russia. It addresses complex interrelated problems of geological setting, magmatism, and ore genesis of Northeastern Transbaikalia, one of the Earth's largest Paleoproterozoic metallogenic provinces. This area comprises three super-large and a number of ordinary deposits and occurrences. First of all, this province hosts 50 Mt of copper in sedimentary and magmatic rocks, including the giant Udokan deposit. Russia's largest reserves of vanadium are concentrated within the Chiney gabbro-anorthosite pluton. And finally, the Katugin deposit, with unique reserves of niobium and zirconium, is also located here.

This work is based on a long study conducted by the authors on geochemistry and mineralogy of rocks and ores from many deposits. The monograph also includes results of predecessors obtained during mapping and prospecting of the Kodar-Udokan area. It shows that the Udokan-Chiney mineral district is a unique metallogenic taxon, combining large and giant deposits of sedimentary, hydrothermal, and magmatic origin in a complex and multistage oremagmatic system associated with the Proterozoic mantle plume. The authors discovered new types of ore in this area, such as hydrothermal sulfide mineralization with platinum group metals and uranium, as well as rare elements in the hydrothermally altered rocks.

However, despite the economic importance and interesting geology of this area, there are no many publications on this unique region. That is why this work may be of great interest to regional and economic geologists, as well as to the students studying geology in the universities across the world.

Academician, Russian Academy of Sciences   Nikolay Laverov (deceased)
Moscow, Russia
September 25, 2016

Comments of the English Text Editor

This monograph by Bronislav Gongalsky and Nadezhda Krivolutskaya is based on their life-long academic studies of mineral deposits, geographically located in Northeastern Transbaikalia, Siberia, Russia. Tectonically, they occur in the west of the Aldan Shield in the Siberian Craton. Its Kodar-Udokan sedimentary basin is internationally most famous to host the Udokan deposit and its satellites, comprising one of the Earth's oldest and largest Paleoproterozoic sediment-hosted copper accumulations. In addition, the basin includes a Katugin Nb-Zr deposit and the Chiney layered mafic-ultramafic pluton, Russia's largest and best explored pluton of this kind with Fe-Ti-V-(Cu-Ni-PGE) mineralization. The understanding of such a metallogenically diverse system is definitely a challenging task. The Chiney pluton was the main focus of the studies by the authors, but they spent a great deal of time to research the metallogenic relationships between all deposit types.

The authors often use traditional Russian terminology, which was directly translated into English in order not to rewrite the entire text. This, for instance, can be seen in such tectonic terms as "geoblock", "trough" and others, which can be found throughout the text. I tried to adapt such terminology, where possible, to international standards, but the Russian flavour is obviously preserved.

Perhaps, most importantly and traditionally for Russian academic studies, the large metal accumulations are described as deposits even if they are not in production. The reader should consider these 'deposits' as a 'mineral endowment' and/or a 'resource'. The latter is used in most cases in accordance with the Russian resource classification, which is not fully compatible with various international standards, such as Australian JORC or Canadian NI 43-101. Despite of this, the metal accumulations are indeed unique and deserve their place on the international geological stage.

The monograph provides useful insights into the diverse geology and metallogeny of the Siberian Craton and I hope that the reader will satisfy his or her curiosity in understanding and knowledge on this remarkable and beautiful region.

Alexander Yakubchuk

30 July 2018

Contents

Abstract

Northeastern Transbaikalia is a region located at the east of Lake Baikal in eastern Siberia. It is one of the largest metallogenic provinces in Russia; it hosts three world-class deposits of the Udokan (26.7 Mt of Cu resource), Chiney (Fe-Ti-V, 30 Gt of ore), and Katugin (2.7 Mt, Nb, and also Ta).

In Russia, there are many mineral provinces with unique concentration of metals in platinoid-copper-nickel deposits in Noril'sk, iron ore deposits in Kursk, apatite and rare earth deposits in the Kola Peninsula, and gold-silver deposits in the Far East. Among them, the Transbaikalian region occupies a special place, whose mineral diversity and scale of mineralization strike the imagination. The mineral resources of Transbaikalia have been developed since Peter the Great times. It is one of the oldest mining regions in Russia, with first gold extracted from silver ores of the Klichka deposit. In 1724, a first large gold medal was cast for Peter the Great from this gold. Since then, many lead-zinc, molybdenum, and gold-silver deposits were discovered here, and Smirnov (1944) outlined main metallogenic zones. Later, uranium, zeolite, and other deposits were discovered in eastern Transbaikalia.

Deposits of Northeastern Transbaikalia were discovered as a result of systematic geological work organized by the USSR Ministry of Geology, especially after the World War II. The Baikal-Amur-Mainline was built in the 1970–1980s exploiting the discovered deposits of iron ore (Sulumat BIF deposit), coal (Apsat deposit), synnyrite (aluminum and potassium raw material in the synnyrite layered mafic-ultramafic complex), sandstone-hosted copper, gold, titano-magnetite, and platinum-copper-nickel. In our work, we consider only three unique deposits (Udokan, Chiney, and Katugin). In addition, we will briefly describe smaller deposits and occurrences.

1.1 General Characteristics of the Area

Northeastern Transbaikalia is a region located at the northeast of Lake Baikal in eastern Siberia (Figs. 1.1 and 1.2). Tectonically, it belongs to the western part of the Aldan Shield in the southeast of the Siberian Craton. Its Kodar-Udokan Basin comprises three world-class deposits (Zadorozhnyi and Bybin 2008, 2012): (i) Udokan, containing more than 26.7 Mt of Cu; (ii) Chiney pluton, containing Fe-Ti-V in 30 Gt of ore; and (iii) Katugin, with 2.7 Mt of Nb and 9 Mt of Zr in ore. In addition, many PGE, Ag, Au, and U deposits and occurrences were discovered in this province. All of them are Paleoproterozoic in age, the most productive time for the concentration of various metals.

1.2 History of Study of Northeastern Transbaikalia

Northeastern Transbaikalia is a remote area. It comprises three major ridges of Kodar and Udokan-Kalar divided by the Chara Basin (Fig. 1.3). The latter is similar in origin to the Baikal Rift, forming its northeastern extension. This area has diverse and very unusual landscapes, such as the world's northernmost sand desert in the Chara Basin (Fig. 1.4), in neighborhood to the glaciers in the Kodar Ridge at the 3000 m elevation.

Geological information on this area was accumulated very gradually, mostly during the twentieth century. Three stages in this work can be recognized.

1. *Regional traversing.* In 1928–1935, Pavlovsky, Efremov, and Arsen'ev completed first regional traverses (Pavlovsky 1933 and unpublished). In the 1930s, a special geological prospecting was carried out in this area for the construction of the Baikal-Amur-Mainline. Its construction started in the early 1940s, but the project was abandoned after the beginning of World War II and

B. Gongalsky, N. Krivolutskaya, *World-Class Mineral Deposits of Northeastern Transbaikalia, Siberia, Russia*, Modern Approaches in Solid Earth Sciences 17, https://doi.org/10.1007/978-3-030-03559-4_1

Fig. 1.1 Position of the Transbaikal area (rectangle) on map of Russia. (Wall Map of Russia. Atlases of Russia, Atlas Print, 2017)

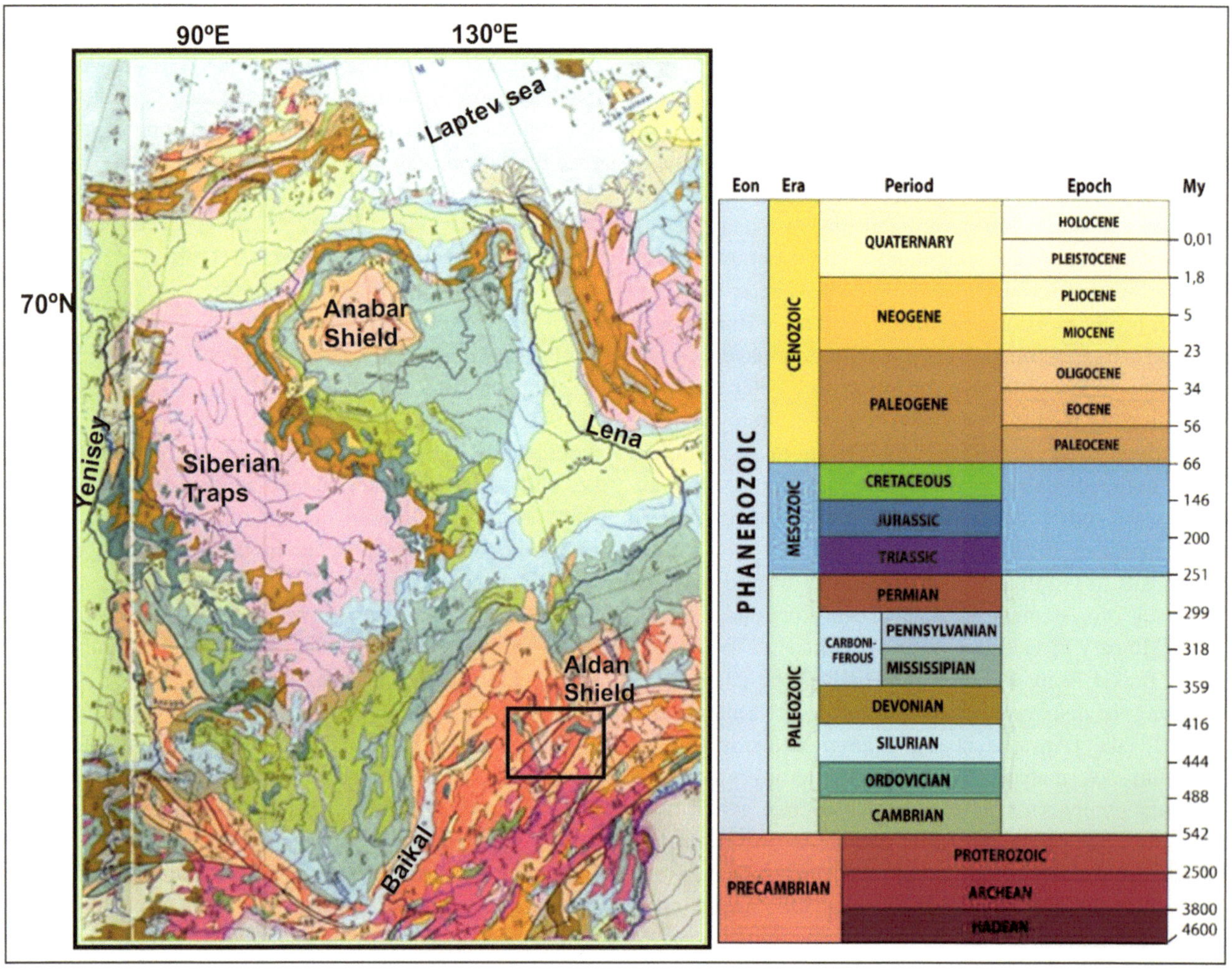

Fig. 1.2 Simplified geological map of eastern Siberia (http://www.myshared.ru/slide/107715/)
Rectangle shows area in Fig. 1.3

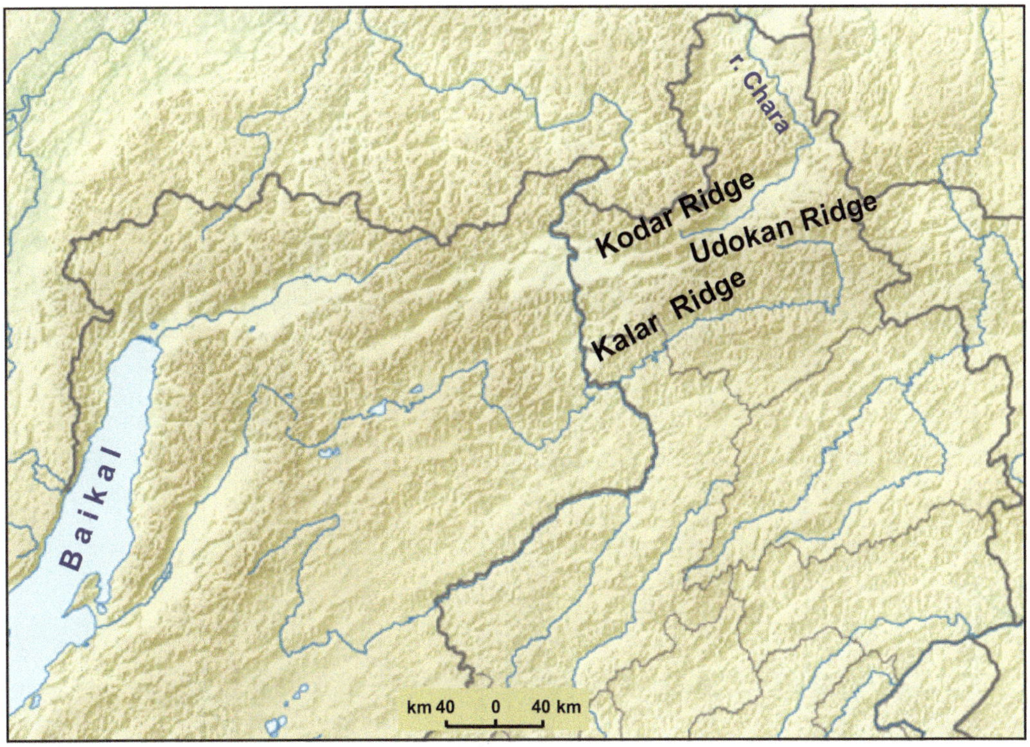

Fig. 1.3 Geographic map of the Kodar-Udokan region (www.google.ru/maps)

was completed only in 1984. In 1938, Mikhail Petrusevich (Fig. 1.5) and Lyudmila Kazik discovered rich titano-magnetite mineralization in the Chiney layered mafic-ultramafic Pluton (Fig. 1.6). Its description was published after the World War II (Petrusevich 1946).

2. *Systematic mapping.* The new phase of geological surveying in Northeastern Transbaikalia began after 1945. In 1949, Elizaveta Burova (Fig. 1.5) and her colleagues from Lesnaya Expedition discovered the sandstone-hosted Udokan copper deposit (Fig. 1.7), during the geological mapping of this area at a scale of 1:200,000. Most intense exploration works were conducted in the 1960s, when the USSR economy recovered after the war. During this period, the entire country, including Northeastern Transbaikalia, was geologically mapped at a scale of 1:1,000,000 and 1:200,000 by various geological organizations (All-Union Geophysical Trust (VAGT), All-Union Geological Institute (VSEGEI)). These maps laid basis for more detailed works at a scale of 1:50,000, 1:25,000, and 1:10,000. In 1954, the Chita Regional Geological Department established the Udokan Expedition for a systematic study of the geology and mineral deposits in this area. Many geologists (EF Grintal, MI Korol'kov, YA Yakimov, VS Chechetkin, KS Kazanov, VK Golev, NG Goleva, LV Sosnovskikh, Y Sosnovskikh, VG Podgorbunskiy, MN Davie, Y Gudyma) contributed to the understanding of local geology. Since that time, many

geoscientists from different organizations (Moscow State Prospecting Institute (MGRI), All-Union Institute for Raw Materials (VIMS), Institute for Geology of Ore Deposits (IGEM), Transbaikal Research Institute (ZabNII)) studied geology, geochemistry, and mineralogy of the discovered deposits. Bakun et al. (1958), Krendelev (1959), and Bakun et al. (1958, 1964, 1966) published first data on the Udokan deposit, based on their work during several years and compilation of the first detailed geological map at a scale of 1:10,000.

In 1963, NB Yusupov discovered the third unique deposit in this area (Katugin deposit) based on rare metal data by VV Arkhangel'skaya.

Salop (1964, 1967) compiled all data and published a fundamental description of this area in two volumes entitled "Geology of the Baikal Mountainous Area." The first complete description of rocks from the Chiney layered mafic-ultramafic complex was carried out by Lebedev (1962) after his 1960 field trip with Oleg Bogatikov to the Kalar and Udokan Ridges.

3. *Detailed study of deposits.* This stage began in 1975, when the USSR government restarted the construction of the Baikal-Amur-Mainline (BAM) and adopted a resolution on exploration and potential exploitation of all mineral deposits within the 100 km corridor near BAM. The

Fig. 1.4 Sand desert in the Chara Trough with Kodar Ridge in the background
Photos: (**a**) Photo A.Savchenko, (**b**) Photo B. Gongalsky

Fig. 1.5 Discoverer of the Chiney Pluto Mikhail Nikolaevich Petrusevich (1908–2003). (Photo B. Gongalsky, 1998) and discoverer of the Udokan deposit Elizaveta Ivanovna Burova (1913–1996). (Photo taken in 1992, from the Burova family archive, copy was made by B. Gongalsky)

USSR Ministry of Geology estimated resources of the titanomagnetite and sulfide ores in the Chiney deposits, rare metal ores in the Katugin deposit, and copper in the Udokan copper deposit. The results were approved by the USSR State Commission for Reserves (GKZ). Geologists from many organizations (VIMS, IGEM, MGRI, CHIPR) worked here during 40 years, studying chemical and mineralogical composition of rocks and ores. Gablina (1983, 1997) studied sulfides from the Udokan deposit. The ZabNII geologists (LF Narkelyun, GAYurgenson, AI Trubachev, VS Salikhov, NA Krivolutskaya) studied Udokan oxide ores (Narkelyun et al. 1987). The Katugin deposit was investigated by Bykov, Lysikov, and many others. Arkhangel'skaya et al. (2004) summarized the geological and mineral data on the Katugin and Udokan deposits.

The Chiney pluton attracted many researches due to the well-developed magmatic layering of the rocks, huge vanadium resources, and sulfide mineralization with noble metals (Lebedev 1962; Kulikov et al. 1980; Gongalsky and Krivolutskaya 1993; Tatarinov et al. 1998; Tolstykh et al. 2008). Konnikov (1986) emphasized a great importance to assimilation of carbonate rocks into mafic magma and its role in ore formation. Golev (unpublished) attributed a complicated structure of the pluton to the heterogeneity of the primary magma.

Today, the licenses to exploit the three deposits belong to the Baikal Mining Company (Udokan), Soyuzmetalresurs (Chiney), and Acropolis Group of Companies (Katugin).

1.3 Methods

The formation parameters of the mineralization remain poorly constrained, and further progress in this respect may be helpful for the discovery of new deposits in the Kodar-Udokan ore-magmatic system, characterized by significant vertical extent and complex relationships. The mineralization, deposited at different depths, is now exposed at the surface and is, therefore, accessible for comprehensive study. In addition to magmatic Fe-Ti-V oxide mineralization in the Chiney and Luktur layered mafic-ultramafic plutons, the post-titanomagnetite copper and noble-metal sulfide mineralization (Rudnoe, Verkhne-Chineyskoe, Skvoznoe, Kontaktovyi, Magnitnyi, and Etyrko deposits) are localized in the endo- and exocontact zones of the plutons. The giant Udokan sandstone-hosted copper deposit is distant from intrusive bodies and hosts sulfide mineralization, also found in the Pravoingamakitskoe, Saku, Unkur, and other deposits. New types of mineralization, e.g., Au-PGE-Cu and REE-U, have been revealed recently in layered plutons and their host rocks (Gongalsky et al. 2009; Makariev et al. 2009, 2010). These manifestations of magmatic activity and ore deposi-

Fig. 1.6 General view of the Chiney pluton (**a**, **b**)

Fig. 1.7 A panoramic view of the Udokan deposit from the Chiney pluton (looking north)
Light gray band is a railway spur leading to the Baikal-Amur-Mainline

tion make up a Kodar-Udokan ore-magmatic system. In this publication, we focus on the geology of ore deposits as parts of this system, on new mineralogical and geochemical data, characterizing the individual deposits, and on their genetic models. A special attention is paid to (1) ore-controlling structures, (2) geochemistry of mafic-ultramafic plutons and their typification, (3) a thorough study of the Chiney Pluton and related mineralization as a reference magmatic deposit, and (4) mineralogy and geochemistry of the Udokan and its satellite deposits.

This monograph is based on our 30-year-long study of geology, petrography, mineralogy, and geochemistry of Ni-Cu-PGE and copper deposits in the Kodar-Udokan mineral district. They were carried out at the Chita Institute of Natural Resources, Siberian Branch of Russian Academy of Sciences (CINR SB RAS) in 1982–1994 and at the Institute for Geology of Ore Deposits, Petrography, Mineralogy, and Geochemistry (IGEM RAS) since 1995 through 2018.

In the course of field works, the sequences of intrusive rocks have been studied in details in large natural outcrops (5.5 km in extent) and deep (down to 1.5 km) boreholes. More than 50 boreholes have been documented and sampled. We collected 10,000 samples and assay duplicates of gabbro rocks from the Chiney Intrusive Complex and carbonate-clastic rocks of the Udokan Supergroup, both from outcrops and drill holes in the Chiney, Udokan, and Pravoingamakitskoe deposits.

The following analytical methods have been used: (1) XRF at CINR SB RAS (analyst NS Baluev) and IGEM RAS (analyst AI Yakushev); (2) ICP-MS at the Institute for Mineralogy, Geochemistry, and Crystal Chemistry of Rare Elements (IMGRE; analyst DZ Zhuravlev) and LA-ICP-MS at the Max Planck Institute for Chemistry, Mainz, Germany (analyst DV Kuzmin); (3) Cameca SX 50 and SX 100 electron microprobe at the Institute of Geochemistry and Analytical Chemistry (GEOKHI RAS; analyst NN

Kononkova) and Lomonosov Moscow State University (analyst NE Sergeeva) and Superprobe Geol JXA 8200 at the Max Planck Institute for Chemistry, Mainz, Germany (analyst DV Kuzmin) and IGEM RAS (analyst EV Kovalchuk); (4) electron microscopy at the Institute of Experimental Mineralogy (IEM RAS; analyst AN Nekrasov), Paleontological Institute, Russian Academy of Sciences (analyst EA Zhegallo), IGEM RAS (analysts LO Magazina and NV Trubkin), and Lomonosov Moscow State University (analyst NN Korotaeva); (5) study of stable isotopes (O, S) at the Geological Institute (GIN RAS; analyst BG Pokrovsky) and Central Institute for Geological Exploration of Base and Precious Metals (analyst SG Kryazhev); (6) study of radiogenic isotopes in rocks: Sm-Nd, IGEM RAS (analyst YuV Goltsman) and U-Pb, A.P. Karpinsky Russian Geological Research Institute (VSEGEI)(analyst AN Timashkov); and (7) determination of PGE and Au in rocks and ore, IGEM RAS (analysts VA Sychkova and VG Belousov) and Institute of Ore Formation, Mineralogy, and Geochemistry, National Academy of Sciences, Ukraine (analyst AA Yushin). Crystallization parameters of magma, eventually solidified into the Chiney pluton, were estimated using PETROTYPE and COMAGMAT-3.5 programs with the assistance of AA Ariskin and GS Nikolaev. All photos were taken by Bronislav Gongalsky, unless specifically indicated.

Acknowledgments We are very grateful to VS Chechetkin, KS Kazanov, LV Sosnovskikh, and Yu Sosnovskikh, NG Goleva, VG Podgorbunsky, MN Devi, and MF Dzyubenko, the geologists from Chitageologiya Expedition, for their assistance during the field trips. We thank analysts NS Baluev, LN Skornyakov, AV Sobolev, DV Kuzmin, OB Kuzmina, VA Sychkova, NE Sergeeva, and EV Kovalchuk for their help. AA Ariskin and GS Nikolaev helped in modeling the crystallization of the Chiney magma. Yuri Safonov, Alexander Volkov, Oleg Bogatikov, and Konstantin Lobanov supported this work for a long time. We are very grateful to Elinor Morrisby for her hard work with English correction. In some cases, Victor Popov helped with English translation, which was then corrected by American Journal

Experts Company (Part I, Chaps. 3 and 4). We are grateful to TB Shlychkova and IV Karlina for the tremendous work with graphic materials.

This work was supported by the Russian Foundation for Basic Research (projects NN 00-05-64507, 07-05-01007, 10-05-10088, 15-05-09250, 15-05-07031 and partially 17-05-01167, 18-05-70094).

References

Arkhangel'skaya VV, Bykhov Yu, Volodin RN, Narkelyun LF, Skursky VC, Trubachev AI, Chechetkin VS (2004) Udokan copper and Katugin rare metal deposits in the Chita region, Russia. Chita, Administration of Chita, 519 p (in Russian)

Bakun NN, Volodin RN, Krendelev FP (1958) The main features of the geological structure of the Udokan deposit of cuprous sandstones and the direction of its further exploration. Geologiya I Razvedka 11:67–83 (in Russian)

Bakun NN, Volodin RN, Krendelev FP (1964) On genesis of the Udokan deposit of copper sandstones (Chita district). Lithology Mineral Dep 3:80–103 (in Russian)

Bakun NN, Volodin RN, Krendelev FP (1966) On genesis of the Udokan cupriferous sandstone deposit (Chita Oblast). Int Geol Rev 8(4):455–466

Gablina IF (1983) Copper accumulation conditions in continental red beds. Moscow, Nauka, p 112 (in Russian)

Gablina IF (1997) Formation of large cupriferous sandstone and shale deposits. Geol Ore Deposits 39(4):320–334

Gongalsky BI, Krivolutskaya NA (1993) The Chiney layered pluton. Novosibirsk, Nauka, p 184 (in Russian)

Gongalsky BI, Makariev LB, Voyakovsky SK (2009) Mesozoic and Cenozoic magmatism of the Udokan–Chiney district and uranium mineralization. In: Gordeev EI (ed) Volcanism and Geodynamics. Institute of Volcanology, Petropavlovsk-Kamchatsky, pp 321–323 (in Russian)

Konnikov EG (1986) Precambrian differentiated mafic–ultramafic complexes in the Transbaikalian Region. Novosibirsk, Nauka, 127 (in Russian)

Krendelev FP (1959) About mineralization of cupriferous sandstones in the Udokan deposit and method of their prospecting. Geologiya i Razvedka 2:107–119 (in Russian)

Kulikov AI, Kryukov VK, Morozova NN, Grechishnikov DN (1980) Ore types of the Chiney titanomagnetite deposits and their compositions. Geol Ore Deposits 22(5):85–88 (in Russian)

Lebedev AP (1962) The Chineygabbro-anorthosite pluton, Eastern Siberia. USSR Academy of Sciences, Moscow, p 100 (in Russian)

Makariev LB, Mironov YB, Voyakovsky SK (2010) The outlook for the discovery of new types of economic uranium deposits in the Kodar–Udokan zone of the Transbaikal territory in Russia. Geol Ore Deposits 52(5):381–391

Makariev LB, Voyakovsky SK, Il'kevich IV (2009) Gold potential of uranium objects in the Kodar–Udokan Trough. Rudy I Metally 6:56–64 (in Russian)

Narkelyun LF, Trubachev AI, Salikhov VS, Kunitsyn VV, Zinov'evYu I, Krivolutskaya NA, Chechetkin VS (1987) Oxidized ores of Udokan. Novosibirsk, Nauka, p 102 (in Russian)

Pavlovskiy EV (1933) Geological sketch of the Upper Chara district (Olekma-Vitim mountainous country). In: Proc Geol Explor Assoc, vol 271, pp 27–35 (in Russian)

Petrusevich MN (1946) The Chiney titanomagnetite deposit. Soviet Geol 10:91–94 (in Russian)

Salop LI (1964) Geology of the Baikal mountainous region. Volume I. Moscow, Nedra, p 515 (in Russian)

Salop LI (1967) Geology of the Baikal Highland: magmatism, tectonics and geological history. Volume 2. Moscow, Nedra, p 699 (in Russian)

Smirnov SS (1944) Metallogeny of Eastern Transbaikalia. Moscow, Gosgeolizdat, p 91 (in Russian)

Tatarinov AV, Yalovik LI, Chechetkin VS (1998) A dynamometamorphic model of the formation of basic layered plutons: a case of the Chiney Pluton in Northern Transbaikalia. Novosibirsk, Nauka, p 120 (in Russian)

Tolstykh ND, Orsoev DA, Krivenko AP, Izokh AE (2008) Noble-metal mineralization in mafic–ultramafic layered plutons in the southern Siberian platform. Novosibirsk, Papallel, p 194 (in Russian)

Zadorozhny VF, Bybin FF (2008) New mining areas of Transbaikalia. Gorny Zhurnal 2:30–34 (in Russian)

Zadorozhny VF, Bybin FF (2012) Udokan deposit in strategy of the Northern Transbaikalia exploration. Geol Miner Resour Siberia 3(11):90–94 (in Russian)

Abstract

The Kodar-Udokan mineral district is spatially constrained to a narrow spur of the western margin of the Aldan Shield in the southeast of the Siberian Craton, sandwiched between the Phanerozoic Baikal-Muya and Mongol-Okhotsk mobile belts. It comprises many deposits of Cu, V, Fe, Ag, Ti, V, Ta-Nb, Zr, and REE-Y of different genetic types in sedimentary and magmatic rocks. The sandstones of the Udokan Supergroup, the gabbro rocks of the Chiney Intrusive Complex, and the alteration assemblages of the Katugin granite are hosts to the main deposits. New geological, mineralogical, and geochemical data on the rocks and mineralization are summarized here.

2.1 General Overview

The Kodar-Udokan mineral district is spatially constrained to a narrow spur of the western Aldan Shield in the southeast of the Siberian Craton, located between the Baikal-Muya and Mongol-Okhotsk mobile belts (Rytsk et al. 2011). According to the current concepts, the Siberian craton was formed in the late Paleoproterozoic (2.0–1.8 Ga) when it was incorporated into the supercontinent Columbia (Khain 2001; Rosen, 2003; Rogers and Santosh 2004; Ernst 2014; Ernst et al. 2016). In its Laurentian portion, there are Paleoproterozoic Ni/Cu deposits of the Superior (including Sudbury), Thompson, and Cape Smith (1.85 Ga) provinces; in Australia, these are deposits of oxide and sulfide ore in the Halls Creek and Pipe Creek districts (1.87–1.81 Ga) (Fraser et al. 2007). The compositionally similar Cu deposits in Africa and China are related to intracontinental rifts that were formed late during amalgamation of the supercontinent (Zhao et al. 2002).

The Paleoproterozoic era worldwide was highly productive for the Fe-Ti-V, Cr, Ni-Cu-PGE, and Au-U mineralization (Irvine 1977; Gruenewald 1977; Dixon 1979; Harn and Gruenewaldt 1995; Vollbrecht et al. 2002; Ames et al. 2008; Cawthorn 2010; Charlier et al. 2015; Lightfoot 2017). The mineralization in Northeastern Transbaikalia province, with large Cu-Ag, Ti-V-(Fe) and Ta-Nb-Zr-REE-Y resources, is also Paleoproterozoic in age (Krendelev et al. 1983; Konnikov 1986; Gongalsky and Krivolutskaya 1993; Gongalsky et al. 1995a, 2009; Arkhangel'skaya 1998, 2014; Chechetkin and Kharitonov 2002; Arkhangel'skaya et al. 2010, 2012). The copper deposits of the Kodar-Udokan mineral district are unique in total endowment (>50 Mt. Cu). The sedimentary rock-hosted Udokan copper deposit is the largest in Russia and also one of the world's largest deposits (Krendelev et al. 1983; Chechetkin et al. 2000; Hedenquist et al. 2012). It is located 30 km away from the Novaya Chara railway station of the Baikal-Amur Mainline (BAM) in the Kalar Administrative District of Zabaikalsky (Transbaikalia) Krai (http://www.bgk-udokan.ru/en/project/deposit/).

The Chiney (Chineysky) pluton is part of the transregional Yenisei-Aldan metallogenic belt, extending along the periphery of the Siberian Platform to the Taymyr Peninsula. Although very different in age, Dodin (2002) speculated that the Ni-Cu-PGE deposits of both the Noril'sk and the Kodar-Udokan mineral districts (Permian-Triassic and Paleoproterozoic, respectively) are interpreted to occur in this belt.

The main metal endowment of the area is contained in the sandstone-hosted chalcopyrite-pyrite and chalcocite-bornite ores of the Udokan deposit and its satellites, as well as in post-magmatic deposits at the contacts of the Chiney pluton. The new types of hydrothermal Cu-Au-PGE (Gongalsky et al. 2007) and U-REE (Makariev et al. 2009, 2010) deposits, hosted in igneous and sedimentary rocks, significantly expand the perspectives for the economic development of this mineral district.

The Aldan Shield (1200 × 270–350 km) is the largest basement inlier of the Siberian craton (Figs. 1.1 and 2.1). In the north, the Archean to Proterozoic rocks plunge beneath

© Springer Nature Switzerland AG 2019

B. Gongalsky, N. Krivolutskaya, *World-Class Mineral Deposits of Northeastern Transbaikalia, Siberia, Russia*,
Modern Approaches in Solid Earth Sciences 17, https://doi.org/10.1007/978-3-030-03559-4_2

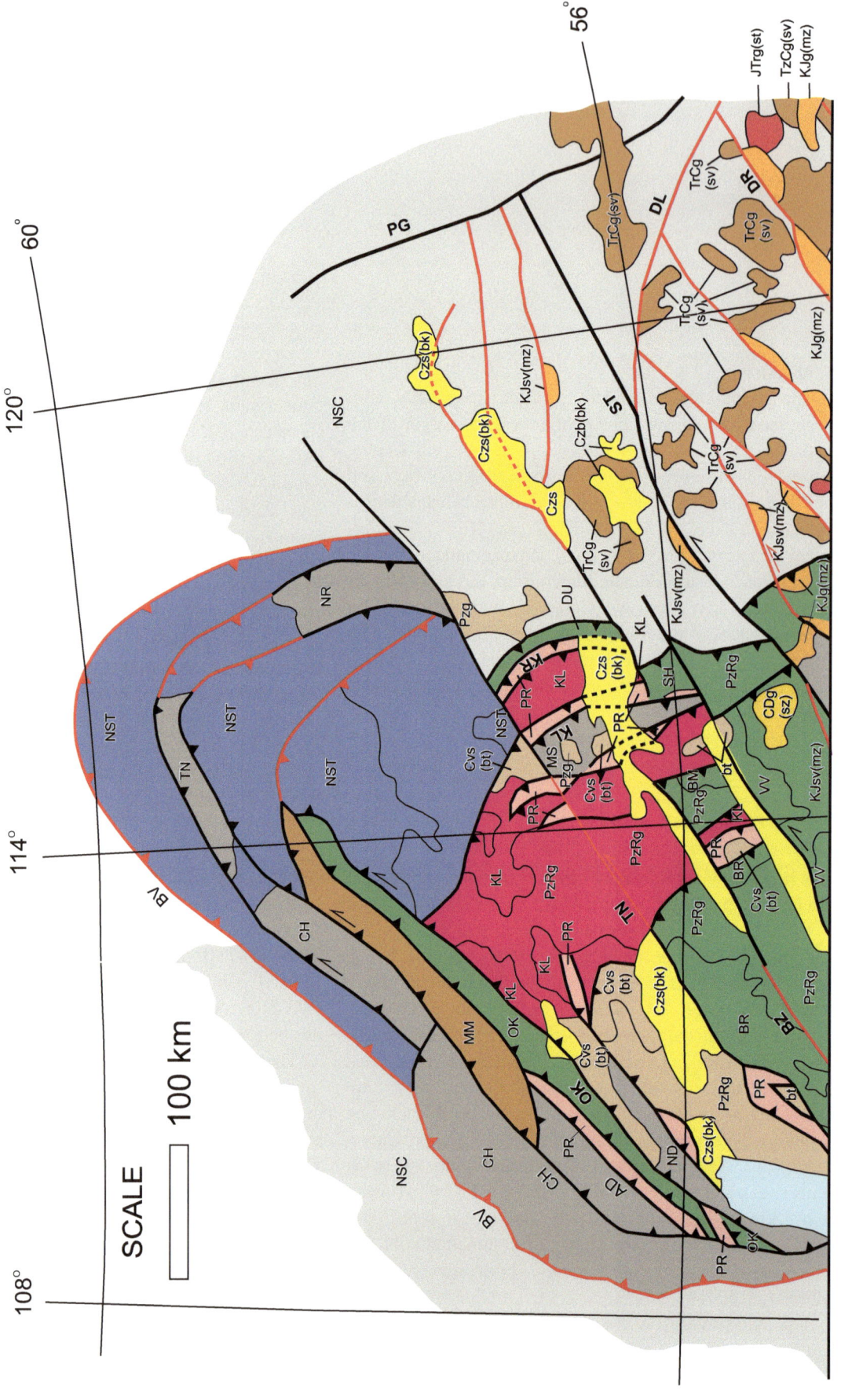

EXPLANATION

Vendian, Cambrian, Mesozoic and Cenozoic Overlap Sedimentary Assemblages and Basinal Deposits

Czs — Sedimentary rocks (Late Cenozoic)

Cenozoic, Mesozoic and Paleozoic Volcanic Sedimentary Assemblages and Basinal Deposits

KJvs — Volcanic and sedimentary rocks (Late Jurassic and Early Cretaceous)

Plutonic Rocks

KJg — Granitoid rocks (Late Jurassic and Early Cretaceous)
JTrg — Granitoid rocks (Late Triassic and Middle Jurassic)
TrCg — Granitoid rocks (Late Carboniferous and Early Triassic)
CDg — Granitoid rocks (Devonian and Early Carboniferous)
Pzg — Granitoid rocks (Paleozoic)
PzRg — Granitoid rocks (Late Riphean and Paleozoic)

Overlap sedimentary, Volcanic and Sedimentary Assemblages, Basinal Deposits and Igneous Arcs (Paleozoic)

bk — Baikal rift system (Late Cenozoic) (Transbaikalia)
bt — Biramin-Tallojin sedimentary basin (Vendian and Cambrian) (Northern Transbaikalia)
mz — Mongol-Transbaikalian sedimentary and volcanic-plutonic belt (Late Jurassic and Early Cretaceous) (Southern Transbaikalia)
sv — Selenga-Vitim sedimentary-volcanic-plutonic belt (Middle Carboniferous and Early Triassic) (Southern Transbaikalia)
sz — Sayn-Transbaikalian sedimentary and volcanic-plutonic belt (Devonian and Early Carboniferous) (Eastern Sayn, Central and Northern Transbaikalia)

Northern Asian Craton and Craton Margin (southern part of Eastern Siberia)

NSC — Nothern Asian Craton (Archean and Early Proterozoic) (southern part of the Nothern Asian craton)
NST — Margin of Nothern Asian Craton (Paton nappe-fold belt) (Riphean) (Between the Nothern Asian Craton and accreted terranes, northeast of Irkutsk region)

Tectonostratigraphic Terranes (arranged alphabetically by map symbol; interpreted tectonic environment and region in parentheses)

BM — Bambui terrane (turbidite basin) (Northern Transbaikalia)
BR — Barguzin terrane (turbidite basin) (Northern Transbaikalia)
CH — Chuji terrane (cratonal) (north of Lake Baikal)
DU — Delum-Uran terrane (turbidite basin) (Northern Transbaikalia)
KL — Kilyan terrane (island arc) (Northern Transbaikalia)
MM — Mama terrane (metamorphic) (north of Lake Baikal)
MS — Muya terrane (craton) (Northern Transbaikalia)
ND — Nerundukan terrane (craton) north of Lake Baikal
NR — Nechera terrane (cratonal) (Northern Transbaikalia)
OK — Olokit terrane (turbidite terrane) (north of Lake Baikal)
PR — Param terrane (oceanic crust) Nothern Transbaikalia
SH — Shaman terrane (turbidite basin) Northern Transbaikalia
TN — Tanad terrane (cratonal) (Northern Transbaikalia)
VV — Verknevitim terrane (turbidite basin) (Central Transbaikalia)

Geological Map Symbols, Contacts, and Faults

Depositional or intrusive contact along margin of overlap assemblage or stitching pluton. Accretionary fault between terranes where not reactivated along post-accretionary fault or where not partly covered by overlap assemblage..

Faults Bounding Terranes

Thrust fault
Strike-slip fault

Major Post-Accretion Faults

Thrust fault
Strike-slip fault
Concealed fault

AD — Abchad strike-slip fault
BV — Baikal-Vituy strike-slip-thrust fault
BZ — Barguzinsky strike-slip fault
CH — Chuy strike-slip fault
DL — Dzheltulak strike-slip fault
DR — Daursky strike-slip fault
KL — Kilyan thrust fault
KR — Karalon thrust fault
OK — Olokit strike-slip fault
PG — Prigiluy strike-slip fault
PR — Pogranichny strike-slip fault
ST — Stanovoy strike-slip fault
TN — Tompuda-Nerpin strike-slip fault

Color for Tectonic Environments

Craton
Craton margin
Cratonal
Island-arc
Oceanic
Turbidite basin

Colors for Overlap Deposits and Postaccretionary Assemblages by Age

Late Cenozoic
Late Jurassic and Early Cretaceous
Late Triassic and Middle Jurassic
Devonian and Early Carboniferous
Vendian and Cambrian
Proterozoic

REFERENCES

This map is compiled from the following references.

Bulgatov, A.N., and Klimuk, V.S., 1998, Structural features of the Dzhida Zone, Caledonides: Geotectonics, no. 1. p. 45-55 (in Russian).

Bulgatov, A.N., Turunkhaev, V.I., 1996, Geodynamics of Central Asia in Late Mesozoic: Doklady Russian Academy of Sciences, v. 349, no. 6. p. 783-785 (in Russian).

Dobretsov, N.L., and Bulgatov A.N., 1991, Geodynamic map of Transbaikalia (concepts of preparation and legend): Novosibirsk: United Institute of Geology, Geophysics and Mineralogy and the Buryat Geological Institute, Siberian Branch, Russian Academy of Sciences, no. 8. 51 p. (in Russian).

Fomin, I.N., Sizich I.V., Cherednichenko, V.P., and Falkin, E.M. 1985, Transbaikalian tectonic complexes and their analogues in the adjacent regions: Tectonics of Siberia, v. 12: Nauka, Novosibirsk, p. 42-52 (in Russian).

Gordienko, I.V., 1997, Major terranes of the Transbaikalian region: Tectonic evolution of the East Asian continent. Short papers for the International Symposium. Seoul, Korea, p. 17-19.

Gordienko, I.V., 1987, Paleozoic magmatism and geodynamics of the Central Asian fold belt: Nauka, Moscow, 240 p. (in Russian).

Kuzmin, M.I., Gordienko, I.V. Almukhamedov, A.I., Antipin, V., Baynov, V.D., and Filimonov A., 1995, Paleo-oceanic complexes: the Dzhida zone of caledonides (Southwestern Transbaikalia): Russian Geology and Geophysics, v. 36, no. 1. p. 1-16.

Parfenov, L.M., Bulgatov, A.N., and Gordienko, I.V., 1995, Terranes and accretionary history of the Transbaikalian orogenic belts: International Geology Review, v. 37, no. 8. p. 73-751.

Fig. 2.1 Tectonic map of Transbaikalia. (after Bulgatov and Gordienko 1999)

the Neoproterozoic to Phanerozoic sedimentary cover of the Aldan-Lena pericratonic graben (Fig. 2.2). In the south, the Stanovoy Thrust Fault separates the shield from the Stanovoi Block, which is sometimes considered as part of the Siberian craton (Bulgatov and Gordienko 1999; Fig. 2.1). The western boundary of the shield extends along the Zhuya Fault System. This important information was obtained in the 1960s during the geological mapping surveys and then integrated by Salop (1964, 1967) and other geologists (Fedorovsky 1972; Krasny 1980; Mitrofanov and Moskovchenko 1985; Chechetkin and Kharitonov 2002). Most recent data were summarized by Mitrofanov (2016) on the geological map of the Russian Federation at a scale 1:1,000,000 (Aldan-Transbaikalia series, Sheets O-50 and O-52), which subdivided the Aldan Shield into the Aldan granulite-gneiss geoblock, the Olekma and Batomga granite-greenstone geoblocks (Chara-Olekma and Batomga geoblocks), and west and east of the Aldan geoblock, respectively.

In contrast to many other shields worldwide, the Aldan Shield and the Baikal-Muya Fold Belt were involved into several episodes of rifting starting from the early Precambrian up until now (Mineeva and Arkhangel'skaya 2007; Bulgatov and Gordienko 1999; Fig. 2.1). The most important tectono-magmatic events in the region are marked by mafic-ultramafic intrusions (Paleoproterozoic Chiney Complex, Neoproterozoic Doros Complex, Mesozoic dikes and volcanic rocks, as well as Neogene-Quaternary basaltic lava plateau and dikes). All of the aforementioned rocks occur at the intersection of northwestern faults as splays of the Syulban Deep Fault and the near-latitudinal structural elements, accompanying the Stanovoi Deep Fault.

The origin of the sutures in the Aldan-Stanovoi geoblock is a matter of debate. Their origin is now referred to the accretion of the Aldan and Stanovoi geoblocks (Kazansky 1982; Duk et al. 1979; Bulgatov and Gordienko 2014a, b). At that time, the Archean protocratons were dissected by large faults, accompanied by sedimentation, metamorphism, and emplacement of mafic-ultramafic plutons. The ancient deep faults defined a general regional structural pattern, including the Chara block and Kodar-Udokan basin, which have angular outlines due to the intersection of rectilinear faults. In the south of the Kodar-Udokan basin (Fig. 2.2), the North Kalar Belt of mafic intrusions and ore-bearing plutons of the Chiney Complex trace the near-latitudinal fault system and coincide with a gravity gradient.

Older faults were rejuvenated during the Phanerozoic reactivation, but the structural pattern of the district remained unmodified. Intense neotectonic processes developed synchronously with the growth of the Baikal rise in the form of the reactivated ENE-trending faults.

2.2   Basement Rocks

The oldest rocks of the Chara-Olekma geoblock comprise the Kurul'ta Group, dominated by tonalite-trondhjemite orthogneisses metamorphosed to high-temperature/moderate pressure amphibolite facies (Berezkin et al. 2007; Smelov et al. 2007). High-alumina gneiss, enderbite, and plagiogneiss are subordinate, occurring as isolated blocks among younger rocks. These are mafic and ultramafic igneous rocks with a minor volume of sedimentary rocks metamorphosed to granulite facies and retrograding into younger amphibolite facies. U-Pb (SHRIMP) zircon age of granulite is 3221 ± 3 Ma (Nutman et al. 1992).

The Chara Metamorphic Complex in the western part of the shield is 10 km thick and is made up of mafic granulites. The BIF (Sulumat deposit) occurs in its central part. Peraluminous schist and plagiogneiss are less abundant and are found as skialiths among igneous rocks. The rocks of the Kurul'ta and Chara Complexes are unevenly affected by granitization and altered. These processes produced gray gneiss of the Olekma Complex. These gray gneisses are thought to be the primary rocks of the crustal granite layer. They correspond to plagiogranite in composition and contain relics of metabasalts and metasediments. The peak metamorphism occurred at granulite facies and was retrograded to amphibolite facies.

The areas of gray gneiss associate with N-S-trending greenstone belts, a few dozen kilometers wide and hundreds of kilometers long. The gray gneiss units are usually thrust over the greenstone belts (principally, Turgincha, Tarynakh, and Olondo) (Fig. 2.2). The supracrustal volcanic and sedimentary rocks in these belts were mapped as the Olondo Group, consisting of orthoamphibolite, metabasalt, talc-chlorite schist, biotite microgneiss, magnetite sandstone, and marble (Glukhovsky 2009).

Metavolcanic rocks are dominant in the Olondo Complex, whereas sandstones dominate among the sedimentary rocks. The volume of the sedimentary rocks increases west- and eastward, including argillites and carbonates. The BIF units are common in the bottom parts of the metasedimentary successions, with large iron deposits known in the Chara-Tokka ore district. The sedimentary sequence becomes coarse-grained upward in the succession, where Molasse units appear in the uppermost layers. Basalt and komatiite dominate at the base of the succession (Fig. 2.3), with andesite being rather sporadic. Upward, these rocks give way to dacite and rhyolite. The volcanic rocks associate with comagmatic mafic and ultramafic intrusions. Apatite-magnetite-bearing pyroxenites with economic iron and phosphorus concentrations are recognized and may constitute IOA (iron-oxides-apatite) style deposits. Tectonic sheets of ultramafic

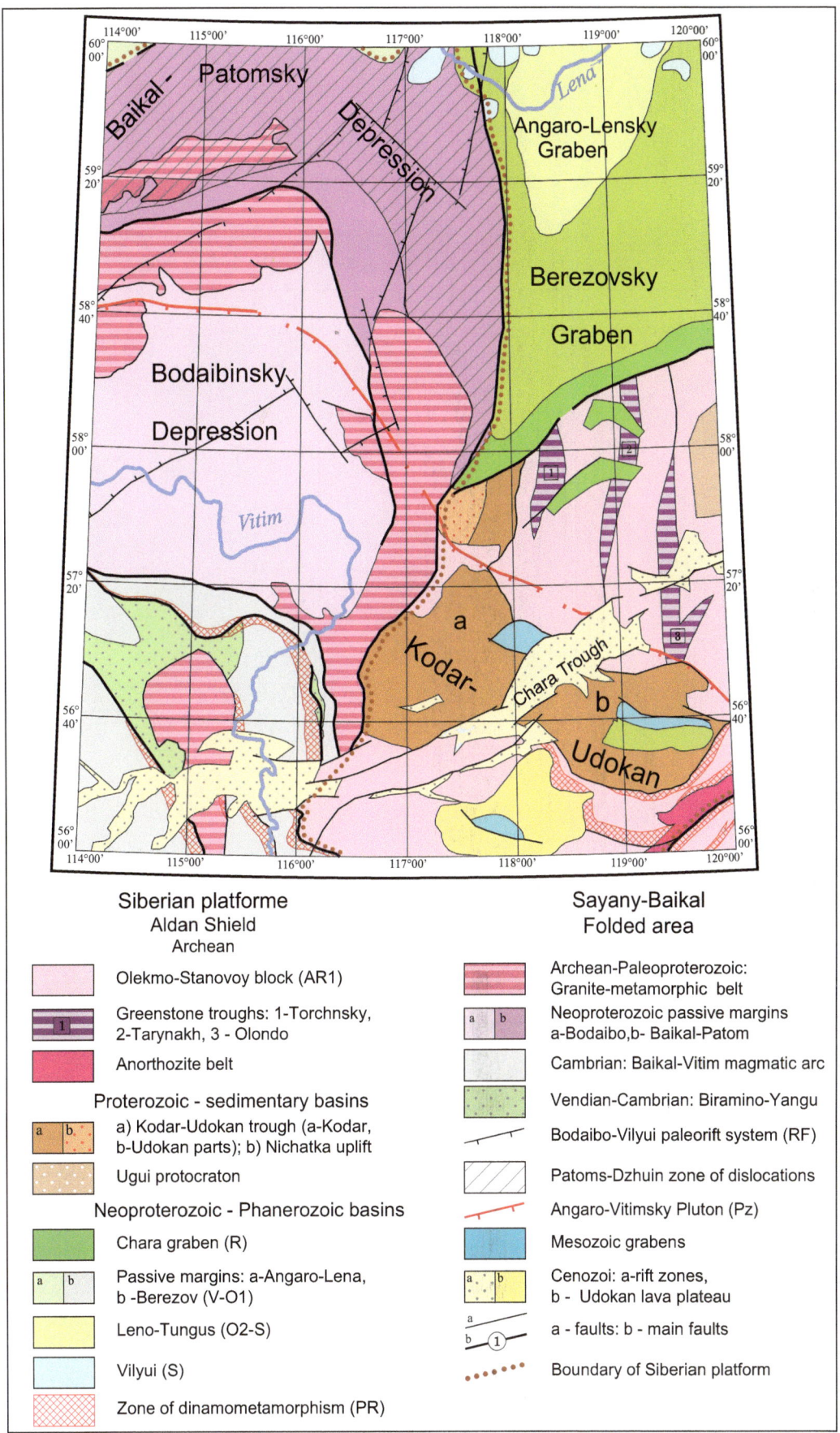

Fig. 2.3 Regional geological map of the Kodar–Udokan Zone

Deposits: 1, Udokan; 2, Chiney pluton; 3, Katugin. Here and in Fig. 2.4, the map is based on A Galyamov,1:100,000 State Geological Maps (Sheets O-49, 50)

rocks are interpreted as protrusions. Formation of greenstone belts was completed after the emplacement of 2.9–3.0 Ga gabbro and plagiogranite (Nutman et al. 1992).

The metamorphic grade in greenstone belts increases from top to bottom and from the center outward. In the central parts of these belts, the grade corresponds to greenschist facies; amphibolite facies rocks are characteristic of the marginal zones, where metamorphism was accompanied by partial melting of the adjacent gray gneiss. The U-Pb isotopic age of zircon from palingenic granite rocks is 2.6–2.7 Ga (Glukhovsky 2009). The granitoids of this age are also widespread beyond the greenstone belts. They make up the Older Stanovoi Complex, which is represented by calc-alkaline granitoids with appreciable amount of K-feldspar. Their formation is frequently accompanied by the growth of granite-gneiss domes. Late-stage alaskite, aplite, and pegmatite veins are abundant. The U-Pb zircon age of metavolcanic rocks is 3.0–2.9 Ga (Nutman et al. 1992). The age of anorthosite, controlled by older suture zones, e.g., the 6000 km^2 Kalar layered mafic-ultramafic complex, remains unconstrained. The scarce isotopic data correspond either to Neoarchean (Larin et al. 2006) or Paleoproterozoic (Glukhovsky et al. 1993) ages. Anorthosites are known along the Stanovoi suture zone with andesitic rocks at its base and labradorite and anorthosite in the upper part.

Thus, the first signs of amalgamation of continental masses appeared approximately 3.0 Ga ago, when sialic and greenstone rocks were tectonically sandwiched into the apparent sequence. The thickness of the continental crust increased at that time as a result of collision of blocks and formation of nappes in the Olondo greenstone belt. Culmination of 2.6–2.8 Ga metamorphic events probably reflects the final stages of the collision (Sklyarov 2006). The Chara-Olekma geoblock was thrust under the Central Aldan geoblock at 2.3–2.1 Ga. This event marks amalgamation of the Siberian craton at 2.0–1.8 Ga (Rosen 2003). The last stages of tectonic evolution were accompanied by extension in relation to the collapse of orogens, induced by the mantle-sourced intrusions of mafic-ultramafic melts. This gave rise to the large intracratonic troughs with the Kodar-Udokan trough (basin) as a striking example (Fig. 2.2), the Akitkan volcanic-plutonic belt, the Chiney layered mafic-ultramafic intrusions (Fig. 2.3), and post-collisional (anorogenic) granitoids of the Kodar and other intrusive complexes (Neymark et al. 1998; Larin et al. 2000; Donskaya et al. 2005).

2.3 Sedimentary Rocks

Carbonate-clastic sequences in the Kodar-Udokan, Ugui, Verkhne-Khani, and other basins represent an older platform cover. The Kodar-Udokan trough, ~300 km long and 60–70 km wide (Fig. 2.3), is filled with Paleoproterozoic

clastic-carbonate rocks of the Udokan Supergroup, reaching 11–14 km in thickness (Fig. 2.4; Krendelev et al. 1983). This supergroup is subdivided into the Kodar, Chiney, and Kemen Groups (macrorhythms) (Salop 1964; Krendelev et al. 1983; Burmistrov 1990; Volodin et al. 1994; Chechetkin et al. 2000), further subdivided into nine formations. In the macrorhythms, the marine sedimentary rocks give way to the continental sediments, interpreted as transgressions and regressions; the intensity of these movements gradually decreased. The style of clastic material regularly changes in succession, with conglomerate and coarse-grained clastic rocks almost always occurring in the lower part of each group. Carbonate rocks (limestone, dolomite) are not abundant and are characteristic of the Butun Formation. The rocks of the Udokan Supergroup are folded, rimming the Chara block. Because of this, the strike of fold axes is variable. The almost west-to-east rather than west-northwestern strike dominates near the Chiney pluton. Three large folds—the Naminga and Katugin brachysynclines and conjugate Chiney anticline—are recognized here (Fig. 2.4).

The Naminga syncline is an 10–12 km-long and 8–10-km-wide asymmetric fold, without parasitic folds. It hosts Udokan sandstone-hosted copper deposit. Its northern limb is steeply dipping and often even overturned as compared to the southern limb, where rocks dip at 35–40°. The Katugin syncline is 12 km wide, striking 85 ESE. Its hinge plunges to the northeast, and the sandstone is mainly contact-metamorphosed by the Kodar granite layered mafic-ultramafic complex. The 10–12 km wide Chiney anticline strikes to 55NE. It is a narrow structural element with numerous second-order folds and faults. Its rocks are intruded by the Chiney Pluton.

The rocks of the Udokan Supergroup are subdivided into three groups, consisting of clastic and carbonate rocks. Copper deposits and occurrences are known within all three groups, mostly in the uppermost Kemen Group (Sakukan and Naminga Formations). The supergiant Udokan copper deposit is hosted in the Sakukan formation, while the Unkur and Burpala deposits are hosted in the lower part of the Sakukan formation. The Pravoingamakitskoe, Krasnoe, and other small deposits occur in the middle part of the Chiney Group (Inyr, Chitkanda, Alexandrov, and Butun Formations). Numerous pyrrhotite- and chalcopyrite-bearing units, containing Ag, Co, and Ni, are localized in the lower formations of the Kodar Group (Bogdanov et al. 1966; Arkhangel'skaya et al. 2004).

The Paleoproterozoic age of the Udokan Supergroup is first of all constrained by the unconformity between the Archean rocks and the overlapping Neoproterozoic sedimentary rocks (with conglomerate at the base). The second is the ages of metamorphism and crosscutting intrusive and palingenic metasomatic rocks of the Kodar and Kuyanda Complexes (Fedorovsky 1972).

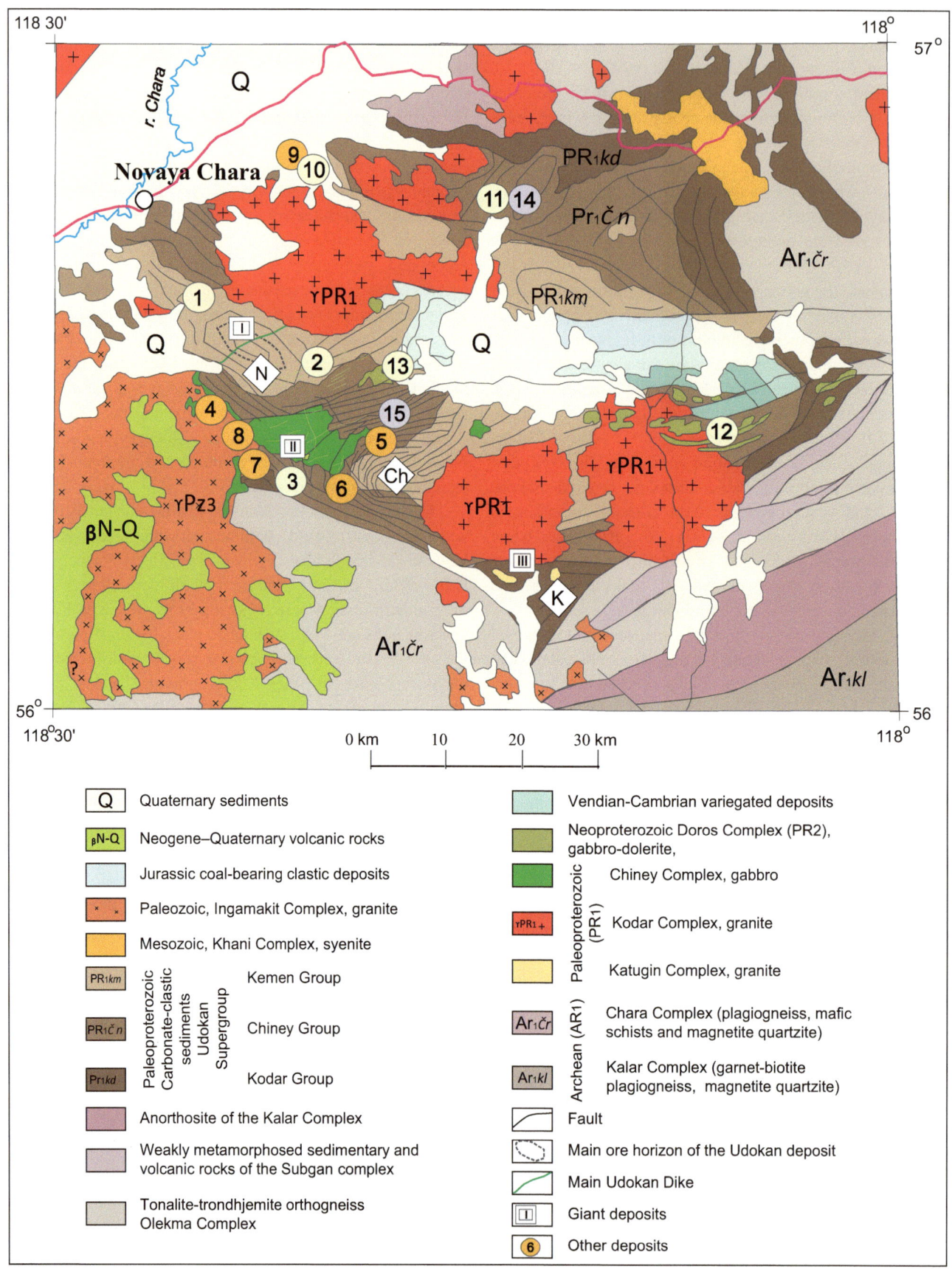

Fig. 2.4 Geological map of the Udokan-Chiney district. (Modified after Chitageologia data)
Superlarge ore deposits: I-Udokan, II-Chiney, III-Katugin; other deposits: (1) Klyukvennoe, (2) Saku, (3) Pravoingamakitskoe, (4) Mylovskiy, (5) Rudnoe, (6) Verkhne-Chineyskoe, (7) Skvoznoe, (8) Kontaktovyi, (9) Luktur, (10) Unkur, (11) Krasnoe, (12) Burpala, (13) Kilcheris, (14) Chitkanda (U), (15) Nizhnechineyskoe (U-REE)

Pokrovsky and Grigoriev (1995) studied the Rb-Sr system in deposits of upper formations. Bulk samples of poorly metamorphosed clastic rocks are rarely used for isotopic dating. The presence of detrital minerals produces a substantial scattering of points on the isochrons or even a mixing line that is not possible to interpret (Clauer 1984; Faure 1986). For dating, a fine fraction (<4 μm or < 2 μm) is usually used, which possibly reflects the age of sedimentation (Clauer 1984) or (more likely) the early stages of diagenesis (Bonhomme 1987). Numerous examples of distorted age were reported for bulk samples of sedimentary rocks (Clauer 1984; Faure 1986). However, they are mainly applied to the Phanerozoic formations. This technique, in our opinion, cannot be applied to Precambrian rocks which are usually strongly metamorphosed. Based only on petrographic criteria, it is extremely difficult to draw a sharp line between the epigenetic and regional metamorphism, which is characterized by the establishment of complete physical-chemical equilibrium in the rock.

As it was noted by Korikovsky (1979), even plagioclase, fully recrystallized during the low-temperature greenschist facies metamorphism, can almost completely retain the shape of the original clastic grains. The data on the oxygen isotopes show that the rocks of the Naminga Formation have been very deeply modified after sedimentation.

The $^{87}Rb/^{86}Sr$ and $^{87}Sr/^{86}Sr$ values in the bulk samples of the Naminga Formation show a direct correlation that can be interpreted as errochron with an age of 1939 ± 101 Ma (Pokrovsky and Grigoriev 1995) and an initial resolution of 0.7092 ± 0.0009 (standard deviation after York 1966). This age practically coincides with the age of metamorphism of the Udokan Supergroup established by other methods. It also coincides (within the error) with Rb-Sr age of metapelites from the Khani Trough (Gorokhov et al. 1989), which is considered to be coeval with the rocks of the Udokan Supergroup (Fedorovsky 1972, 1985). The errochrons (Gorokhov et al. 1989; Pokrovsky and Grigoriev 1995) have close MSWD −16.40 and 17.35, respectively, although the rocks of the Khani Trough, judging from the descriptions, have experienced a stronger metamorphism than the Udokan rocks. It should be noted that the data obtained for the rocks of the Naminga Formation (Pokrovsky and Grigoriev 1995) can be also interpreted for two independent isochrons: (1) with an age of 2102 ± 23 Ma ($^{87}Sr/^{86}Sr)_0$ = 0.70880 ± 0.00014 (for rocks with K_2O content of <2 wt%) and −(2) with an age of 1992 ± 143 Ma ($^{87}Sr/^{86}Sr)_0$ = 0.7052 ± 0.0048 (for rocks with K_2O content >2 wt%). This illustrates a more or less obvious fact that the rocks were modified many times. The relatively low initial value ($^{87}Sr/^{86}Sr)_i$ on the younger isochron, which seems paradoxical at first glance, can be explained by the contamination of strontium from the underlying carbonate rocks of the Kodar Group (Table 2.1). The conclusion that the carbonate material was very intensely dissolved and

redeposited is driven by the similarity of the C isotopes in the carbonate cement of the Naminga Formation and the carbonates of the underlying rocks, as well as by strong fluctuations in the thicknesses of the carbonate rocks of the Butun Formation (Fedorovsky 1972). The age of metamorphism of the uppermost formation in the Udokan Supergroup is constrained by an errochron of 1939 ± 101 Ma (Pokrovsky and Grigoriev 1995). This value is almost identical to the ages obtained for the lower formations of the Udokan Supergroup (Gorokhov et al. 1989). From this, it can be concluded that either the entire 10-km-thick sequence was formed within a very short time interval (2.1–2.25 m.y.) or all available data are indicative of the age of metamorphism.

Rocks of the Udokan Supergroup were metamorphosed to epidote-chlorite or biotite greenschist facies but locally attain amphibolite grade (Bogdanov et al. 1966; Volodin et al. 1994; Abramov 2011). Elsewhere in the region, granulite-facies peak metamorphism was reached at ~1.9–1.8 Ga (Glebovitsky et al. 2008). Furthermore, the Nd isotopic signature of the metasedimentary rocks from the Udokan Supergroup is similar to that of island arc-type metavolcanic sequences in the central Aldan Shield, dated at 2006 ± 3 Ma, implying a possibility of contemporaneous derivation from this source region (Podkovyrov et al. 2006).

The isotopic age of the Udokan sedimentary rocks, constrained using the K-Ar method, is 1832–1980 Ma (Chechetkin and Kharitonov 2002). Perelló et al. (2017) separated titanite crystals from a sample collected in the Medny site of the deposit, where typical high-grade disseminated and veinlet chalcocite-bornite mineralization is well exposed. Titanite crystals from the disseminated and veinlet fractions were separately dated by the ID-TIMS U-Pb method, with the disseminated fraction returning a concordia age of 1895.3 ± 9.7 Ma and the veinlet fraction a concordia age of 1896.7 ± 7.8 Ma. The combination of both titanite fractions produced a concordia age of 1896.2 ± 6.2 Ma, interpreted as synorogenic mineralization.

The stratigraphic unconformity between the Chiney and Kemen Groups (Tombasov and Sinitsa 1990) suggests a younger age of the latter. Indeed, the type of the Udokan biota suggests a Riphean to Vendian age of the Kemen rocks (Vil'mova 1990; Sinitsa 1996; Salikhov 2010). The data on detrital zircons indicate a much older age of the Chitkanda Formation of the Chiney Group (2180 ± 50 Ma) in comparison with the Kemen Group (Berezhnaya et al. 1988).

The aforesaid has substantial implications for the timing of ore formation in this district, because it was assumed that the magma, responsible for the formation of the Chiney pluton and related ore, intruded only rocks of the Chiney Group (Burmistrov 1990) and thus predated the Kemen Group. Meanwhile, new evidence for the age of the Chiney pluton has been obtained recently using different specific isotopic systems, and this is explained not only by analytical uncer-

Table 2.1 U-Pb age of gabbronorite from Chiney pluton (sample 0305–1)

Analyt. point	Concentration					Age* and error, Ma			Isotope ratios (**)		
	U, ppm	Th, ppm	$^{206}Pb_R$, ppm	$^{206}Pb_C$, %	$^{232}Th/^{238}U$	$^{206}Pb/^{238}U$	$^{207}Pb/^{206}Pb$	D, %	$^{207}Pb^*/^{235}U$ / ±%	$^{206}Pb^*/^{238}U$ / ±%	EC
0305-1.3.1	6898	8918	774	1.09	1.34	783 ± 3	1539 ± 13	97	1.70/0.8	0.1292/0.5	0.57
0305-1.4.1	1635	1066	238	0.77	0.67	1001 ± 3	1650 ± 13	65	2.35/0.7	0.1680/0.3	0.38
0305-1.2.1	517	31	217	0.03	0.06	2568 ± 34	2722 ± 5	6	12.67/1.6	0.4894/1.6	0.98
0305-1.5.1	139	48	62	0.05	0.36	2681 ± 15	2767 ± 19	3	13.71/1.3	0.5157/0.7	0.50
0305-1.1.1	74	47	35	0.05	0.65	2807 ± 21	2849 ± 12	1	15.25/1.2	0.5456/0.9	0.79

Note. Errors are given for 1σ except for specified cases. * with correction for common lead based on measured ^{204}Pb; $^{206}Pb_R$ and $^{206}Pb_C$ are radiogenic and common lead, respectively. D, discordance, %: D=100 {[age ($^{207}Pb/^{206}Pb$)]/[age($^{206}Pb/^{238}U$)]–1}
Error correlation of $^{207}Pb/^{235}U$–$^{206}Pb/^{238}U$ determination

tainties but also by the multiple and diachronous injections of mafic melts forming this pluton. The study of Sm-Nd isotopic system (Tables 2.1 and 2.2; Fig. 2.5) yields a rather uncertain age of the Chiney pluton of 1850 ± 90 Ma (Gongalsky et al. 2008). The study of U-Pb system in zircons allowed us to constrain the age of particular rock groups: 1858 ± 17 Ma for the high-Ti gabbro rocks of the Group 2 and 1811 ± 27 Ma for low-Ti rocks of the Group 3 (Gongalsky et al. 2012), which post-dates the age of mineralization at Udokan constrained by Perelló et al. (2017).

We collected ~1-kg-samples 0303 and 0305 from low-Ti gabbro and sample 0306 from high-Ti gabbro. Zircon grains suitable for further study (SHRIMP II, VSEGEI, analyst AN Timashkov) have been selected from these samples. Three types of zircons are distinguished based on their morphology and internal zoning. The first type dominates in all samples. These are pink and light pink subhedral prismatic crystals, 80–400 μm in size. The cathodoluminescent image (CL) of the first-type zircon displays a complex internal structure (Fig. 2.6). The grains consist of a vaguely zonal and partly recrystallized core surrounded by one or two outer shells (black and white). This type of internal structure is typical for zircons from high-temperature metamorphic rocks. The grains of this type are occasionally very dark and reveal vague zoning with a bright thin outer rim. The U and Th contents and Th/U ratio in the first-type zircon vary in a wide range (58–610 ppm U, 31–406 ppm Th; Th/U = 0.03–0.80).

The second-type zircon grains are represented by large fragments of prismatic crystals, commonly 200–300 μm in size (sample 0303). In CL images, these zircons are dark, almost black, and reveal both fine and rough zoning. In sample 0305, the second-type zircons occur as euhedral prismatic crystals, 150–300 μm long, which are uniformly black in CL (Fig. 2.6). Figure 2.7 shows their age. Second-type zircons were also detected in sample 0306. These are large turbid fragments and roughly zonal long-prismatic crystals, which are dark and vaguely zonal in CL (Fig. 2.8). Second-type zircons from samples 0303 and 0305 (low-Ti gabbro) are characterized by anomalously high U (838 to 1179–6898 ppm) and Th (660 to 1066–8918 ppm); the Th/U ratio is 0.74–1.09 (Table 2.1). The second-type zircons from sample 0306 (high-Ti gabbro) are characterized by lower U content ranging from 229 to 652 ppm (Table 2.2), and the Th content is 144–385 ppm, with Th/U = 0.61–0.65. The second-type zircon was probably formed during the post-metasomatic stage of gabbro intrusions. The third-type zircon is distinguished from the first- and second-type crystals in containing a low U concentration (13–383 ppm) and no Th; Th/U = 0. The third-type zircon was formed during the recrystallization of the gabbro.

The age of first-type zircons from the high- and low-Ti gabbro was estimated at 2872–2919 Ma (Figs. 2.7 and 2.9). The data points of second- and third-type zircons from the high-Ti gabbro (sample 0306) lie on discordia with an upper intercept with the concordia at 1858 ± 17 Ma (Fig. 2.9). The second- and third-type zircons from the low-Ti gabbro (samples 0303 and 0305) yielded a concordant age of 1811 ± 27 Ma (Fig. 1.6); the lower intercept with the discordia corresponds to 205 ± 32 Ma. The above data allowed us to constrain the age of high-Ti gabbro of the second group at 1858 ± 17 Ma. The age of the third group (low-Ti gabbro) is most likely close to 1811 ± 27 Ma. The lower intercept between the discordia and concordia corresponds to 305 ± 32 Ma and indicates the emplacement time of the Ingamakit granite pluton, which affected the gabbro of the Chiney pluton.

Our results are generally consistent with the data of other researchers, who obtained a 1867 ± 3 Ma U–Pb zircon age for the marginal facies of the Chiney pluton (Popov et al. 2009) and a 1880 ± 16 Ma Ar/Ar mica age (Polyakov et al. 2006). These age estimates are close to the age of the Udokan Supergroup and thus show that no temporal gap existed between the formation of the Chiney and Kemen Groups and related deposits. This is also constrained by the age of the Kodar Granite (1875 Ma, Larin et al. 2002) that cuts through the rocks of the Kemen Group.

In addition to the large Kodar-Udokan trough, Paleoproterozoic clastic rocks were found in a number of smaller, tectonically bound, and deeply eroded Ugui, Verkhne-Khani, and Oldogsa Troughs.

The Upper Kalar Basin in the southwestern part of the Aldan Shield is filled with Vendian to Lower Paleozoic sedimentary rocks, which are defined as the Upper Kalar Group. The Vendian to Lower Cambrian coarse clastic rocks overlie with an unconformity of the Paleoproterozoic rocks of the Udokan Supergroup and granite rocks that cut through the latter. The Cambrian carbonate rocks (>1000 m), largely dolomite with sporadic siltstone and claystone beds followed by Ordovician carbonate-terrigenous rocks, occur in the upper part of the sequence. In general, the fill of this basin makes up a common transgressive-regressive sequence.

In the Jurassic and Cretaceous, the Aldan Shield was tectonomagmatically reactivated. Continental coal-bearing clastic rocks, up to 4 km thick, were deposited in the newly formed basins. The thickness of coal seams in the South Yakutian Basin reaches 50 m. Mafic volcanic rocks were recorded in some basins. Chul'man, Usmun, Tokko, and Apsat are the largest basins. The youngest stratified rocks are Neogene-Quaternary basalts of the Udokan lava plateau, just 16–2 Ka in age (Stupak 1987).

Table 2.2 U-Pb age of zircon in gabbronorite from the Chiney pluton (sample 0306)

Analytical point	$^{206}\mathrm{Pb}_c$, %	U, ppm	Th, ppm	$^{232}\mathrm{Th}/^{238}\mathrm{U}$	$^{206}\mathrm{Pb}^*$, ppm	Age, Ga $^{206}\mathrm{Pb}/^{238}\mathrm{U}$ (1)	(2)	(3)	$^{207}\mathrm{Pb}/^{206}\mathrm{Pb}$ (1)	$^{208}\mathrm{Pb}/^{232}\mathrm{Th}$ (1)	D, %
0306.1.1	0.10	141	40	0.29	65.9	2.799 ± 48	2.777 ± 68	2.801 ± 49	2.848 ± 13	2.730 ± 83	2
0306.2.1	0.22	139	104	0.77	61.7	2.681 ± 48	2.653 ± 64	2.685 ± 52	2.755 ± 16	2.628 ± 66	3
0306.3.1	0.06	157	56	0.37	70.2	2.695 ± 46	2.627 ± 60	2.697 ± 48	2.864 ± 12	2.624 ± 91	6
0306.4.1	0.12	383	1	0.00	98.4	1.684 ± 29	1.665 ± 32	1.686 ± 29	1.858 ± 15	−3.220 ± 2400	9
0306.5.1	1.20	610	345	0.58	249	2.479 ± 42	2.374 ± 51	2.465 ± 45	2.811 ± 67	2.703 ± 320	12
0306.6.1	0.00	528	406	0.80	241	2.745 ± 43	2.712 ± 59	2.739 ± 48	2.825.80 ± 6.3	2.821 ± 98	3
0306.7.1	0.02	229	144	0.65	64.5	1.831 ± 32	1.830 ± 36	1.838 ± 35	1.843 ± 17	1.749 ± 39	1
0306.8.1	0.03	652	385	0.61	185	1.842 ± 30	1.839 ± 34	1.845 ± 33	1.859 ± 10	1.796 ± 36	1
0306.9.1	0.02	482	16	0.03	215	2.700 ± 42	2.684 ± 58	2.701 ± 42	2.743.50 ± 6.9	2.613 ± 82	2
0306.10.1	5.91	82	0	0.00	25.3	1.881 ± 44	1.895 ± 44	1.885 ± 54	1.776 ± 240		−6
0306.11.1	0.04	559	176	0.32	236	2.571 ± 41	2.534 ± 53	2.648 ± 42	2.681.80 ± 8.0	150 ± 11	4
0306.12.1	–	276	188	0.70	113	2.516 ± 41	2.434 ± 51	2.571 ± 45	2.770.50 ± 9.4	1.785 ± 38	9
0306.13.1	0.22	13	0	0.00	3.09	1.608 ± 62	1.558 ± 66	1.588 ± 62	2.053 ± 87	23.260 ± 7500	22
0306.14.1	0.14	106	0	0.00	30.8	1.874 ± 35	1.872 ± 40	1.875 ± 36	1.889 ± 24	−11.250 ± 5400	

Total $^{238}\mathrm{U}/^{206}\mathrm{Pb}$/±%	Total $^{207}\mathrm{Pb}/^{206}\mathrm{Pb}$/±%	$^{238}\mathrm{U}/^{206}\mathrm{Pb}^*$/±%	$^{207}\mathrm{Pb}/^{206}\mathrm{Pb}^*$/±%	$^{207}\mathrm{Pb}^*/^{235}\mathrm{U}$/±%	$^{206}\mathrm{Pb}^*/^{238}\mathrm{U}$/±%	Corr.Err
1.837/2.1	0.2036/0.76	1.839/2.1	0.203/0.78	15.2/2.2	0.544/2.1	0.938
1.935/2.2	0.1935/0.92	1.939/2.2	0.192/0.99	13.62/2.4	0.516/2.2	0.910
1.926/2.1	0.2052/0.72	1.927/2.1	0.205/0.73	14.65/2.2	0.519/2.1	0.944
3.345/2	0.1147/0.75	3.349/2	0.114/0.83	4.678/2.1	0.2986/2	0.921
2.107/1.9	0.2084/2.8	2.132/2	0.198/4.1	12.82/4.6	0.469/2	0.443
1.884/1.9	0.2/0.39	1.884/1.9	0.2/0.39	14.64/1.9	0.531/1.9	0.980
3.043/2	0.1128 / 0.95	3.044/2.0	0.113/0.96	5.1/2.2	0.3285/2	0.902
3.023/1.9	0.1139/0.55	3.024/1.9	0.114/0.57	5.18/2	0.3307/1.9	0.958
1.922/1.9	0.1903/0.42	1.922/1.9	0.19/0.42	13.64/2	0.5203/1.9	0.977
2.776/2.3	0.1606/3.6	2.951/2.7	0.109/13	5.07/13	0.3389/2.7	0.203
2.04/1.9	0.1836/0.48	2.04/1.9	0.183/0.48	12.38/2	0.4901/1.9	0.970
2.094/2	0.1933/0.57	2.094/2	0.193/0.57	12.73/2	0.4775/2	0.960
3.52/4.3	0.1285/4	3.53/4.3	0.127/5	4.95/6.6	0.283/4.3	0.660
2.96/2.2	0.1167/1.3	2.964/2/2	0.116/1.3	5.38/2.6	0.3374/2/2	0.853

Note. Common Pb corrected using measured $^{204}\mathrm{Pb}$; (2) common Pb corrected by assuming $^{206}\mathrm{Pb}/^{238}\mathrm{U}$–$^{207}\mathrm{Pb}/^{235}\mathrm{U}$ age concordance; (3) common Pb corrected by assuming $^{206}\mathrm{Pb}/^{238}\mathrm{U}$–$^{207}\mathrm{Pb}/^{206}\mathrm{Pb}$ age concordance; D, discordance, %; error in standard calibration is 0.77%; errors are 1-sigma; Pb_c and Pb^* are common and radiogenic lead, respectively

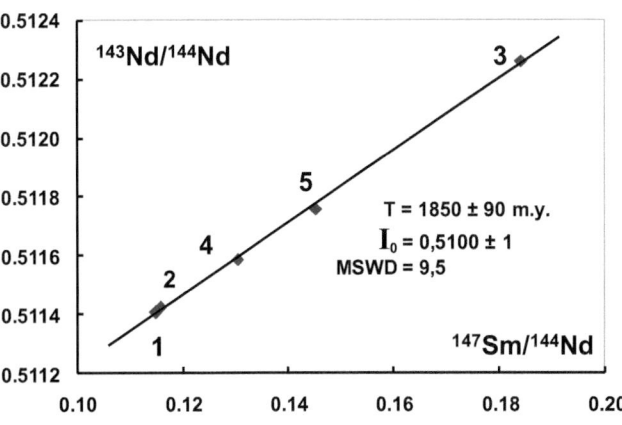

Fig. 2.5 ^{143}Nd/^{144}Nd versus ^{147}Sm/^{144}Nd for rocks of the Chiney pluton
See Table 2.1 for sample numbers

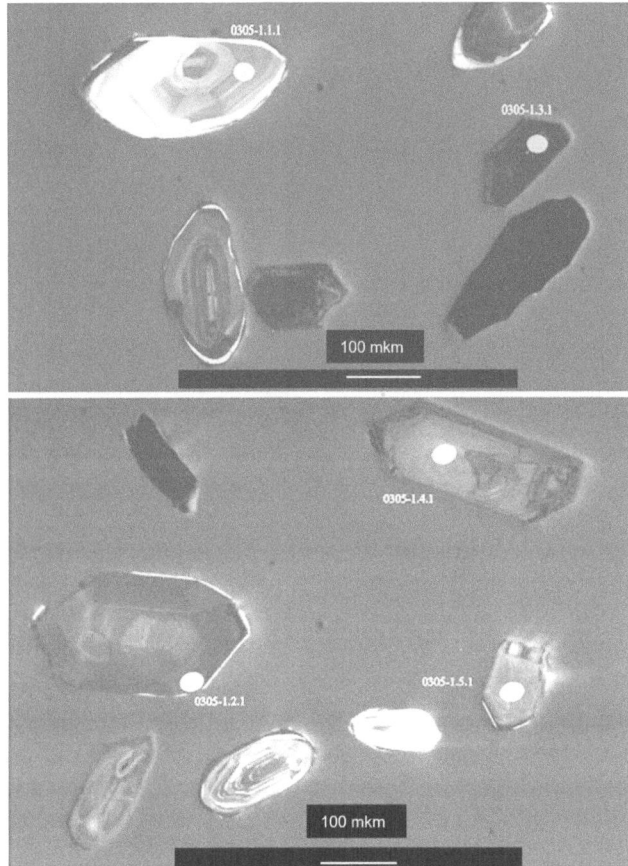

Fig. 2.6 CL image of zircons from low-Ti gabbro with position of analytical points. (sample 0305)
See Table 2.1 for sample numbers

2.4 Intrusive Rocks

The available geological and geochemical data on the known Proterozoic platinum-bearing ultramafic-mafic in the south of Siberia compare ore-bearing complexes of different ages (Mekhonoshin et al. 2016). For example, platinum bearing of southern Siberia, combined into three age groups, namely, late Paleoproterozoic (Chiney Complex), late Mesoproterozoic (Srednecheremshansky layered mafic-ultramafic complex), and Neoproterozoic (Kingash and Yoko-Dovyren and some of the central East Sayan), has their equivalents of the large igneous provinces (LIP) in northern Canada: 1880–1865 Ma Ghost-Mara River-Morel LIP, 1270–1260 Ma Mackenzie LIP, and 725–720 Ma Franklin LIP (Ernst 2014). The PGE-Cu-Ni bearing Verkhniy Kingash, Yoko-Dovyren in Siberia and Tartay, Zhelos, and Tokti-Oy in the East Sayan might have probably associated with the Franklin LIP. Polyakov et al. (2013), based on new data on mineralogy, geochemistry, and age of Precambrian ultramafic-mafic complexes on the southern periphery of the Siberian Craton, outlined the East Siberian PGE-Cu-Ni province. It includes the Yenisei Ridge, Kan Uplift, and Alkhadyr terrane with the adjacent Biryusa block, northern Baikal region (Yoko-Dovyren), and should also include the Kodar-Udokan area.

The rocks of the Udokan Supergroup are intruded by bodies of diverse magmatic complexes (Fig. 2.4). The oldest ones are two small (3 and 18 km^2) alkali granite plutons of the Katugin Complex, to which the Katugin rare-metal deposit is related. These plutons extend nearly west-east along the southern margin of the Kodar-Udokan trough, near the Stanovoi suture. The granite rocks are dated as 2066 ± 6 Ma (U–Pb zircon age; Larin et al. 2002). A series of large granite plutons, consisting of two intrusive phases, is related to the Kodar Complex. Vodovozov and Zverev (2015) and Vodovozov et al. (2016), based on a 5-year study of the Paleoproterozoic complexes in the Olekma block, concluded that the main intrusive complexes of these areas were approximately coeval to post-collisional complexes: 1873+/−3 Ma and 1877+/−4 Ma granite of the Kodar Complex, 1867+/−3 Ma gabbroids of Chiney, and 1863+/−9 Ma Kuranakh Complexes. The mean paleomagnetic pole for three complexes is lat = −21.7° long = 96.1° A95 = 11.6°. This pole coincides, as expected, to the beginning of the Paleoproterozoic APWP of Siberia (Didenko et al. 2015). During opening of the Vilyui rift it shifted

Fig. 2.7 Data for zircons
from sample 0305 plotted on
a diagram with concordia

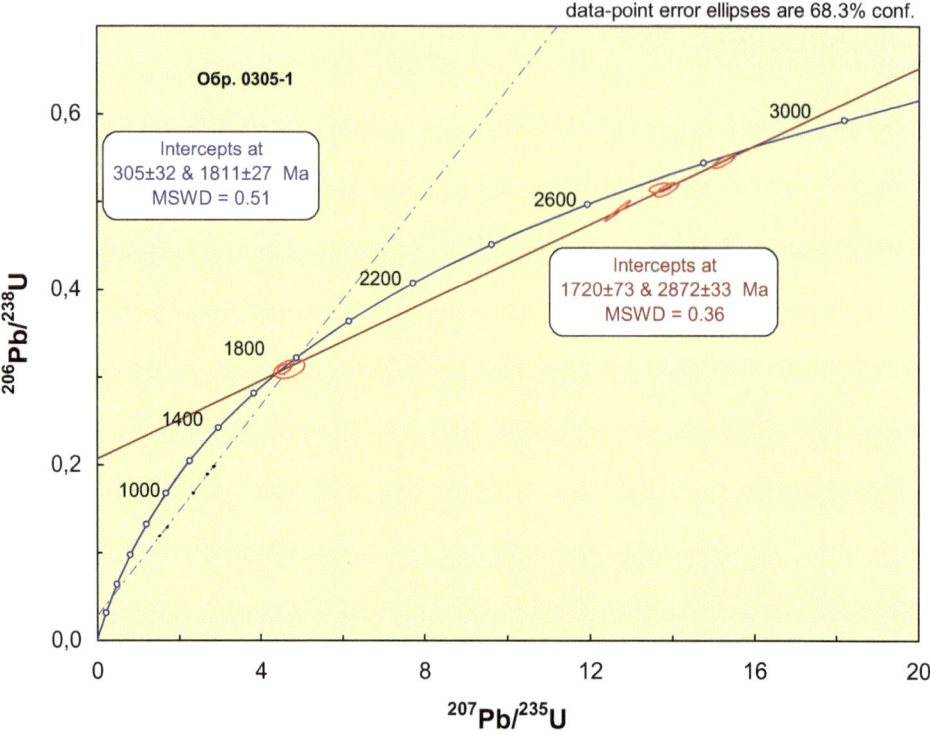

toward younger poles. Comparing the data on the Paleoproterozoic poles of southern Siberia, one can conclude that craton was consolidated near 1.86 Ga.

Dating of the Chiney and Luktur constrained their ages in the range of 860–1880 Ma (Tolstykh et al. 2008; Polyakov et al. 2006; Gongalsky et al. 2009, 2012). These data are consistent with formation of the Kalar-Nimnyr dike swarm (1865 Ma), which is considered by several researchers (Gladkochub et al. 2010; Ernst 2014; Ernst et al. 2016) to belong to the same LIP. Moreover, the dike swarms (Ghost swarm) and the sills (Morel sills, Mara River), close in age (1880–1870 Ma), were also recorded in the Slave Craton, northern Canada (Buchan et al. 2010; Ernst et al. 2016; Fig. 2.10). Considering that paleomagnetic reconstructions admit a close spatial location of southern Siberia and northern Laurentia in the late Paleoproterozoic (Pisarevsky et al. 2008; Didenko et al. 2009), it can be assumed that the Kalar-Nimnyr dike swarm, the Chiney Intrusive Complex, and the Slave dike swarm could be fragments of a single LIP.

West of the Chiney pluton, gabbro, gabbronorite, and gabbrodolerite are exposed along the northern and eastern contacts of the Lurbun granite pluton and in the deep canyon of the River Chukchudu. Their petrography and geochemistry, Fe-Ti-V and Cu sulfide mineralization, as well as the pattern of magnetic and gravity fields provide grounds to assign these rocks to the Chiney Complex. The emplacement of the late Paleozoic granite of the Lurbun pluton was associated with the fragmentation of the gabbro body. Its major part was overlain by granite, which was emplaced along a gently dip-

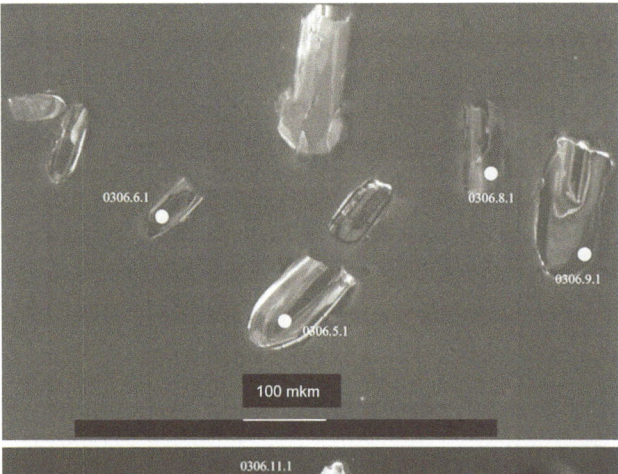

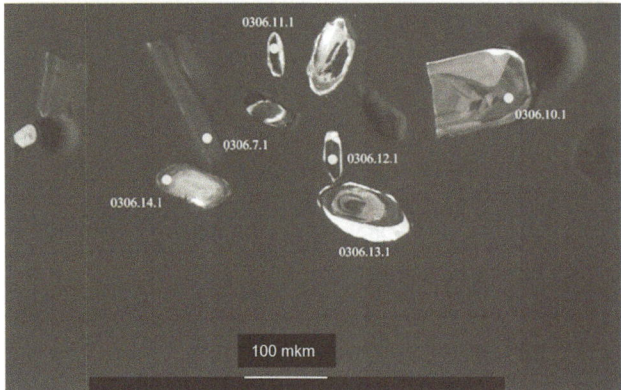

Fig. 2.8 CL image of zircons from high-Ti gabbro with position of analytical points (sample 0306). Sample numbers correspond to Table 1.1

Fig. 2.9 Data for zircons from sample 0306 plotted on a diagram with concordia

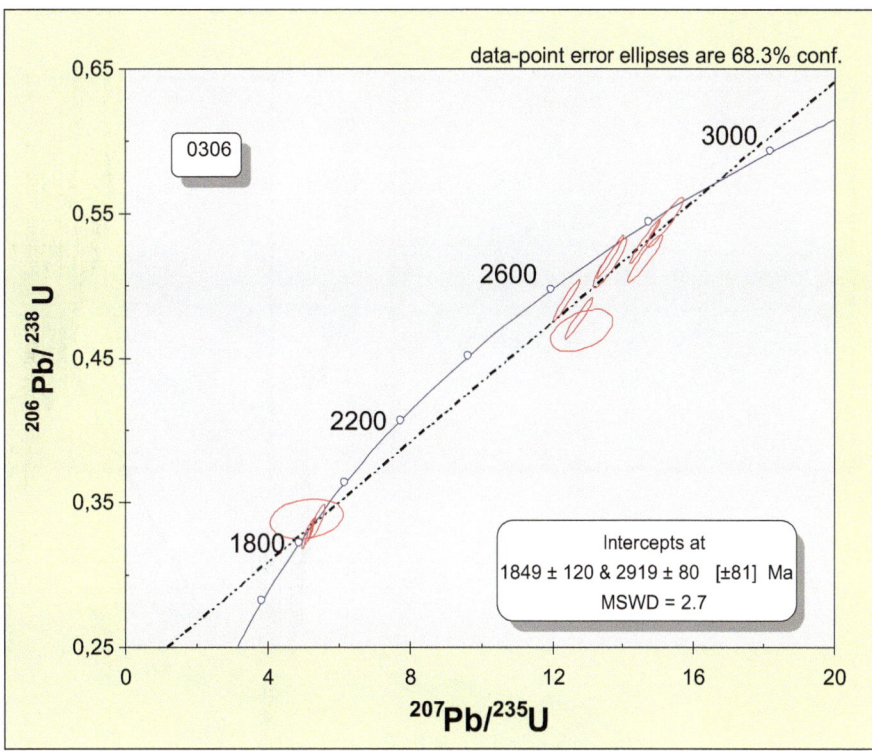

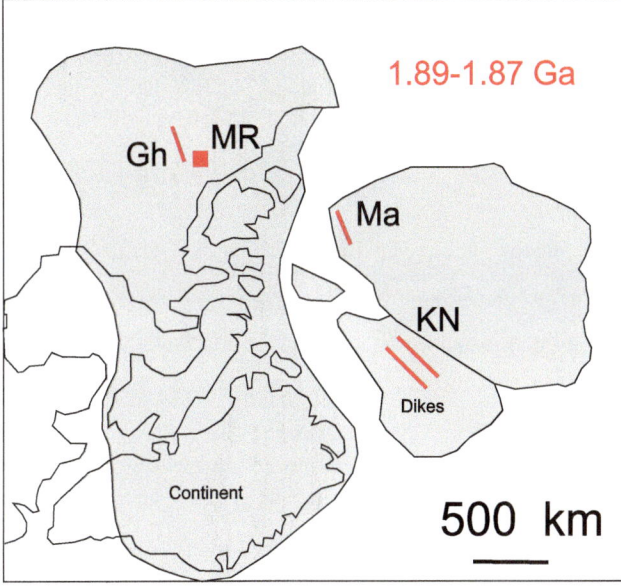

Fig. 2.10 The 1.85–1.89 Ga LIP event of southern Siberia and northern Laurentia. (after Ernst et al. 2016).
Ma Malozadoisky dike, *KN* Kalar-Nimnyr swarm, *Gh* Ghost dolerite dikes, *MR* Mara River

ping fault, and its minor part was uplifted, forming the Chiney pluton in its present-day configuration.

At regional scale, the Kodar-Udokan trough, a first-order structure, coincides with a large gravity minimum (ΔgRres ~0 mGal), reflecting a greater depth of the Archean basement (Arkhangel'skaya et al. 2004). The Kemen and Ingamakit

ring units with similar geophysical pattern (Fig. 2.11) are structures of the second order (Gongalsky et al. 1995a, b). They are characterized by zonal structure: their central zones correspond to the lowest Δg values and coincide with granite plutons of the Kodar (Paleoproterozoic) and Ingamakit (Paleozoic) Complexes. The intermediate zones reveal positive Δg and negative ΔT gradients and correspond to the hydrothermally altered rocks; the marginal zones are noted due to stronger gravity field with local Δg and ΔT maxima above the exposed (Chiney, Luktur) and buried layered mafic plutons (Fig. 2.11).

These were mapped based on the limited outcrops of gabbro on the periphery of the late Paleozoic Lurbun granite pluton (Fig. 2.12). The authors, who compiled magnetic and gravity maps in the 1960s, assumed identity of the anomaly-inducing masses with gabbro rocks of the Chiney pluton, which is exposed nearby, but this assumption was not taken into account on geological maps.

It should be emphasized that the attribution of gabbro rocks that occur in granite plutons, to the first phase of the same intrusive complex, was a customary approach used by mapping geologists in the Olekma, Amanan, and Amudzhikan Intrusive Complexes in the eastern Transbaikalia region (Pavlinov et al. 1976; Gavrikova et al. 1976; Gavrikova and Gongalsky 1980). Following this approach, the gabbro rocks, spatially associated with granitoids of the Ingamakit Complex, have been grouped into the first intrusive phase of this complex. The recently applied geochemical methods distinguished separate intrusive phases and complexes in

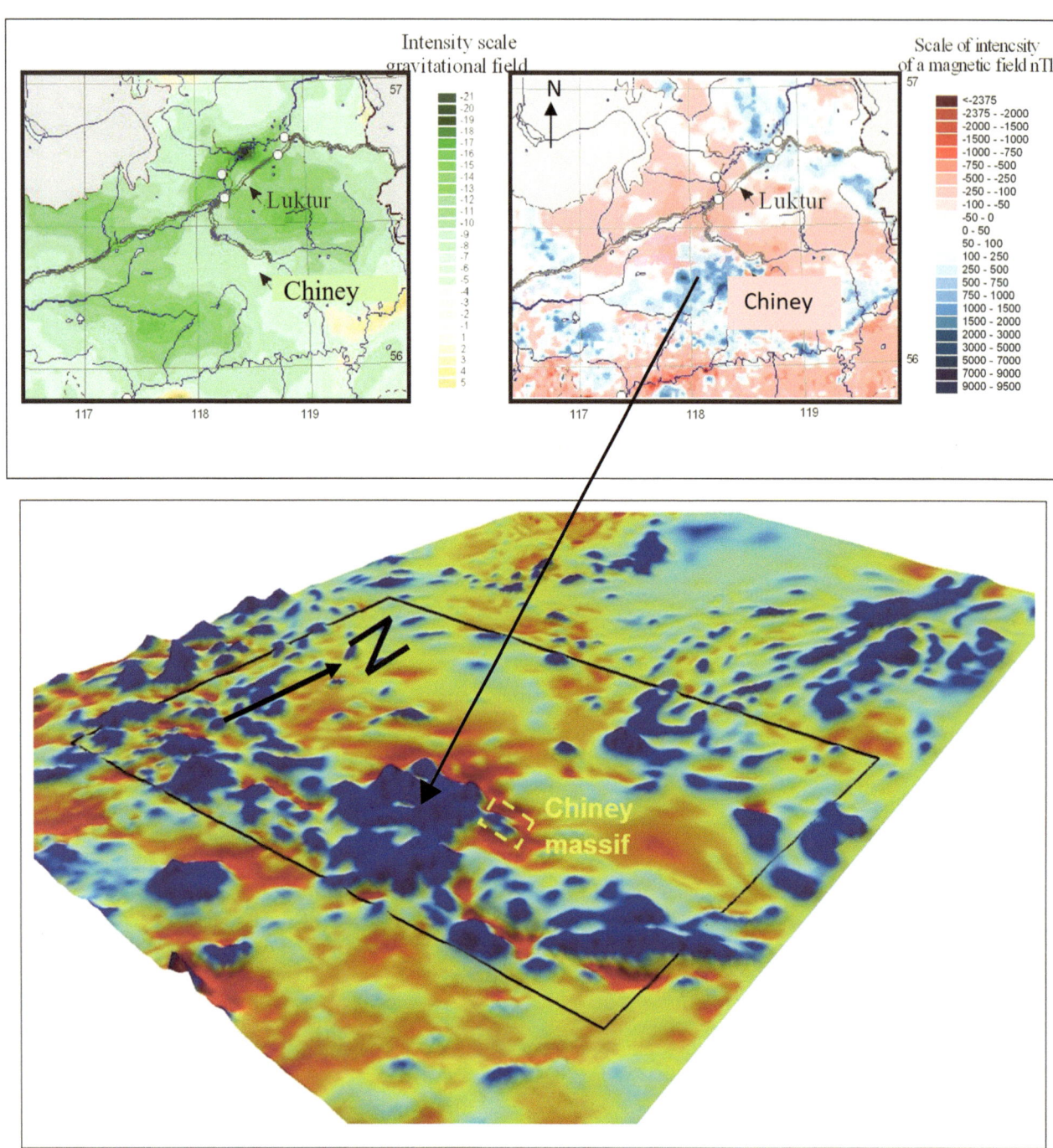

Fig. 2.11 Gravity (in arbitrary units (**a**)) and magnetic maps (ΔT (**b**)). (http://www.vsegei.ru/ru/info/gisatlas/sfo/zabaykalsky_kray/index.php)

large magmatic bodies, and the Mylovskiy pluton of the Chiney Complex has been recognized on this basis (see below).

Location of plutons pertaining to the Chiney Complex is indicated by arrows on the gravity and magnetic maps of northeastern Transbaikalia (unpublished data). On the gravity map, these plutons are distinguished based on positive anomalies, contrasting to the lower background values above the Kodar-Udokan trough filled with Paleoproterozoic carbonate-clastic rocks of the Udokan Supergroup. On the 1:200,000 map of anomalous magnetic field (Sokol 1978, unpublished), the Chiney pluton corresponds to intense local positive anomaly. West of the Chiney pluton is the magnetic anomaly over a much larger area (Fig. 2.13), corresponding to the Mylovskiy pluton of the Chiney Complex, which is overlain by late Paleozoic granite. According to the interpretation of the gravity data, the top of the Mylovskiy pluton is 0.8–1.5 km deep (VM Kravchenko, pers. comm.).

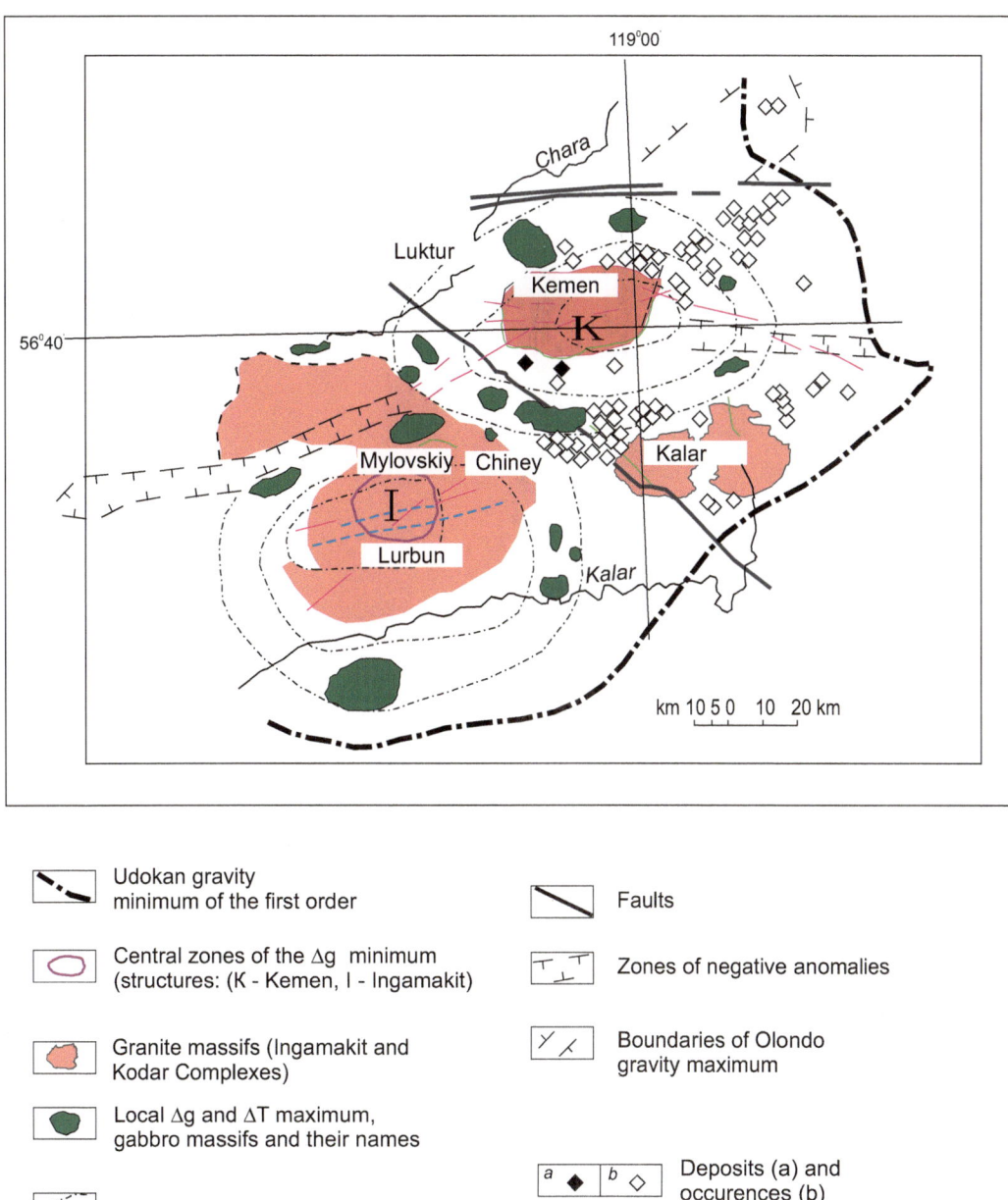

Fig. 2.12 Geological-geophysical scheme of the Kodar-Udokan district. (after Gongalsky et al. 1995a, b)

The intensity of magnetic anomalies above the late Paleozoic granite rocks of the Ingamakit Complex is comparable to that above the Fe-Ti-V ore in the Chiney pluton, and we use this as an evidence for genetic links of sporadic gabbro outcrops in the marginal part of the granite pluton with the gabbro of the Chiney Complex. The gabbrodolerite dikes radiate from the inferred gabbro layered mafic-ultramafic complex, such as the Main Udokan Dike. They are related to the Chiney Complex or to younger intrusive complexes.

Important evidence for these intrusive rocks, belonging to the magmatic system of the Kodar-Udokan district, is their geochemical similarity. The distribution of trace elements, in particular, topology of their normalized patterns, which are independent of magma fractionation, is especially revealing. In addition to the Chiney and Mylovskiy plutons, gabbro rocks of the Luktur pluton and the Main Udokan Dike that intersects the deposits in its central part, as well as a number of smaller intrusive bodies, were examined. All of these rocks are characterized by elevated Ti concentrations, similar K_2O/Na_2O ratios, regardless of the MgO concentration, and similar configurations of the normalized multi-elemental patterns (Fig. 2.14). This figure displays the distribution of trace elements normalized to the primitive mantle (Hofmann 1988) in gabbro rocks of the Luktur pluton and titanomagne-

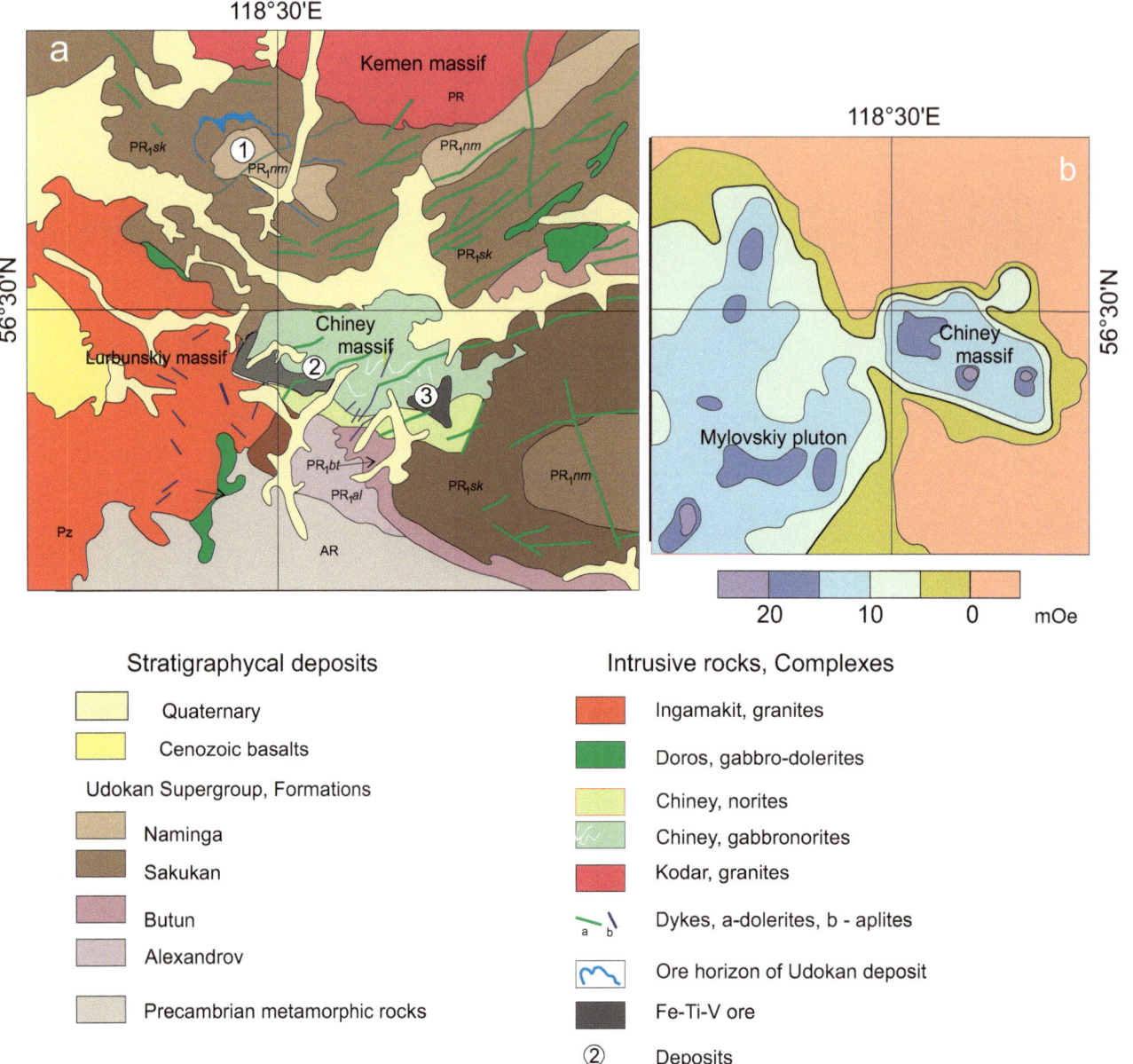

Fig. 2.13 Schematic geological (**a**) and magnetic maps (**b**) of the Chiney-Mylovskiy pluton

tite gabbro of the Mylovskiy pluton, as compared with the chemical composition of the Chiney pluton. The latter is characterized by representative analyses of slightly fractionated gabbro of the Ti-magnetite-bearing gabbro and norite series that differ from each other only in positive Ti anomaly in the former. The monzodiorite of the Chiney pluton and biotite-bearing gabbronorite of the Luktur pluton contain the highest concentrations of all elements as a result of magma fractionation in the upper part of the intrusive bodies. The titanomagnetite gabbro of the Mylovskiy pluton is distinguished by depletion in all trace elements, especially LILE, and by positive Ti and Sr anomalies (Table 2.3).

Despite the aforementioned differences in mineralogical specialization, the trace-element patterns for all rocks are quite similar, with pronounced negative Ta and Nb anomalies, positive Pb anomalies, and similar La/Sm and Gd/Yb ratios, which indicate their similar origin (Fig. 2.14). All rocks display evidence of crustal contamination, and this is confirmed by negative ε_{Nd} = −4.4 to −5.0 (Table 2.4). In addition to petrographic and geochemical similarities between gabbro from the Chiney, Luktur, and Mylovskiy plutons, these rocks contain titanomagnetite and sulfides with elevated PGE and Au concentrations.

Fig. 2.14 Spider diagram for rocks of the Chiney Complex normalized to primitive mantle (PM). (After Hofmann 1988)
Here and below Cr, element content in rock; C_{PM}, element content in primitive mantle. Massifs, lines: green, Chiney; blue, Luktur; red, Mylovskiy

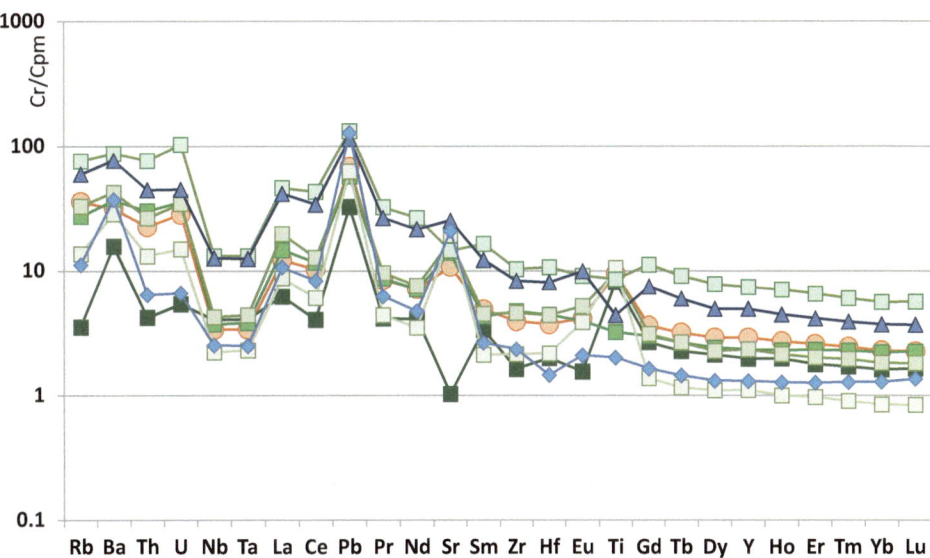

Table 2.3 Chemical composition of rocks from the Chiney pluton, wt. %

No	1	2	3	4	5	6	7	8	9	10	11	12	13	14
SiO_2	45.86	53.15	37.48	50.43	44.33	44.81	49.78	43.2	44.65	41.42	35.49	30.81	21.41	17.02
TiO_2	2.11	0.39	3.79	0.34	2.26	2.38	1.12	2.61	2.29	3.36	4.67	5.74	8.13	9.05
Al_2O_3	23.94	26.16	7.18	13.41	12.64	14.26	13.43	14.11	14.22	7.21	6.81	6.03	4.33	3.4
Fe_2O_3	13.48	3.53	30.63	12.82	20.04	20.41	14.71	21.32	17.95	10.12	15.11	19.41	26.47	29.92
FeO_2	n.a.	n.a.	n.a.	n.a.	n.a.	n.a.	n.a.	n.a.	n.a.	16.89	19.84	21.95	27.77	29.62
MnO	0.06	0.03	0.29	0.2	0.18	0.17	0.2	0.17	0.17	0.26	0.28	0.3	0.32	0.31
MgO	0.72	0.69	8.93	13.04	6.49	5.95	9.31	5.72	6.09	9.96	8.81	9.12	7.05	6.91
CaO	9.29	9.68	8.95	7.91	11.21	8.53	8	9.42	11.1	9.35	7.76	5.3	3.27	2.14
Na_2O	3.82	4.71	0.98	1.17	2.05	2.32	2.52	2.72	3.06	0.98	0.94	1.01	0.99	1.43
K2O	0.36	1.22	0.14	0.27	0.34	0.66	0.83	0.73	0.54	0.37	0.26	0.3	0.24	0.19
P_2O_5	0.09	0.14	0.03	0.04	0.07	0.07	0.07	0.05	0.08	0.08	0.03	0.03	0.03	0.01
Total	99.73	99.7	98.4	99.63	99.61	99.56	99.97	100.05	100.15	100	100	100	100.01	100

Note. Rocks with known Sm-Nd isotopic data (1–5): (1) anorthosite, (2) chineite, (3) titanomagnetite gabbronorite, (4) gabbronorite, (5) titanomagnetite gabbro; ordinal numbers correspond to numbers of points in isochron plot. XRF of rock-forming elements were carried out at IGEM RAS, analyst AI Yakushev; (6–11) average weighted composition: (6) all rocks of 987 analyses: (7) gabbronorite series (16 analyses), (8) titanomagnetite gabbro series (31 analyses), (9) leucogabbro series (20 analyses), (10–11) rhythm from Borehole 11 from a depth of 622 m (10 analyses)

Table 2.4 Sm-Nd isotopic composition of rocks from the Chiney pluton

Sample	Sm, ppm	Nd, ppm	$^{147}Sm/^{144}Nd \pm 2\delta$	$^{143}Nd/^{144}Nd \pm 2\delta$	ε_{Nd}	$t_{mod, Ma}$
1	0.78	4.1	0.1149 ± 2	0.511405 ± 10	−4.6	2290
2	1.68	8.8	0.1156 ± 2	0.511424 ± 9	−4.5	2270
3	0.54	2.5	0.1303 ± 3	0.511587 ± 13	−4.8	2400
4	2.05	8.5	0.1453 ± 3	0.511757 ± 8	−5.0	2600

Note (1) anorthosite, (2) chineite, (3) gabbronorite, (4) titanomagnetite gabbro
Analyses were performed at Institute of Geology of Ore Deposits, Petrography, Mineralogy, and Geochemistry, Russian Academy of Sciences (IGEM RAS), analyst YuV Gol'tsman

Authors with the aid of AV Petrov and GD Pavlovich used geophysical data for modeling of the internal structure of upper crust in the Udokan area and created a 3D model (Fig. 2.15) of gravitational and magnetic masses within the standard Russian 1:200,000 map sheets (O-XXXV (Tombasov et al. 2004a) and O-XXXVI (Tombasov et al. 2004b) designed in the CosCAD 3D software (Trusov and Petrov 2000; Demoura and Petrov 2014, http://coscad3d.ru/). These 3D models were imported into RockWorks 16 (https://www.rockware.com/) and Voxler 3 (http://www.goldensoftware.com/) software packages, where the distribution of gravitational and magnetic masses was interpreted.

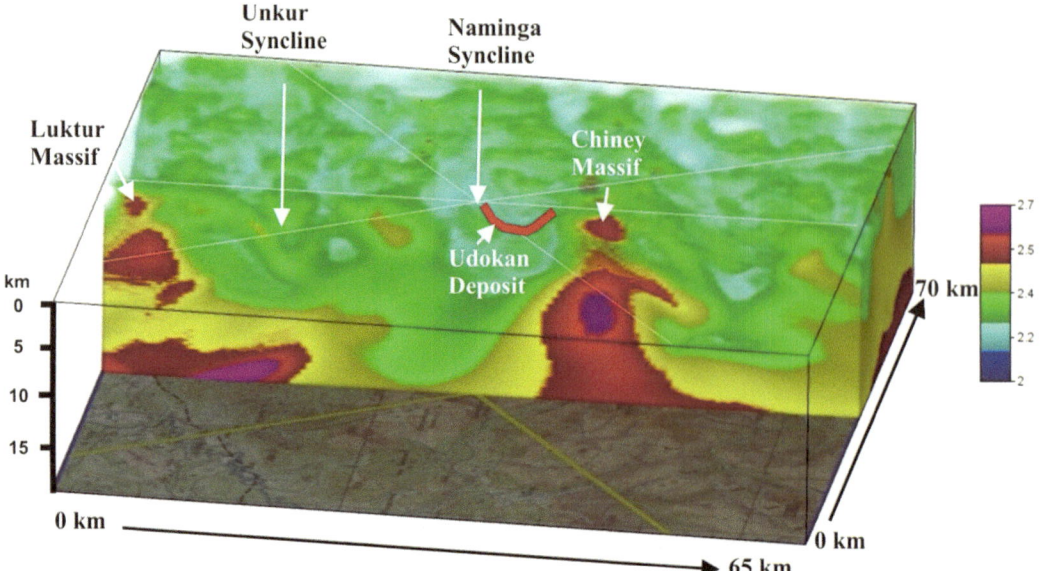

Fig. 2.15 A 3D gravity model of the Udokan subzone. Scale for gravity field is given in arbitrary units

The 3D model (Fig. 2.15) reflects main gravity structures where anorthosite-gabbronorite of Chiney and Luktur form positive (anticline) structures, between which there are the synclines (Unkur, Naminga). The latter contains the Udokan Cu-Ag-Fe and Unkur Cu-Ag deposits. The layered intrusions exposed on the surface represent the upper zones of high-density columns. The columns can be traced to a depth of about 20 km, making branches of a large magma chamber at a depth of about 20 km. The magmatic columns are zonal, gabbros, and gabbronorites dominating in the upper parts. Ultramafic rocks can be found at the lower levels of magmatic columns.

The younger intrusive rocks are minor mafic intrusions of the Neoproterozoic Doros Complex, which are clustered into swarms, cutting through the Vendian rocks. Scarce intrusions of the Khani Complex, a counterpart of the Synnyr Complex in the northern Baikal region, are composed of ultrapotassic rocks (synnyrites), whose K_2O content reaches 15–17 wt %.

The late Paleozoic intrusive rocks of the Ingamakit Complex are widespread in the Baikal-Vitim region. The plutons, up to 1000 km^2 in area, consist of four intrusive phases varying in composition from monzonite and monzo-diorite to granite. The isotopic age of the Ingamakit Complex corresponds to the Early Permian (Chechetkin and Kharitonov 2002).

The Jurassic-Cretaceous intrusive rocks of the Aldan complex group into clusters, belts, and fields, which are controlled by faults. Three rock associations are localized in various parts of the Aldan Shield. The association of leucite and alkali syenites occurs in the western and central parts of the shield; granite rocks are abundant in the eastern and southern parts;

monzonite and syenite are exposed throughout the shield except its extreme western and eastern portions.

Gold mineralization of the Aldan province (Kuranakh, Lebedinskoe, Elkon) is related to Mesozoic felsic rocks. The rare metal and charoite occurrences associate with alkaline rocks (134 Ma, Wang et al. 2014). Potentially diamondiferous dunite, kimberlite, and lamproite of the same age (2.96–3.0 Ga) can be also expected (Smelov et al. 2012).

2.5　Conclusions

1. The Kodar-Udokan mineral district is spatially constrained to a narrow spur of the western margin of the Aldan Shield on the southeastern periphery of the Siberian craton between the Baikal-Muya and Mongol-Okhotsk mobile belts.
2. The Northeastern Transbaikalia province, with large Fe-Cu-Ag, Ti-V-Fe, and Ta-Nb-Zr-REE-Y resources, is of Paleoproterozoic age. The copper deposits of the Kodar-Udokan basin form the largest accumulation of copper (>50 Mt. Cu) in Russia.
3. The Ingamakit and Kemen ring structures play an important role in regional geology and localization of the Paleoproterozoic deposits of diverse genetic types.
4. Available data on zircon ages of gabbro in the Chiney pluton (1858 ± 17 Ma, 1811 ± 11 Ma) show that there was no appreciable time gap between the deposition of clastic rocks of the Chiney and Kemen Groups and no time gap occurred between formation of copper sandstone and magmatic deposits related to the Chiney Complex.

5. The Chiney pluton is a fragment of the deep Paleoproterozoic magmatic system. The Mylovskiy pluton, which is cut through by late Paleozoic granite rocks of the Ingamakit Complex and is overlain by Neogene to Quaternary volcanic rocks of the Udokan lava plateau, is part of this system.

6. The 1.9 Ga U-REE occurrences and deposits, recently discovered in the Chiney pluton and its host rocks, are coeval with the leading types of mineralization.

References

Abramov BN (2011) Formation conditions and ore potential of black shales of the Udokan Group, eastern Siberia. Lithol Mineral Dep 46(4):353–362

Ames DE, Davidson A, Wodicka N (2008) Geology of the giant Sudbury polymetallic mining camp, Ontario, Canada. Econ Geol 103:1057–1077

Arkhangel'skaya VV (1998) Cryolite. Moscow, Geoinformmark, 26 (in Russian)

Arkhangel'skaya VV (2014) Deposits of the Baikal-Amur Mainline zone. VIMS, Moscow, p 226 (in Russian)

Arkhangel'skaya VV, Ryabtsev VV, Shuriga TN (2012) Geological structure and mineralogy of tantalum deposits. VIMS, Moscow, p 191 (in Russian)

Arkhangel'skaya VV, Bykhov Y, Volodin RN, Narkelyun LF, Skursky VC, Trubachev AI, Chechetkin VS (2004) Udokan copper and Katugin rare metal deposits in the Chita region, Russia. Administration of Chita, Chita, p 519 (in Russian)

Arkhangel'skaya VV, Getmanskaya TI, Pechenkin IG, Shuriga TN (2010) Cryolite of Russia. In: Mashkovtsev AG (ed) Exploration and conservation of mineral resources. VIMS, Moscow, pp 26–31 (in Russian)

Berezhnaya KK, Bibikova EV, Sochava VB, Kirnozova TI, Makarov VA, Bogomolov YS (1988) Isotopic age of the Chiney subformation, Udokan Group from the Kodar–Udokan Trough. DoklAkadNaukUSSR 302(2):1209–1212 (in Russian)

Berezkin VI, Timofeev VF, Smelov AP, Postnikov AA, Timoshina ID, Zedgenizov AN, Popov NV (2007) Geology and petrology of Paleoproterozoic lower Khani graben-syncline (Aldan–Stanovoi shield): to the problem of search for traces of ancient life on the earth. Otech Geol 5:62–71 (in Russian)

Bogdanov YV, Kochin GG, Kutyrev EI, Travin LV, Feoktistov VP (1966) Geology, formation conditions, and distribution of cupriferous sandstones in northeastern Olekma-Vitim mountain province. Int Geol Rev 8(11):1305–1315

Bonhomme MC (1987) Type of sampling comparison between K-Ar and Rb-Sr isotopic dating of fine fractions from sediments in attempt to date young diagenetic events. Chem Geol Isotope Geosci Sect 65(3–4):209–222

Buchan KL, Ernst RE, Bleeker W, Davies W, Villeneuve M, van Breemen O, Hamilton M (2010) Proterozoic magmatic events of the Slave Craton, Wopmay Orogen and environs. Geological Survey of Canada, p 26

Bulgatov AN, Gordienko IV (1999) Terrane and overlap assemblage map of Transbaikalia and Eastern Sayan Region, Southern Siberia, Russia. Scale 1:5,00 0,000. U.S. Geological Survey, 1 sheet

Bulgatov AN, Gordienko IV (2014a) Fold systems of Sayan-Baikal Mountain area. In: Leonov YuG, Petrov OV, Pospelov II (eds) Explanatory note to the tectonic map of Northern, Central and Eastern Asia. Scale 1:2,500,000. VSEGEI, St-Petersburg, pp 53–58 (in Russian)

Bulgatov AN, Gordienko IV (2014b) Tectonic map of Northern-Central-Eastern Asia and adjacent areas. Scale 1:2,500,000. Petrov OV, Leonov YuG (eds) VSEGEI, St-Petersburg, p 15 (in Russian)

Burmistrov VN (1990) Structure and composition of the Kemen Group of the Udokan Complex of eastern Siberia. Soviet GeolGeofiz 31(3):23–30 (in Russian)

Cawthorn RG (2010) The platinum group element deposits of the bushveld complex in South Africa. Platinum Metals Rev 54:205–215

Charlier B, Namur O, Latypov R, Tegner C (eds) (2015) Layered intrusions. Springer, Dordrecht, p 732

Chechetkin VS, Kharitonov YuF (2002) Geology and mineral deposits of the Chita Segment of BAM. Chita, p 63 (in Russian)

Chechetkin VS, Yurgenson GA, Narkelyun LF, Trubachev AI, Salikhov VS (2000) Geology and ore of the Udokan copper deposit: a review. Russ GeolGeophys 41(5):710–722

Clauer N (1984) New approach to Rb-Sr method for sedimentary rocks. In: Chernyshov IV, Gorochov IM (eds) Isotope geology. Nedra, Moscow, pp 40–62 (in Russian)

Demoura GV, Petrov AV (2014) Physical-geological modeling (simulation) and anizotropy geomagnetic tomography. Russ Geophys J 6:18–24. in Russian

Didenko AN, Vodovozov VY, Pisarevsky SA, Gladkochub DP, Donskaya TV, Mazukabzov AM, Stanevich AM, Bibikova EV, Kirnozova TI (2009) Palaeomagnetism and U-Pb dates of the Palaeoproterozoic Akitkan Group (South Siberia) and implications for pre-Neoproterozoic tectonics. In: Reddy SM, Mazumder R, Evans DAD, Collins AS (eds) Palaeoproterozoic supercontinents and global evolution. Special Publication 323. Geological Society of London, pp 145—163

Didenko AN, Vodovozov VY, Peskov AY, Guryanov VA, Kosynkin AV (2015) Paleomagnetism of the Ulkan massif (SE Siberian platform) and the apparent polar wander path for Siberia in late Paleoproterozoic – early Mesoproterozoic times. Precambrian Res 259:58–77

Dixon CJ (1979) The Witwatersrand gold-uranium deposits — South Africa. In: Atlas of economic mineral deposits. Springer, Dordrecht, pp 26–27

Dodin D (2002) Metallogeny of the Taimyr–Noril'sk Region, the north of Central Siberia. Nauka, St.-Petersburg, p 822 (in Russian)

Donskaya TV, Gladkochub DP, Kovach VP, Mazukabzov AM (2005) Petrogenesis of early Proterozoic post-collisional granitoids in the southern Siberian Craton. Petrology 13(3):229–252

Duk LV, Kitsul VI, Berezkin VI (1979) Structures and metamorphism of the early Precambrian in the Aldan shield in the Timpton and Sutam River Valleys. In: Pushcharovsky YM (ed) Geodynamic studies. Nauka, Moscow, pp 7–29 (in Russian)

Ernst RF (2014) Large igneous provinces, vol 653. Cambridge University Press, Cambridge

Ernst RF, Hamilton MA, Söderlund U, Hanes JA, Gladkochub DP, Okrugin AV, Kolotilina TB, Mekhonoshin AS, Bleeker W, LeCheminant AN, Buchan KL, Chamberlain KR, Didenko AN (2016) Long-lived connection of Siberia and northern Laurentia in the Proterozoic. Nat Geosci 9:464–469

Faure G (1986) Principles of isotope geology. Willey, New York, p 589

Fedorovsky VS (1972) Lower Proterozoic stratigraphy of the Kodar and Udokan Ridges (East Siberia). Nauka, Moscow, p 130 (in Russian)

Fedorovsky VS (1985) The lower Proterozoic of the Baikal mountainous area. Nauka, Moscow, p 200 (in Russian)

Fraser GL, Huston DL, Gibson GM, Neumann NL, Maidment D, Kositsin N, Skirrow RG, Jaireth S, Lyons P, Carson C, Cutten H, Lambeck A (2007) Geodynamic and metallogenic evolution of Proterozoic Australia from 1870–1550 Ma: a discussion. Geosci Aust Rec 16:76

Gavrikova SN, Gongalsky BI (1980) Alkalinity and acidity of granitoids from the Olekma Stanovik ridge. Geokhimiya 18(10):1481–1500 (in Russian)

Gavrikova SN, Pavlinov VN, Orlov VN, Rachkov VS, Gongalsky BI, Fedchuk VY, Surikova EP (1976) Mesozoic tectonomagmatic reactivation of the southwestern Olekma Stanovik ridge, eastern Transbaikalian region. Izv Vuzov Geologiya I Razvedka 10:12–19 (in Russian)

Gladkochub DP, Wingate MTD, Pisarevsky SA, Donskaya TV, Mazukabzov AM, Ronomarchuk VA, Stanevich AM (2010) Proterozoic mafic magmatism in Siberian craton: an overview and implications for paleocontinental reconstruction. Precambrian Res 183:660–668

Glebovitsky VA, Khil'tova VY, Kozakov IK (2008) Tectonics of the Siberian Craton: interpretation of geological, geophysical, geochronological, and isotopic geochemical data. Geotectonics 42:8–20

Glukhovsky MZ (2009) Paleoproterozoic thermotectogenesis: a rotation–plume model of the formation of the Aldan shield. Geotectonics 43(3):226–250

Glukhovsky MZ, Moralev VM, Sukhanov MK (1993) Tectonic position of Paleoproterozoic anorthosites and granitoids of the Aldan shield and zoning of thermotectogenesis. Geotekt Forsch 28(3):69–81 (in Russian)

Gongalsky BI, Krivolutskaya NA (1993) The Chiney layered pluton. Novosibirsk, Nauka, 184 (in Russian)

Gongalsky BI, Krivolutskaya NA, Goleva NG (1995a) Ore deposits of the Chiney pluton. In: Laverov NP (ed) Mineral deposits of the Transbaikalian region, vol 1. Geoinformmark, Moscow, pp 20–28 (in Russian)

Gongalsky BI, Golovatyi AS, Abushkevich SA (1995b) Zonal ring structures of the Udokan range. Doklady Earth Sci 343(1):80–82 (in Russian)

Gongalsky BI, Safonov YG, Krivolutskaya NA, Prokof'ev VYu, Yushin AA (2007) A new type of copper–noble metal mineralization in Northern Transbaikalia. Dokl Earth Sci 414(5):645–648

Gongalsky BI, Sukhanov MK, Holtzman YV (2008) Sm-Nd isotope system of the Chiney anorthosite-gabbronorite pluton (eastern Transbaikalia). In: Chernyshev ID (ed) Problems of geology of ore deposits, mineralogy, petrology and geochemistry. IGEM, Moscow, pp 57–60 (in Russian)

Gongalsky BI, Makariev LB, Voyakovsky SK (2009) Mesozoic to Cenozoic magmatism of the Udokan–Chiney District and uranium mineralization. In: Gordeev EI (ed) Volcanism and geodynamics. Institute of Volcanology, Petropavlovsk-Kamchatsky, pp 321–323 (in Russian)

Gongalsky BI, Timashkov AN, Voyakovsky SL (2012) U-Pb dating results on Paleoproterozoic zircons from intrusions of the Udokan-Chiney ore district (Russia). In: Laverov NP (ed) Abstracts of materials from Russian conference on isotope geochemistry. IGEM, Moscow, pp 110–112 (in Russian)

Gorokhov IM, Timofeev VF, Bisunok MB, Kuznetsov AB (1989) Rb-Sr systems in metasedimentary deposits of the Khani trough (Olekma granite-greenstone area). In: Levsky LK, Levchenkov JA (eds) Isotope geochronology of the Precambrian. Nauka, Leningrad, pp 110–125 (in Russian)

Gruenewaldt GV (1977) The mineral resources of the bushveld complex. Minerals Sci Eng 9:83–95

Harn DMV, Gruenewaldt GV (1995) Ore-forming processes in the upper part of the bushveld complex, South Africa. J Afr Earth Sci 20:77–89

Hedenquist JW, Harris M, Camus F (eds) (2012) Geology and genesis of major copper deposits and districts of the world: attribute to Richard H. Sillitoe. SEG Spec Publ 16:618

Hofmann AW (1988) Chemical differentiation of the earth: relationships between mantle, continental crust and oceanic crust. Earth Planet Sci Lett 90:297–314

Irvine TN (1977) Origin of chromitite layers in the Muskox intrusion and other stratiform intrusions: a new interpretation. Geology 5:273–277

Kazansky VI (1982) Faults in the Baikal–Amur region and related endogenic mineralization. In: Kuznetsov VA (ed) Faults and endogenic mineralization of the Baikal–Amur region. Nauka, Moscow, pp 5–14 (in Russian)

Khain VE (2001) Tectonics of continents and oceans, vol 604. Scientific World, Moscow (in Russian)

Konnikov EG (1986) Precambrian differentiated mafic–ultramafic complexes in the Transbaikalian region. Nauka, Novosibirsk, p 224 in Russian

Korikovsky SP (1979) Facies of metamorphism for metapelites, vol 263. Nauka, Moscow (in Russian)

Krasny LI (1980) Geology of the Baikal–Amur railroad region. Nedra, Moscow, p 159 (in Russian)

Krendelev FP, Bakun NN, Volodin RN (1983) Udokan copper sandstone, vol 248. Nauka, Moscow (in Russian)

Larin AM, Kovalenko VI, KotovAB S'n EB, Kovach VP, Makariev LB, Timashkov AN, Berezhnaya NG, Yakovleva SZ (2000) New data on the age of granite of the Kodar and Tukuringra complexes, eastern Siberia: geodynamic constraints. Petrology 8(3):238–248

Larin AM, Kotov AB, Sal'nikova EB, Kovalenko VI, Kovach VP, Yakovleva SZ, Berezhnaya NG, Ivanov VE (2002) Age of the Katugin Ta–Nb deposit, Aldan–Stanovoi shield: evidence for the identification of the global rare-metal metallogenic epoch. Dokl Earth Sci 383:336–339

Larin AM, Kotov AB, Sal'nikova EB, Glebovitsky VA, Sukhanov MK, Yakovleva SZ, Kovach VP, Berezhnaya NG, Velikoslavinsky SD, Tolkachev MD (2006) The Kalar complex, Aldan–Stanovoi shield, an ancient anorthosite–mangerite–charnockite–granite association: geochronologic, geochemical and isotopic–geochemical characteristics. Petrology 14(1):2–20

Lightfoot PC (2017) Nickel Sulfide Ores and Impact Melts: Origin of the Sudbury Igneous Complex. Elsevir 687

Makariev LB, Voyakovsky SK, Il'kevich IV (2009) Gold ore potential of uranium objects in the Kodar–Udokan trough. Rudy I Metally 6:56–64 (in Russian)

Makariev LB, Mironov YB, Voyakovsky SK (2010) The outlook for the discovery of new types of economic uranium deposits in the Kodar–Udokan zone of the Transbaikalian territory in Russia. Geol Ore Deposits 52(5):381–391

Mekhonoshin AS, Ernst RE, Söderlund U, Hamilton MA, Kolotilina TB, Izokh AE, Polyakov GV, Tolstykh ND (2016) Relationship between platinum-bearing ultramafic-mafic intrusions and large igneous provinces (exemplified by the Siberian craton). Russ Geol Geophys 57(5):1043–1057

Mineeva IG, Arkhangel'skaya VV (2007) New direction in methodology of uranium and gold deposits on shields and in the Precambrian folded areas. Explor Conserv Subsoil 11:18–25 (in Russian)

Mitrofanov GL (ed) (2016) Explanatory note to geological map of Russian Federation, scale 1:1,000,000 (third generation). VSEGEI, St-Peterbourg (in Russian)

Mitrofanov GL, Moskovchenko NI (eds) (1985) The early Precambrian of the Aldan shield and its framework. Nauka, Leningrad, p 184 (in Russian)

Neymark LA, Larin AM, Nemchin AA, Ovchinnikova GV, Ritsk EY (1998) Anorogenic nature of magmatism in the northern Baikal volcanic belt: evidence from geochemical, geochronological (U–Pb) and isotopic (Pb-Nd) data. Petrology 6(2):124–148

Nutman AP, Chernyshev IV, Baadsgaard H, Smelov AP (1992) The Aldan shield of Siberia, USSR: the age of its Archean components and evidence for widespread reworking in the mid-Proterozoic. Precambrian Res 54(4):195–209

Pavlinov VN, Gavrikova SN, Gongalsky BI, Orlov VN, Rachkov VS (1976) Pre-Mesozoic tectonomagmatic evolution of the south-western Olekma Stanovik ridge, eastern Transbaikalian region. GeologiyaiRazvedka 8:10–24 (in Russian)

Perelló J, Sillitoe RH, Yakubchuk AS, Valencia VA, Cornejo P (2017) Age and tectonic setting of the Udokan sediment-hosted copper-silver deposit, Transbaikalia, Russia. Ore Geol Rev 86:856–866

Pisarevsky SA, Natapov LM, Donskaya TV, Gladkochub DP, Vernikovsky VA (2008) Proterozoic Siberia: a promontory of Rodinia. Precambrian Res 160:66–76

Podkovyrov VN, Kotov AB, Larin AM, Kotova LN, Kovach VP, Zagornaya NY (2006) Sources and provenances of lower Proterozoic terrigenous rocks of the Udokan group, southern Kodar–Udokan depression: results of Sm–Nd isotopic investigations. Doklady Earth Sci 408:518–522

Pokrovsky BG, Grigoriev VS (1995) New data on age and geochemistry of isotopes of Udokan Supergroup, lower Proterozoic, eastern Siberia. Lithol Miner Dep 3:273–283 (in Russian)

Polyakov GV, Isokh AE, Krivenko AP (2006) Platiniferous ultramafic–mafic formations of mobile belts of Central and Southeastern Asia. Russ Geol Geophys 47:1227–1241

Polyakov GV, Izokh AE, Krivenko AP (2006) Platiniferous ultramafic–mafic formations of mobile belts of central and southeastern Asia. Russ Geol Geophys 47(12):1227–1241

Polyakov GV, Tolstykh ND, Mekhonoshin AS, Izokh AE, Podlipskiy MY, Orsoev DA, Kolotilina TB (2013) Ultramafic-mafic igneous complexes of the Precambrian east Siberian metallogenic province (southern framing of the Siberian craton): age, composition, origin and ore potential. Russ Geol Geophys 54(11):1689–1704

Popov NV, Kotov AB, Postnikov AA, Sal'nikova EB, Shaporina MN, Larin AM, Yakovleva SZ, Plotkina YV, Fedoseenko AM (2009) Age and tectonic position of the Chiney layered massif, Aldan shield. Dokl Earth Sci 424(1):64–67

Rogers JJ, Santosh M (2004) Continents and supercontinents. Oxford University Press, New York, pp 5–22

Rosen OM (2003) Siberian Craton: tectonic zoning and stage of evolution. Geotectonics 3:3–21

Rytsk EY, Kovach VP, Yarmolyuk VV, Kovalenko VI, Bogomolov ES, Kotov AB (2011) Isotopic structure and evolution of the continental crust in the east Transbaikalian segment of the central Asian Foldbelt. Geotectonics 45(5):349–377

Sklyarov EV (2006) Exhumation of metamorphic complexes: basic mechanisms. Russ Geol Geophys 47(1):71–75

Salikhov VS (2010) Problems of age for the Udokan group. In: Kurilenko AV (ed) Geology and Minerageny of Transbaikalia. ZabNII, Chita, pp 77–83 (in Russian)

Salop LI (1964) Geology of the Baikal highlands, vol 1. Nedra, Moscow, p 515 (in Russian)

Salop LI (1967) Geology of the Baikal highlands: magmatism, tectonics and geological history. Moscow, Nedra, vol 2, p 699 (in Russian)

Sinitsa SM (1996) Issue of the Udokan Biota of the Kodar-Udokan Area in the Transbaikal Region. In: Yurgenson GA (ed) Problems of ore formation and prospecting and evaluation of raw minerals, Novosibirsk, pp 177–181 (in Russian)

Smelov AP, Kravchenko AA, Berezkin VI, Dobretsov VN (2007) Geology and geochemistry of basic-ultrabasic complexes of central part of the Aldan Shield. Otechestennaya Geologiya 7:53–62

Smelov AP, Shatsky VS, Ragozin AL, Reutsky VN, Molotkov AE (2012) Archean diamondiferous rocks of the Olondo green-stone belt (western Aldan-Stanovoy shield). Russ Geol Geophys 10:1012–1022

Stupak FM (1987) Cenozoic Volcanism of the Udokan Range. Novosibirsk, Nauka, 169 (in Russian)

Tolstykh ND, Orsoev DA, Krivenko AP, Izokh AE (2008) Noble-metal mineralization in mafic–ultramafic layered plutons in the southern Siberian Platform. Parallel, Novosibirsk, p 194 (in Russian)

Tombasov IA, Sinitsa SM (1990) Stratigraphy of rocks from Udokan Complex in Ikabya-Chitkanda region. In: Karsakov LP (ed) Stratigraphy of lower precambrian of the far east. Vladivostok, pp 56–61 (in Russian)

Tombasov IA, Shemelina SF, Afonin GA, Drozdov SA (2004a) Governmental geological map of the Russian Federation, scale 1: 200,000. 2nd edn. Udokan series. O-50-XXXVI (Katugin). Maps. Factory of A. P. Karpinsky Russian Geological Research Institute, St-Petersburg, Russia (in Russian)

Tombasov IA, Shemelina SF, Afonin GA, Drozdov SA (2004b) Governmental geological map of the Russian Federation, scale 1:200,000. 2nd edn. Udokan series. O-50-XXXV (Naminga). Maps. Factory of A. P. Karpinsky Russian Geological Research Institute, St-Petersburg, Russia (in Russian)

Trusov AA, Petrov AV (2000) Computer technology of statistical and spectral-correlation analysis of three-dimensional geoinformation: COSCADE 3D. Russ Geophys J 4: 29–33 (in Russian)

Vil'mova ES (1990) A possible reconstruction of Udokania colonies in Proterozoic sedimentary rocks of southern Transbaikalian Region. In: Yurgenson GA (ed) Topical Problems of Geosciences. Chita, pp 33–38 (in Russian)

Vodovozov VYu, Zverev AR (2015) Paleomagnetism of Early Proterozoic formations of the south of the Siberian craton (Udokan and Kodar Ridges). In: Shecherbakov VP (ed) Paleomagnetism and magnetism of rocks: theory, practice, experiment. Materials of the All-Russian School-Seminar on Paleomagnetism and Magnetism of Rocks. Filigran, Yaroslavl, pp 27–33 (in Russian)

Vodovozov VYu, Zverev AR, Filev EA (2016) Paleomagnetism of the Early Proterozoic complexes of the Olekma block, Siberian craton. In: Abstract of the 11th International Conference "Problems of geocosmos", St.- Petersburg, p 183

Vollbrecht A, Oberthur T, Ruedrich J, Weber K (2002) Microfabric analyses applied to the Witwatersrand gold- and uranium-bearing conglomerates: constraints on the provenance and post-depositional modification of rock and ore components. Mineral Deposita 37:433–451

Volodin RN, Chechetkin VS, Bogdanov YV (1994) The Udokan copper sandstone deposit, eastern Siberia. Geol Ore Deposits 36(1):3–30

Wang Y, Hy H, Ivanov AV, Zhu R, Lo C (2014) Age and origin of charoitite, Malyy Murun Massif, Siberia, Russia. Int Geol Rev 56:1007–1019

York D (1966) Least-squares fitting of a straight line. Can J Phys 44:1079–1086

Zhao GC, Cawood PA, Wilde SA, Sun M (2002) Review of global 2.1–1.8 Ga orogens: implications for a pre-Rodinia supercontinent. Earth Sci Rev 59(1/4):125–162

Part II

Mineral Deposits in Sedimentary Rocks

Abstract

The cupriferous sandstones occur within the Paleoproterozoic Kodar-Udokan basin, superimposed onto Archean crystalline rocks of the Aldan Shield. In addition to the giant Udokan deposit, the basin hosts many smaller satellite deposits and occurrences. The Paleoproterozoic Sakukan Formation, hosting cupriferous sandstone units, and the overlaying Naminga Formation are principal hosting stratigraphic units of the Udokan deposit, forming a Naminga brachysyncline. The ore-bearing stratum was traced for 25 km along strike and includes four mineral levels in the vertical sequence. The Udokan deposit contains a huge endowment of Cu (>25 Mt), Fe (10 Mt), Ag (12 kt), and modest Au (13 t). The mineralization is subdivided into sulfide (70–100% sulfides), oxide (0–30% sulfides), and mixed (30–70% sulfides) ores. The ore principally consists of bornite-chalcocite and less frequently of pyrite-chalcopyrite. New data on mineral composition of ores and its sulfur isotopes are summarized here in order to constrain the genesis of the Udokan deposit.

3.1 Stratigraphic Position of Cupriferous Sandstones

There are numerous occurrences of cupriferous sandstones in the Siberian Platform (Fig. 3.1). They are typically located near outcrops of Archean basement. There are six metallogenic provinces inside the Siberian Platform, (I) Igarka-Yenisei, (II) Sayan, (III) Pribaikal, (IV) Aldan, (V) Verkhoyansk, and (VI) Anabar, and one province outside the platform, on the Severnaya Zemlya Archipelago (Narkelyun et al. 1983; Fig. 3.1). The deposits occur in the Proterozoic, Vendian, Early Cambrian, Middle Cambrian, and Ordovician strata.

The Aldan metallogenic province includes the Aldan Shield and adjacent area to west (Fig. 3.1). The deposits of cupriferous sandstones occur in synclines and grabens. The largest copper accumulation was discovered in the Kodar-Udokan basin (Fig. 2.2), which comprises a giant Udokan deposit and many satellites. It consists of the clastic Proterozoic rocks of the Udokan Supergroup (Fig. 3.2). The Cu-Ag-Fe Udokan deposit is part of the Naminga copper cluster (Fig. 3.3), also comprising the Pravoingamakitskoe, Unkur, and Saku deposits and many occurrences (Fig. 2.4).

Copper mineralization is recorded at different levels of the Paleoproterozoic stratigraphic sequence. The best endowed cupriferous sandstone unit of the Udokan deposit is related to the Upper Subformation of the Sakukan Formation. The deposit was described in many publications (Bakun et al. 1958, 1964, 1966; Grintal, 1968; Krendelev 1959, 1987; Reznikov 1965; Bogdanov et al. 1966, 1973, Yurgenson et al. 1968; Gablina 1983, 1997; Narkelyun et al. 1977, 1983, 1987; Konnikov 1986; Konnikov et al. 1986; Volfson and Arkhangel'skaya 1987; Volodin et al. 1994; Chechetkin et al. 2000; Yurgenson and Abramov 2000; Ptitsyn et al. 2003;Gongalsky and Krivolutskaya 1993, 2009; Gongalsky 2010, 2015; Fat'yanov et al. 2000; Soloviev 2010; Trubachev 2010; Trubachev et al. 2016).

The Udokan deposit was discovered by Elizaveta Burova in 1949, and since that time it remains undeveloped. Since its discovery, the deposit is a subject of comprehensive research, but until now, its origin is a matter of debate. In the twentieth century, its sedimentary origin was widely accepted. It seemed reasonable that ore deposition continued during several million years, when sulfide minerals were accumulated at stratigraphic levels within a depth interval of 2–3 km (Narkelyun et al. 1977; Krendelev et al. 1983; Gablina 1983, 1997). The Sakukan Formation was recognized as the most productive stratigraphic unit for copper mineralization in clastic rocks of the Udokan Supergroup (Fig. 3.2).

© Springer Nature Switzerland AG 2019
B. Gongalsky, N. Krivolutskaya, *World-Class Mineral Deposits of Northeastern Transbaikalia, Siberia, Russia*,
Modern Approaches in Solid Earth Sciences 17, https://doi.org/10.1007/978-3-030-03559-4_3

Fig. 3.1 Map of the Siberian
Platform and its Cu-sandstone
provinces. (After Narkelyun
et al. 1983)
I-VIII – Cu metallogenic
provinces: I, Igarka-Yenisei;
II, Sayan; III, Baikal; IV,
Aldan; V, Verkhoyansk; VI,
Anabar; VII, Severnaya
Zemlya

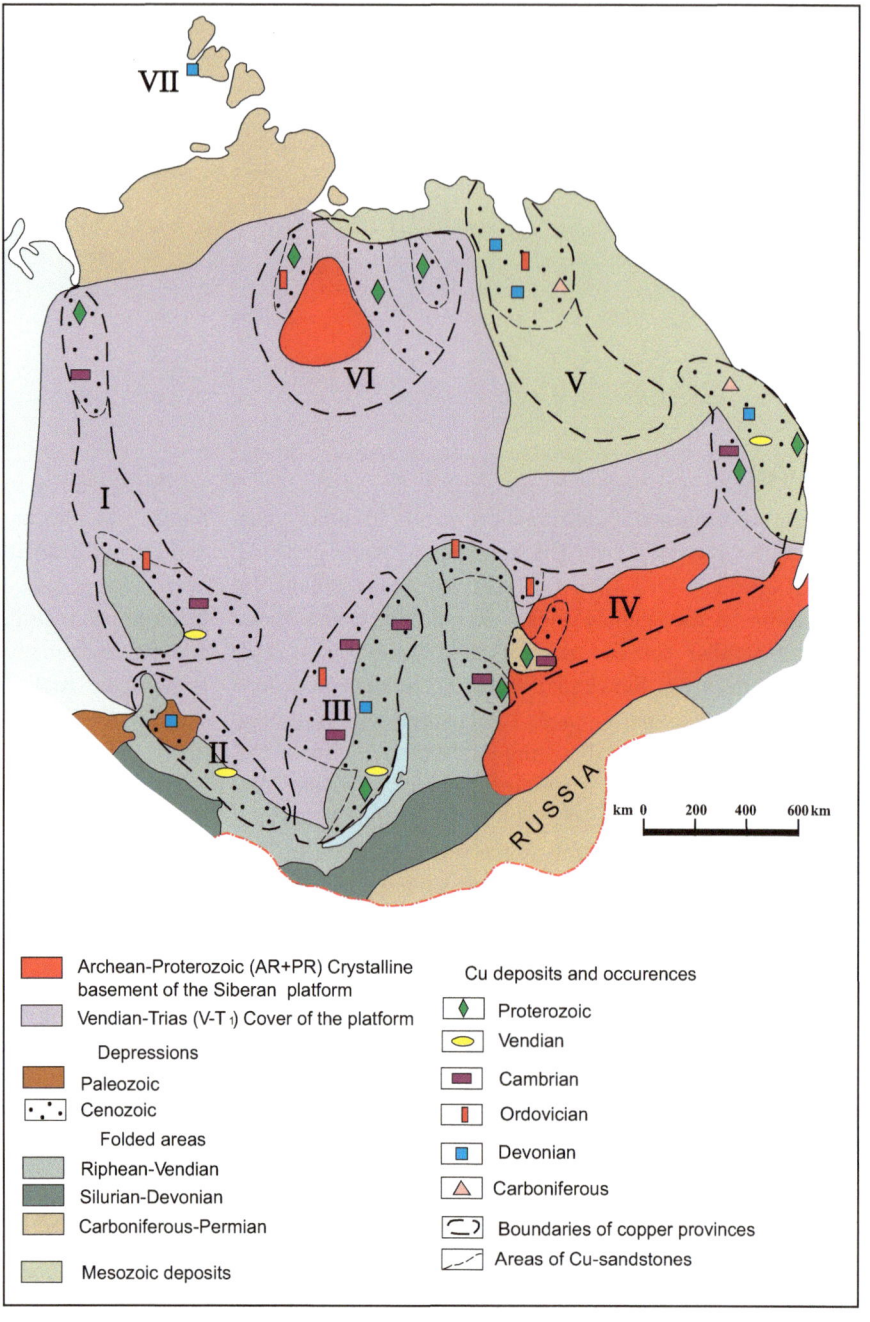

Archean-Proterozoic (AR+PR) Crystalline
basement of the Siberan platform

Vendian-Trias (V-T₁) Cover of the platform

Depressions

Paleozoic

Cenozoic

Folded areas

Riphean-Vendian

Silurian-Devonian

Carboniferous-Permian

Mesozoic deposits

Cu deposits and occurences

◆ Proterozoic

⬯ Vendian

▬ Cambrian

▯ Ordovician

◼ Devonian

△ Carboniferous

⊂⊃ Boundaries of copper provinces

⟋ Areas of Cu-sandstones

km 0   200   400   600 km

The two uppermost formations of the Udokan
Supergroup—the Sakukan Formation, including cupriferous
sandstone units, and the overlying Naminga Formation—are
the hosts of the Udokan deposit. They constitute a Naminga
brachysyncline (Fig. 3.4) in the axial zone of the Kodar-
Udokan mineral district. In plain view, the brachysyncline is
10 × 15 km in size, with elongation in the WNW direction.

The rocks at the northern, eastern, and western limbs dip
inward the syncline core at 10–12° to 35–40°. The southern
limb is more complexly shaped (Fig. 3.4). In its western part,
the strata dip normally to the north at 25–30°, whereas in the
middle and eastern parts, they are overturned, steeply dip-

ping at 45–50° to the south-southwest. At depth, the strata
become vertical and then acquire a normal northward dipping. In the core of the syncline, the rocks lie almost
horizontally.

The Naminga syncline is trans-sected by crush zones,
reverse and reverse-strike-slip faults with displacements of
0.1–15 m. The larger reverse-strike-slip fault zones are filled
with dikes of variable composition. An offset along one of
such faults is filled by a gabbrodolerite dike, which reaches
150 m in thickness. The bedding-parallel and crosscutting
crush zones, 5–25 cm thick, are spaced 30–70 m apart and
can be traced for hundreds of meters or, less frequently, even

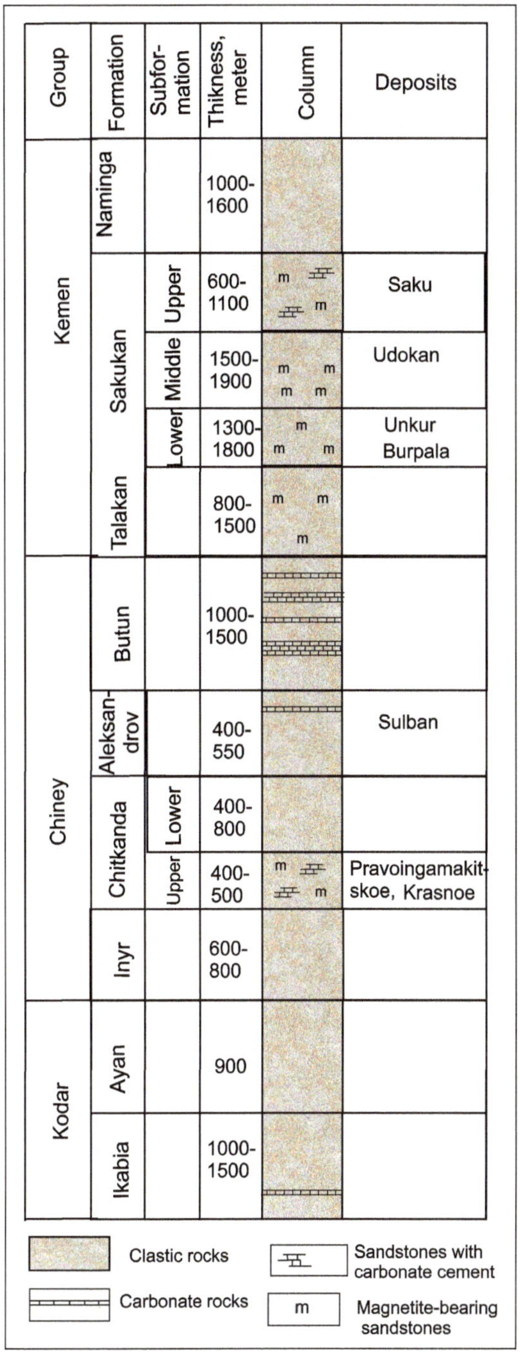

Group	Formation	Subformation	Thickness, meter	Column	Deposits
Kemen	Naminga		1000–1600		
Kemen	Sakukan	Upper	600–1100		Saku
Kemen	Sakukan	Middle	1500–1900		Udokan
Kemen	Sakukan	Lower	1300–1800		Unkur Burpala
Kemen	Talakan		800–1500		
Chiney	Butun		1000–1500		
Chiney	Aleksandrov		400–550		Sulban
Chiney	Chitkanda	Upper / Lower	400–800		
Chiney	Chitkanda	Upper	400–500		Pravoingamakitskoe, Krasnoe
Chiney	Inyr		600–800		
Kodar	Ayan		900		
Kodar	Ikabia		1000–1500		

Legend: Clastic rocks; Carbonate rocks; Sandstones with carbonate cement; Magnetite-bearing sandstones (m).

Fig. 3.2 Stratigraphic chart of the Udokan Supergroup. (Modified after Krendelev et al. 1983)

and, south of the brachysyncline, by a gabbro pluton of the Chiney Complex (Salop 1967; Krendelev et al. 1983). The gabbrodolerite dikes that are widespread throughout the whole brachysyncline are related to the Chiney and Doros complexes.

The rocks of the Sakukan Formation are subdivided into the lower, middle, and upper subformations (Fig. 3.2) (Narkelyun et al. 1968; Chechetkin et al. 2000). The Cu-bearing units are localized in the middle part of the Upper Sakukan subformation, which is about 650 m thick. The Upper Sakukan subformation is, in turn, subdivided into the sub-ore, ore, and supra-ore units. The sub-ore unit (250 m thick) consists of gray and pinkish gray fine- to medium-grained sandstone with sericite-quartz and calcite cement and is distinguished due to numerous lenticular cross-bedded calcareous sandstone layers, 0.05–1.5 m thick, which reoccur at intervals of 1–5 m. The low-grade copper sulfide dissemination in this unit is only of mineralogical interest.

The ore unit and its rhythms show an increase in the amount of clastic material upward in the stratigraphic sequence and a simultaneous decrease in the size of clasts. The thickness of the ore unit does not exceed 20 m at the eastern closure of the fold; it is 140 m at the southern limb; and it reaches 330 m at the northern limb. The supra-ore unit of the Upper Sakukan subformation (Fig. 3.4) reaches 100 m in thickness and consists of pale gray and pinkish gray, mainly fine-grained, horizontally and wavy-bedded quartz-feldspar sandstone and less abundant siltstones without copper mineralization.

3.2 Copper-Bearing Units

The total thickness of the copper-bearing unit at the Udokan deposit varies from a few to 300 m (Krendelev et al. 1983; Narkelyun et al. 1983; Chechetkin et al. 2000). The ore-bearing horizon was traced for 25 km along the current perimeter of the Naminga syncline (Fig. 3.4). There are four ore-bearing levels in the vertical sequence (Chechetkin et al. 1995), labeled as 0, I, II, and III (multiple levels). Their thicknesses vary from 50–100 m to 250–270 m (Fig. 3.5). Fifty-three orebodies were recognized within the ore-bearing horizons. Their typical parameters are as follows: the strike length is 300–2000 m, with 400–2500 m downdip. The thickness is 16–52 m. The shapes of the orebodies vary from tabular and lenticular to ribbonlike. They are characterized by a complex structure (Volodin et al. 1994) and abundant copper sulfide disseminations (Fig. 3.6).

The orebodies are stratiform or lenticular, frequently en echelon lodes, which are sometimes traced for as much as 2–3 km. The orebodies generally dip to the southwest. The thickness of the orebodies is markedly reduced at the southern limb. The internal structure of the bodies is characterized

for 2–3 km. The limbs of the brachysyncline have parasitic folds of higher orders and small flexures. The widespread shear fractures and tension cracks are occasionally filled with quartz and calcite and also with bornite, chalcopyrite, and chalcocite, where they intersect cupriferous sandstone.

Northeast of the Naminga brachysyncline, the metasedimentary rocks of the Udokan Supergroup are cut through by a large Paleoproterozoic granite pluton of the Kodar Complex

a

b

Fig. 3.3 The Naminga site (**a**) and small open pit at the Zapadnyi site (**b**)

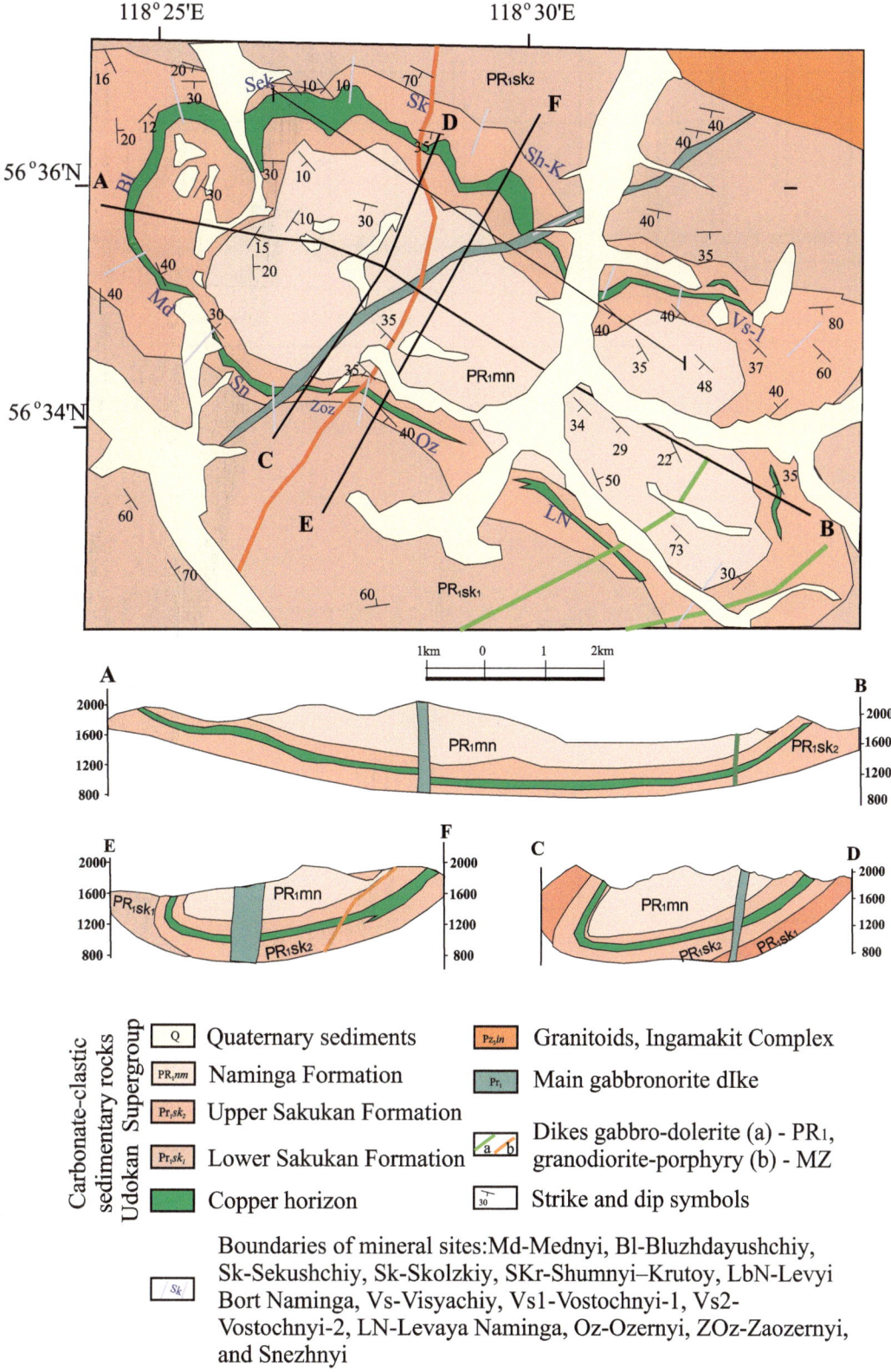

Fig. 3.4 Geological map of the Udokan deposit. (Modified after Krendelev et al. 1983; Chechetkin et al. 2000)

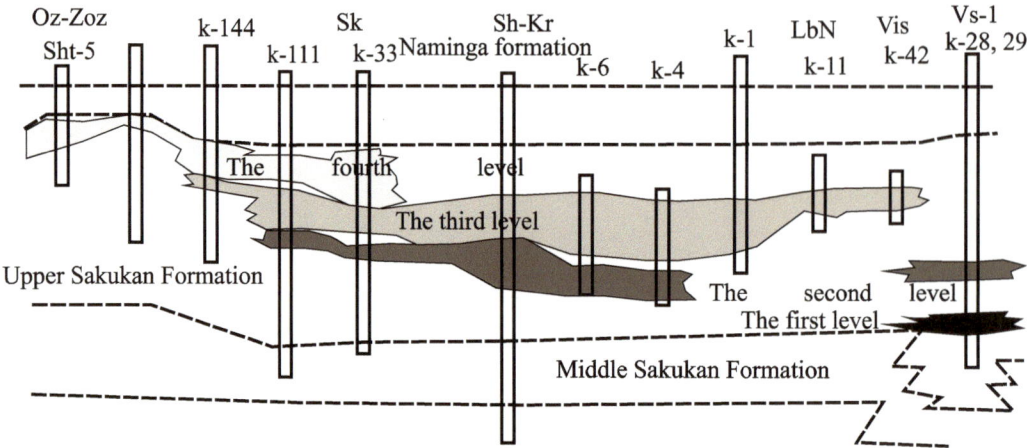

Fig. 3.5 A model, showing distribution of the copper-bearing lenses within the main ore stratum at the Udokan deposit. (After Chechetkin et al. 1995)

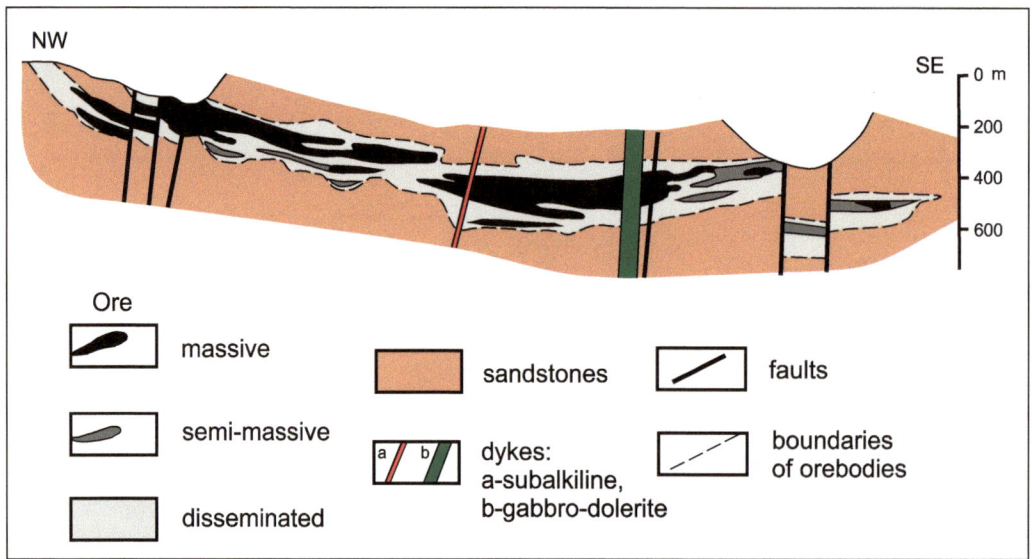

Fig. 3.6 A NW-SE section of the Udokan deposit, showing distribution of massive, semi-massive, and disseminated orebodies. (After Volodin et al. 1994) Line I-I in Fig. 3.4

by frequently alternating and mutually grading layers with mineralization intensity varying downdip, along the strike, and in the direction normal to bedding (Fig. 3.7a). As a result, the orebodies look like custard pies. They are frequently represented by closely spaced veinlets and large sulfide pockets (Fig. 3.7b).

The units are composed of claystone, siltstone, silty sandstone (Fig. 3.8), sandy limestone, and conglomeratic breccia, with intricate mutual transitions seen both laterally and downdip (Chechetkin et al. 2000). The rhythmically repeating beds (from bottom to top) are conglomeratic breccia – sandstone-siltstone, as seen in some vertical sections. The sandstone constitutes >90% of the ore host (Fig. 3.9a–d), whereas clayey rocks collectively account for no more than 3–5%.

Mildly calcareous and noncalcareous varieties are dominant in the sandstone, and less than 40% of the ore volume occurs in the sandstone with calcareous cement (Fig. 3.9e–h). Carbonate rocks are rare (Fig. 3.9i–l). The quartz sandstone is gray, fine- to medium-grained, and inequigranular. In addition to strongly dominating clastic quartz grains, fragments of albite-oligoclase, saussuritized plagioclase, microcline, quartzite, micropegmatite, and felsic volcanic rocks are abundant. The accessory minerals are magnetite, titanomagnetite, ilmenite, zircon (Fig. 3.10), tourmaline, apatite, hematite, and titanite. The sericite-quartz cement with minor calcite dominates in the sandstone. The composition of rock-forming minerals is shown in Table 3.1.

The calcareous medium-grained and noncalcareous fine-grained and equigranular sandstone occurs in the lower part

Fig. 3.7 (**a**) Outcrops of
cupriferous unit at the
Sekushchiy site.
(Photo by K. Murashev).
(**b**) quartz-sulfide veinlets in
sandstone. (Photo
B. Gongalsky)
Cc chalcocite, *Bn* bornite,
Q quartz

of the sequence as thin (~2.5 m) layers and lenses that are
less than 300 m in lateral extent. They dominate in the lower
part of the sequence. Fine-grained and inequigranular quartz
sandstone is widespread in its upper part. These aforementioned rocks are the most typical hosts of copper mineralization at the Udokan deposit. In their outer portions, they are
fused light gray rocks with clastic grains often cemented by
chalcocite and bornite. All sandstone types reveal cross and
wavy bedding (Figs. 3.11 and 3.12). The cross-bedded rocks
are mainly localized in the lower part of this section. The
northern portion of the Naminga syncline is mineralized
most intensely.

The average rock composition of the Udokan deposit is
given in Table 3.2. The distribution of rare elements in studied samples is shown in Table 3.3 and Fig. 3.13.

3.3 Sulfide Ore

3.3.1 Chemical Composition

The orebodies of the Udokan deposit contain a huge
resources (JORC complaint) (https://www.bgk-udokan.ru)
of Cu (>26.7 Mt), Fe (10 Mt), Ag (12 kt), and modest Au
(13 t) (Chechetkin et al. 2011). The ores are subdivided
into sulfide (70–100% sulfides), oxidized (0–30% sulfides), and mixed (30–70% sulfides) types. In the mineral
endowment, the sulfides constitute 43%, with 40% mixed
and 17% oxide ores. Table 3.4 shows proportions of various ore types at different parts of the deposit, both at surface and at depth.

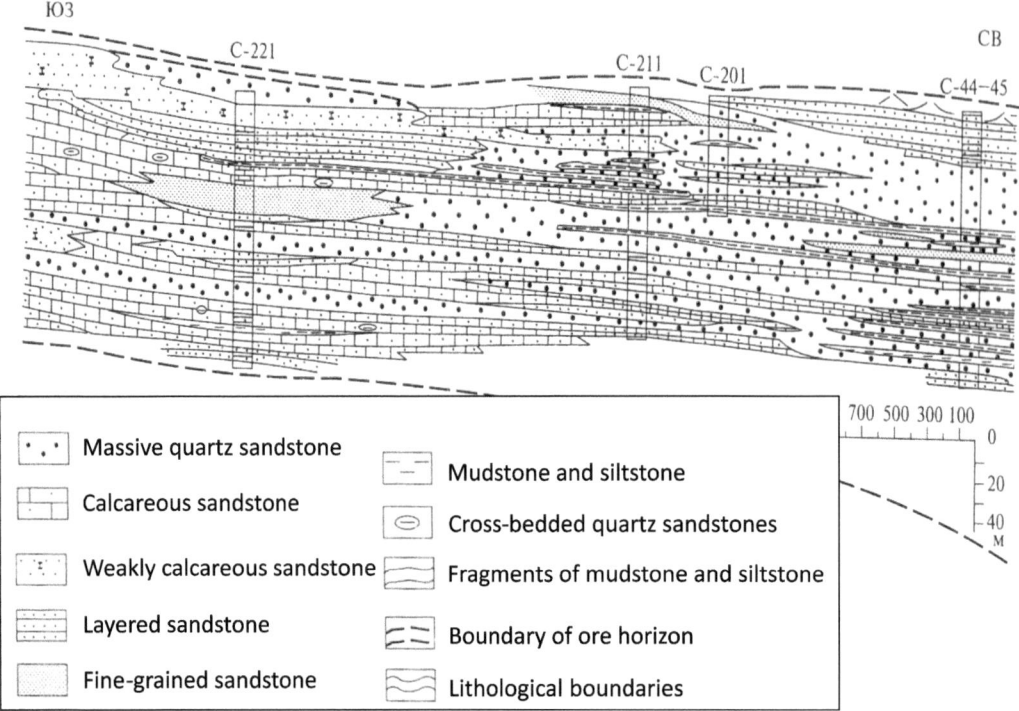

Fig. 3.8 Location of ore units at the Udokan deposit. (After Chechetkin et al. 2000)

The Udokan deposit hosts extended and lenticular tabular orebodies, composed of chalcocite-bornite (67.5%), chalcopyrite (6.5%), and malachite-brochantite (26.0%) (Arkhangel'skaya et al. 2004; Volodin et al. 1994; Chechetkin et al. 2000). The ores of the Udokan deposit contain Ni, Co, Zn, Mo, and other trace elements in addition to the main Cu-Ag-Fe components (Table 3.3). The trace element patterns (Fig. 3.13) of samples collected from Adit 9 (in the western part of the deposit) display elevated U concentrations and very low concentrations of Ta, Nb, Ti, and Cr. The bornite-chalcocite ore reveals elevated values of Ag, Bi, Mo, Re, Pb, and Sb. The pyrite-chalcopyrite ore is enriched in Zn, Co, Ni, Se, and Te.

3.3.2 Structure and Texture

The Udokan ores can be subdivided into disseminated and massive varieties according to their structure and texture (Fig. 3.14). The disseminated ores are dominant in the deposit, and the stringer-disseminated ore is limited in abundance (Krendelev et al. 1983; Gablina 1995). Most orebodies are characterized by disseminated structure. However, alternation of laminae, containing mineralization of various intensities, makes the layered structure rather widespread. The thickness of the laminae with intense sulfide disseminations varies from 1 mm to 1 cm. The morphological varieties of this kind of ore structure can be correlated with cross-bedding, fibrous, and other types of bedding, inherent to the host rocks. The cement texture of the rocks is the most typical (Figs. 3.15 and 3.16), and the bornite-chalcocite aggregates are often characterized by a graphic texture.

The pyrite-chalcopyrite ore (Fig. 3.17) occurs less frequently than the bornite-chalcocite ore, and these types of ore are distinctly separated from each other in the vertical section of the ore unit. The former ore type occurs in gray, dark gray, with greenish hue in cross-bedded fine-grained sandstone and siltstone that contain fine pyrite and chalcopyrite disseminations. The grains range from a few hundredths to, less frequently, a few tenths of a millimeter in size. The sulfide disseminations are distributed in sandstone nonuniformly, and their localization in siltstone is controlled by the bedding-plane cleavage. The breccia ores (Fig. 3.18) are widespread as well. In rare cases, sulfide minerals (mostly chalcocite) were found in late quartz veins (Fig. 3.19).

3.3.3 Mineralogy

The mineralogical composition of the ores varies very little (Narkelyun et al. 1987; Gablina and Ermilov 1990; Gablina 1997, 2014; Krasnokutskaya 2012; Gongalsky et al. 2017). Chalcocite, bornite, and chalcopyrite are major copper minerals. Magnetite is rather widespread. Pyrite and hematite are present as well (Table 3.5). Valleriite, molybdenite, wittichenite, pyrrhotite, sphalerite, marcasite, tennantite, poly-

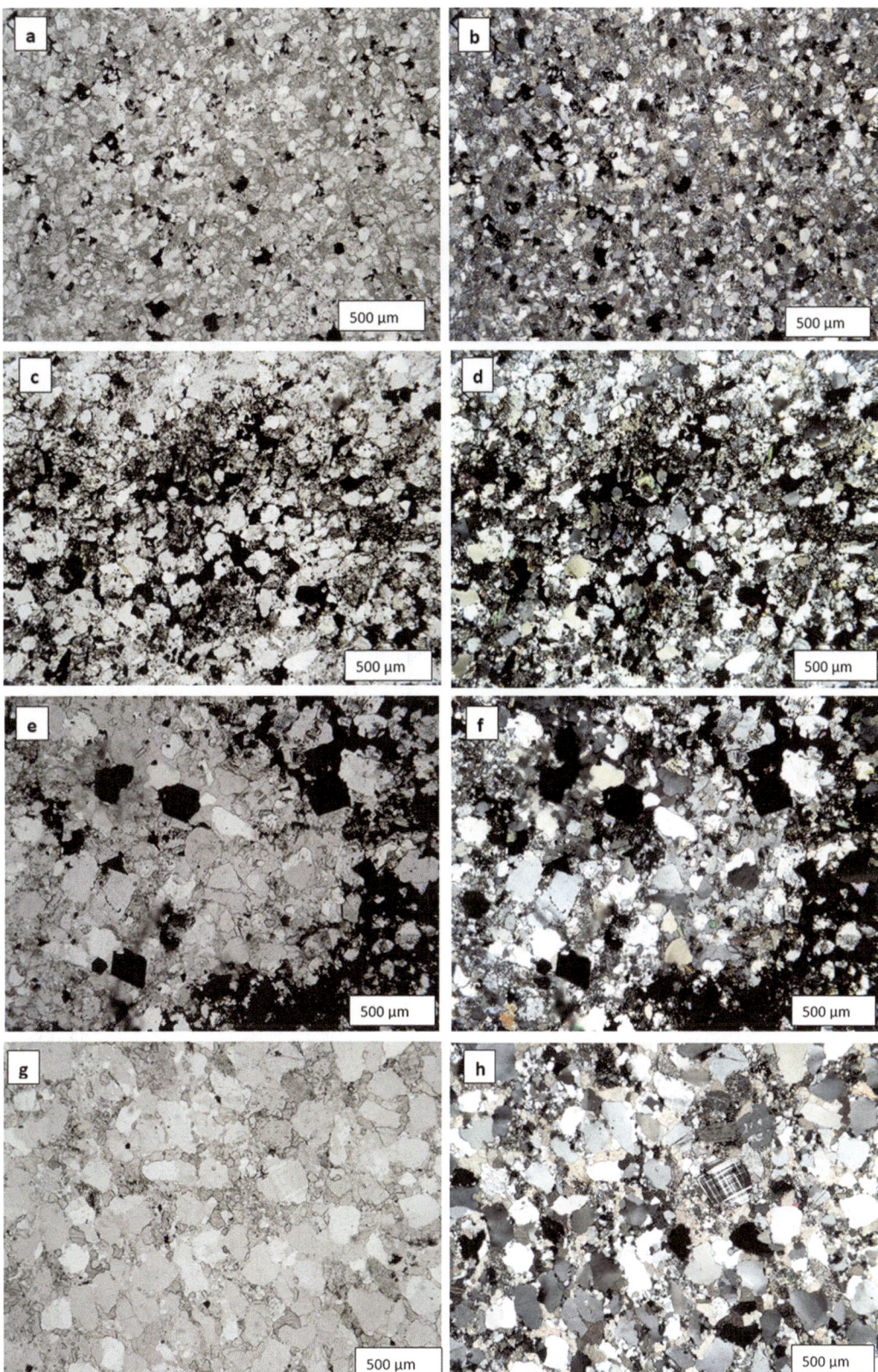

Fig. 3.9 Microphotographs of rocks from the Udokan deposit
(**a**–**d**) Fine- and medium-grained sandstone, (**e**–**h**) medium-grained and coarse-grained sandstone with calcareous cement, (**i**–**l**) carbonate rocks (**a**, **c**, **e**, **i**, **k**, one nicol; **b**, **d**, **h**, **j**, **l**, two nicols)

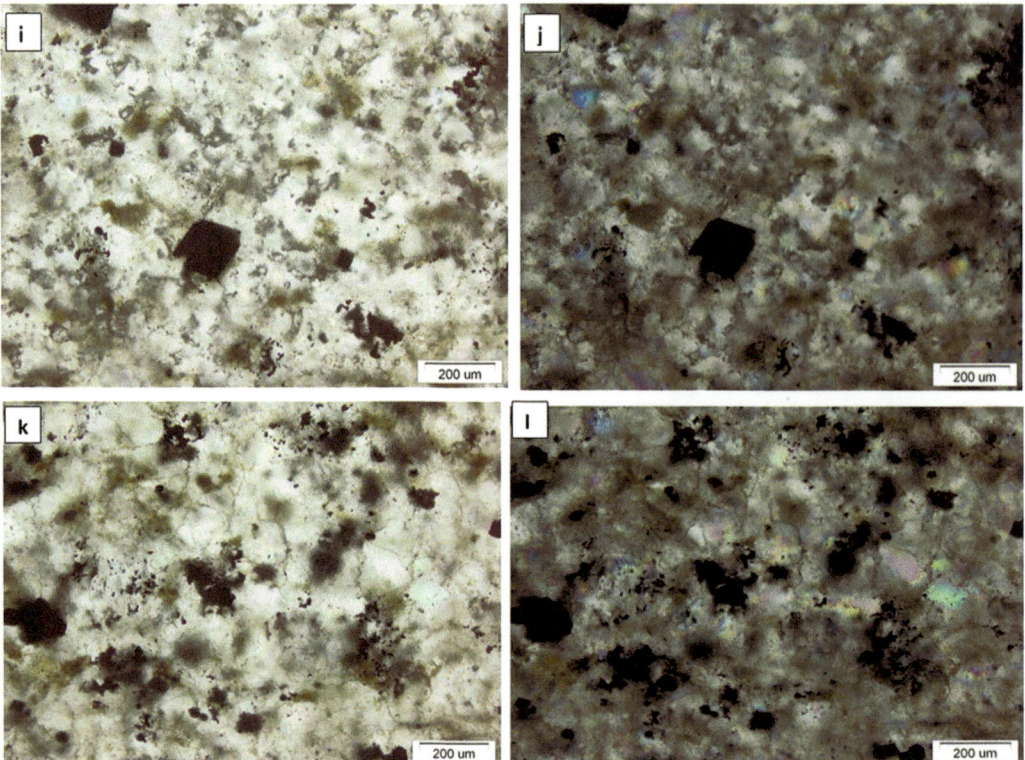

Fig. 3.9 (continued)

dymite, cobaltite, stromeyerite, native silver, and native gold are rare. In the sulfide ore, 65% of copper is contained in chalcocite, 20–25% in bornite, and 10–15% in chalcopyrite. In the oxidation zone, malachite, azurite, gypsum, goethite, and limonite are the most abundant; and tenorite, cuprite, delafossite, chalcanthite, brochantite, antlerite, chrysocolla, melanterite, and jarosite occur less frequently. The three dominant paragenetic mineral assemblages, (i) bornite-chalcocite, (ii) chalcopyrite-bornite, and (iii) pyrite-chalcopyrite, are recognized. The bornite-chalcopyrite ore is transitional between assemblages (i) and (iii).

The zonal distribution of mineral types in the ore can be clearly seen in the plan view and vertical section of the Cu-bearing sequence. The distribution of chalcocite-bornite and pyrite-chalcopyrite mineralization reflects a hypogene zonation, which is typical for many sediment-hosted Cu deposits worldwide.

The northern limb of the brachysyncline is noted for the dominance of bornite-chalcocite ore. The thickness ratio of pyrite-chalcopyrite to bornite-chalcocite ore varies from 1:2 to 1:20. Toward the bottom part of the brachysyncline (i.e., southward), the amount of bornite-chalcocite ore decreases, while the amount of pyrite-chalcopyrite ore increases. At the southern limb, the proportion of bornite-chalcocite ore increases again. However, the ratio of pyrite-chalcopyrite to bornite-chalcocite ore reaches merely 1:1 to 1:3. In the

Cu-bearing sequence at the northern limb of the brachysyncline, the vertical zoning is expressed in the upward transition of the pyrite-chalcopyrite ore into bornite-chalcocite ore. A symmetric zoning is found in the center of the syncline, where the central part of the lode is composed of pyrite ore, which gives way to pyrite-chalcopyrite-bornite-chalcocite ore at the base and top of the lode. At the southern limb, an inverse zoning is recorded. In contrast to the northern limb, the pyrite-chalcopyrite ore occurs at the top of the lode at the southern limb, and the bornite-chalcocite ore is localized below.

The main orebodies at the Udokan are composed of chalcocite and bornite (Figs. 3.15 and 3.20). Their grains with exsolution textures are also very common for chalcopyrite (Fig. 3.21). Bornite and chalcocite always occur in association with magnetite (Fig. 3.15). Chalcocite forms rims around bornite grains, whose morphology changes and depends on the interstitial space (Fig. 3.22a–f, i). Figure 3.23 demonstrates relationships of sulfide minerals with rock-forming minerals. However, veinlets and veins of chalcocite-bornite are also mostly oriented conformably to the bedding (Fig. 3.18a–c). Pyrite also occurs there. The composition of chalcocite minerals in the western part of the Udokan deposit is given in Table 3.6 and Fig. 3.24. Chalcocite and djurleite with elevated Fe content were identified. Among the rare minerals, native

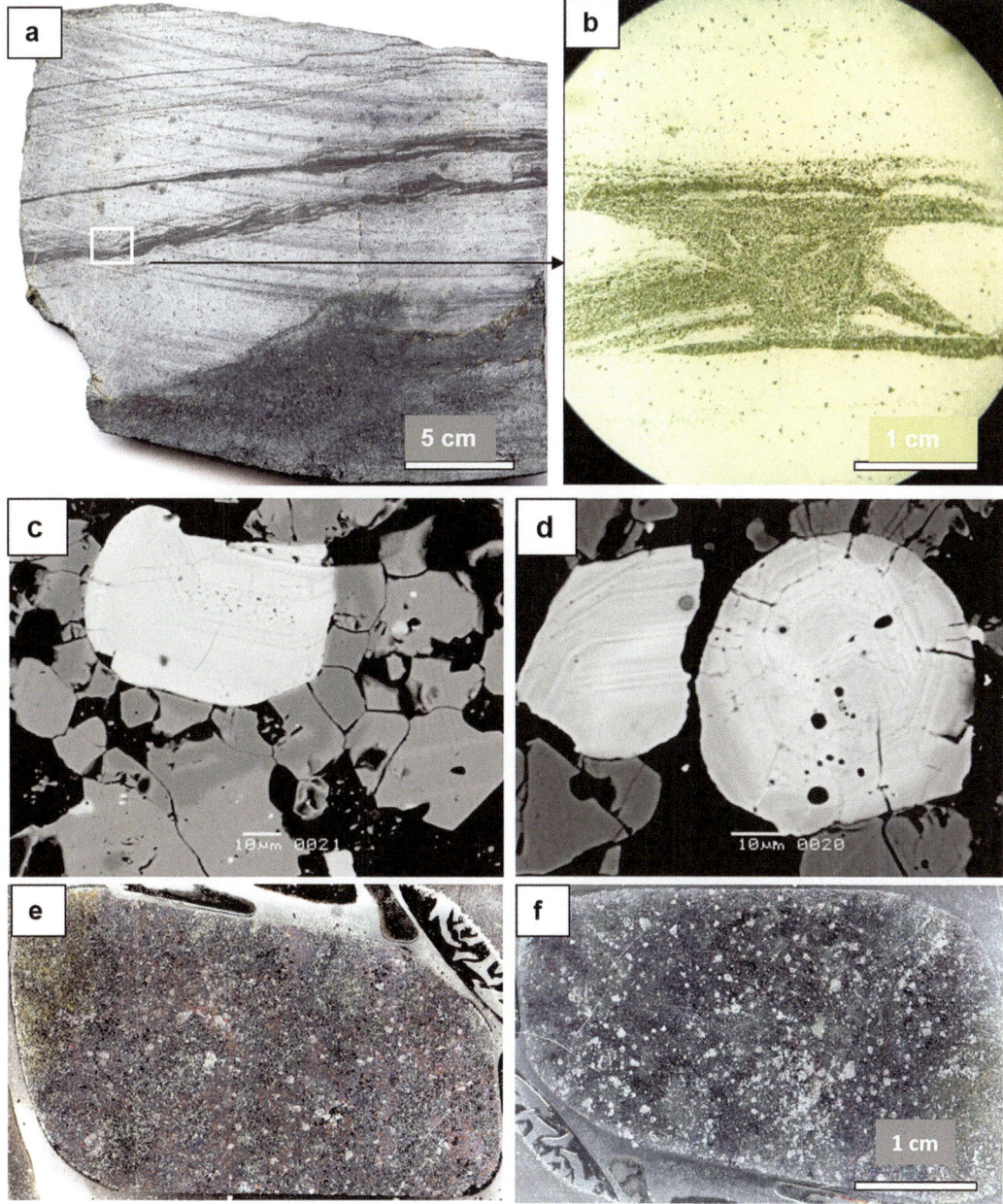

Fig. 3.10 Magnetite layers with traces of erosion and redeposition (**a, b**), grains of zoned zircons (**c, d**) and magnetite-chalcocite-bornite ore (**e, f**). Sample U-9. Concentrations of zircon in magnetite layers (**a, b**) reach 8–10% vol. In the lower part of the section (**a**), massive chalcocite-bornite ore with superimposed impregnation and clusters of these sulfides between magnetite layers; (**e, f**) – massive magnetite-chalcocite-bornite ores

silver and carrollite were identified most frequently (Fig. 3.25).

Iron oxides, predominantly magnetite, and sulfides are found in several orebodies. The magnetite content in the disseminated and massive chalcocite-magnetite ore frequently reaches 50 vol %, with 10–15 vol % on average. Magnetite occurs as round grains and euhedral crystals (Fig. 3.15), which are either products of recrystallization due to superimposed hydrothermal alteration (Gablina and Ermilov 1990) or a newly formed mineral. Magnetite is often martitized

with the relict core of the magnetite grains surrounded by hematite rims. The composition of oxide minerals from the Udokan deposit is presented in Table 3.7. Thin laminae of martitized magnetite often occur in the upper part of the deposit. In Adit 2 in the Naminga River valley, these laminae also contain zircon, apatite, chromite, hematite, and round to angular clastic zircon grains with fine zoning. They are thought to be of volcanic origin. Layered magnetite rims are also found around chalcocite-bornite segregations. In addition to newly formed magnetite, there is another line of evi-

Table 3.1 Chemical compositions of rock-forming minerals in disseminated ore

No	SiO$_2$	TiO$_2$	Al$_2$O$_3$	FeO	MnO	MgO	CaO	Na$_2$O	K$_2$O	Cr$_2$O$_3$	Total	Mineral
1	63.21	0.037	18.05	0.025	0.001	0.026	–	0.31	15.683	–	97.35	Kfs
2	63.41	0.028	18.28	0.012	–	–	–	0.698	15.071	–	97.50	Kfs
3	67.19	0.006	18.81	0.046	–	0.011	0.062	11.415	0.098	0.008	97.64	Ab
4	46.56	0.203	28.57	4.712	0.039	2.263	0.02	0.192	10.532	0.014	93.10	Mic
5	66.92	0.014	18.95	0.19	0.009	–	0.239	11.463	0.07	–	97.85	Ab
6	66.94	0.006	19.75	0.09	–	–	0.858	11.166	0.098	–	98.90	Ab
7	61.27	0.044	21.83	0.993	0.009	0.34	0.208	9.973	2.047	0.013	96.73	Ab
8	0.78	0.097	0.42	91.672	0.073	0.018	0.064	0.058	0.004	0.024	93.20	Mag
9	0.47	0.063	0.62	92.029	0.078	0.11	0.125	0.068	0.011	0.038	93.61	Mag
10	49.32	0.41	27.05	4.65	0.028	1.837	0.026	0.907	10.116	0.011	94.36	Mic
11	0.34	0.12	0.15	88.004	0.037	–	0.206	0.054	0.017	0.028	88.96	Mag
12	46.71	0.328	28.58	4.023	0.037	2.294	0.032	0.21	10.348	0.087	92.64	Mic
13	67.09	0.017	19.73	0.12	–	–	0.724	11.158	0.098	0.002	98.94	Ab
14	47.89	0.623	28.61	4.054	0.027	2.147	0.011	0.254	10.529	0.025	94.17	Mic
15	67.66	0.014	19.13	0.062	0.01	–	0.054	11.464	0.094	–	98.49	Ab
16	66.91	0.011	19.20	0.114	0.003	0.01	0.333	11.626	0.133	0.007	98.35	Ab
17	66.74	0.05	19.84	0.429	–	0.114	0.228	11.54	0.675	–	99.61	Ab
18	65.66	0.015	19.91	0.134	0.003	–	0.807	11.069	0.287	0.005	97.88	Ab
19	66.90	0.014	19.63	0.075	0.008	–	0.379	11.622	0.155	–	98.78	Ab
20	65.54	0.025	20.52	0.054	–	0.009	0.937	11.599	0.064	0.001	98.75	Ab
21	0.22	0.053	0.09	94.629	0.102	0.072	0.001	0.007	0.013	–	95.19	Mag
22	0.37	0.21	0.32	88.696	0.093	–	0.067	–	0.006	0.01	89.77	Mag
23	62.76	0.029	18.15	0.089	–	0.009	–	0.318	15.638	–	96.99	Kfs
24	56.38	0.211	23.45	3.193	0.035	1.568	0.244	3.828	8.071	0.013	96.99	Mic
25	66.27	0.024	19.67	0.126	–	0.068	0.489	11.037	0.062	0.009	97.75	Ab
26	0.30	0.183	0.15	91.263	0.077	0.097	0.03	0.065	0.002	0.042	92.21	Mag
27	0.35	0.256	0.15	90.466	0.073	–	0.058	–	0.005	0.037	91.40	Mag
28	61.40	0.014	21.38	0.401	0.002	0.053	0.317	9.286	1.678	–	94.53	Ab
29	0.24	0.101	0.18	89.457	0.08	–	0.043	0.048	0.003	0.007	90.16	Mag
30	0.73	0.133	1.04	71.056	2.247	–	0.062	0.067	0.005	17.25	92.58	Cr
31	0.20	0.17	1.96	61.235	3.818	–	–	0.02	0	28.1	95.51	Cr
32	67.98	0.015	19.05	0.037	–	–	0.14	11.638	0.074	–	98.93	Ab
33	64.07	0.01	17.83	0.264	–	–	–	0.303	16.134	–	98.62	Kfs
34	65.92	0.013	18.51	0.115	0.002	–	0.25	9.334	1.656	–	95.80	Ab
35	37.66	1.942	14.98	18.765	0.409	10.28	0.039	0.019	8.697	0.072	92.87	Bi
36	67.66	0.006	19.54	0.151	0.009	–	0.094	11.752	0.075	–	99.29	Ab
37	68.22	0.011	18.66	0.115	0.002	0.024	0.498	10.67	0.222	–	98.42	Ab
38	62.37	0.055	19.23	0.808	–	0.361	0.463	8.23	1.475	0.001	92.99	Ab
39	67.56	0.007	19.35	0.08	–	0.028	0.143	11.729	0.111	0.003	99.01	Ab
40	68.30	0.003	19.41	0.075	0.005	–	0.144	11.767	0.127	0.012	99.84	Ab
41	68.14	0.036	19.19	0.225	0.002	0.077	0.154	11.355	0.544	0.004	99.74	Ab
42	66.72	0.017	19.34	0.115	0.005	0.009	0.072	11.834	0.164	0.009	98.29	Ab
43	48.94	0.693	25.45	4.468	0.039	2.318	0.024	0.171	10.711	0.053	92.87	Mic
44	67.55	0.017	19.32	0.05	–	–	0.198	11.637	0.059	–	98.83	Ab
45	68.45	0.023	19.06	0.031	–	–	0.096	11.483	0.093	–	99.23	Ab
46	67.97	0.004	19.35	0.047	–	–	0.055	11.63	0.105	0.008	99.17	Ab
47	67.76	0.006	19.21	0.039	0.007	0.027	0.23	11.54	0.129	–	98.95	Ab
48	68.77	0.006	18.75	0.034	–	–	0.133	11.626	0.071	0.001	99.39	Ab
49	65.81	0.012	19.09	0.035	–	–	0.06	8.571	5.069	0.003	98.65	Ab
50	63.84	0.013	18.08	0.038	0.012	–	–	0.472	15.929	–	98.38	Kfs
51	64.45	0.017	17.91	0.044	0.001	–	–	0.615	15.692	–	98.73	Kfs

Notes: *Kfs* K-feldspar, *Ab* albite, *Mag* magnetite, *Cr* chromite, *Mic* mica. Oxide contents are given in wt%. Dash denotes that element was not analyzed. Analyses were carried out at IGEM RAS, analyst E.V. Kovalchuk

Fig. 3.11 Cross-bedding in cupriferous sandstone of Sekushchiy (**a**) and Medny (**b**) sites

Fig. 3.12 Cross-bedding in sample U-22

dence for hydrothermal alteration of the ore. This is demonstrated most distinctly by the occurrence of chalcopyrite-pyrite or monomineral pyrite veinlets and larger veins that are commonly oriented at an angle to the general bedding, emphasized by the distribution of sulfide minerals. At the lower level of the Left Naminga site at a depth of 420 m, boreholes intersected pyrite and pyrite-chalcopyrite mineralization in the upper part of the western flank of the deposit. This implies that sedimentary rocks at the Left Naminga site are overturned. Table 3.8 and Appendix 1 show composition of sulfide minerals.

3.4 Oxide Ore

The zone of oxidation plays a very important role for the Udokan deposit. It is widespread at the surface and inside deep fault zones (Narkelyun et al. 1987). Oxidized and mixed ore constitute ~50% of total ore in the deposit (Table 3.9). Its distribution is shown in Fig. 3.26. Oxidized ore dominates at the surface (Fig. 3.27), where minerals of sulfate group occur (Fig. 3.28). Their volume (as mixed ores) decreases toward the deep zones of the deposit (Fig. 3.29). There are many supergene minerals in this zone as well, with the most important (in order of dominance), being sulfates, carbonates, Fe-oxides, and silicates (Table 3.10). Yurgenson (1973), Yurgenson and Bezrodnykh (1966), Yurgenson et al. (1968), Trubachev and Narkelyun (1968), Trubachev (1981), Narkelyun et al. (1987), Krivolutskaya (1989), and Belogub et al. (2015) diagnosed them on the basis of optic data and infrared spectra (Fig. 3.30), thermograms (Fig. 3.31), and X-ray powder diffraction.

3.4.1 Minerals from the Oxidation Zone

Sulfates

Brochantite $Cu_4(SO_4)(OH)_6$ is a main mineral of oxidation zone in the Udokan deposit. It is a very common secondary copper hydroxyl sulfate of green emerald to light green color. It forms crusts and fillings; it was found in the voids in the form of individual prisms and acicular and elongated crystals, given: at times in the forming radiating aggregates and earthy masses. It forms pseudomorphs after malachite and azurite and altered to chrysocolla. According to Yurgenson (1973), the specific gravity of brochantite is 3.966–3.819 g/cm^3, and its hardness is 3.5, with Ng = 1.795 Np = 1.725 Nm = 1.770. X-ray powder diffraction records the following characteristic

Table 3.2 Chemical composition of the main types of ore-bearing rocks from the Udokan deposit, wt%

Rock, type	n	SiO_2	TiO_2	Al_2O_3	Fe_2O_3	FeO	MgO	CaO	Na_2O	K_2O	CO_2	Cu	S
1	49	68.86	0.53	11.60	1.76	3.12	0.88	2.12	2.93	2.88	0.78	2.09	0.54
2	16	54.48	0.45	8.84	0.99	2.76	0.93	10.41	2.29	2.26	6.30	2.31	0.96
3	24	59.56	0.68	16.60	4.45	2.83	2.00	1.18	1.88	5.57	0.24	1.23	0.36

Note. Rock type: 1 – quartzite-like sandstone, 2 – calcareous sandstone, 3 – argillite, siltstone; *n* number of analyses; data after (Ptitsyn et al. 2003)

Table 3.3 Trace element contents in chalcocite-bornite ore and host sandstone at the Udokan deposit (Western site, adit 9)

Oxide, element	USHT 9–41	USHT 9–42	USHT 9–44	Element	USHT 9–41	USHT 9–42	USHT 9–44
K_2O	3.67	2.18	н/a	Ti	1914.4	1750.4	124.5
CaO	0.65	2.56	н/a	Eu	0.37	1.10	0.80
MnO	0.05	0.06	н/a	Gd	1.98	3.44	1.63
Fe_2O_3	3.09	4.52	н/a	Tb	0.33	0.53	0.27
P_2O_5	0.12	0.18	н/a	Dy	2.11	2.96	1.61
S	0.30	5.73	н/a	Y	14.7	17.3	10.2
Cs	3.37	2.90	0.28	Ho	0.46	0.60	0.39
Rb	127.7	92.6	3.8	Er	1.29	1.64	1.45
Ba	940.8	320.7	36.6	Tm	0.20	0.25	0.27
Th	6.24	7.67	3.50	Yb	1.22	1.37	2.24
U	2.24	4.53	27.89	Lu	0.18	0.18	0.52
Nb	7.24	6.34	0.58	Be	2.13	2.12	1.33
Ta	0.56	0.53	0.06	V	200	100	43
La	14.4	42.4	5.1	Mn	364	477	366
Ce	24.5	79.7	9.4	Co	9.5	6.5	2.0
Pb	2.3	5.4	89.9	Cr	60.0	44.1	9.5
Pr	2.91	9.39	1.61	Ni	30.0	47.3	14.8
Mo	9.25	4.00	4.21	Zn	33.6	26.5	22.8
Sr	120.6	374.5	218.7	Ga	13.5	8.7	8.6
Nd	10.2	33.0	6.9	Cu	5305	130	202
Sm	1.9	5.3	1.9	Au	0.01	0.10	0.30
Zr	103.6	81.9	41.0				
Hf	2.8	2.2	0.6				

Analyses were carried out at IEM RAS, analyst V.K. Karandashev

lines (hereinafter d-space/intensity): 6.95 (100), 6.45 (100), 5.39 (100), 3.92 (100), 2.72 (90), 2.54 (100), and 1.75 (90). Infrared spectra demonstrate absorption bands at 426, 475, 489, 514, 602, 631, 650, 736, 775, 851, 873, 936, 987, 1089, 1118, 3265, 3380, 3360, and 3581 cm $^{-1}$ (Fig. 3.30–3). The thermograms of brochantite show records of four endothermic effects at 180, 460, 740, and 1000 °C, which correspond to the stages of transformation and destruction of the mineral by heating (allocation of water, sulfuric anhydrite). Brochantite (Fig. 3.32) is one of the latest minerals in the zone of oxidation of the Udokan deposit.

Antlerite $Cu_3(SO_4)(OH)_4$ is less common than related and sometimes visually similar brochantite, with which it may be confused. Emerald green to blackish green and also light green crystals are commonly thick tabular {010}(Fig. 3.33a–c), also equant or short prismatic [001] (Fig. 3.34). It also occurs as cross-fiber veinlets or friable interlaced aggregates of acicular or fibrous, felt-like, and granular crystals. It has perfect cleavage on {010}, with Ng = 1.726, Np = 1.738, Nm = 1.790, and 2V = 53. Three endothermic effects are recorded on the thermograms of antlerite at 520–540 °C (release of water, the weight loss of 10%), 850–870 °C (separation of SO_3), and 1050–1070 °C (destruction of the mineral). The X-ray diffraction analysis shows that antlerite resemblance to its reference mineral is 5.40 (50), 4.85 (90), 3.61 (40), 2.69 (80), 2.562 (100), and 2.128 (100). Obtained absorption bands in infrared spectra are very characteristic for antlerite (Fig. 3.30–4).

Chalcanthite $CuSO_4 \cdot 5H_2O$ is a water-soluble copper sulfate of green, green blue, light blue, or dark blue color. It is characterized by a glass luster (quickly becoming dim in the air), blue color, low hardness (2.5), and high fragility. It does not form huge aggregates. It was found as crusts and columnar fibrous aggregates, as well as massive (3–4 mm), granular filling of veins (cross-fiber veinlets). Sometimes, it forms short prismatic crystals. The mineral is biaxial (−), Ng = 1.512, Np = 1.550, Nm = 1.540, and 2V = 55°. It can be diagnosed very well using the thermoweight analysis (Fig. 3.31). The thermograms of chalcanthite (Fig. 3.31–1,2) give noticeable exothermic (120–140, 180–200, 315, 505, 750, 320,

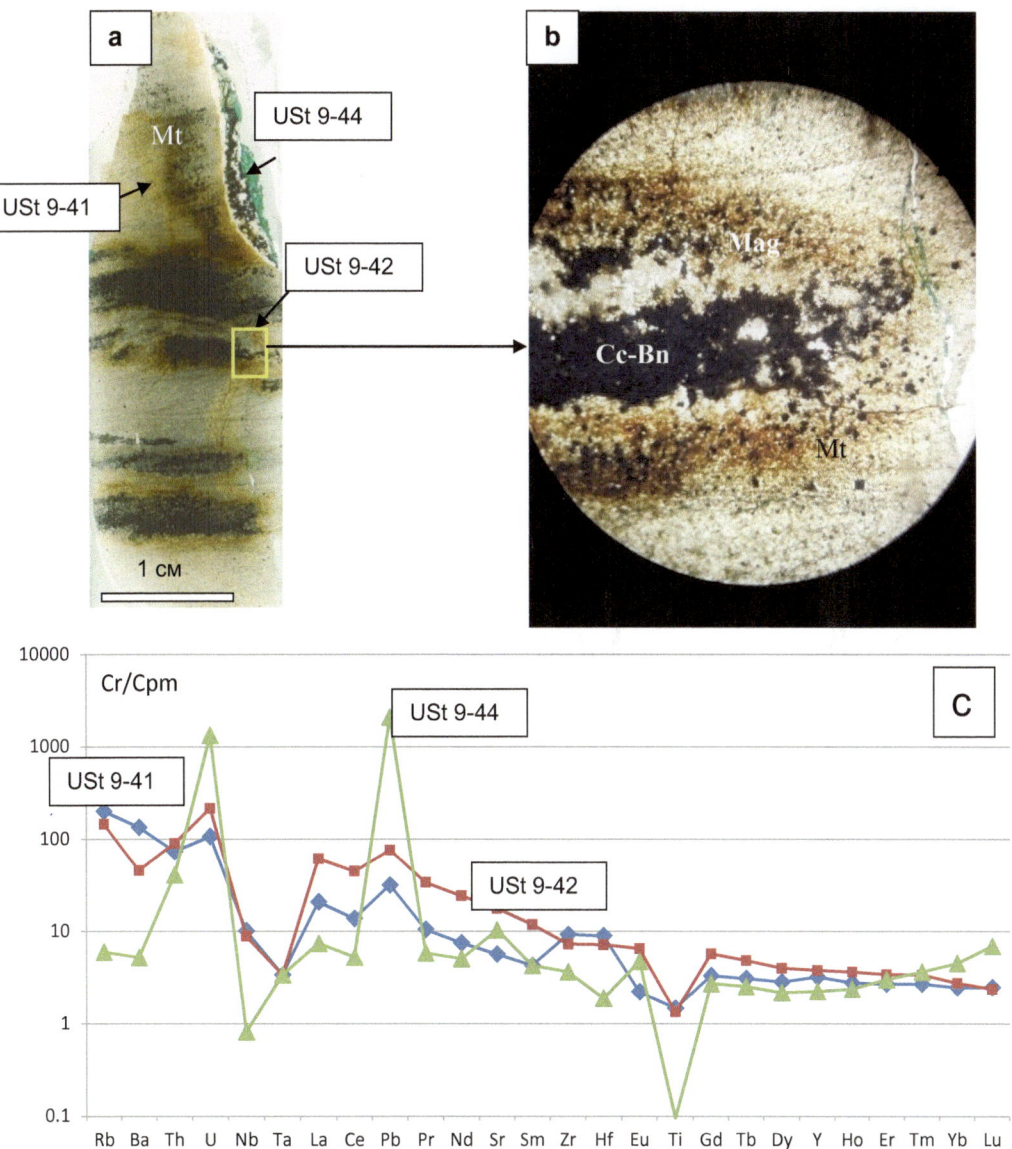

Fig. 3.13 Sandstone sample with magnetite veinlets and spider diagram for the rocks of the Udokan deposit. Crosscutting veinlet (**a**, green line, point USht 9–44), nearly concordant lenses (**b**, red line, point USht 9–42), and **c** – spider diagram of trace element concentrations in sandstone (blue line, point USht 9–41 Fig. 3.13a), normalized to primitive mantle (Hofmann 1988). See data in Table 3.2. Cr/C$_{PM}$_ rock/element in primitive mantle, *Cc* chalcocite, *Bn* bornite, *Mag* magnetite

Table 3.4 Proportions of ore types at the Udokan deposit

Level, m	Site	Average degree of oxidation, %	Sulfide	Mixed	Oxidized
Surface	Left bank of River Naminga	43.2	41	56	3
	Shumnyi-Krutoi	70.1	–	26	74
	Sekushchiy	76.0	–	37	63
	Skolzkiy	57.8	3	74	26
	Bluzhdayushchiy-Medny	77.6	–	44	56
−150	Adits 11, 12, 7	48.4	29	67	4
−260	Adits 3, 9	32.8	45	47	8
−310	Adit 1	20.8	73	11	16
−470	Adit 2	19.2	82	18	–
−670	Boreholes	18.2	85	15	–

Note. Data after Narkelyun et al. (1987)

Fig. 3.14 Massive (**a**) and banded (**b**) chalcocite-bornite ore

890, 930, 1040–1090 °C) and endothermic (130–150, 250–260, 400, 600, 700, 800–840, 950–960, 1050) effects. The X-ray diffraction analysis confirms the identification of the mineral as chalcanthite: 4.72 (100), 3.97 (90), 3.68 (80), 3.29 (60), 3.04 (60), 2.43 (100), 2.06 (100), and 1.63 (100). There are following absorption bands typical for infrared spectra of chalcanthite: 633, 855, 1120, 1678, and 3485 cm^{-1} (Fig. 3.30–8). Chalcanthite is associated with gypsum, antlerite, and brochantite. It was noted on the aluminum wire. This proves that this mineral was formed very recently in the permafrost (Yurgenson and Bezrodnykh 1966). Usually, chalcanthite is a secondary mineral that forms in arid climate or in rapidly oxidizing copper deposits. It usually forms on mine walls due to the action of surface acidic waters on copper veins. Generally, it develops in arid regions, where it is not dissolved by groundwater. The presence of chalcanthite at the Udokan deposit can be explained by permafrost that keeps it on the surface without dissolution.

Gypsum $CaSO_4*2H_2O$ is a very common sulfate mineral at the Udokan deposit found as both massive material and clear crystals. Usually it forms thin crusts. Also it was found as granular masses, massive beds, and elongated [010] small (<1 mm) crystals, colorless to white, light yellow, often tinged to other hues due to impurities. Rosette-like clusters of lenticular crystals are common in thin crusts. Ng = 1.520 Np = 1.521 Nm = 1.530. The thermograms record two endothermic peaks at 180–200 °C and 840–860 °C and an exothermal peak at 360–400 °C. Gypsum is well confirmed by X-ray analysis on the reference lines, 7.61 (100), 4.21 (100), 3.83 (50), 3.03 (80), 2.85 (50), 2.66 (50), 2.08 (90), 1.77 (40), 1.62 (30), 1.37 (20), 1.23 (20), 1.200 (20), and 1.085 (10), and by infrared spectra (Fig. 3.30–9).

Melanterite $FeSO_4*7H_2O$ forms a thin crust, small lamellar crystals (up to 0.1 mm) and a powdery mass in pyrite-chalcopyrite ore in association with goethite, jarosite, halotrichite, and other minerals. It often contains minor copper, which causes its bluish color. Melanterite is a hydrated iron sulfate formed after decomposition of pyrite or other iron minerals due to the action of surface waters. It was found on the walls in galleries 9 and 12. Optic axes are Ng = 1.470, Np = 1.477, and Nm = 1.486. Melanterite has the following X-ray lines: 4.89 (100), 3.07 (100), 2.71 (20), 2.060 (20), and 1.462 (10).

Halotrichite $FeAl_2 (SO_4)_4*22H_2O$ has white, yellowish, greenish color, or colorless. Usually it forms aggregates of hair-like crystals and spheroids. It was recognized by X-ray powder diffraction (4.79 (100); 4.29 (30); 3.974 (20); 3.50 (80)) and optical features (Ng = 1.48, Np = 1.49, Nm = 1.49, 2 V = 40°).

Jarosite $KFe^{3+}_3(SO_4)_2(OH)_6$ is a Fe analogue of alunite. The color of mineral is yellow; it is dark straw-golden in color. It was found as granular crusts. It also forms nodules or fibrous masses. Jarosite associates with hydrogoethite and goethite. In very rare cases, jarosite forms smallest rhombohedral crystals (<0.1 mm). It is diagnosed in X-ray powder diffraction: 3.07 (30), 2.29 (50), and 1.524 (30). The thermograms have three endothermic effects at 130, 360–470, and 650 °C. It forms as a result of reaction of diluted sulfuric acid in groundwater, derived from the oxidation of pyrite, with gangue minerals and wall rock at the deposits.

Dolerophanite $Cu_2(SO_4)O$ is a very rare mineral of dark brown color. It was distinguished by Yurgenson (1968) in thin sections in the form of fine-grained aggregates, pleochroic from greenish yellow to brownish on Ng and yellow on Np. The larger grains are with cleavage. Antlerite forms very often pseudomorphs after dolerophanite.

Carbonates

Malachite $Cu_2(CO_3)(OH)_2$ is a very common green (bright green, even vary dark to nearly black) secondary copper mineral with a widely variable habit (Figs. 3.28 and 3.35). It is developed in calcareous sandstone. It is found as crystalline aggregates or crusts (several centimeters in diameter), often banded, around sulfide grains (the first generation). The

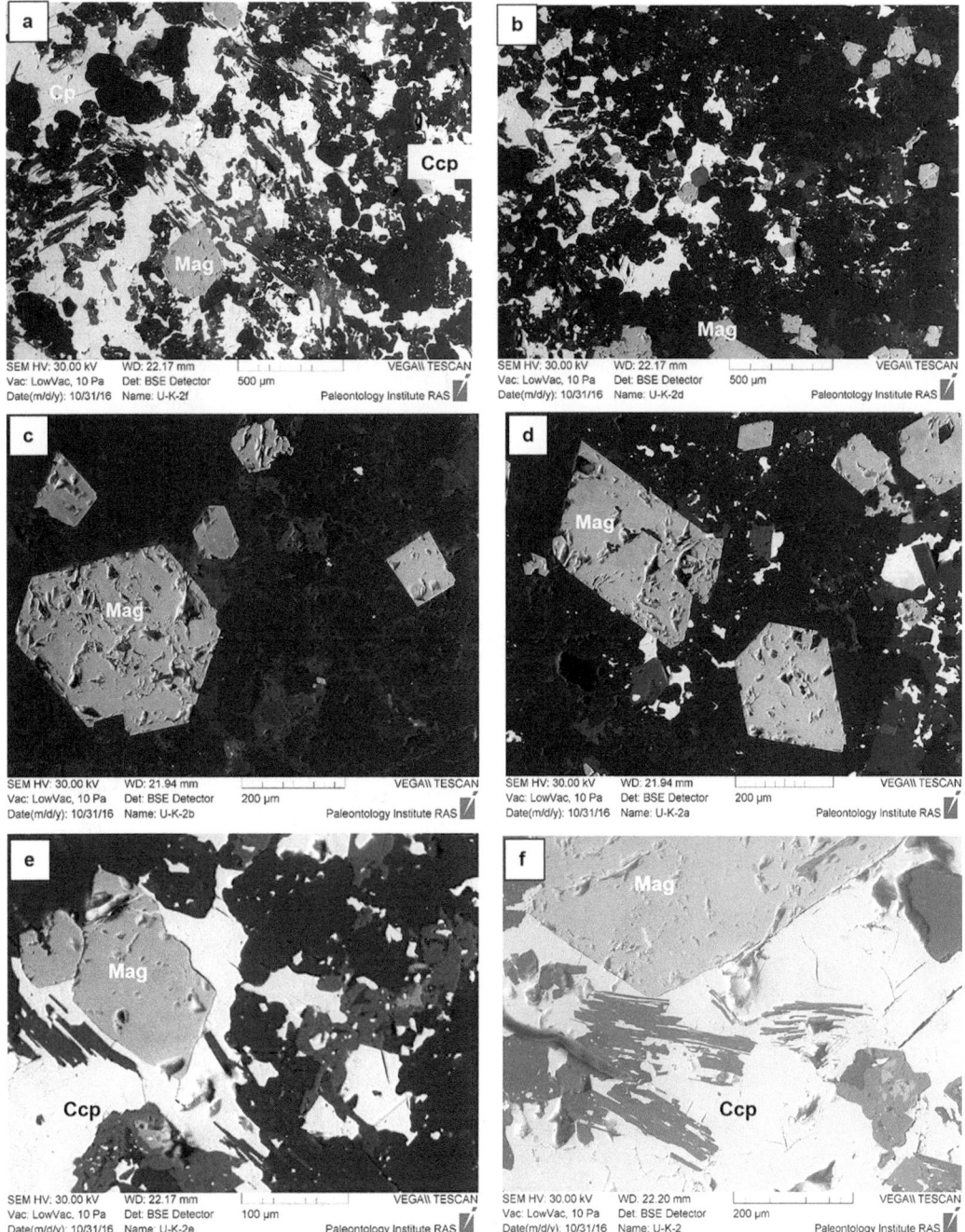

Fig. 3.15 Disseminated ore from the Bluzhdayushchiy site, Udokan deposit

BSE images of *Ccp* chalcopyrite, *Bn* bornite, *Mag* magnetite (Images were taken at the Paleontology Institute, Russian Academy of Sciences. Analyst EA Zhegallo). (**a, b**) – Disseminated sulfides in rock; (**c-d**) – large idiomorphic magnetite crystals in silicate matrix; (**e-f**) – large magnetite crystals surrounding by sulfides

veinlets are very common as well. Botryoidal aggregates of radiating fibrous crystals are more common (the second generation). Single crystals were found as a pseudomorph after azurite crystals, which are generally more tabular in shape. The refractive indices are as follows: Np = 1.664 and Ng = 1.906. 2V = 40°. The thermal curve confirms this mineral as malachite, with endothermic effects at 395 and 1100 °C. The characteristic X-ray lines for the malachite of the first generation are as follows: 5.07 (70), 3.70 (90), 2.86 (100), 2.78 (90), 1.69 (80), 1.42 (80), 1.16 (70), and 1.07 (60). Malachite of the second generation has some extra lines: 3.92 (80), 3.70 (70), 3.20 (80), 2.86 (100), 2.686 (70), 2.533 (80), 2.078 (5), 1.478 (4), and 1.318 (4). The crystals are typically twinned on {100}. Infrared spectroscopy

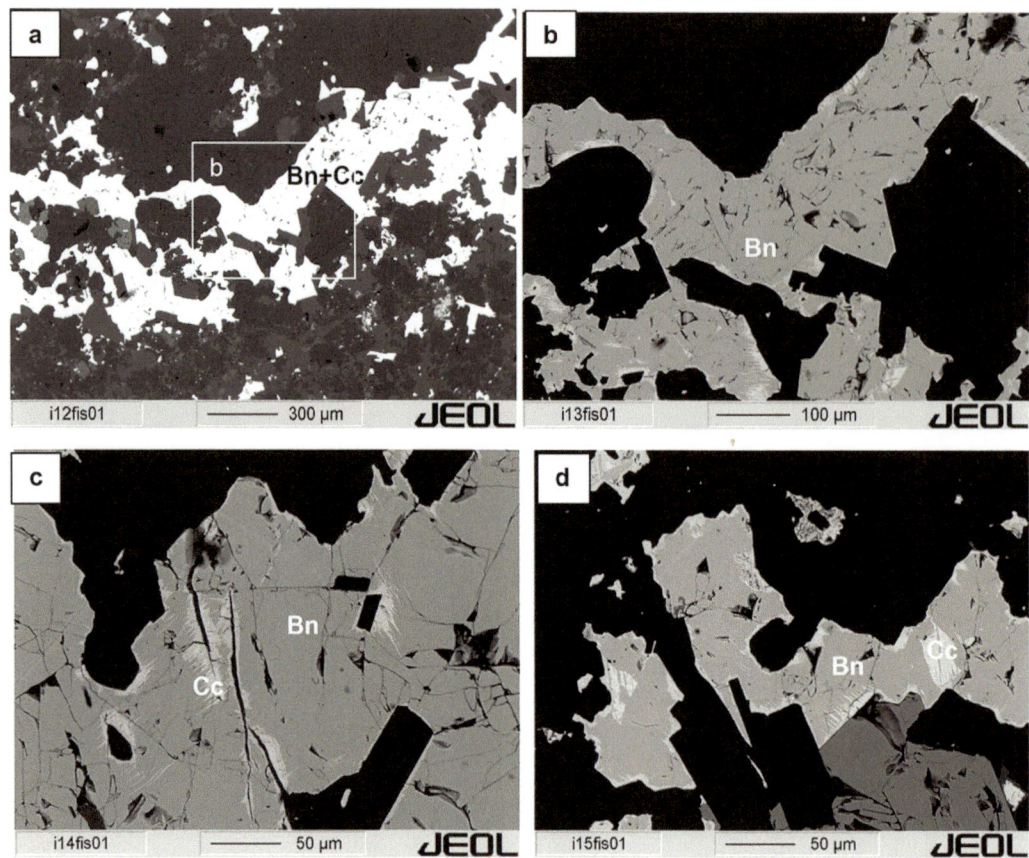

Fig. 3.16 Disseminated ore from the Ozernyi site, Udokan deposit
BSE images were taken at IGEM RAS. (Analyst EV Kovalchuk) Sample U-20: (**a**) general view, (**b**) detail of Fig. 3.16a, (**c**, **d**) sample U-23. *Cc* chalcocite, *Bn* bornite

Fig. 3.17 Strata-parallel (**a**) and crosscutting (**b**, **c**) chalcopyrite-pyrite mineralization

Fig. 3.18 Polished sections of breccia ores: (**a**) lamina and crosscutting veinlets of pyrite-chalcopyrite ore, (**b**) thin magnetite lamina, (**c**) close-up fragment of image **b**, (**d**) massive chalcocite-bornite

method confidently diagnosed malachite by characteristic absorption spectra (Fig. 3.30–1,2): 400, 434, 488, 510, 529, 577, 587, 715, 752, 779, 824, 877, 1050, 1100, 1390, 1420, 3310, and 3400 cm^{-1}. In addition, the spectra 1500, 1800, and 3490 cm^{-1} can distinguish early malachite from the needle malachite of the second generation.

Azurite $Cu_3(CO_3)_2(OH)_2$ is carbonate with additional anions without H_2O. It is rarer than malachite. It is developed in calcareous sandstone as well. The color is azure blue (Fig. 3.28), light blue or dark blue, and light blue in transmitted light. Azurite is typically found as tabular to prismatic crystals (Fig. 3.36) of Ng =1.730, Np = 1.757, Nm = 1.838, and 2V = 65°. The thermal curve of azurite shows exothermic (at 280 °C) and endothermic (at 340–345, 1025–1110 °C) effects of water and carbonated acid at 180–340 °C. Calculated radiographs indicate standard lines for azurite: 2.53 (100), 2.29 (100), 1.46 (80), 1.86 (60), 1.81 (60), 1.62 (60), 1.59 (60), and 1.44 (60). Azurite can be clearly determined by

infrared spectroscopy (Fig. 3.31–7); there are many visible characteristic absorption bands: 411, 466, 501, 607, 668, 750, 775, 825, 843, 958, 1100, 1156, 1421, 1621, 3425, and 3555 cm^{-1}.

Cerussite $PbCO_3$ is a very rare mineral in the Udokan deposit. It forms massive, granular, or dense compact aggregates in galena grains. It is of white or gray color.

3.4.1.1 Oxides

Hydroxides of iron (Fe_2O_3-wH_2O) are widely distributed in the form of scabs, crusts, veinlets, and oolitic aggregates in cracks at the outcrops of orebodies at surface and inside the faults. Under the microscope, one can see them as veinlets in pyrite, chalcopyrite, bornite, and other sulfides. They are less likely to occur as disseminations in cement of sandstone. According to Krendelev et al. (1983) and Narkelyun et al. (1987), these minerals were recorded at many locations at the Udokan deposit. Iron hydroxides are represented by goe-

Fig. 3.19 Quartz vein with oxidized sulfide mineralization at the Levyi Bort Naminga site

Table 3.5 Minerals of sulfide ore from the Udokan deposit

Main	Major	Rare
Chalcocite CuS	Hematite F_e2O_3	Carrollite $Cu(Co, Ni)_2S_4$
Djurleite $Cu_{31}S_{16}$	Digenite Cu_9S_5	Pyrrhotite $Fe_{1-x}S$
Bornite Cu_5FeS_4	Covellite CuS	Wittichenite $Cu_3(BiS_3)$
Chalcopyrite $CuFeS_2$		Marcasite FeS_2
Pyrite FeS_2		Idaite Cu_5FeS_6
Magnetite Fe_3O_4		Galena PbS
		Sphalerite ZnS
		Native Ag
		Native Au
		HgS
		Molybdenite MoS
		Pentlandite $(Ni,Fe)_9S_8$
		Cobaltite CoAsS
		Arsenopyrite FeAsS
		Acanthite Ag_2S
		Valleriite $Fe^{2+}, Cu)_4(Mg, Al)_3S_4(OH,O)_6$
		Polydymite $Ni^{2+}Ni_2^{3+}S_4$
		Stromeyerite AgCuS
		Ilmenite $Fe^{2+}TiO_3$
		Sulvanite Cu_3VS_4

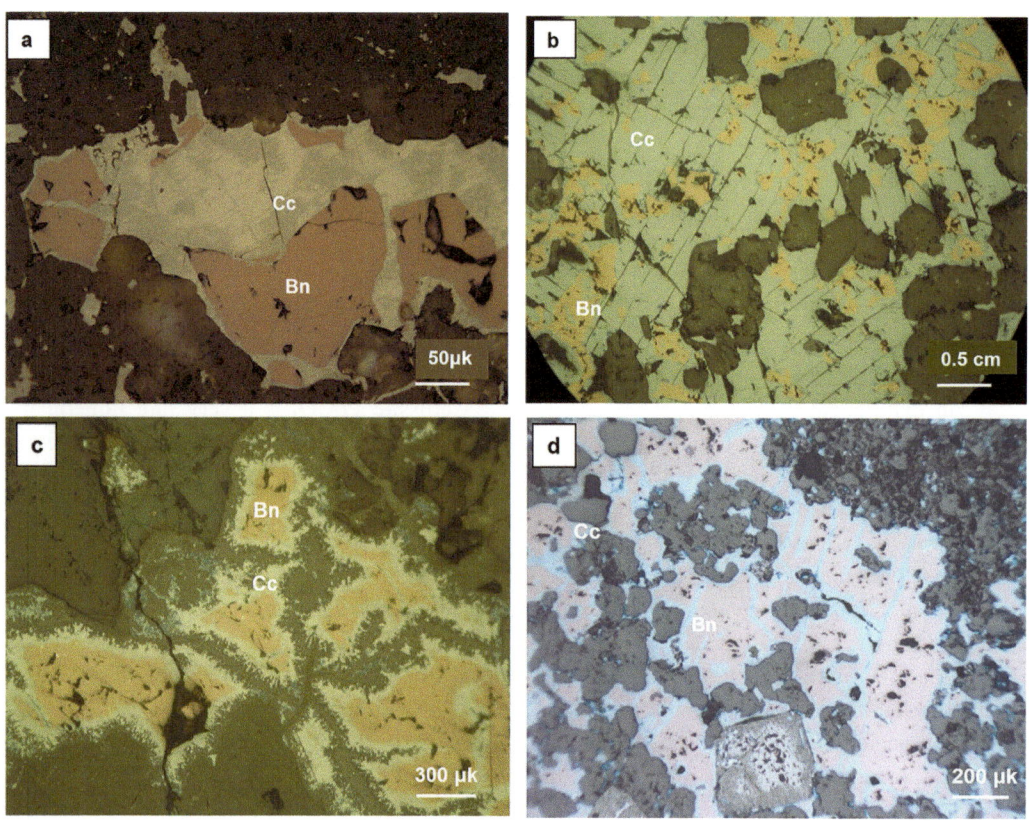

Fig. 3.20 Microphotographs of chalcocite-bornite (Cc-Bn) ore in reflected light
Mineral sites: (**a**) Medny, (**b**) Sekushchiy, (**c**) Levyi Bort Naminga, (**d**) Ozernyi

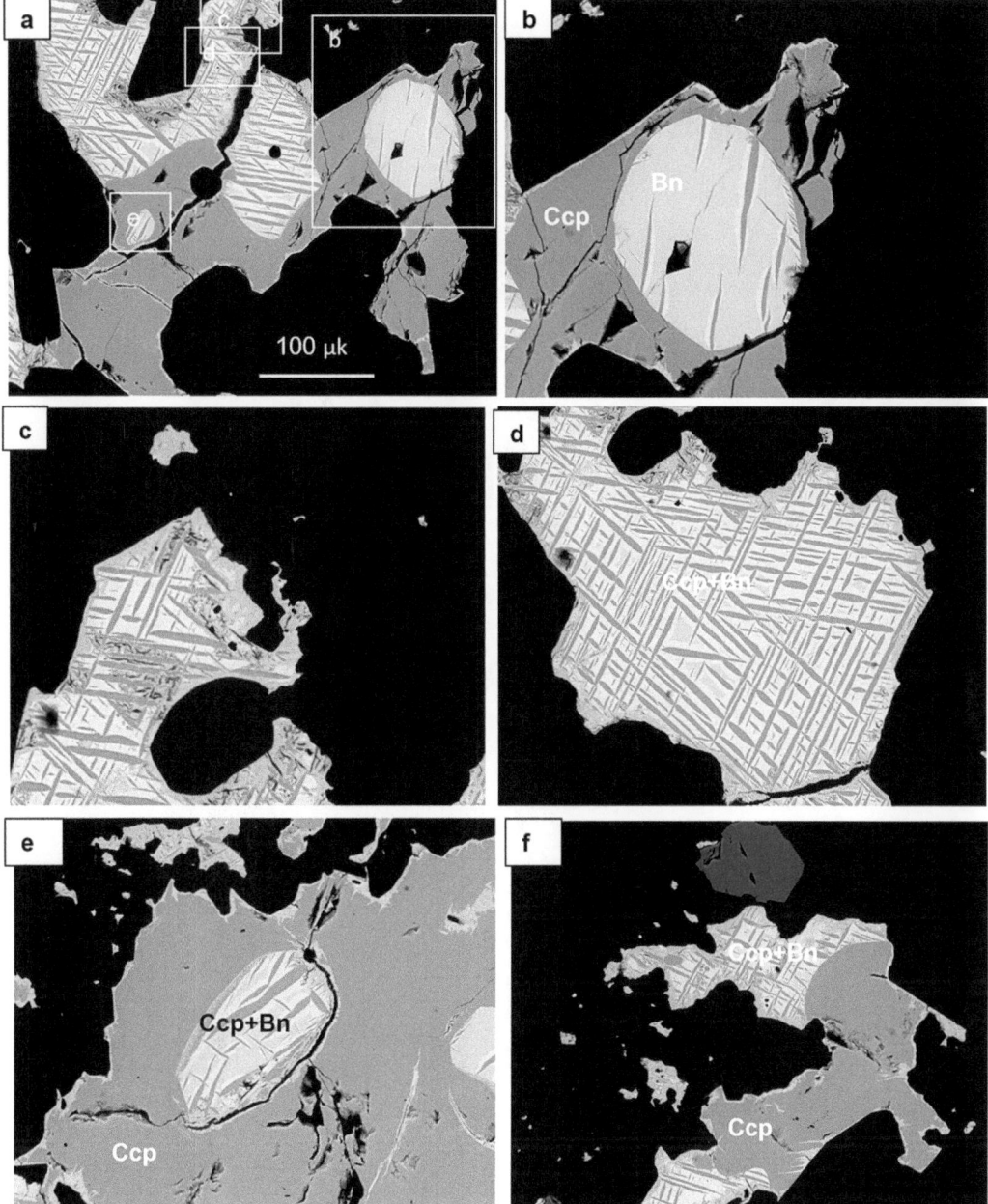

Fig. 3.21 Exsolution textures in bornite-chalcopyrite (Bn-Ccp) grains. BSE images were taken at IGEM, Russian Academy of Sciences (analyst LO Magazina). Sample U-9/1, Bluzhdayushchiy mineral site. (**a**) general view, (**b–d**) detailed textures shown in Fig. 3.20a. Dark gray, chalcopyrite; light gray, bornite

thite, hydrogoethite, and hydrohematite. Goethite often substitutes chalcopyrite, while hydrogoethite is present among the bornite aggregates.

Goethite α-$Fe^{3+}O(OH)$ is the most common iron hydroxide, which is often found in fractures. It was recorded at considerable depth (1700 m) and is widely developed on the surface. Goethite forms black, reddish, brown prismatic small (<1 mm) crystals, but it is more often as oolitic, reniform, and botryoidal aggregates. It has gray color in polished sections with bright yellow-brown internal reflexes. The thermograms have weak endothermic effects at 280–300 °C. The presence of goethite is well confirmed in X-ray powder diffraction data: 4.27 (100), 2.72 (100), 2.46 (100), 2.27 (40), 2.20 (40), 1.72 (90), 1.509 (70), and 1.448 (60).

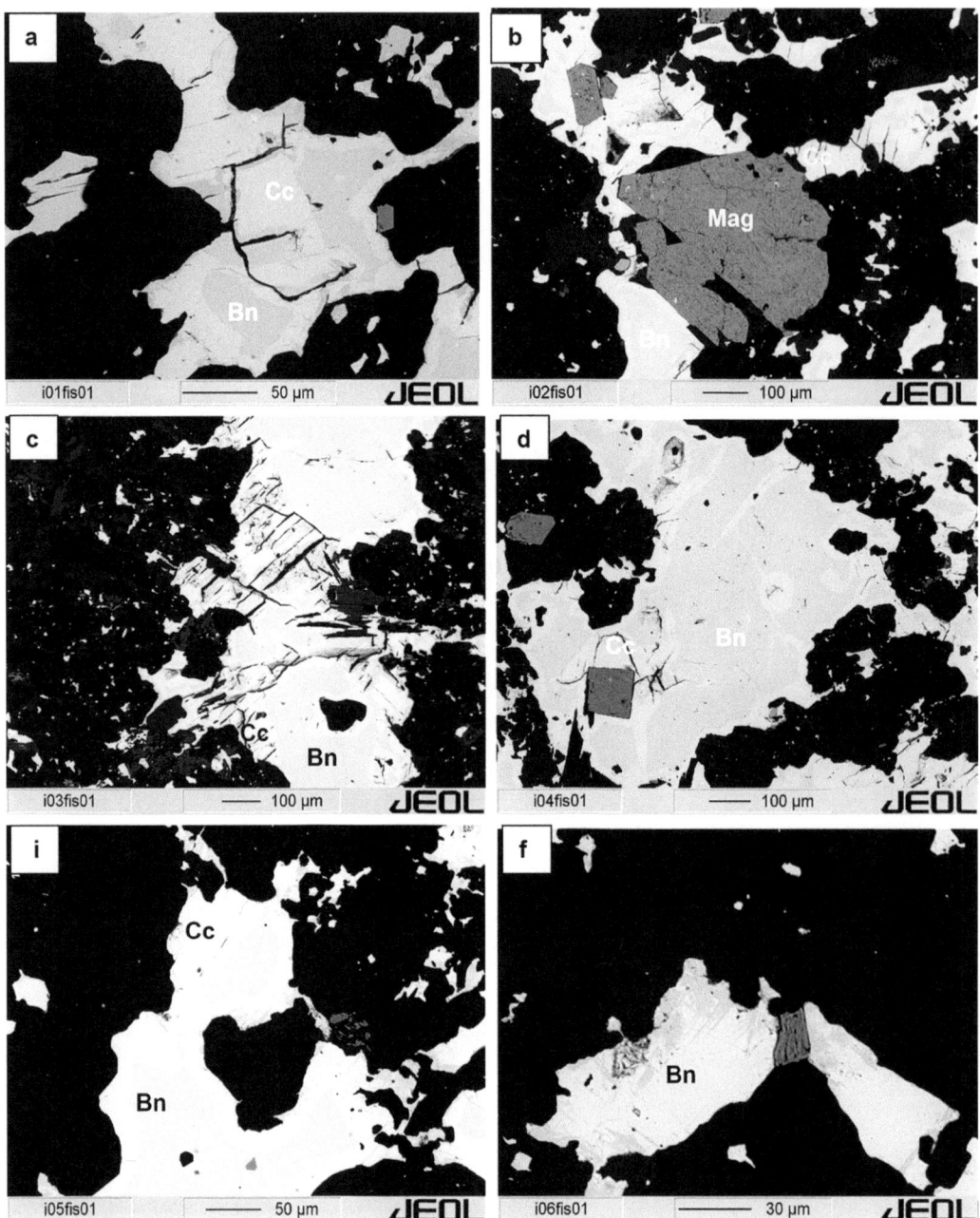

Fig. 3.22 Ore minerals from the Naminga site (Sample U-2) (BSM images were taken at IGEM RAS. Analyst EV Kovalchuk). *Cc* chalcocite, *Bn* bornite. (**a**) Replacement of bornite by chalcocite, (**b**) magnetite crystal with sulfide minerals in disseminated ore, (**c**) bornite with thin rim of chalcocite, (**d**) large bornite grain with chalcocite in thin fractures, (**e**) replacement of bornite by chalcocite, (**f**) grains of bornite

Goethite is most often found at the Medny, Skolzkiy, and Shumnyi mineral sites of the Udokan deposit. This mineral was formed near chrysocolla and malachite.

Hydrogoethite $3Fe_2O_3 \cdot 4H_2O$ is often found with goethite on the surface. It replaces goethite and sulfides. Very often, it forms a crust of yellowish, brown color on the surface of goethite. X-ray analysis confirms its presence in oxide ore:

4.29 (90), 2.46 (90), 2.22 (80), 1.638 (90), 1.567 (60), and 1.45 (70).

Lepidocrocite γ-$Fe^{3+}O$ (OH) is very common in association with goethite and hydrogoethite. It forms red-brown, deep red aggregates, sometimes small rosettes (<0.5 mm). X-ray powder diffraction data are 3.40 (100), 2.46 (100), 1.937 (30), and 1.721 (90).

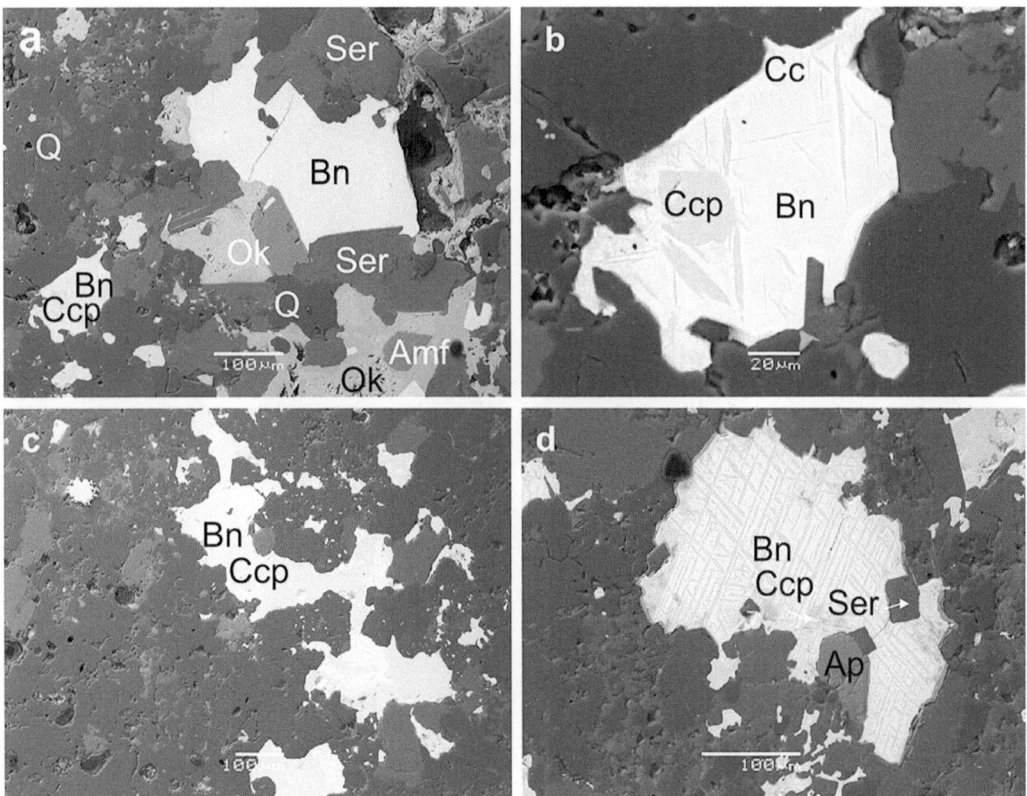

Fig. 3.23 Rock-forming and ore minerals from the Naminga mineral site (Sample U-2/1). (BSM images were taken at IGEM RAS. Analyst LO Magazina). Minerals: *Ccp* chalcopyrite, *Cc* chalcocite, *Bn* bornite, *Q* quartz, *Ser* sericite, *Amf* amphibole, *Ap* apatite. (**a**) – Bornite in association with seritsite, quartz, amphibole; (**b**) – chalcopyrite inside bornite-chalcocite grain; (**c**) – chalcopyrite-bornite association; (**d**) – chalcopyrite-bornite solid solution structure

Table 3.6 Composition of sulfide minerals from sample U-6 (Fig. 3.9), wt%

No	As	S	Fe	Ni	Co	Se	Sb	Cu	Total
1	0.05	35.17	30.27	0	0	0.09	0.001	34.93	100.51
2	0.10	24.95	6.43	0.002	0.007	0.094	0.011	72.68	104.27
3	0.01	34.93	28.97	0	0	0.08	0.005	36.40	100.39
4	0.08	34.80	30.43	0.001	0	0.106	0.012	34.94	100.37
5	0.10	24.46	3.34	0	0.011	0.109	0.008	74.93	102.96
6	0.05	34.71	29.23	0	0	0.115	0.004	36.26	100.38
7	0.07	23.80	1.41	0	0.011	0.135	0	76.83	102.26
8	0.02	24.49	3.41	0	0.008	0.145	0	73.45	101.53
9	0.04	31.65	17.99	0	0	0.097	0	49.62	99.40
10	0.03	34.41	0.36	0.003	0.011	0.122	0.028	66.13	101.08
11	0.06	41.86	0.69	2.392	36.6	0.085	0.004	19.28	101.06
12	0.04	34.81	30.58	0	0	0.073	0	34.69	100.19
13	0.09	24.79	7.30	0.002	0.008	0.123	0.007	71.39	103.71
14	0.06	34.87	29.36	0.002	0	0.071	0	36.13	100.50
15	0.04	24.06	5.24	0.003	0.013	0.06	0.015	72.24	101.67
16	0.01	34.92	30.41	0.004	0	0.106	0.014	34.76	100.23
17	0.09	27.11	10.86	0	0.003	0.124	0	61.77	99.96
18	0.09	26.27	10.23	0	0	0.116	0	66.44	103.15
19	0.04	31.54	23.10	0	0	0.087	0.004	44.50	99.27
20	0.06	23.26	2.40	0	0.013	0.08	0	76.60	102.42
21	0.09	24.87	7.12	0	0.01	0.127	0.016	69.38	101.62
22	0.05	34.77	30.55	0.002	0	0.102	0	34.66	100.12
23	0.04	30.75	21.97	0.004	0	0.097	0.003	45.83	98.70
24	0.09	25.36	9.39	0.004	0.007	0.071	0.009	67.75	102.68
25	0.07	22.83	1.52	0	0.014	0.085	0	78.04	102.55

Note. № 1, 3, 4, 6, 12, 14, 16, 22 – chalcopyrite; 2, 5, 7, 8, 10, 11, 13, 15, 17, 18, 20, 21, 24 –minerals of chalcocite group, 17, 18, 19, 24 – bornite, 11– carrollite. Analyses were carried out at GEOKHI RAS, analyst NN Kononkova

Fig. 3.24 Disseminated ore
from the Levyi Naminga site
(**a**) Thin section; BSM images
of chalcocite: (**b**) general
view, (**c**) detail with analytical
points from Table 3.6

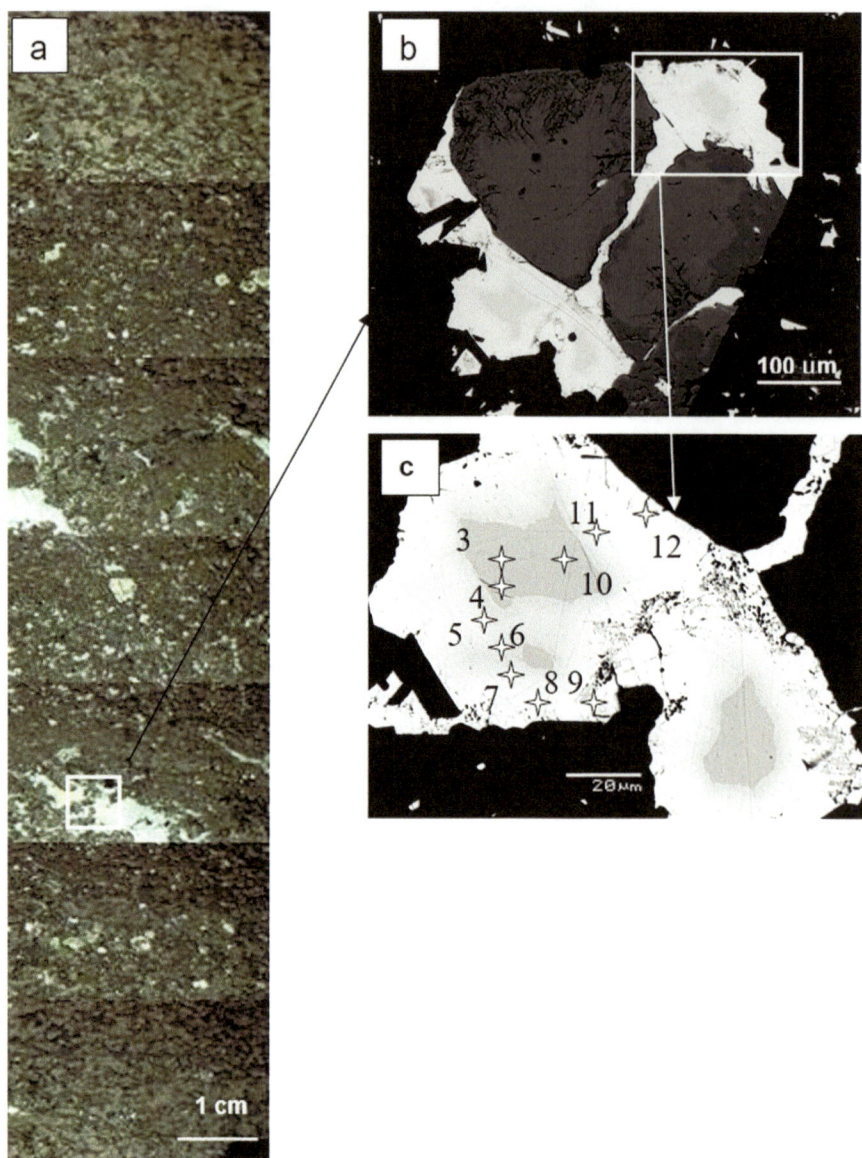

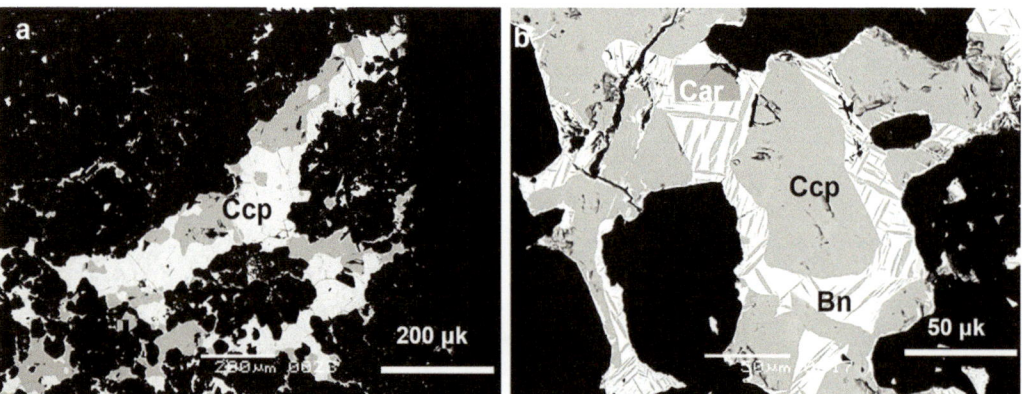

Fig. 3.25 Back scattered electron images of carrollite (Car, $Cu(Co,Ni)_2S_4$), chalcopyrite (Ccp), and bornite (Bn) in disseminated ore from the Udokan deposit in Sample Un-6. See Table 3.5 for composition of minerals. (**a**) – chalcopyrite-bornite veinlet, (**b**) – carrolite in grain with chalcopyrite-bornite solid solution structure

Table 3.7 Composition of Fe oxide minerals, wt%

N	TiO_2	FeO	MnO	V_2O_3	Nb_2O_5	Cr_2O_3	Al_2O_3	Total
1	4.01	84.68	0.00	0.02		0.08		88.90
2	3.72	85.81	0.02	0.15		0.07		89.86
3	3.24	85.41	0.00	0.00		0.00		88.76
4	3.11	85.07	0.01	0.08		0.01		88.38
5	4.36	85.23	0.03	0.04		0.06		89.82
6	3.97	84.46	0.06	0.03		0.07		88.65
7	3.31	83.84	0.03	0.02		0.11		87.38
8	2.15	87.41	0.04	0.00		0.00		89.67
9	3.76	84.88	0.00	0.06		0.06		88.95
10	3.83	85.05	0.02	0.13		0.09		89.23
11	3.27	85.24	0.05	0.00		0.04		88.66
12	3.99	85.36	0.00	0.18		0.05		89.65
13	4.31	84.79	0.06	0.10		0.07		89.38
14	83.40	13.43		0.15	0.14	0.07	0.06	97.39
15	99.13	0.54		0.08	0.03	0.02	0.01	99.89

Note. Analyses were carried out in IGEM RAS. Analyst EV Kovalchuk. Analyses 1–13, magnetite; 14–15, rutile. Empty cell, element was not analyzed

Psilomelane ($Mn_2O*MnO*nH_2O$) is widespread at the Udokan deposit. It forms iron-black dendritic aggregates, films, crusts, and veinlets in calcareous sandstone and siltstone. Yurgenson (1968) guessed that it forms at different stages in development of oxidation zone, because it was found as relics in copper sulfates and carbonates, and at the same time, it develops on them. Shcherbina (1972) indicated a marked tendency of Mn oxides precipitation from alkaline solutions. This conclusion is confirmed by the presence of psilomelane in the calcareous sandstone.

Cuprite Cu_2O is widely developed at the Bluzhdayushchiy, Medny, and Sekushchiy mineral sites of the Udokan deposit. It is rare in other places. Usually it was observed in bornite-chalcocite ore. It is often associated with blue supergene chalcocite and almost never occurs in pyrite-chalcopyrite ore. This feature can be easily explained using diagrams of oxide ratios and sulfides of copper in the system Cu-S-O, which clearly shows that cuprite is in equilibrium only with chalcocite, covellite, and native copper. Analysis of pH-Eh diagram shows that it needs high pH (6–12) for the deposition of cuprite and tenorite (Garrels and Christ 1965). Cuprite occurs in the form of powdery granular masses, thin films, and dendritic aggregates. It is recognized by dark red color and diamond-like brilliance, sometimes almost black. Under the microscope in transmitted light, it is orange-red with high relief. It is light gray with a blue shade in reflected light. R = 25–30% is weakly anisotropic along the edges of the grains with ink internal reflexes. Associated with native copper are malachite, azurite, and limonite. X-ray diffraction gives characteristic lines: 3.09 (100), 2.48 (10), 2.30 (20), 1.52 (20), and 0.97 (30).

Tenorite CuO is less common mineral in oxide ore than cuprite. It forms a pulverulent black mass, smears, and incrustations on the walls of cavities and fine cracks in pyrite-chalcopyrite. It is less common in bornite-chalcocite ore. In reflected light, it is of yellowish-brown color, with clear anisotropy, blue internal reflections, and visible polysynthetic twins in individual grains.

Magnetite (Fe_3O_4) is a very rare mineral. It mainly forms rims and small grains on the periphery of bornite. It is present as individual small euhedral grains along with chalcocite, reaction hematite, and chalcopyrite. The microhardness is 402–645 kg/mm^2.

Hematite Fe_2O_3 is formed after magnetite. It is developed as streaks, rims, and veinlets. Hematite is rather variable in its appearance. It can be reddish brown; very often it forms ocherous masses. Internal reflections are red, and it has weak pleochroism. X-ray powder diffraction data: 3.66 (30), 2.72 (90), 2.52 (70), 1.84 (50), 1.72 (50), and 1.49 (30).

Delafossite $CuFeO_2$ is a very rare trigonal multiple oxide. It forms botryoidal crusts of black color. It is brown-white in reflected light.

Silicates

Chrysocolla $Cu_{2-x}Al_x(H_{2-x}Si_2O_5)(OH)_4*nH_2O$ (x < 1) is a rare mineral at the Udokan deposit, but sometimes its concentrations are high in oxide ore. For example, at the Skolzkiy site, it forms big clusters of buds, crusts, and earthy masses. There are two different types of chrysocolla: (1) glassy blue or bright blue (under the microscope with low birefringence, Np = 1.574, Nm = 1.598, Ng = 1.608) crystalline to amorphous botryoidal aggregates and crusts on surface of sandstone. Yurgenson (1968) was able to obtain some X-ray data, from which it follows that chrysocolla in some samples shows a high degree of crystallinity: 4.49 (70), 2.91 (100), 1.626 (80), 1.475 (100), and 1.315 (30). The thermograms of chrysocolla from different areas of the Udokan deposit (Fig. 3.31–3,4,5) demonstrate exothermic (150, 300–310, 600–620, 730–750, 950 °C) and endothermic (240–300, 350–400, 680–700, 810–820, 960–980 °C) effects, sometimes different from the standard, perhaps, because of the admixture of other minerals. However, it is possible to recognize that water adsorption exists in the temperature range 100–300 °C, 300–600 °C (crystalline hydrate), and in 600–700 °C range (water hydroxyl). Infrared spectra from chrysocolla (Fig. 3.31–5, 6) clearly demonstrate the following absorption lines: 437, 475, 503, 676, 780, 800, 1033, 1635, 3400, and 3620 cm^{-1}. The bands of 600–1200 cm^{-1} indicate clear links Si-O in the silicon-oxygen tetrahedra, and the band 676 cm^{-1} is related to fluctuating Si-O-Cu connections. The broadband of valence fluctuations 3200–3500 cm^{-1} and the line 1653 cm^{-1} of deformational vibrations indicate the presence of molecular water. The band 3620 cm^{-1} of stretching vibrations is a result of coordinated H_2O molecules. Therefore, the IR spectra and DTA curves of chrysocolla show molecular (adsorption) water and hydroxyl water.

Table 3.8 Composition of ore minerals from the Udokan deposit, wt%

№	As	S	Fe	Ni	Co	Se	Sb	Cu	Total
1	0	23.16	7.08	0.006	0.001	0.25	0.013	71.31	101.83
2	0.067	23.37	2.56	0	0.008	0.216	0	75.00	101.21
3	0.066	24.60	6.52	0	0.008	0.108	0.005	72.07	103.39
4	0.026	24.75	6.44	0	0.008	0.124	0	70.58	101.93
5	0.094	22.78	1.27	0	0.013	0.139	0	77.03	101.32
6	0.053	22.57	1.08	0	0.013	0.066	0.009	76.52	100.30
7	0.026	22.51	0.62	0.003	0.015	0.149	0.006	76.52	99.86
8	0.097	22.62	0.67	0.006	0.015	0.062	0	77.70	101.17
9	0.045	24.04	6.61	0	0.012	0.08	0	72.58	103.37
10	0.134	22.38	0.46	0	0.014	0.103	0	78.82	101.92
11	0.021	20.90	0.37	0.005	0.02	0.131	0	78.82	100.27
12	0.169	23.30	3.05	0.003	0.014	0.097	0	75.57	102.20
13	0.076	25.25	8.74	0	0.006	0.062	0.018	68.65	102.80
14	0.054	22.37	0.82	0.003	0.012	0.114	0.005	78.33	101.70
15	0.045	24.16	5.91	0	0.006	0.149	0.006	73.21	103.49
16	0.107	23.92	5.82	0.001	0.01	0.128	0.012	73.61	103.61
17	0.043	24.39	6.36	0.006	0.011	0.082	0	72.18	103.07
18	0.059	22.43	0.19	0.001	0.009	0.136	0.01	78.95	101.79
19	0.147	22.05	0.53	0	0.015	0.116	0.01	78.39	101.26
20	0.029	23.70	3.64	0.003	0.012	0.064	0.005	74.30	101.75
21	0.101	24.23	5.78	0	0.006	0.142	0	73.47	103.74
22	0.05	35.17	30.27	0	0	0.09	0.001	34.93	100.51
23	0.10	24.95	6.43	0.002	0.007	0.094	0.011	72.68	104.27
24	0.01	34.93	28.97	0	0	0.08	0.005	36.40	100.39
25	0.08	34.80	30.43	0.001	0	0.106	0.012	34.94	100.37
26	0.10	24.46	3.34	0	0.011	0.109	0.008	74.93	102.96
27	0.05	34.71	29.23	0	0	0.115	0.004	36.26	100.38
28	0.07	23.80	1.41	0	0.011	0.135	0	76.83	102.26
29	0.02	24.49	3.41	0	0.008	0.145	0	73.45	101.53
30	0.04	31.65	17.99	0	0	0.097	0	49.62	99.40
31	0.03	34.41	0.36	0.003	0.011	0.122	0.028	66.13	101.08
32	0.06	41.86	0.69	2.392	36.6	0.085	0.004	19.28	101.06
33	0.04	34.81	30.58	0	0	0.073	0	34.69	100.19
34	0.09	24.79	7.30	0.002	0.008	0.123	0.007	71.39	103.71
35	0.06	34.87	29.36	0.002	0	0.071	0	36.13	100.50
36	0.04	24.06	5.24	0.003	0.013	0.06	0.015	72.24	101.67
37	0.01	34.92	30.41	0.004	0	0.106	0.014	34.76	100.23
38	0.09	27.11	10.86	0	0.003	0.124	0	61.77	99.96
39	0.09	26.27	10.23	0	0	0.116	0	66.44	103.15
40	0.04	31.54	23.10	0	0	0.087	0.004	44.49	99.27
41	0.06	23.26	2.40	0	0.013	0.08	0	76.60	102.42
42	0.09	24.87	7.12	0	0.01	0.127	0.016	69.38	101.62
43	0.05	34.77	30.55	0.002	0	0.102	0	34.65	100.12
44	0.04	30.75	21.97	0.004	0	0.097	0.003	45.83	98.70
45	0.09	25.36	9.39	0.004	0.007	0.071	0.009	67.75	102.68
46	0.07	22.83	1.52	0	0.014	0.085	0	78.04	102.55

Note. Analyses were carried out at IGEM RAS, analyst EV Kovalchuk

Minerals: 1–21, 23, 26, 28, 29, 34, 36, 41, 46 – chalcocite group; 22, 24, 25, 6, 27, 33,37,43– chalcopyrite; 38, 39, 42, 45 – bornite, 32 – carrollite

Table 3.9 Proportions of ore types at the Udokan deposit

Level, m	Site	Average oxidation, %	Sulfide	Mixed	Oxidized
Surface	Left bank of River Naminga	43.2	41	56	3
	Shumnyi-Krutoi	70.1	–	26	74
	Sekushchiy	76.0	–	37	63
	Skolzkiy	57.8	3	74	26
	Bluzhdayushchiy-Medny	77.6	–	44	56
−150	Adits 11, 12, 7	48.4	29	67	4
−260	Adits 3, 9	32.8	45	47	8
−310	Adit 1	20.8	73	11	16
−470	Adit 2	19.2	82	18	–
−670	Boreholes	18.2	85	15	–

Note. Data after Narkelyun et al. (1987)

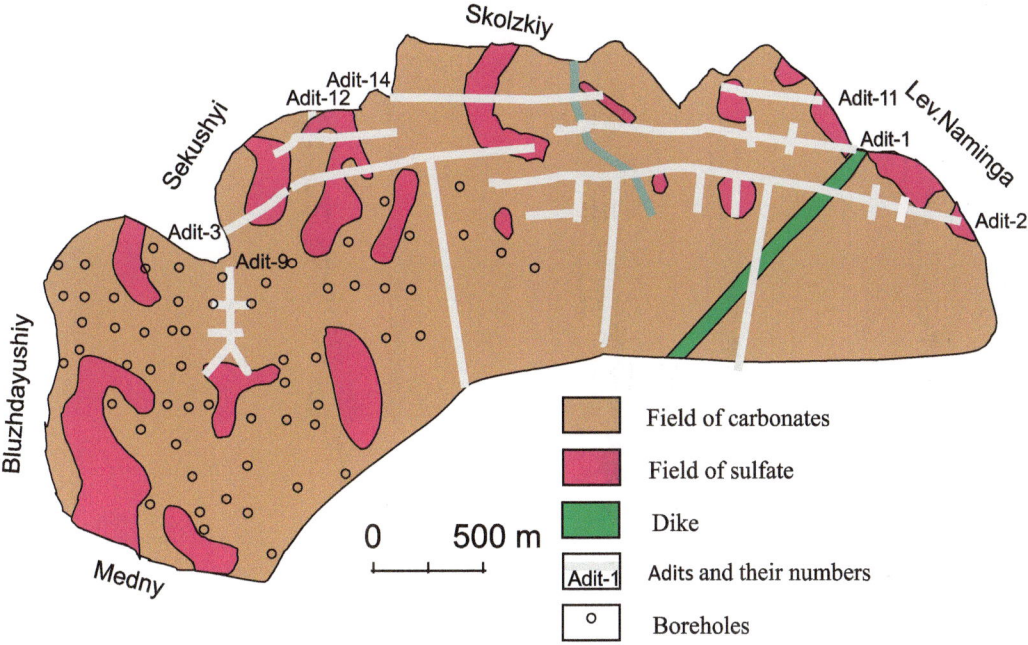

Fig. 3.26 Plan view distribution of oxidized ore at the Udokan deposit. (After Narkelyun et al. 1987)

Phosphates

Cornetite $Cu_3(PO_4)(OH)_3$ was found by Demidovich as small crusts (0.05–0.2 mm) in cracks in sulfide ore. It has dark blue to green blue color. It was recognized by its optical features: biaxial (−), Ng = 1.768, Np = 1.811, Nm = 1.820, and 2V = 40°.

Two very rare minerals were discovered by Krivolutskaya and Gongalsky at the Bluzhdayushchiy site of the Udokan deposit. One of them is very close in morphology and optical features to chalcophyllite (Dana et al. 1954), but it has different chemical composition and represents a new mineralogical species. Preliminary it was named krendelevite (pending the decision of IMC).

Krendelevite is a preliminary named new P-mineral. We suggested to name this mineral after Fedor Petrovich Krendelev, one of the first geologists who studied the Udokan deposit in 1956–1959 and who founded the Institute of Natural Resources, Siberian Branch, Russian Academy of Sciences in Chita in 1981. The mineral is very similar to chalcophyllite $Cu_{18}Al_2$ $(SO_4)_3(AsO_4)_3(OH)_{27*}33H_2O$, an extremely rare mineral from oxidation zone of copper deposits. It is an aqueous basic sulfate-arsenate of copper and aluminum. As pointed out by John Dana in System of Mineralogy, it was described by many researches under different names starting from 1801. The modern name "chalcophyllite" was given by Breithaupt in 1841. In Russia, it was first described by Koksharov in 1866 at the Mednorudyanskoe deposit (Nizhny Tagil area) in the Urals. There were no other findings in the USSR. That is why the findings of similar mineral at the Udokan deposit are extremely interesting. However, the discovered mineral is a phosphate-sulfate complex. Conventionally, we regard it as phosphorus analogue of chalcophyllite (P-chalcophyllite).

Fig. 3.27 Zone of oxidation at the Udokan deposit
(**a**) Medny mineral site, (**b–d**) Bluzhdayushchiy mineral site

Fig. 3.28 Samples of oxide ores
Cc chalcocite, *Bn* bornite, *Az* azurite, *Mh* malachite, *Br* brochantite, *Chn* chalcanthite, *Cal* calcite

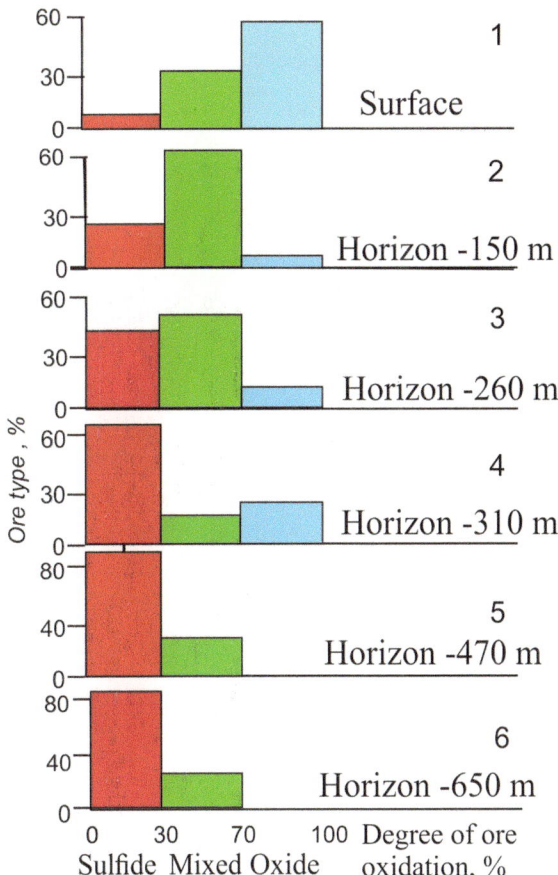

Fig. 3.29 Distribution of ore types at different horizons of the Udokan deposit. (After Narkelyun et al. 1987)

Krendelevite forms subtle streaks (0.1–2 mm thick) in sandstone (Figs. 3.37 and 3.38). Typically, it forms hexagonal plates (up to 5 mm), which form the sockets (1.5 cm in diameter) in cracks. Massive aggregates are less abundant (fine-grained aggregates). As usual, P-chalcophyllite forms monomineral crusts. It is rarely observed in association with antlerite. Some samples contain sockets, about 1 cm in diameter, with concentric zonal structure (from the center to the periphery): P-chalcophyllite-antlerite-P-chalcophyllite. In rare cases, it occurs in association with gypsum.

Transparent emerald green P-chalcophyllite acquires a bluish-green hue and becomes translucent after 2 days of storage in dry environment. The shine changes from glass to pearl. According to some researchers (Dana and Dana 1951), this is due to the partial dehydration of the mineral. The hardness is 2, and the cleavage is very perfect on {001}; it can be dissolved in HCl with a barely noticeable bubbling. The color of mineral in immersion is green, uniaxial negative. No = 1,620 ± 0,003, Ne could not be measured, because the mineral is broken into too thin leaves (No for chalcophyllite 1618). We obtained infrared spectra (ZabNII, analyst A Kozachenko) for sample No 130 from the Udokan deposit

(Fig. 3.31 –11) and sample No 22278 from Cornwall, England (a sample from the Mineralogical Museum collection of the Mining Institute, St. Petersburg, Russia; Fig. 3.31-10). These spectra were compared with the standard spectrum of chalcophyllite (Menke 1970). Referring to identification of absorption bands in the IR spectra and chemical composition of chalcophyllite, we can say the following: absorption at 1000–1100 cm^{-1} due to the anion SO_4 (sulfate) is present in all three compared spectra; absorption in the region 800–900 cm^{-1} due to the anion AsO_4 (arsenate) is noted only in two spectra (sample No 22278 and the pattern from the Atlas of Menke). This band is absent in the spectrum of the Udokan chalcophyllite (or very weakly manifested). The nature of bands at 1640 and 3300–3400 cm^{-1} is identical in all three cases, which indicates the presence of water in all samples in the form of H_2O. All the abovementioned lead to the following conclusions: sample No 22278 from Cornwall corresponds to chalcophyllite described in literature; sample No 130 according to the infrared spectrum mostly belongs to water sulfate and phosphate rather than arsenate.

Therefore, the Udokan Chalcophyllite is different mineral space in comparing with Cornwall one (usually $SO_4:AsO_4$ is less or equals to 1:1). This also follows from chemical analysis of the studied mineral (in wt%, analyst M Kozakova): CuO 46.56, Al_2O_3 1.00, AsO_4 0.18, SO_4 10.97, H_2O ± 16.49, [H_2O] 0.99, P_2O_5 9.82, LOW 15.35, and total 101.37. Repeated identification of arsenic in Udokan chalcophyllite yielded 2.94 wt%. The study of these minerals by microprobe method (Moscow State University, analyst N Krivolutskaya) showed that arsenic in the Udokan chalcophyllite was not detected, whereas in the reference chalcophyllite from Cornwall, arsenic values exceed 10 wt%. Considering the above data, we can conclude that the mineral from the Udokan deposit is a low-As mineral species (0.18–2.94% As according to qualitative chemical analysis). It is possible that we deal with an isomorphic range, whose end members are chalcophyllites from Udokan and Cornwall.

Fluellite $Al_2(PO_4)F_2(OH)*7H_2O$. In addition to P-chalcophyllite, another rare mineral was found (the second finding in Russia). It occurs on the surface of chalcophyllite as fine aggregates (0.1 mm) of the smallest isometric grains. The mineral is colorless; its spherical aggregates are white to light gray. It was not investigated in detail due to insignificant amount. Its identification is based on the X-ray analysis. Powder patterns are similar to the ASTM standard. The main lines of powder patterns are 6.50 (100), 3.23 (80), 2667 (80), 2153 (80), and 1703 (70).

Sulfides

Chalcocite Cu_2S is blue black to gray. It forms pseudomorphs after bornite and chalcopyrite in the form of streaks, stains, and borders. It differs from the hypogene white chalcocite by many characteristic features: microhardness, optical data,

Table 3.10 Minerals from zone of oxidation at the Udokan deposit

Class of Mineral	Main	Major	Rare
Sulfates	Brochantite $Cu_4(SO_4)(OH)_6$	Melanterite $FeSO_4 \cdot 7H_2O$	Jarosite $KFe^{3+}{}_3(SO_4)_2(OH)_6$
	Antlerite $Cu_3(SO_4)(OH)_4$		Dolerophanite $Cu_2(SO_4)O$
	Gypsum $CaSO_4 \cdot 2H_2O$		Halotrichite $FeAl_2(SO_4)_4 \cdot 22H_2O$
	Chalcanthite $CuSO_4 \cdot 5H_2O$		
Carbonates	Malachite $Cu_2(CO_3)(OH)_2$	Azurite $Cu_3(CO_3)_2(OH)_2$	Cerussite $PbCO_3$
Oxides	Hematite Fe_2O_3	Tenorite CuO	Magnetite $FeFe_2O_4$
	Cuprite Cu_2O	Lepidocrocite γ-$Fe^{3+}O(OH)$	Delafossite $CuFeO_2$
	Psilomelane $(Mn_2O*MnO*nH_2O)$	Hydrogoethite $3Fe_2O_3 \cdot 4H_2O$	
	Goethite α-$Fe^{3+}O(OH)$		
Silicates	Chrysocolla $Cu_{2-x}Al_x(H_{2-x}Si_2O_5)(OH)_4 \cdot nH_2O$ $(x < 1)$		
Phosphates			Atacamite $Cu_2(OH)_3Cl$
			Cornetite $Cu_3(PO_4)(OH)_3$
			Fluellite $Al_2(PO_4)F_2(OH) \cdot 7H_2O$
			P-Chalcophyllite $Cu_{18}Al_2(SO_4)_3(PO_4)_3(OH)_{27}$ x $33H_2O$
Sulfides	Chalcocite Cu_2S secondary	Covellite CuS	
		Chalcopyrite $CuFeS_2$	
		Bornite Cu_5FeS_4	
Native metals	Copper Cu		
	Gold Au		

Note: after Narkelyun et al. (1987) and Gongalsky and Krivolutskaya (1993)

color, associations, and conditions of formation (Trubachev and Narkelyun 1968). In the northern part of the ore horizon, it is confined to near-surface areas in the fracture zones (Trubachev 1981).

Covellite CuS. Its "extended" formula is $Cu^+{}_4Cu^{2+}{}_2(S_2)_2S_2$ (Goble 1985). Usually it forms indigo-blue massive metallic material. It develops after chalcocite and bornite and to a lesser degree on chalcopyrite in the form of plates, up to 0.005–0.01 mm. Relatively large aggregates of small scales of covellite were detected by distinct anisotropy in the reflected light. Most often, it forms rims around grains of chalcocite and bornite (0.03 mm thick). It was observed in chalcopyrite and bornite in the form of 0.05 mm veinlets. The mineral is usually accompanied by iron hydroxides, malachite, and rarely cuprite. Paragenetic associations of covellite and its wide presence in oxidized ore prove its supergene origin. X-ray powder diffraction data are 3.25 (30), 3.05 (60), 2.80 (100), and 2.72 (60).

Bornite Cu_5FeS_4. Supergene bornite is a mineral from oxidation zone. It is rare, developing along the periphery of the grains of chalcopyrite and its cracks. Typically, it occurs as massive metallic material. It has a copper-red color on fresh exposures, which quickly tarnish into an iridescent purple color after exposure to air and moisture. It has low microhardness (89 kgs/mm²) in comparison with hypogene

bornite (110 kg/mm²), and it associates with supergene minerals.

Chalcopyrite CuFeS₂. Supergene chalcopyrite develops as rims around chalcocite, also forming a plate-shaped lattice in the decomposition zone during the oxidation of bornite (Bezrodnykh et al. 1968). Volodin identified three generations of supergene chalcopyrite in oxidized ore (Krendelev et al. 1983).

Native Metals

Copper Cu is a rare mineral in oxidation zone. It is usually located on the surface in association with cuprite. It was noted with iron hydroxides not only on the surface but also at the deep levels. For example, it was found in borehole 204 at a depth of 320–370 m. Native copper forms long platy grains and dendritic 1–5 mm crystals.

3.4.2 Mineral Associations and Their Origin

Age relationships between supergene minerals are very complex and not always certain due to their separation in space. For this reason, various researchers identified different mineral parageneses formed at different stages of oxidation at Udokan. Pramzintsev (1966) identified five associations, and many minerals in ore constitute 2–3 generations. He believed

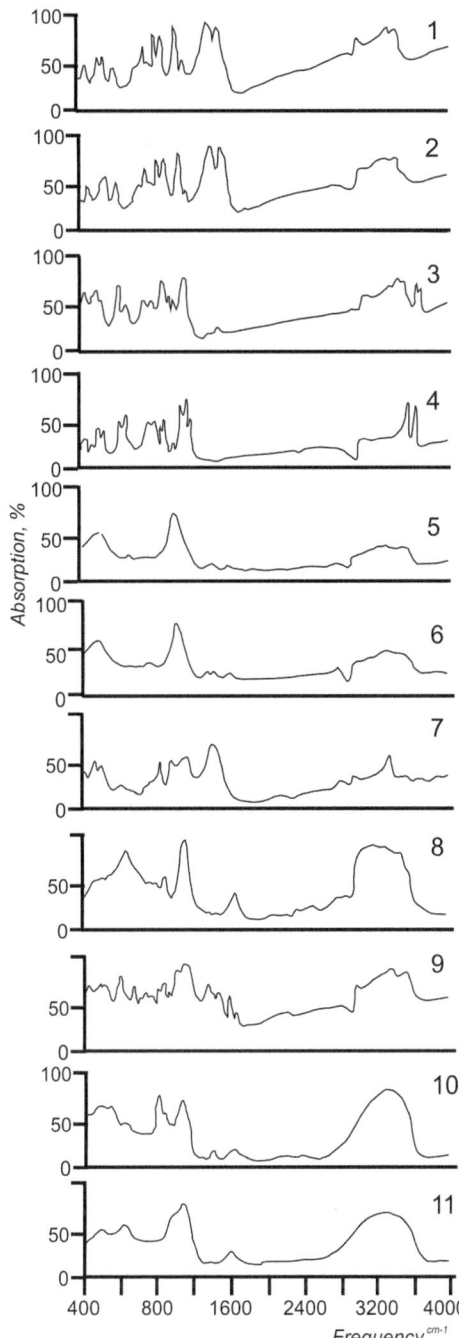

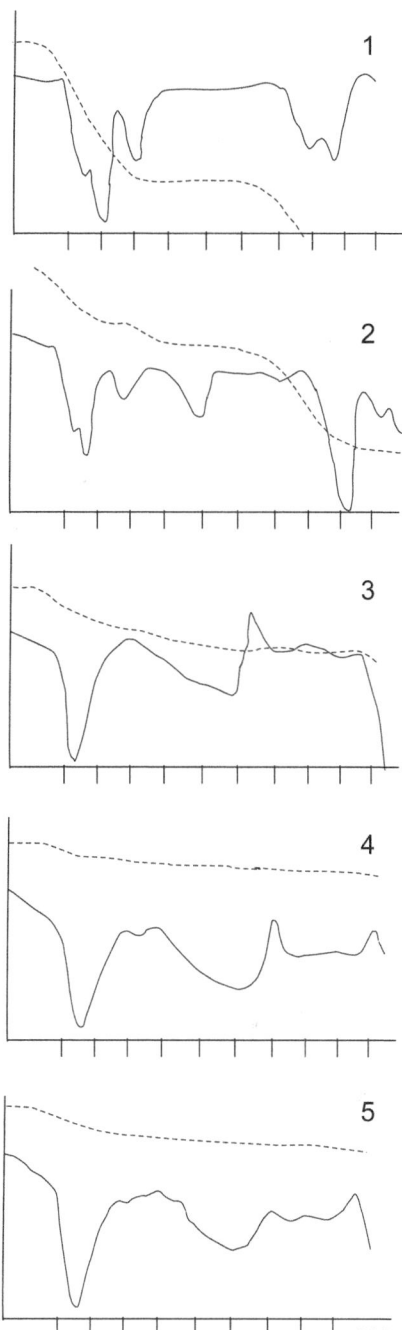

Fig. 3.30 Infrared spectra of minerals from oxidation zone of the Udokan deposit
1–2, malachite (1, crystalline; 2, needle); 3, brochantite; 4, antlerite; 5–6, chrysocolla (5, blue; 6, green); 7, azurite (+gypsum); 8, chalcanthite; 9, gypsum (+malachite); 10, chalcophyllite from Cornwall; 11, krendelevite (The data here and in Fig. 3.31 are after Narkelyun et al. 1987)

Fig. 3.31 Thermograms and curves of weight loss of minerals from oxidation zone at the Udokan deposit
1–2, chalcanthite; 3–5, chrysocolla

that the main association is malachite-brochantite. Yurgenson et al. (1968) identified several mineral parageneses irregularly distributed in the Udokan deposit. Sulfate and carbonate are two main mineral associations. There is a following

tendency in formation of supergene minerals: earlier formation of chrysocolla, iron hydroxides, covellite, supergene chalcocite, bornite, and chalcopyrite followed by sulfates. At the same time, radiating malachite crystallized after sulfates. Pramzintsev explained a combination of various associations in oxide ores by evolution of chemical composition of water. Subsequently, a completely different model of supergene ore formation was proposed. It was shown that the main factors

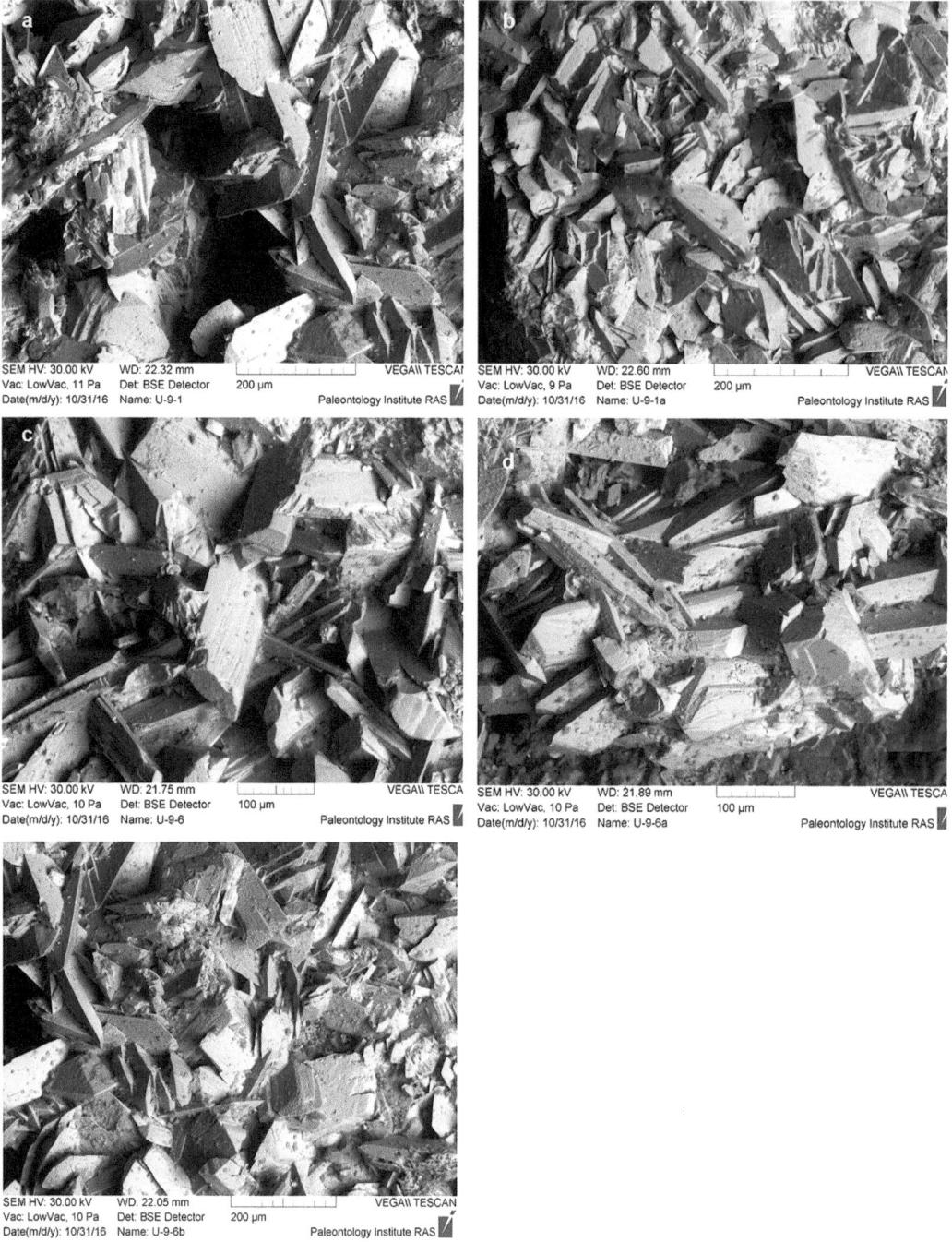

Fig. 3.32 BSE images of brochantite. Here and in Figs. 3.33, 3.34, 3.35, 3.36, and 3.38, images were taken at the Paleontology Institute, Moscow (analyst EA Zhegallo) (**a–d**) Sample numbers: (**a**) U-9-1, (**b**) U-9-1a, (**c**) U-9-6, (**d**) U-9-6a

controlling the formation of supergene minerals are the hypogene sulfide minerals, lithological composition of ore-bearing rocks, geomorphology, as well as tectonic and climatic (permafrost) conditions.

In general, it was concluded that the Udokan oxidation zone was formed in two main phases: (i) during the preglacial period in warm humid climate and (ii) under the specific permafrost conditions. During the first phase, copper from

sulfides passed into water and formed soluble sulfates that easily migrated within the ore horizon. The modern forms of Cu migration on the surface are given in Table 3.11. These solutions produced carbonates in the areas of calcareous rocks and cuprite-tenorite in silica-rich rocks. The reaction between Si-bearing and Cu-bearing solutions produced chrysocolla. At the same time, there were other oxidation processes: formation chalcocite after bornite and

Fig. 3.33 Minerals from oxidation zone of the Udokan deposit
Samples of NA Krivolutskaya and BI Gongalsky; photo by MA Bogomolov. (**a**, **b**) Crystals of antlerite (Ant), (**c**) krendelevite (Kr), (**d**) gypsum (Hyp, white) and brochantite (Br, green)

chalcopyrite, martitization of magnetite, formation of plate-lattice structures of chalcopyrite, and some other features.

This special process in permafrost was named cryomineralization (Yurgenson 1996). It is characterized by <0 °C temperatures, lack of liquid water, and isothermal conditions of the mineralization processes.

Bezrodnykh (1969) for the first time calculated the reactions in the frozen column and showed a large role of solution concentration and pH for the deposition of sulfates at the Udokan deposit. It was found that crystallization of melanterite and chalcanthite needs the highest concentration of solutions, while antlerite was formed at lower concentrations, whereas brochantite crystallized at the lowest concentrations. This fact explains the prevalence of brochantite.

During formation of permafrost, mainly sulfate minerals were formed under the influence of chemical, electrochemical, and microbiological processes, with atmospheric oxygen as the most real agent of sulfide oxidation. Thermodynamic calculations and experimental studies (Pitulko 1977; Bezrodnykh 1969) were conducted in the northeast of the USSR and at the Udokan deposit. They demonstrated that

only sulfates could be energetically formed in frozen ore, while all other supergene minerals were formed mainly at >0 °C temperatures. Indeed, malachite coexists with sulfates, growing on brochantite. Its appearance can be explained by thermal anomalies during the oxidation of sulfides. More complex relationships of copper sulfates and carbonates with other supergene minerals are probably due to the long existence of oxidation zone. Pitulko (1977) recognized at least four stages: ancient oxidation (formation of supergene sulfides), early sulfation (forming low-H_2O sulfates), late oxidation (formation of iron and manganese hydroxides, gypsum, phosphates, arsenates, etc.), and finally, modern sulfation (produced the main volume of sulfate).

Hydrogeochemical activity in cryogenic zone is due to the presence of liquid aqueous phases. There are two mechanisms to decrease the temperatures for the aqueous solutions. The first one is a cryogenic concentration of solution controlled by the temperature before freezing; it has influence on the amount of solution but not on its concentration (Ptitsyn et al. 2003). Another mechanism is the formation of film fluids, chemically or physically associated with the surface of

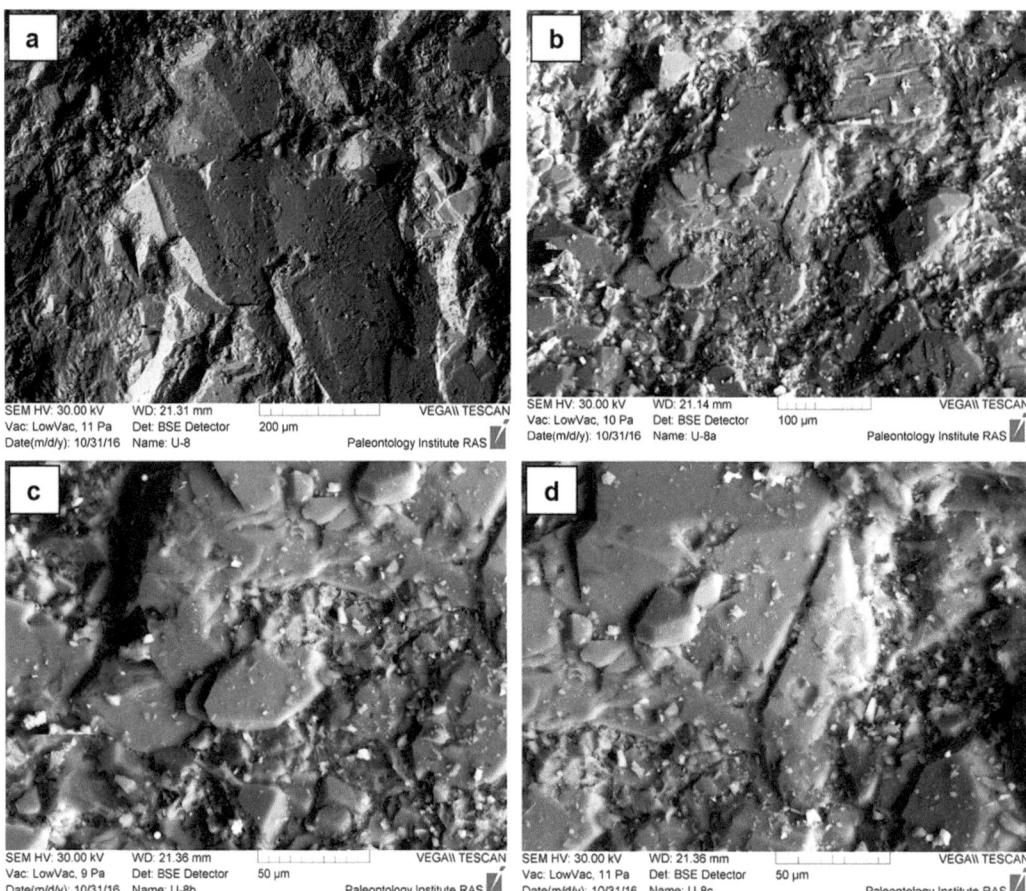

Fig. 3.34 BSE images of antlerite
(**a–d**) Sample numbers: (**a**) U-8, (**b**) U-8a, (**c**) U-8c, (**d**) U-8c

the minerals. In this case, the aqueous liquid phase can be maintained at very low temperatures (e.g., down to 140 K on channels of zeolites). Thus, the existence of liquid water phase is possible in almost any real subzero temperatures in the Earth's crust.

Ptitsyn (1992) conducted experiments on the leaching of Udokan oxide ore with sulfurous solution at −18 °C. The 0.5% sulfuric acid solution was frozen. A tablet of oxide ore (malachite) was placed on the ice surface. Thereafter, the same sulfuric acid solution was frozen slowly on the top of the tablet. After 5 months, a layer of nonfreezing solution was formed at the bottom of the container, wherein the concentration of Cu reached 22 g/l. During the experiment, copper was recovered from the ore almost completely. According to the estimates of Ptitsyn (1992), the speed of gravitational migration of solution through the ice was 1.5–2 cm per month. At such rate, a 100-m-thick cryogenic oxidation zone over a sulfide deposit could be formed during 2000 years. According to experimental data, temperature variations (even within the negative temperature range) might have accelerated all leaching processes up to 3–5 times.

Takenaka et al. (1992) published some interesting data. Based on their experiments, they argued that speed of oxidation for nitrogen compounds at low temperatures is five orders of magnitude higher than at the normal temperature. In more detail, these results were republished later (Takenaka et al. 1996).

The low temperatures determine the specifics of migration and formation of stable forms of compounds in cryogenic zone. The oxidation took place at a relatively constant temperature in the frozen zones (−1 to −7 °C), reflecting the mode of daily and seasonal temperature fluctuations. In the latter case, abnormally active melt water (Shvartsev 1998) and gravitational movement of solutions in frozen rocks (Ptitsyn and Sysoeva 1988) might have taken place, with a large number of ionic compounds crystallizing from brines into crystalline hydrates.

This is a possible oxidation reaction of chalcocite to form hydroantlerite:

$$2Cu_2S + (2+1)H_2O + SO_2 = CuS + Cu_3SO_4(OH)_4^* nH_2O$$

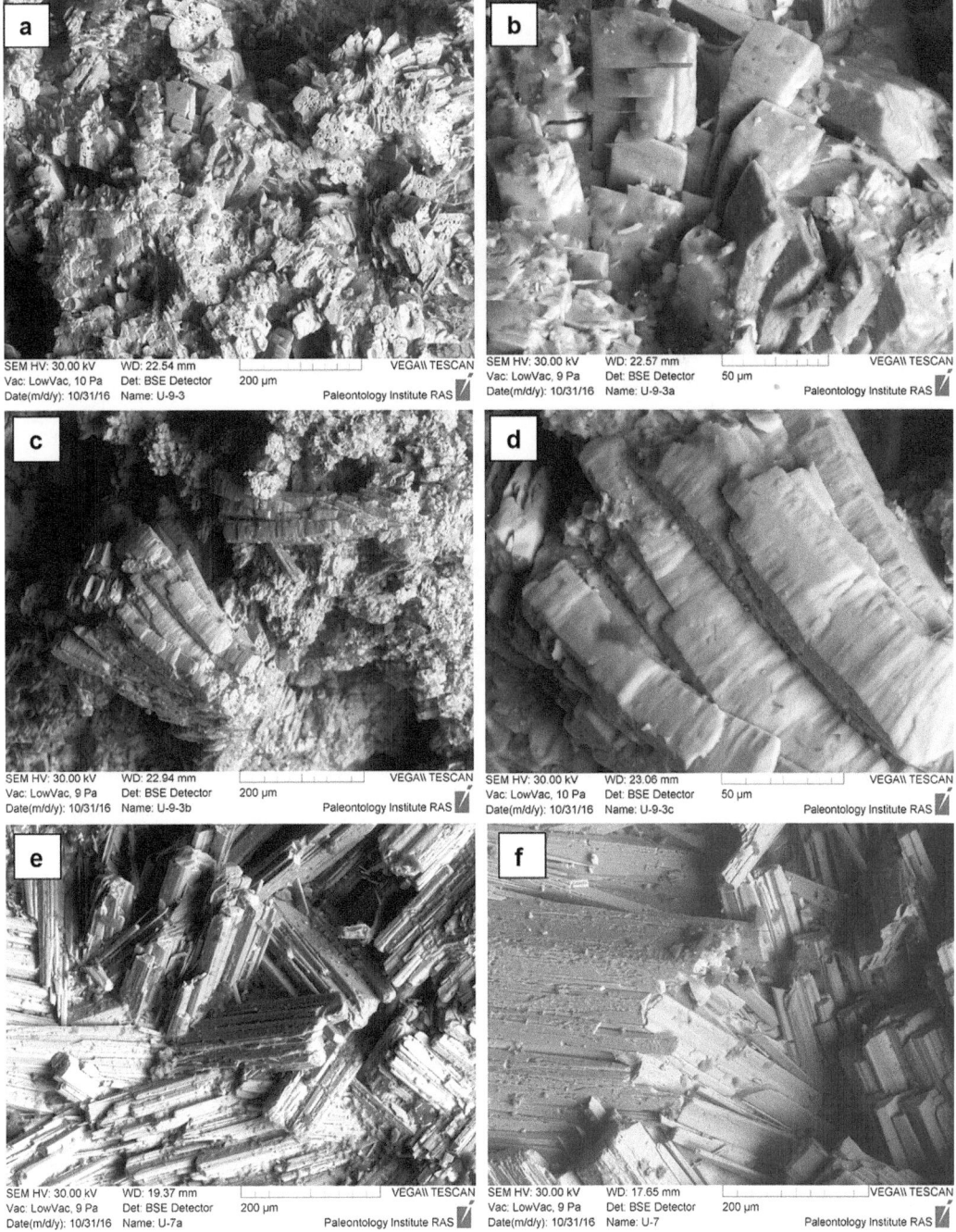

Fig. 3.35 BSE images of malachite
Sample numbers: (**a**) U-9-3, (**b**) U-9-3a, (**c**) U-9-3b, (**d**) U-9-3c, (**e**) U-7, (**f**) U-7a

The change in the Gibbs free energy of this reaction is estimated as -830 kJ/mol (for standard conditions). The low-temperature components do not change much the potential assessment since the majority of the binary eutectic temperature of sulfates does not fall below $-5\ °C$.

The rocks under permafrost conditions are therefore in the active area of physical-chemical processes for sulfate formation at the Udokan deposit, and the permafrost factor plays an important role in redistribution of minerals in the oxidation zone. It is important to emphasize that domination of brochantite $Cu_4(SO_4)(OH)_6$ and antlerite $Cu_3(SO_4)(OH)_4$ is a specific feature of supergene mineralization at the Udokan deposit that distinguishes it from other copper deposits of Russia and elsewhere. The quantitative estimates of relationships between the minerals in oxidation zone revealed their following ratio (wt%): carbonate (12–14%), sulfate (25–30%), pyrite-chalcopyrite (3%), and bornite-chalcocite (53–60%). Spatial distribution of ore types

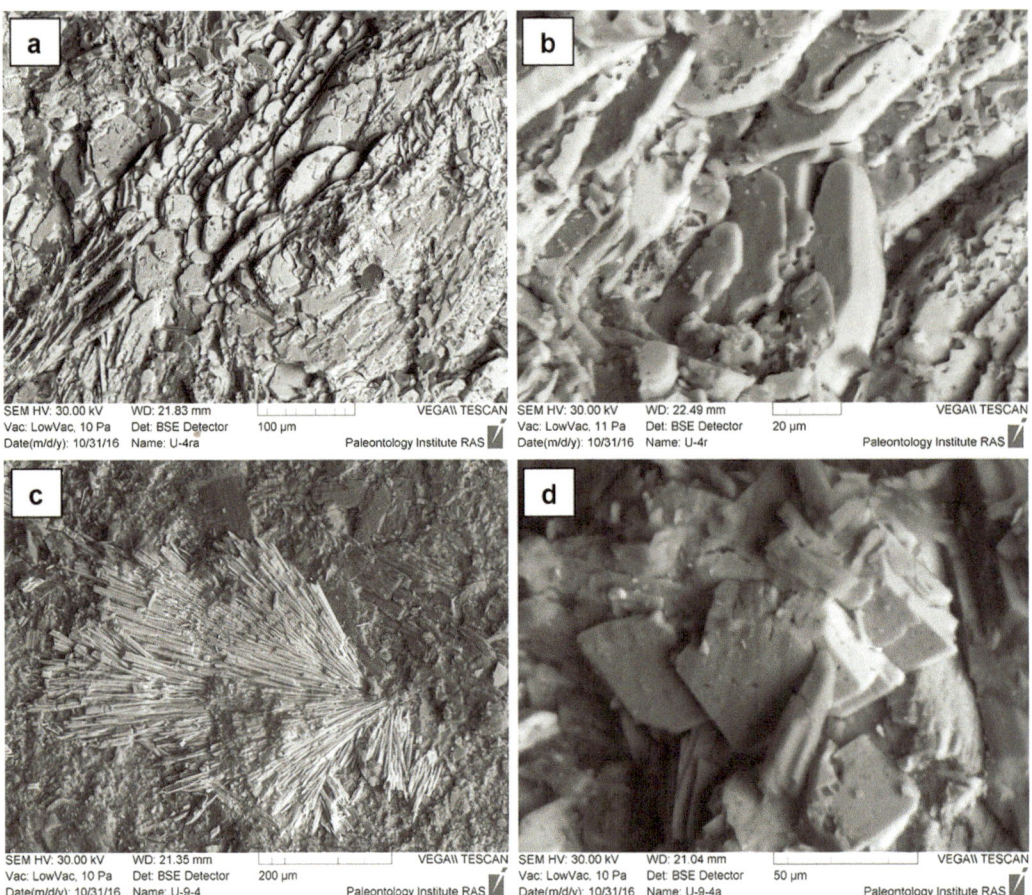

Fig. 3.36 BSE images of azurite
Sample numbers: (**a**) U-4r, (**b**) U-4ra, (**c**) U-9-4, (**d**) U-9-4a

Fig. 3.37 Photo of krendelevite

(Fig. 3.29) is controlled by altitude and copper content. At the same altitude, the poor ore is deeply oxidized, while the rich ore is generally slightly oxidized. The high rate of the oxidation by biogenic process (by several orders of magnitude higher abiotic) plays important role in formation of basic sulfate process. This is a main reason for their rapid accumulation and crystallization.

3.5 Metamorphism of Rocks and Ore

The rocks and the ore of the Udokan deposit were affected by regional and, to a lesser extent, contact metamorphism. Contact metamorphism of the copper sandstone is generally evident from the recrystallization of the sandstone and siltstone cement and slight regeneration of the sulfides.

The host rocks retain their original look, and alteration can be seen only under a microscope. The recrystallized cement of the sandstone acquires lepido- and granoblastic structures. The siltstones and claystones are characterized by low-grade greenschist facies metamorphism. Several researchers assume that metamorphism was associated with the transformation of the iron hydroxides into magnetite. As

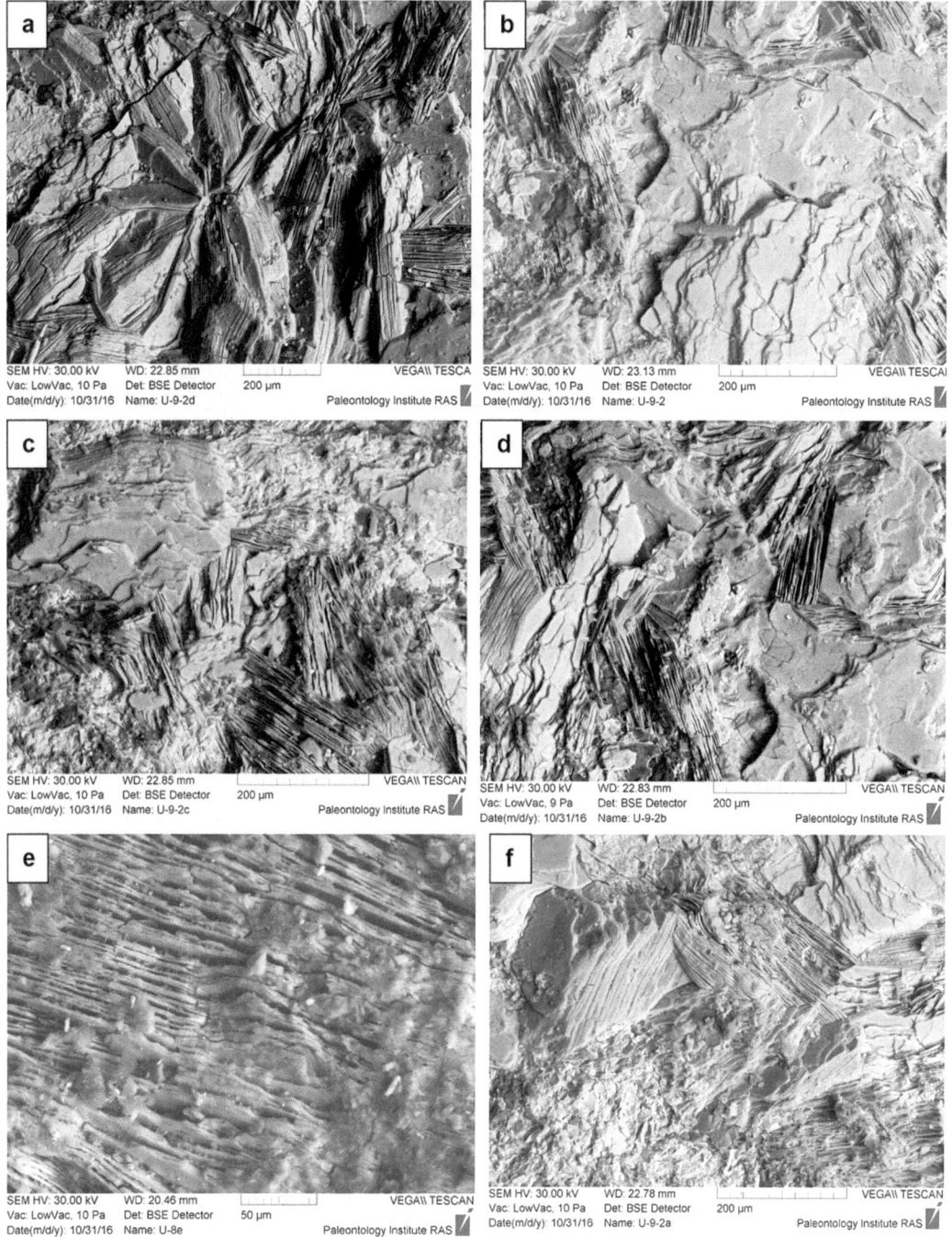

Fig. 3.38 BSE images of krendelevite
Sample numbers: (**a**) U-9-2d, (**b**) U-9-2, (**c**) U-9-2c, (**d**) U-9-2b, (**e**) U-8e, (**f**) U-9a

a result, the primary red color of some of the beds was modified. The most significant metamorphism is discernible in the copper sandstone. The sulfide disseminations are commonly accompanied by newly formed quartz, chlorite, muscovite, and epidote. The primary bedding is complicated. Some laminae locally lost rectilinear boundaries due to the appearance of a network of tiny sulfide and thin quartz and quartz-sulfide veinlets and offsets. The metamorphosed ore reveals corrosion of the clastic grains and displays the development of interstitial structures. More significant alterations of copper sandstone are related to contact metamorphism near the crosscutting dikes. Toward a 150 m-thick gabbrodolerite dike, the degree of recrystallization of the cement and clastic grain gradually increases; the bedding of rocks is disturbed;

Table 3.11 Forms of Cu migration in water in scattering fluxes of the Bluzhdayushchiy mineral site, mol/l

Component	K	pH = 6,41		3D	pH = 7,02		pH = 7,5	
		Probe №			(N 86)		(N 106)	
		Content, wt%			Content, wt%		Content, wt%	
Cu^{2+}	–	$4.2 \cdot 10^{-6}$		82.4	$1.7 \cdot 10^{-6}$	34.1	$7.1–10^{-8}$	5.9
$CuOH^-$	6.66	$5.3 \cdot 10^{-7}$		13.4	$8.8 \cdot 10^{-7}$	17.7	$9.9 \cdot 10^{-8}$	8.2
$Cu(OH)_2$	13.91	$2.6 \cdot 10^{-7}$		1.2	$1.7 \cdot 10^{-6}$	33.9	$5.8 \cdot 10^{-7}$	48.2
$Cu(OH)_3^-$	16.69	$4.3 \cdot 10^{-12}$		<0.1	$9.8 \cdot 10^{-11}$	<0.1	$1.1 \cdot 10^{-10}$	<0.1
$Cu(OH)_4^-$	16.47	$7.1 \cdot 10^{-20}$		<0.1	$6.0 \cdot 10^{-18}$	<0.1	$2.1 \cdot 10^{-17}$	<0.1
CuF^+	1.54	$2.1 \cdot 10^{-12}$		<0.1	$1.7 \cdot 10^{-10}$	<0.1	$2.0 \cdot 10^{-11}$	<0.1
$CuCl^+$	0.74	$1.1 \cdot 10^{-10}$		< 0.1	$4.8 \cdot 10^{-11}$	<0.1	$3.5 \cdot 10^{-11}$	< 0.1
$CuCl_2^0$	0.69	$4.5 \cdot 10^{-16}$		<0.1	$2.0 \cdot 10^{-16}$	<0.1	$2.8 \cdot 10^{-15}$	<0.1
$CuBr^+$	1.13	$3.4 \cdot 10^{-11}$		<0.1	$1,5 \cdot 10^{-11}$	<0.1	$4.3 \cdot 10^{-14}$	<0.1
$CuHCO_3^+$	2.4	$4.4 \cdot 10^{-8}$		1.1	$5.6 \cdot 10^{-8}$	1.1	$1.2 \cdot 10^{-8}$	1.0
$Cu(HCO_3)_2^0$	4.2	$1.2 \cdot 10^{-10}$		<0.1	$4.4 \cdot 10^{-10}$	<0.1	$5.9 \cdot 1O^{-10}$	<0.1
$CuCO_3^0$	6.75	$6,8 \cdot 10^{-8}$		1.4	$6.5 \cdot 10^{-7}$	13.0	$4.4 \cdot 10^{-7}$	36.5
$Cu(CO_3)_2^{2-}$	9.92	$9.1 \cdot 1O^{-13}$		<0.1	$5.9 \cdot 10^{-11}$	<0.1	$2.6 \cdot 10^{-10}$	<0.1
$CuSO_4^0$	2.34	$2.2 \cdot 10^{-8}$		0.4	$9.9 \cdot 10^{-9}$	0.2	$3.5 \cdot 10^{-10}$	< 0.1
$Cu(SO_4)_2^{2-}$	3.4	$6.0 \cdot 10^{-12}$		<0.1	$2.7 \cdot 10^{-12}$	<0·1	$2.9 \cdot 10^{-14}$	<0.1
Total		$5.0 \cdot 10^{-6}$		100	$5.0 \cdot 10–6$	100	$1.2 \cdot 10^{-6}$	100

Note. Data after Savenko and Shatalov (1998); $K = (SO_4^{2-})/(HCO_3)_2^-$

the quartz-feldspar sandstone is transformed into quartzite; and calcareous sandstone is transformed into epidote hornfels, which often contains actinolite and garnet.

Evidence of the redeposition of the primary sulfide and changes in their phase and chemical compositions were detected by Krendelev et al. (1983), who recognized four types of bedding-parallel and crosscutting quartz-carbonate veinlets with chalcocite-bornite mineralization, whose origin was explained by regional or contact metamorphism. Gablina (Gablina and Vasilovskaya 1989; Gablina and Lur'e 2001; Gablina 2008) suggested that the rocks of the Udokan Supergroup lost their red color due to metamorphism. Less attention was paid to the veins nearly concordant to the bedding, which are often brecciated and contain fragments of quartz and host rocks. These veins were found by Petrovsky (1985, 2003) who studied the cross-bedding. The en echelon lenses of sulfides with quartz and the crosscutting chalcocite-bornite veinlets at the Zapadnyi, Ozernyi, and Naminga sites provide evidence for ore deposition from hydrothermal solutions or copper redeposition after sedimentation. Thick (a few tens of meters) zones of brecciated host sandstone and claystone cemented by gangue quartz with chalcocite, bornite, and chalcopyrite were also found. The quartz-sulfide veinlets are rimmed with metasomatic magnetite, which differs in morphology and composition from the magnetite in the sedimentary laminas. The veinlets reveal up to 0.3 ppm Au, whereas the gold grade in the barren sandstone does not exceed 0.01 ppm. In addition to the crosscutting quartz-carbonate veins with sulfides, which were documented by all researchers, stratabound and crosscutting veins, complex in structure, often brecciated (Fig. 3.18) and with large sulfide

segregations, are widespread in the upper part of the productive sequence. Fragments of host rocks and gangue quartz are cemented in central parts of the veins with coarse crystalline bornite and, to a lesser extent, by chalcocite. Sulfides of the cement often penetrate into the zone of finely disseminated chalcocite ore, which, in turn, gives way to the zone of magnetite disseminations.

Furthermore, the bedding-parallel and crosscutting sulfide and quartz-sulfide veins, together with the adjacent host rocks, are characterized by the following zoning. Coarse-crystalline bornite, replaced by chalcocite-group minerals in grain margins and along fractures, is dominant in the veins. The next zone is composed of finely disseminated chalcocite with much lower abundance of bornite; and the outer zone contains magnetite disseminations. The Fe content in the minerals varies oppositely to the zoning. At one locality, laminae with clastic magnetite, titanite, and zircon were found in the outer zone. Single euhedral magnetite crystals typical of the outer zone are also noted here. The intermediate zone in the crosscutting veinlets is expressed less distinctly, but the magnetite content is much higher.

Some authors reported similar data with another interpretation of veins at the Udokan deposit. The detailed study of veins hosted in sandstone demonstrated two main genetic types of veins (Ptitsyn et al. 2003). The first one is a core of Alpine type, which was produced during epigenesis and regional metamorphism. They were formed by the in situ recrystallization of materials along minor cracks, with the migration of ore minerals. The second type is characterized by clearly sharp contacts, later formation (as compared with the surrounding rocks), and localization near contacts of sandstone with magmatic rocks. These features correlate them to contact metamorphism. Along with a polygenic genesis of the veins, it has been established that they were formed in several stages. The earliest minerals are calcite, quartz-calcite veins, and Cu- and Fe-sulfides (chalcopyrite, bornite, chalcocite), hosted in both calcite and quartz.

3.6 Genesis of Ore

Since the discovery of the Udokan deposit in 1949, different points of view on the genesis of Udokan ores have been proposed. Despite the dominance of the hydrothermal hypothesis on the origin of copper mineralization in sandstone in the 1950s, many geologists (Korol'kov, Grintal, Chechetkin, Bykov, Koshelev, Mel'nichenko, Volodin, Bakun, Domarev, Bogdanov, Feoktistov, Narkelyun, Bezrodnykh, Yurgenson, Trubachev, and others) explained formation of the Udokan deposit via sedimentary processes. There are currently five hypotheses regarding the origin of the Udokan ore.

3.6.1 Hypotheses on Origin of the Udokan Ores

The sedimentary hypothesis. According to numerous researchers (Krendelev 1959, 1987; Krendelev et al. 1983) and results of geological exploration, this model is based on the facts presented below. We present only the most important facts here and recommend the detailed descriptions in the work by Ptitsyn et al. (2003). The main supporting facts of the sedimentary hypothesis are as follows:

1. The presence of copper-bearing horizons at all stratigraphic levels of the Udokan Supergroup.
2. Cupriferous sandstones occupy a particular stratigraphic position and have a pronounced rhythmic style.
3. The horizons of cupriferous sandstone are part of the folded structure.
4. The copper-bearing horizon occurs in the particular lithofacies, reflecting conditions of their formation. There is a clear link between the bornite-chalcopyrite ore type and deltaic facies, while pyrite-chalcopyrite mineralization is found in lagoon facies.
5. Cross-bedded ores correlate with cross-bedding of sandstone; the scale of ore length and the inclination of sulfide disseminations show the formation of ore in the beds of the most active water flows where heavy sulfide fraction has settled.
6. There is no correlation between the intensity of mineralization and metamorphism; the most intense metamorphism is recorded under the ore layers.

7. The ore lenses and elongated bodies have a general west or south-west direction that reflects the material flow and mineral composition of the ore-bearing strata and eroded rocks of the Aldan Shield.

The *hydrothermal-magmatic-metasomatic hypothesis* (Reznikov 1965; Volfson and Arhangel'skaya 1972, 1987; partially Petrovsky and others) is based on the following assumptions:

1. Mineralization zones are controlled by deep faults serving as pathways for the penetration of magmatic melts and fluids since the Late Archean.
2. The presence of typical hydrothermal veins in the lower structural zone of the Udokan Supergroup (similar to the veins in the Kemen Group).
3. The epigenetic (transgressive) nature of the ore relative to the host rocks.
4. Localization of orebodies within (a) folds and (b) fault sunders the claystone strata.
5. The vein, veinlet, and breccia nature of some ores.
6. Control of the hydrothermally altered rocks and orebodies by the same faults.
7. The presence of copper sulfides not only in sandstone but also in the intrusions and dikes.
8. The relatively high temperature of ore formation (above 90 °C, when cubic chalcocite was formed).

Hydrothermal-sedimentary hypothesis was suggested by Sochava (1979, 1981, 1986), based on the following observations and assumptions:

1. Copper and other metals are imposed into the alluvial plain in the form of metal-bearing solutions, which contained H_2S, resulting in certain amount of hydrogen sulfide.
2. Mechanical redeposition of sulfides in the process of erosion by river flows, which led to the formation of ribbon-shaped lodes with high ore component as a natural concentrate.
3. Regeneration of sulfide grains during diagenesis and catagenesis with the loss of their primary clastic structures.
4. Partial dissolution of sulfides and formation of new deposits by groundwater in fluvial sediments enriched inorganic matter, followed by a new mechanical redeposition.
5. Redistribution of sulfides during the deformations and related late catagenesis and metamorphism, leading to formation of ore veins.
6. The main factors in mineralization are the heterogeneous geochemical parameters, i.e., the hydrothermal solutions themselves or mechanically reworked sulfides.

Hydrogenic hypothesis was developed by Gablina (1994, 1995, 1997) and partially by Petrovsky (1985, 2003). It is based on the following assumptions:

1. Mineralization is controlled by the change in the oxidation conditions, which regulated copper mobility and its deposition at the geochemical barrier.
2. Geochemical barriers were lenses and bands of pyrite-bearing rocks, around which mineralization had concentric zonation, with pyrite in the center and chalcopyrite-bornite-digenite-chalcocite on the periphery.
3. Each stage of formation and transformation of rocks has typomorphic sulfide minerals and parageneses.
4. Copper mineralization is located in the zones, changing primarily red-colored formations to gray-colored (at the stage of green schist facies) as a result of hematite transformation into magnetite due to the presence of reducing agents in the rocks, such as hydrocarbons, hydrogen, carbon monoxide, and sulfides.
5. As a result of solid-phase reactions in the system, the Cu-S role of pyrite and chalcopyrite grows, while the role of bornite and chalcopyrite and chalcocite reduces.
6. The history of copper sandstone formation based on analysis of rock-forming mineral associations and sulfide paragenesis suggests the following stages:
   (a) Diagenesis—early catagenesis in the transition areas with red rocks changing into gray-colored rocks; sulfides begin to form ore mineralization at the hydrogen barriers. The presence of copper in the pore solutions of red rocks is necessary.
   (b) Late catagenesis—metagenesis, with redeposition of earlier sulfides by metamorphic water.
   (c) Composition of sulfide was changed during the epidote-amphibolite facies metamorphism, with copper partially dissipated, and appearance of the chalcopyrite-pyrite-pyrrhotite (±bornite) paragenesis.
   (d) The supergene stage oxidized copper sulfide, with iron sulfides being displaced.

The *rift-related hypothesis of ore fluid* (Salikhov 1995) is based on the following observations and assumptions:

1. Udokan is located within the back-arc rift zone, in which there were repeated deformations, accompanied by dynamic effects in fault zones. The central fault of spreading corresponds to the Main Udokan Dike.
2. The development of mineralized cleavage fractures, obliquely oriented to the layering, creates a false impression of cross-bedding.

3. Saturation of this area by magmatic dike and intrusive complex is related to faults.
4. The source of ore is a fluid-magmatic system.

3.6.2 Isotopic Data

3.6.2.1 Copper Isotopes

New data on the Udokan deposit were published recently. Unfortunately, they were often not incorporated into genetic models. First, copper isotopes give additional information on the genesis of metal in ore (Belogub et al. 2015). The first attempts to use the copper isotope ratios in nature employed thermal ionization mass spectrometers (TIMS). They did not give sufficient results due to low accuracy; there was an incomplete transfer of copper from the sample into plasma (Shields et al. 1965). In the end-twentieth century, multicollector mass spectrometers with inductively coupled plasma (MC-ISP-MS) made it possible to determine the isotopes of many metals, including copper.

There are two stable isotopes ^{63}Cu and ^{65}Cu in nature, with representation of 69.1 and 30.9%, respectively. Deviations of the isotopic composition of the sample from the international standard (NIST 976, 0.4456) are calculated as follows: $\delta^{65}Cu = ((^{65}Cu/^{63}Cu$ sample)/$(^{65}Cu/^{63}Cu$ standart) $- 1) \times 1000$. $\delta^{65}Cu$ for magmatic sulfides varies from -0.10 to -0.20, which is regarded as an evidence of the isotopic homogeneity of the source (Zhu et al. 2002; Larson et al. 2003). Indeed, Malitch et al. (2014, 2015) obtained data for many magmatic sulfides from different deposits of the Noril'sk area, in addition to the Urals, Stillwater, and other locations. Most sulfide samples, including the ores from the Talnakh intrusion at Noril'sk and the Stillwater Complex, have a restricted range of $\delta^{65}Cu$ values from $-1.1‰$ to $0‰$, whereas those from the Kharaelakh and Noril'sk 1 intrusions reveal greater variations of $\delta^{65}Cu$ values from $-2.3‰$ to $1.0‰$. Samples from the three economic deposits at Noril'sk have distinct mean $\delta^{65}Cu$ values of $-1.56 \pm 0.27‰$ for Kharaelakh, $-0.55 \pm 0.41‰$ for Talnakh, and $0.23 \pm 0.28‰$ for Noril'sk 1, consistent with those for carbonaceous chondrite and iron meteorites. The $\delta^{65}Cu$ data are very close, although the authors believe that the variations of $\delta^{65}Cu$ are significant and can be interpreted as a result of magmatic fractionation of Cu isotopes. Assimilation from an external source cannot be ruled out, particularly in the case of the Kharaelakh ores.

Sulfides from the Udokan deposit have a greater range of $\delta^{65}Cu$ (Table 3.12), which varies from -6.0 to $+1.46$ (Belogub et al. 2015). Thus, they have no typical magmatic source, although one could exist during ore deposition. Similar data are typical for hydrothermal deposits that are characterized

Table 3.12 Isotope compositions of copper and sulfur for mineral from the Udokan deposit

№	Type of preparation	Mineral	δ⁶⁵Cu ‰, NBS	δ³⁴S ‰, CDT
1	1	Ccp	−6.00	−8.9
2	1	Cc	−0.32	−19.8
3	1	Cc	−0.04	−4.2
4	1	Bn	−3.30	−10.96
5	1	Ccp	−2.31	−12.3
6	2	Bn-Cc	0.82	−4.6
7	2	Ccp	−1.02	−14.2
8	2	Cc-Bn	0.40	−16.6
9	2	Bn-Cc	−0.34	−10.0
10	2	Bn-Cc	−1.33	−23.9
11	2	Ccp	1.86	−21.3
12	2	Cc-Bn	−0.61	2.0
13	1	Ant	−0.70	+1.7
14	1	Bn	0.90	+3.8
15	1	Mh	−0.36	n.d.
16	1	Mh	−1.06	n.d.
17	1	Ant	0.98	−7.3
18	1	Bn	1.73	−1.8
19	1	Ant	1.12	−0.05
20	1	Cc	1.46	−8.9
21	1	Mh	−0.56	n.d.
22	1	Bn	−0.27	n.d.
23	1	Bn	−0.27	+1.7

After Belogub et al. (2015)

Note. Type of preparation: 1, sample; 2, concentrate. Minerals: *Cc* chalcocite, *Ccp* chalcopyrite, *Bn* bornite, *Ant* antlerite, *Mh* malachite. *n.d.* not detected, δ⁶⁵ Cu were analyzed at Dzhuanita college by R. Mantur, σ = ~0.06‰, δ³⁴S were analyzed at Institute of Mineralogy by S.A. Sadykov

Table 3.13 Sulfur isotope composition of ore minerals from the deposits of the Kodar-Udokan district

Sample number	Mineral	δ³⁴S$_{CDT}$, ‰
Pravoingamakitskoe		
5045	Chalcopyrite + pyrite	+2.1
3038-3	Chalcopyrite + pyrite	+4.2
900/2[a]	Chalcopyrite	+2.6
900/3[a]	Chalcopyrite	+3.5
900/4[a]	Chalcopyrite	+2.0
Rudnoe		
п-11-06	Chalcopyrite	+1.7
9-197.5	Chalcopyrite + pyrite	+2.6
45-38.2	Chalcopyrite	+2.5
20-149	Pyrrhotite	+3.9
20-145.9	Pyrrhotite	+4.4
44-71.1	Pyrrhotite	+3.5
45-38.2	Pyrrhotite	+2.6
Verkne-Chineyskoe		
kan-01	Chalcopyrite	+3.0
64-247	Pyrrhotite	+3.6
Kontaktovyi		
53-890	Pyrite	+3.3
Udokan		
zap-06	Chalcocite + bornite	−2.7
zap-07	Chalcocite + bornite	−8.6
nam-03-1	Chalcocite + bornite	−5.3
nam-03-3	Chalcocite + bornite	−6.2
zap-05	Chalcocite + bornite	−16.0
STD-N 1	Chalcopyrite + pyrite	−12.9
STD-N -2	Chalcopyrite + pyrite	−13.1
928-405	Pyrite	−27.2
Saku		
VII-1[a]	Pyrrhotite	+1.89
VII-3[a]	Pyrrhotite	+2.87

Note. Analyses were carried out at TsNIGRI, analyst SG Kryazhev; [a]After Konnikov et al. (1986)

by a wide variation in δ⁶⁵Cu (Larson et al. 2003; Maher et al. 2003; Mason et al. 2005) due to multiple remobilizations of copper during hydrothermal activity. The values of δ⁶⁵Cu for the chalcopyrite pipes from active modern "black smokers" are different from those of much older buildings, suggesting a possibility of isotopic exchange between the solution and the deposited sulfide (Zhu et al. 2002).

The most extensive δ⁶⁵Cu variations are associated with supergene and accompanying redox reactions (Larson et al. 2003). Moreover, the possibility of "facilitating" the relict sulfides of copper from oxidized ore, compared to primary ore, was evaluated based on the hydrothermal deposits of Schwartzvald (Markl et al. 2006). The distribution of copper isotopes can also be influenced by complicated bacteria sorption processes and other factors. It has been proven experimentally that under non-biogenic conditions, the isotopic ratio of copper in the solution corresponds to those in the dissolved sulfide. The presence of *Thiobacillus ferrooxidans* results in copper in the solution that is lighter, while a heavier isotope is concentrated in the amorphous hydroxides forming a shell around the bacterial cells (Mathur et al. 2005).

3.6.2.2 Sulfur Isotopes

The second important question is the source of sulfur in the Udokan ore. Sulfur isotopes have been studied in Udokan ore by many geologists. According to Belogub et al. (2015), sulfides and sulfates (Table 3.12) are light in composition (mostly −9 to +3.8, except 3 analyses). But our data (Table 3.13) and the data of Volodin et al. (1994) and Konnikov (1986) correspond to even lighter values δ³⁴S = −20–24‰ (up to −27‰). This difference can be explained by different origin of the studied minerals. The first group represents oxidized ore while the second group from primary sulfide minerals.

It was shown that the biogenic processes of sulfate reduction lead to the most significant isotope fractionation (Faure

Fig. 3.39 Fluid inclusions in quartz from the Udokan deposit (**a**) General view of the sample with inclusions, (**b, c**) inclusions: (**b**) three phases, (**c**) two phases

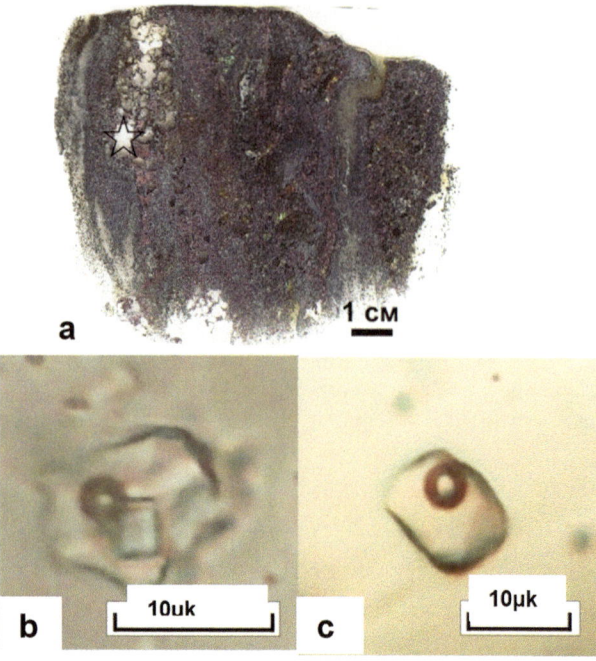

1986). Sulfate-reducing bacteria are a diverse group of eubacteria, the representatives of which are distinguished by the ability to obtain energy by the oxidation of hydrogen in anaerobic conditions. Sulfate-reducing bacteria play a major role in the formation of the light isotopic composition of sulfur in sulfides, because the primary electron acceptor during their respiration is the sulfate ion. Some species can, in addition to sulfates, use more sulfur compounds (thiosulfate, sulfite, molecular sulfur). The product recovery of the sulfate ion is a sulfide as a result of the intermediate step informa-tion of the sulfite: $4H_2 + SO_4^{2-} + 2H^+ \rightarrow H_2S + 4H_2O$ ($\Delta G = -152$ kJ/mol). There is *Thiocapsa thiozimogenes*, which is able to grow using the natural energy of sulfites and thiosulfates. It is known by 220 species of sulfate-reducing bacteria (Korneeva 2015).

It was suggested that sulfate-reducing bacteria are the oldest on Earth (*Archaea* bacteria). *Archaeoglobus* (type species *A. fulgidus*) was found in 1987 (Gusev and Mineeva 2010) in hot marine sediments near hydrothermal submarine sources of "black smokers" of the Mid-Atlantic Ridge, in the oil of the North Sea, and under the frozen surface of the Northern Slope of Alaska. The optimal temperature for growth is around 110 °C. Although bacteria were dated as far back as 3.5 Ga, most of them do not have the characteristic morphological features, and therefore fossilized forms cannot be identified as the remains of *Archaea*. At the same time, chemical residues of the unique lipid for the *Archaea* are more informative since these compounds are not found in other organisms. In some publications, it has been indicated that the lipids of *Archaea* and eukaryotes are present in the 2.7 Ga rocks. These lipids have also been found in younger Precambrian formations. The oldest of these residues were found in the greenstone belt in West Greenland, where the world's oldest sedimentary rocks were found. Typically, these bacteria inhabit the bottom of marine sediments (in contrast to methanogens, which usually grow in the sediments of fresh water lakes) or occur in waters that are rich in decaying organic matter. Representatives of the group are an important link in the global sulfur cycle.

3.6.2.3 Oxygen and Carbon Isotopes

Stable isotopes were studied in the rocks inside the Udokan sub-zone of the Kodar-Udokan Basin (Pokrovsky and Grigoriev 1995; Pokrovsky 2002). The analyzed rocks comprise samples from the upper part of the Udokan Supergroup, including the Alexandrov, Butun, Sakukan and Naminga Formations (Fig. 3.2).

Thick layers of carbonate rocks occur in the Butun and par-tially in the Alexandrov formations (150 m), while the thick-ness of the Butun Formation varies from 400 to 900 m. They contain stromatolites *Conophiton* и *Stratoconophiton* (Fedorovsky 1972) and *Udokania problematica* of unknown origin (Leites 1965). Sandstone dominates in the Sakukan Formation, including ore-bearing rocks. The Naminga Formation consists of siltstone with layers of sandstone. The metamorphism changes from the bottom of the Udokan Supergroup, reaching amphibolite facies in the Kodar Group and decreasing to the greenschist facies in the Kemen Group, at the top (Fig. 3.39).

Carbon. Carbon-bearing rocks of the Udokan Supergroup (Table 3.14) are characterized (Pokrovsky and Grigoriev 1995) by a very monotonous isotope composition of carbon within a range of $\delta^{13}C$ from -3.6 to $0.5‰$. The only exception is a small (3–5 m) lens of marble in which $\delta^{13}C$ is -11.3 to $-11.4‰$, and the participation of carbonic acid of organic origin is suggested. In the thickest strata of limestone and dolomite from the Butun Formation, located in the northern area of the Butun Creek (this sequence was tested from top to bottom without significant breaks), the variations in $\delta^{13}C$ do not exceed $0 \pm 0.8‰$.

One study (Salop 1964) suggested that salt deposits could be originally present in the Butun Formation and were then dissolved during metamorphism. For carbonates associated with evaporate, carbon isotope anomalies are very typical, both positive and negative (Casanova and Hillaire-Marcel 1987; Pierre and Rouchy 1988; Sanders et al. 1988). They were not recognized in the Butun Formation, which was formed, probably, under more or less normal conditions. There are no significant deviations from the "normal" values (close to zero of $\delta^{13}C$) in the carbonate material from the clastic rocks of the Naminga and Sakukan formations, including sandstone enriched in copper sulfides (Pokrovsky and Grigoriev 1995). This is an unexpected result. Detailed studies have shown that sulfides of the Udokan Supergroup display wide variations of $\delta^{34}S$ (from -27.8% to $+22.7\%$ *) (Sidorenko and Borshchevsky 1977). At least in part, their isotopes were isotope-heavy sulfates of seawater ($\delta^{34}S = \sim 20‰$). The reduction of sulfates during diagenesis and catagenesis usually occurs during interaction with organic matter (Vinogradov 1980). This leads to a large amount of isotopically light carbon dioxide, which is manifested in carbonates with low values of $\delta^{13}C$, for example, in Mn diagenetic deposits (Kuleshov and Dombrovskaya 1988, 1990). The absence of such carbonates in the Sakukan and Naminga formations shows that reduction of sulfates did not occur in situ, but it could have rather taken place in other parts of the sequence.

The main sources of hydrogen sulfide fluids were most likely associated with the Kodar Group, in which black carbonaceous shales (Sochava 1986) and carbonates with low values of $\delta^{13}C$ (Table 3.14) were found. The source of the carbonate material was, apparently, the lower limestones and dolomites of the Hallas Subgroup dissolved in the process of

catagenesis and metamorphism (Kholodov 1982). In general, the isotopic data support the model of epigenetic formation of the Udokan ores, which was proposed by Belyi (1973) and later developed by Apolsky (1992).

Oxygen. Variations of $\delta^{18}O$ in carbonates are much greater (from 11.9 to 20.0‰) than variations of $\delta^{13}C$ (Pokrovsky and Grigoriev 1995). The highest values of $\delta^{18}O$ were detected in thick layers of limestone and dolomite from the Butun Formation, while the lowest were found in the intensely altered carbonates and clastic-carbonate rocks. This implies, on the one hand, that the difference in $\delta^{18}O$ is due to the post-sedimentation $\delta^{18}O$ isotope exchange with the carbonate-rich catagenic and metamorphic fluids, which were balanced with the bulk of the silicate species. Correlations of $\delta^{13}C$ and $\delta^{18}O$ are not observed.

Silicate (clastic) rocks of the Udokan Supergroup are subdivided into two groups based on their oxygen isotope composition. They differ from each other by the sharply varying intensity of post-sedimentary alteration. The first group includes samples from the Naminga syncline (Sakukan, Ayan, and Ikabiya formations) characterized by values $\delta^{18}O = 9.3$–$14.9‰$. The second one combines altered samples collected near the Katugin rare metal deposits (Ayan and Ikabiya formations), with low (2.3–$5.1‰$) values of $\delta^{18}O$ (Table 3.14).

Variations in the oxygen isotopic composition of rocks from the Naminga, Sakukan, Ayan, and Ikabiya formations (outside the Katugin deposit) were controlled by the composition of the initial deposits, on the one hand, and interaction in the "water-rock" system during diagenesis, catagenesis, and metamorphism, on the other hand (Pokrovsky and Grigoriev 1995). The highest values of $\delta^{18}O$ (13.3–$14.9‰$) are recorded in the siltstone of the Naminga Formation. These rocks with fine-grained basic fabric consist of quartz, feldspar, sericite, chlorite, and calcite that cement detrital quartz grains and significantly rarer feldspar grains (the amount of detrital grains reaches 40 vol%). Isotope data show, however, that the degree of secondary transformations of these rocks is much higher than can be concluded based on petrographic observations.

The most informative in this sense is the ratio of $\delta^{18}O$ and K_2O. Unbalanced clastic sediments represent a mixture of mineral fragments: predominantly quartz from the rocks of the crystalline basement ($\delta^{18}O = 8$–$10‰$) and low-temperature clay minerals ($\delta^{18}O = \sim 15$–$25‰$) formed during the weathering stages, sedimentation, and diagenesis. Such sediments are characterized by the direct correlation of $\delta^{18}O$ and the contents of potassium; sometimes, it is kept in metapelites (Hoernes and Hoffer 1985; Hoernes and van Reenen 1992). In the layers of the Udokan Supergroup, the opposite trend can be seen. The values $\delta^{18}O$ decrease with

Table 3.14 Isotope composition of oxygen and carbon in carbonate and terrigenous rocks of the Udokan Supergroup

No sample	Rocks	Place of selection	Silicates $\delta^{18}O,\%o$	Carbonates $\delta^{13}C,\%o$	$\delta^{18}O,\%o$
		Naminga Formation			
1/92	Siltstone	R. Prav. Naminga	13.8	–	–
2/92	Calcareous siltstone	»	14.7	−1.5	13.7
3/92	»	»	14.9	−1.4	12.5
4/92	»	»	13.4	–	–
5/92	»	»	14.4	–	–
6/92	»	»	13.7	−1.6	14.4
7/92	»	»	13.70	−2.4	15.1
8/92	»	»	13.7	−2	14.4
9/92	»	»	13.9	0.4	15.1
*1/90	Calcareous sandstone with copper sulfides	Sakukan Formation R. Naminga	–	−1.6	12
5/90	»	»	–	−2.2	12.5
7/90	Siltstone	»	11.7	–	–
9/90	Carbonate breccia	»	–	−0.7	11.9
12/90	Calcareous sandstone	»	–	−0.6	12.4
44/90	»	»	–	−3.9	13.8
10/92	»	»	13.2	–	–
Л/92	»	»	12.5	−3.1	12.1
12/92	Sandstone		10.9	–	–
		Butun Formation			
37/90	Calcareous sandstone	R. China	–	−3.6	17.4
39/90	Sandstone	»	12.6	–	–
40/90	Calcareous sandstone	»	–	−0.7	16.7
41/90	»	»	–	−3.2	13.8
95/90	Dolomite	R. Butune	–	0.6	15.9
96/90	Brominated dolomite	».	–	0.5	16.2
97/90	Limestone	»	–	0.5	16.1
99/90	Dolomite siltstone	»	–	−2	12
100/90	Calcareous dolomite	»	–	0	17
101/90	»	»	–	0.8	17.7
102/90	»	»	–	−3.4	14.5
103/90	»	»	–	0.4	12.8
108/90	»	»	–	−0.6	18.6
109/90	Siltstone	»	–	0	20
U-1	»	R. Prav. Ingamakit	–	0.6	14.8
		Alexandrov Formation			
105/901	Siltstone	R. Butune	–	14.2	–
	The Kodar Group (the Ikabiya and Ayan Formation)				
Ж-00/87	Quartz-biotite-feldspar shale	R. Saku	12.2	–	–
112/90	»	»	12.6	–	–
21/92	»	»	11	–	–
22/92	»	»	9.3	–	–
25/92	Marble	»	–	−11.4	13.3
26/92	»	»	–	−11.3	13.1
28/92	Quartz-biotite-feldspar shale	»	13.4	–	–
29/92	»	»	13.3	–	–
30/92	»	»	11.4	–	–
31/92	»		13.4	–	–
36/92	»	»	13.4	–	–
37/92	»	»	13.3	–	–
2406	Migmatite	Sakukan-Katuginsky watershed	13	–	–
2407	»	»	11.6	–	–
2409	»	»	11.9	–	–

Note: Data after Pokrovsky and Grigoriev (1995)
– not analyzed

increasing K_2O (Pokrovsky and Grigoriev 1995). The maximum value of $\delta^{18}O$ is observed in the sample with the richest SiO_2 and poor K_2O, which estimates the isotopic composition of quartz in siltstone with $\delta^{18}O$ values equal to $15 \pm 17‰$. These are much higher than in quartz from the rocks of the Aldan Shield (according to our data $\delta^{18}O = 9.5–10.5‰$) and are close to newly formed quartz from the metapelites, studied in detail from the Nam (Namibia) Group (Hoernes and Hoffer 1985; Hoernes and van Reenen 1992), metamorphosed to the greenschist facies. Quartz with similar isotopic composition can be formed in sediments in equilibrium with seawater at a temperature of the order of 150 °C (Friedman and O'Neil 1977). Thus, quartz in the rocks of the Naminga Formation sharply (not less than 4–5 times) prevails over the clastic one. Feldspar is considerably less stable than quartz, so we do not take into account other minerals that apparently were completely recrystallized.

The lower values of $\delta^{18}O$ in the rocks of the Kodar Group are obviously related not to the type of the initial substrate but to the increase in temperatures and, possibly, the quantities of metamorphic solutions that passed through the rocks. The established relationships between the isotopic composition of oxygen and the content of potassium can be interpreted as an evidence of enrichment of rocks in potassium at the stage of catagenesis and metamorphism. High probability of this phenomenon for Precambrian metapelites was noted by Garrels and MacKenzie (1971). However, it is not always taken into account in petrochemical modeling.

Inverse correlation $\delta^{18}O$ and K_2O is also observed in the metasandstone and metasiltstone of the Sakukan Formation (Pokrovsky and Grigoriev 1995). These rocks also undoubtedly experienced profound transformations under the influence of post-sedimentation solutions. Some depletion of the $\delta^{18}O$ rocks of the Sakukan Formation in comparison with the rocks of the Naminga Formation can be explained both by a higher fraction of quartz and different composition of post-sedimentation solutions. A difference in the isotopic composition of hydrogen seems to confirm the latter assumption. However, there is no enough data to make a definitive conclusion.

Similar geothermal systems are widely distributed in the areas of modern volcanism, and they are accompanied by the introduction of many granite intrusions (including Precambrian), as isotope studies show (Sturchio and Muehlenbachs 1985; Taylor 1977). The problem is that the granites of the Katugin ore field, according to a widespread view (Arkhangel'skaya et al. 2004), are not intrusive rocks, interpreted to be rather as products of metasomatic transformation of rocks of the Udokan Supergroup in the fault zone. One of the most important parameters of geothermal systems is the water/rock ratio. The water/rock ratio calculated for the geothermal system of the Katugin ore field (Pokrovsky and Grigoriev 1995) ranges from 0.6 to 1.2 for closed and from 0.5 to 0.8 for open systems. Such high ratios are typical for the most powerful modern geothermal systems, such as Mutnovsky in Kamchatka, Russia. The necessary condition for the emergence of such systems is the high permeability of rocks and anomalously high heat flux. This hydrothermal system cannot be related to the Kodar granite, which does not demonstrate obvious isotopic signs of a hydrothermal process. Moreover, this granite is apparently much younger than the Katugin deposit (Tauson et al. 1983; Larin et al. 2000, 2015).

Indeed, abundant new data have appeared regarding the Katugin deposit that suggest the metasomatic origin of ore. Sklyarov et al. (2016) suggested that the host rocks of the Katugin ore are granite (see Chap. 9). Nevertheless, the volume of these rocks cannot form such a large hydrothermal system.

Thus, the stable isotope studies (Pokrovsky and Grigoriev 1995) led to some important conclusions:

1. The siltstone and sandstone of the Udokan Supergroup were seriously transformed. As a result, the isotopic composition of oxygen in the rocks was almost completely balanced with the catagenetic and metamorphic fluids. It is very likely that there was post-sedimentation redistribution of carbonate material, as well as potassium, whose concentrations correlate with the isotopic composition of oxygen.

2. Carbonate rocks have a normal isotope composition close to the modern one. This is very important because it was suggested that Paleoproterozoic carbonates had an anomalously heavy isotope composition of carbon due to different compositions of the ancient atmosphere (Shidlowski et al. 1975, 1983; Baker and Fallick 1989; Yudovich et al. 1990; Pokrovsky and Melezhik 1992). Consequently, there are no grounds for discussing the significant global changes in composition of the atmosphere at this stage.

3. Of particular interest is the light isotopic composition of oxygen in the hydrothermally altered rocks and rare metal granite of the Katugin field. This indicates that a powerful geothermal ore-forming system with an atmospheric feed might have existed in the south of the Udokan zone (before or during the Udokan Supergroup metamorphism). In general, such systems can be initiated by large intrusions, which are not recorded at the Katugin ore field. Perhaps, this is a rare example of a geothermal system, the emergence of which was due to fault formation and associated increase in heat flow and permeability of rocks.

Table 3.15 Trace elements in copper ore of the Udokan-Chiney area

№	Deposit	№ sample	Mineral composition of ore	Mo	Sb	Bi	Pb	Co	Ni	Re
				(%)	(%)	(%)	(%)	(%)	(%)	ppm
1	Pravoingamakitskoe	5038-1	Pyrite-chalcopyrite	0.0006	<0.0005	0.0008	0.0138	0.0054	0.0115	0.1220
2	Pravoingamakitskoe	5038-2	Pyrite-chalcopyrite	0.0005	<0.0005	<0.0005	0.0057	0.0093	0.0115	0.0510
3	Pravoingamakitskoe	5038-3	Pyrite-chalcopyrite	0.0007	<0.0005	0,0019	0.0024	<0.0005	0.0145	0.0830
4	Udokan	Ush 2-07	Chalcocite-bornite	0.0004	<0.0005	<0.0005	0.0022	0.0083	0.0220	0.0250
5	Udokan	Ush 2-07	Chalcocite-bornite	0.0009	<0.0005	<0.0005	0.0051	0.0063	0.0340	0.0230

Note. Analyses were carried out at IMGRE, samples of BI Gongalsky

3.6.3 Discussion

What are the differences, contradictions, and similarities in the abovementioned hypotheses?

The main question of genesis is the source of copper and other metals. On this basis, all the hypotheses can be divided into two groups: (a) deep sources of metals brought by fluids, magma, and saline solutions in different forms (chlorides, true solutions, colloids, sulfides, carbonates, etc.) from different crustal or even mantle depths (Volfson and Arkhangel'skaya 1972, 1987; Gongalsky 1993) and (b) surface sources, mainly from the areas of denudation with clastic material (Yurgenson and Bezrodnykh 1966; Yurgenson 1968; Bezrodnykh et al. 1968; Narkelyun and Yurgenson 1968; Yurgenson and Abramov 2000). Some researchers (Ptitsyn et al. 2003) suggested several sources of copper and other metals in the Udokan deposit because of the presence of (a) products of weathering of rocks sourced from the Aldan Shield and Chara block, (b) erosion of copper-bearing horizons in the Chiney Groups of the Udokan Supergroup, (c) volcanic products, and (d) metals of magmatic origin (Pt, Co, Ni, Mo, Sn, etc.).

Despite many similarities in structure and mineral composition between the ores of Udokan and Dzhezkazgan deposits in Kazakhstan (Narkelyun 1962; Narkelyun et al. 1983; Satpaeva 1985, 1995), Udokan has specific features. Firstly, the Dzhezkazgan ore is enriched in Re (0.1 ppm; concentrated in dzhezkazganite $ReMoCu_2PbS_6$; https://www.mindat.org/min-6908.html and in sulfides), while the concentrations of this metal are much lower in the Udokan ore (Table 3.15). Secondly, Satpaeva discovered many new minerals in the Dzhezkazgan ores: solid solutions of chalcopyrite-bornite-digenite range, high-temperature cubic modification of chalcopyrite-putoranite (primarily discovered at the PGE-Cu-Ni Talnakh deposit, Putorana Plateau, Russia; Filimonova et al. 1974, 1980) and mercury silver (kongsbergite (Ag, Hg)), the silver Domenic (Kutenait), new minerals discovered for the first time in the Dzhezkazgan ores, as well as djurleite, stromeyerite, nickel cobaltite, nickel-skutterudite, and loellingite.

Chalcocite-bornite disseminated and vein mineralization is related to gray sandstone, argillite, and siltstone, which interlayered with red rocks. So, the idea that Udokan sandstone lost its red color during metamorphism is untenable

because sulfides with a light sulfur isotope composition could not be crystallized under oxidation conditions.

All of the above helps us to better understand the origin of Udokan ores. It is obvious that sulfide minerals were formed "in situ" in specific water conditions in a reducing environment and enriched in SO_4 ion. It might have happened during sedimentation or immediately after it. Copper could be transported by deep fluids in the form of complex compounds rather than transported as solid fragments. There is no evidence of Cu-bearing rocks around the Kodar-Udokan basin. Porphyry copper occurrences of the Aldan Shield near the Kodar-Udokan area have $\delta^{34}S = -3 -/+ 3‰$ and cannot be a source of sedimentary sulfides for Udokan ore. The presence of different metals, including Au, Ag, and sometimes PGE, indicates that fluid composition and sources could be different. This idea is supported by oxygen isotope data that demonstrate the hydrothermal origin of quartz (Pokrovsky and Grigoriev 1995).

3.7 Conclusions

1. The Udokan Cu-Ag-Fe deposit is part of the large Naminga mineral cluster. Cu-sandstone is related to the Upper Sakukan subformation (about 650 m thick), part of the Udokan Supergroup.

2. The Udokan ores can be subdivided into disseminated and massive types according to their structure and texture. The ores are represented by sulfide (70–100% sulfides), oxide (0–30% sulfides), and mixed (30–70% sulfides). The identified copper endowment consists of 43% sulfide, 40% mixed, and 17% oxide ores. The bornite-chalcocite ore reveals elevated concentrations of Ag, Bi, Mo, Re, Pb, and Sb, and the pyrite-chalcopyrite ore is enriched in Zn, Co, Ni, Se, and Te.

3. Isotopic composition of sulfide ore is very specific for $\delta^{34}S = -20–22‰$, and $\delta^{18}O = 15 \pm 17‰$ is specific for quartz from sandstone.

4. Oxidation zone is a very important part of the Udokan deposit. It is widespread on the surface and inside deep fault zones. Oxidized and mixed ores constitute around 50% of ore in the Udokan deposit. They are present in all parts of the deposit. There are two main oxide mineral associations: sulfate and carbonate. The earlier supergene

minerals are chrysocolla, iron hydroxides, covellite, supergene chalcocite, bornite, and chalcopyrite followed by sulfates. Udokan oxidation zone was formed in two main phases: (i) in the preglacial period with warm and humid climate and (ii) under specific conditions during formation of the permafrost.

5. Copper and sulfur could be source from deep fluids that supplied diverse metals to the basins. The precipitation of ore took place in areas of bacterial activity. Where they were absent, the metals were injected into deltaic facies. Hence, this lithofacial zoning was noted by all researchers.

References

Apolsky OP (1992) About genesis of Cu-sandstone from Kodar-Udokan zone. Doklady Earth Sci 324:1–5

Arkhangel'skaya VV, Bykhov Yu, Volodin RN, Narkelyun LF, Skursky VC, Trubachev AI, Chechetkin VS (2004) Udokan copper and Katuginskoe rare metal deposits in the Chita region, Russia. Administration of Chita, Chita, p 519 (in Russian)

Baker AJ, Fallick AR (1989) Heavy carbon in two-billion-year-old marbles from Lofoten-Vesteralen, Norway: implications for the Precambrian carbon cycle. Geochim Cosmochim Acta 53:1111–1115

Bakun NN, Volodin RN, Krendelev FP (1958) The main features of the geological structure of the Udokan deposit of cuprous sandstones and the direction of its further exploration. Geologiya I Razvedka 5:67–83 (in Russian)

Bakun NN, Volodin RN, Krendelev FP (1964) About genesis of Udokan Cu-sandstone deposit (Chita area). Lithol Miner Dep 3:89–103 (in Russian)

Bakun NN, Volodin RN, Krendelev FP (1966) Genesis of Udokan cupriferous sandstone deposit (Chita Oblast). Int Geol Rev 8(4):455–466

Belogub EV, Matur R, Sadykov SA, Novoselov KA (2015) The first data on the isotope composition of copper and sulfur in minerals from the ores of the Udokan cupriferous sandstone deposit (Transbaikalia). In: Maslennikov VV, Melekevtseva IY (eds) Metallogeny of ancient and modern oceans-2015. IMIN UB RAS, Miass, pp 35–42 (in Russian)

Belyi VM (1973) Behavior of S isotope in formation of U- and Cu sandstone. In: Vinogradov VI (ed) Role of S isotopes in studying of stratiform deposits. GIN, Moscow, pp 38–48 (in Russian)

Bezrodnykh YuP (1969) Distribution and conditions of accumulation of silver, gold and other impurities in copper sandstones and shale. Author on the competition Degrees PhD. Chita 27.

Bezrodnykh YP, Narkelyun LF, Trubachev AI (1968) On structures such as the decomposition of a solid solution in the ores of the Udokan cupriferous sandstone deposit. Irkutsk. Trudy Irkutsk Polytec Institute 42:36–45 (in Russian)

Bogdanov YuV, Bur'yanova EZ, Kutyrev EI, Feoktistov VP, Trifonov NP (1973) Stratiform copper deposits of the USSR. Nedra, Leningrad, p 310 (in Russian)

Bogdanov YV, Kochin GG, Kutyrev EI, Travin LV, Feoktistov VP (1966) Geology, formation conditions and distribution of cupriferous sandstones in northeastern Olekma-Vitim mountain province. Int Geol Rev 8:1305–1315

Casanova J, Hillaire-Marcel C (1987) Chronology et paleohydrology des hautsniveauxquaternaires du bassin Natron-Magadi (Tanzanie-Kenya) d'apres la composition isotopique (^{18}O, ^{13}C, ^{14}C, U/Th) des stromatolites littoraux. Bull Sci Geol 10:121–134

Chechetkin VS, Narkelyun LF, Trubachev AI, Volodin RN, Bykov YuV (1995) Udokan copper deposit. Deposits of Transbaikalia. Moscow 3–27

Chechetkin VS, Kharitonov SF, Chaban NN (2011) Mineral resources of Transbaikalia. Perspectives of development. Gornyi Zhurnal 3:18–23 (in Russian)

Chechetkin VS, Yurgenson GA, Narkelyun LF, Trubachev AI, Salikhov VS (2000) Geology and ore of the Udokan copper deposit: a review. Rus Geol Geophys 41:710–722

Dana JD, Dana ES (1951) Chalcophyllite. The system of mineralogy, 7th edn, 2(2):452–454

Dana DD, Dana ES, Pelach C (1954) The system of mineralogy, vol 2, p 592

Fat'yanov AV, Yurgenson GA, Glotova YV (2000) Effect of features of the mineral composition and conditions of formation of oxidized copper ores at the Udokan deposit and their beneficiation technology. Gornyi Zhurnal 36:185–192 (in Russian)

Faure G (1986) Principles of isotope geology. Wiley, New York, p 589

Fedorovsky VS (1972) Paleoproterozoic stratigraphy of the Kodar and Udokan Ridges. Nauka, Moscow, p 130 (in Russian)

Filimonova AA, Muravieva IV, Evstigneeva TL (1974) Minerals from chalcopyrite group in cu-Ni ores of the Noril'sk deposits. Geol Ore Deposits N5:31–47 (in Russian)

Filimonova AA, Evstigneeva TL, Laputina IP (1980) Putoranite and Ni-putoranite – new minerals in chalcopyrite group. Zap Vses Min Obsh 109(3):335–341 (in Russian)

Friedman I, O'Neil JR (1977) Compilation of stable isotope fractionation factors of geochemical interest. In: Fleischer M (ed.) Data of geochemistry. US Print, Washington, DC, p 440

Gablina IF (1983) Copper accumulation conditions in continental red beds. Nauka, Moscow, p 112 (in Russian)

Gablina IF (1994) Regularities in localization of Cu ore at the Udokan deposit. Lithol Miner Dep 3:53–67 (in Russian)

Gablina IF (1995) Role of metamorphic and supergene processes in formation of the Udokan copper deposit. In: Pasava J, Kribek B, Zak K (eds) Mineral deposits. Balkema, Rotterdam, pp 863–864

Gablina IF (1997) Formation conditions of large cupriferous sandstone and shale deposits. Geol Ore Deposits 38(4):320–334

Gablina IF (2008) Sulfides of Cu and Fe as indicators of conditions of ore formation and recrystallization. International Fedorovsky conference. St. Petersburg, pp 32–34 (in Russian)

Gablina IF (2014) Structural properties of low-temperature chalcocite Cu_2S. Abstract international Fedorovsky conference. St. Petersburg, pp 28–29 (in Russian)

Gablina IF, Vasilovskaya LV (1989) Primary indicators of associations of rocks of the Udokan series. Doklady Earth Sci 301:99–102

Gablina IF, Ermilov VV (1990) New data on magnetite from ore-bearing rocks of the Udokan deposit. Lithol Miner Dep 1:119–123

Gablina IF, Lur'e AM (2001) Formation and alteration of the stratiform copper deposits. In: Pieszyński A (ed) Mineral deposits in the beginning of the 21st century. Balkema, Tokyo, pp 231–234

Garrels RM, Christ CL (1965) Solutions, minerals and equilibria. Harper & Row, New York

Garrels PM, MacKenzie F (1971) Evolution of sedimentary rocks. Norton WW, New York, p 397

Goble RJ (1985) The relationship between crystal structure, bonding and cell dimensions in the copper sulfides. Can Mineral 23:61–76

Gongalsky BI (1993) Origin of copper ore in sedimentary and igneous rocks (Chiney Pluton). Nedra Vostoka 2:2–4 (in Russian)

Gongalsky BI (2010) Sulfide ores from the Udokan–Chiney mineral district. In: Proceedings of the 21st international conference to 100th Anniversary of Academician V.I. Smirnov. MSU, Moscow, pp 272–289 (in Russian)

Gongalsky BI (2015) Deposits of unique metallogenic province of Northern Transbaikalia. VIMS, Moscow, p 248 (in Russian)

Gongalsky BI, Krivolutskaya NA (1993) Chiney layered pluton. Nauka, Novosibirsk, p 184 (in Russian)

Gongalsky BI, Krivolutskaya NA (2009) The Udokan-Chiney ore-magmatic system, Russia. Northwestern Geol 42:180–184

Gongalsky BI, Krinov DV, Magazina LO, Krivolutskaya NA, Kovalchuk EN (2017) Mineralogy and geochemistry of the Cu-sandstone deposits from the Kodar-Udokan area, South Siberia, Russia. In: Ore deposits of Asia: China and beyond. Inc. SEG 2017 Conference P010

Gusev MV, Mineeva LA (2010) Microbiology: textbook for university students in the direction of "Biology" and biological specialities. MSU, Moscow, p 381 (in Russian)

Grintal EF (1968) Some regularities in distribution of ore mineralization inside Udokan deposit in contex of sedimentary-hydrothermal theory of genesis. Lithol Miner Dep 3:42–50 (in Russian)

Hoernes S, Hoffer E (1985) Stable isotope evidence for fluid-present and fluid-absent metamorphism in metapelites from the Damara Orogen, Namibia. Contrib Mineral Petrol 90(4):322–330

Hoernes S, van Reenen DD (1992) The oxygen-isotopic composition of granulites and retrogressed granulites from Limpopo Belt as a monitor of fluid-rock interaction. Precambrian Res 55(1):353–364

Hofmann AW (1988) Chemical differentiation of the Earth: relationship between mantle, continental crust, and oceanic crust. Earth Planet Sci Lett 90:297–314

Kholodov VN (1982) New data in understanding the catagenesis. Lithol Miner Dep 5:15–32 (in Russian)

Konnikov EG (1986) Relationship of Cu sandstones of the Kodar–Udokan zone to Precambrian basic magmatism. Soviet Geol Geofiz 27:23–27

Konnikov EG, Truneva MF, Kaviladze MS (1986) Genetic relationships of stratiform and magmatic copper mineralization in the Kodar–Udokan zone. Izv Akad Nauk USSR, Ser Geol 10:102–110 (in Russian)

Korneeva VA (2015) Biodiversity of sulfate-reducing bacteria in O-bearing water of Black and Barents seas. Dissertation, MSU, Moscow, p 178 (in Russian)

Krasnokutskaya AV (2012) Ore-forming minerals of oxidized copper ores of the Udokan deposit, Siberia. In: Zaykov VV (ed) Metallogeny of ancient and modern oceans. Institute of Mineralogy, Miass, Ural Branch of RAS 18: 147–150

Krendelev FP (1959) About mineralization of cu sandstones of Udokan deposit and method of its prospecting. Geologiyai Razvedka 2:107–119 (in Russian)

Krendelev FP (1987) About genesis of sulfide ore in the Udokan Cu sandstone deposit. Rus Geol Geophys 8:133–134 (in Russian)

Krendelev FP, Bakun NN, Volodin RN (1983) Udokan cupriferous sandstone. Nauka, Moscow, p 248 (in Russian)

Krivolutskaya NA (1989) Mineralogical and geochemical features and the genesis of copper ores from the Chineiskoye deposit (Northern Transbaikalia). Author on the competition Degrees PhD. Moscow 27

Kuleshov VN, Dombrovskaya ZhV (1988) Isotopic composition and formation conditions of Mn and Carbonate in Nikopol ore. In: Shukolukov YuA (ed) Isotope geochemistry of the process of ore formation. Nauka, Moscow, pp 233–258 (in Russian)

Kuleshov VN, Dombrovskaya ZV (1990) Isotopic composition and origin of MangyshlakMn ore. Lithol Miner Dep 2:50–62 (in Russian)

Larin AM, Kovalenko VI, Kotov AB, Sal'nikova EB, Kovach VP, Makar'ev LB, Timashkov AN, Berezhnaya NG, Yakovleva SZ (2000) New data on the age of granites of the Kodar and Tukuringra complexes, Eastern Siberia: geodynamic constraints. Petrology 8(3):238–248

Larin AM, Kotov AB, Vladykin NV, Gladkochub DP, Kovach VP, Sklyarov EV, Donskaya TV, Veklikoslavinskii SD, Zagornaya NY,

Sotnikova IA (2015) Rare metalgranites of the Katugin complex (Aldan Shield): sources and geodynamic formation settings. Dokl Earth Sci 464:889–893

Larson P, Maher K, Ramos FC, Chang Z, Gaspar M, Meinert LD (2003) Copper isotope ratios in magmatic and hydrothermal ore-forming processes. Chem Geol 201:337–350

Leites AM (1965) The lower Proterozoic of NE Olekma-Vitim mountain area. Nauka, Moscow, p 184 (in Russian)

Maher K, Ramos F, Larson P (2003) Copper isotope characteristics of the Cu (±Au, Ag) skarn at Coroccohuayco, Peru. Geological Society of America, Annual meeting, paper, pp 211–244

Malitch KN, Latypov RM, Badanina IY, Sluzhenikin SF (2014) Insights into ore genesis of Ni-Cu-PGE sulfide deposits of the Noril'sk Province (Russia): evidence from copper and sulfur isotopes. Lithos 204:172–187

Malitch KN, Badanina IYu, Belousova EA, Griffin W (2015) Radiogenic and stable isotope study of the Vologochan and Mikchangda ore-bearing intrusions of the Noril'sk Province: implications for exploration. In: The 9th international conference on the analysis of geological and environmental materials. Geoanalyses. Austria, Leoben p 59

Markl G, Lahaye Y, Schwinn G (2006) Copper isotopes as monitors of redox processes in hydrothermal mineralization. Geochim Cosmochim Acta 70:4215–4228

Mason TFD, Weiss DJ, Chapman JB, Wilkinson JJ, Tessalina SG, Spiro B, Horstwood MSA, Spratt J, Coles BJ (2005) Zn and Cu isotopic variability in the Alexandrinka volcanic–hosted massive sulfide (VHMS) ore deposit, Urals, Russia. Chem Geol 221:170–187

Mathur R, Ruiz J, Titley S, Liermann L, Buss H, Brantley S (2005) Cu isotopic fractionation in the supergene environment with and without bacteria. Geochim Cosmochim Acta 69:5233–5246

Menke W (1970) Far ultraviolet circular dichroism and infrared absorption of thylakoids. Z Naturforsch 25:849–885

Narkelyun LF (1962) Geology and ore of the Dzhezkazgan deposit. IGEM, Moscow, p 128 (in Russian)

Narkelyun LF, Yurgenson GA (1968) Copper sources in the formation of deposits of the cupriferous sandstone type. Lith. and Mineral Ressources. Cons Bur, New York 6:739–747 (in Russian)

Narkelyun LF, Bezrodnykh Yu P, Trubachev AI, Yurgenson GA (1968) Geology and genesis of the Udokan copper sandstone deposit. In: Narkelyun LF (ed) Geology of some deposits in the Transbaikalia region, pp 70–90 (in Russian)

Narkelyun LF, Bezrodnykh YuP, Trubachev AI, Salikhov VS (1977) Copper sandstone and shale in the southern Siberian Platform. Nedra, Moscow, p 223 (in Russian)

Narkelyun LF, Salikhov VS, Trubachev AI (1983) Cupriferous sandstones and shales of the World. Nedra, Moscow, p 414 (in Russian)

Narkelyun LF, Trubachev AI, Salikhov VS, Kunitsin VV, Chechetkin VS, Zinoviev YuI, Krivolutskaya NA (1987) Oxidized ore of the Udokan deposit. Nauka, Novosibirsk, p 102 (in Russian)

Petrovsky PP (1985) Tectono-magmatic factors of ore formation at the Udokan copper sandstone deposit. In: Kuznetsov YuA (ed) Metasomatism and ore formation in Transbaikalian Region. Nauka, Novosibirsk, pp 66–73 (in Russian)

Petrovsky PP (2003) Boundary disturbances and their influence on the geological structure of the Udokan deposit. In: Kuznetsov VV (ed) Proceedings of the fourth scientific and technical conference of the mining. Chita State Technical University, Chita, pp 122–125 (in Russian)

Pierre C, Rouchy JM (1988) Carbonate replacements after sulfate evaporites in the middle Miocene of Egypt. J Sediment Petrol 55(3):446–456

Pitulko VM (1977) Secondary halo dispersion in the cryolithozone. Leningrad: Nedra. Leningrad detachment 197 (in Russian)

Pokrovsky BG (2002) Crustal contamination of mantle magmas based on isotope geochemistry. Nauka, Moscow, p 194 (in Russian)

Pokrovsky BG, Melezhik VA (1992) Carbon isotope anomalies and composition of Earth atmosphere in lower Proterozoic. Abstracts XIII Symposium on isotope geochemistry, pp 144–145 (in Russian)

Pokrovsky BG, Grigoriev SV (1995) New data on the age and geochemistry of isotopes of the Udokan series from the Lower Proterozoic of eastern Siberia. Lithol Miner Dep 30(3):243–283

Pramzintsev FB (1966) On the oxidation zone of the Udokan field. Geol Mines Field 2:113–115. (in Russian)

Ptitsyn AB, Sysoeva EI (1988) Some issues of the geochemistry of copper in the cryolithozone Udokan deposit. Geol Geophys 12:54–61

Ptitsyn AB, Zamana LV, Yurgenson GA, Abramov BN, Bashurova NF, Vilmova EV, Eremin OV, Zheleznyak II, Malchikova IYu, Petrovsky PP, Sinitsa SM, Trubachev AI, Turanova TK, Usmanov MT, Shesternev DM, Chechel AP (2003) Udokan: geology, mineralization, conditions of exploration. Nauka, Novosibirsk, p 160 (in Russian)

Reznikov IP (1965) Genesis of the Udokan deposit. Lithol Miner Dep 2: 85–94 (in Russian)

Salikhov VS (1995) Genetic principles of stratiform sedimentation. Abstract of dissertation for doctor geol.-min. sciences. Irkutsk, IZK, p 53 (in Russian)

Salop LI (1964) Geology of the Baikal mountainous region. Moscow Nedra I: 515 (in Russian)

Salop LI (1967) Geology of the Baikal highlands:magmatism, tectonics and geological history. Nedra, Moscow, p 699 (in Russian)

Sanders JA, Prikryl JD, Posey HH (1988) Mineralogic and isotopic constraints on the origin of strontium-rich cap rock, Tatum Dome, Mississippi, USA. Chem Geol 74:137–152

Satpaeva MK (1985) Dzhezkazgan ores and conditions of their formation. Nauka, Alma-Ata, p 208 (in Russian)

Satpaeva MK (1995) Mineralogy and zoning of Dzhezkazgan lodes based on material from the Annensky ore district. Bylym, Almaty, p 122 (in Russian)

Savenko VS, Shatalov IA (1998) Dissolution of atakamite and physical-chemical conditions for copper in marine water. Geochem Int 8:842–851

Shcherbina VV (1972) Basics of Geochemistry 296 (in Russian)

Shidlowski M, Eichmann R, Jung CE (1975) Precambrian sedimentary carbonates; carbon and oxygen isotope geochemistry and implications for terrestrial oxygen budget. Precambrian Res 2:1–69

Shidlowski M, Hayes JM, Kaplan IR (1983) Isotopic inference of ancient biochemistries: carbon, sulfur, hydrogen and nitrogen. In: Earliest biosphere: its origin and evolution. Princeton University Press, New York, pp 149–186

Shields WR, Goldich SS, Garner EL, Myrphy TJ (1965) Natural variations in the abundance ratio and weight of copper. L Geophis Res:479–491

Shvartsev SL (1998) Hydrogeochemistry of hypergenesis zone. Nedra Moscow 128 (in Russian)

Sidorenko AV, Borshchevsky YuA (1977) General tendency of isotope composition change for carbonates in the Precambrian. Dokl AN USSR 234:892–894 (in Russian)

Sklyarov EV, Gladkochub DP, Kotov AB, Starikova AE, Sharygin VV, Velikoslavinsky SD, Larin AM, Mazukabzov AM, Tolmacheva EV, Khromova EA (2016) Genesis of the Katugin rare metals deposit: magmatism contra metasomatism. Rus J Pac Geol 10:155–167 (in Russian)

Sochava VB (1979) Red-colored formations of the Precambrian and the Phanerozoic. Nauka, Leningrad, p 207(in Russian)

Sochava VB (1981) Problems of the Precambrian and Phanerozoic ecology and fauna of ancient basins. Nauka, Leningrad, p 209(in Russian)

Sochava VB (1986) Petrochemistry of Upper Archean and Proterozoic of the western Vitim-Aldan Shield. Nauka, Leningrad, p 144 (in Russian)

Soloviev SG (2010) Iron oxide copper-gold and related mineralisation of the Siberian Craton, Russia: 2 – iron oxide, copper, gold and uranium deposits of the Aldan Shield, South-Eastern Siberia. In: Porter TM (ed) Hydrothermal iron oxide copper-gold and related deposits: a global perspective, Advances in the understanding of IOCG deposits, vol 4. PGC Publishing, Adelaide, pp 515–534

Sturchio VC, Muehlenbachs K (1985) Origin of low-^{18}O metamorphic rocks from a late Proterozoic shear zone in the Eastern Desert of Egypt. Contrib Mineral Petrol 91(2):188–195

Takenaka N, Ueda A, Maeda Y (1992) Acceleration of the rate of nitrite oxidation by freezing in aqueous solution. Nature 358:736–738

Takenaka N, Ueda A, Daimon T (1996) Acceleration mechanism of chemical reaction by freåzing. The reaction of nitrous acid with dissolved oxygen. J Phys Chem 100(3):13874–13884

Tauson LV, Sobachenko VN, Plyusnin GS, Sandimirova GP (1983) Rb-Sr age of rapakivi granite and metasomatite of Katugin-Ayan zone. (NE Transbaikalia) Dokl AN USSR 273(5):1233–1236 (in Russian)

Taylor HP (1977) Water-rock interaction and origin of H,0 in granitic batholiths. J Geol Soc L 133 (6):509–558

Trubachev AI (1981) Minerals of oxidation zone at the Udokan deposit and regularities of their distribution. Geol Geophys 5:80–90 (in Russian)

Trubachev AI, Narkelyun LF (1968) About origin of chalcocite of Udokan deposit. In: Narkelyun LF (ed) Geology of some deposits of Transbaikalia. ZabNII, Chita, pp 12–17 (in Russian)

Trubachev AI (2010) Genetic models of mineralization of cupriferous sandstones and copper schists. Chita, Vestnik Chit GU 64:106–113 (in Russian)

Trubachev AI, Sekisov AG, Salikhov VS, Manzirev DV (2016) Commercial components in cupriferous sandstone ores of the Kodar-Udokan Zone (Eastern Transbaikalia) and their extraction technologies. Izvestiya of the Siberian Branch of the Earth Sciences Section of the Russian Academy of Natural Sciences, Ser Geology, prospecting and exploration of ore deposits 54:9–19 (in Russian)

Vinogradov VI (1980) Role of sedimentary cycle in geochemistry of sulfur isotopes. Nauka, Moscow, p 192 (in Russian)

Volfson FI, Arhangel'skaya VV (1972) On the formation conditions of cupriferous sandstone deposits. Lithology and Mineral Dep 3:11–25 (in Russian)

Volfson FI, Arhangel'skaya VV (1987) Stratiform deposits of nonferrous metals. Moscow, Nedra,p 255 (in Russian)

Volodin RN, Chechetkin VS, Bogdanov YV, Narkelyun LF, Trubachev AI (1994) The Udokan copper sandstone deposit, Eastern Siberia. Geol Ore Deposits 36(1):1–25

Yudovich YE, Makarikhin VV, Medvedev PV, Sukhanov NV (1990) Isotopic carbon anomalies in carbonates of Kola peninsula. Geochem Int 7:972–987

Yurgenson GA (1968) Mineralogy and petrography of the ore-bearing stratum of the Udokan deposit and its connection with copper mineralization. Author dis Cand geol-min sciences. Kazan 32. (in Russian)

Yurgenson GA (1973) About unusual brochantite of Udokan deposit. Zap Vses Min Ob-va 102:103–106 (in Russian)

Yurgenson GA (1996) Problems of mineralization, exploration and evaluation of mineral resources. Publishing House of SB RAS, Novosibirsk, pp 127–160 (in Russian)

Yurgenson GA, Bezrodnykh YuP (1966) About oxidation zone of Udokan deposit and its role in formation of temperature field of permafrost rocks. Geocryological conditions of northern Transbaikalia. Nauka, Moscow, pp 53–55 (in Russian)

Yurgenson GA, Smirnova NG, Karenina LA (1968) Specific features of oxidation zone of Udokan Cu deposit. Vestnik of Transbaikalia branch Geographic Soc USSR. Chita, pp 3–9 (in Russian)

Yurgenson GA, Abramov BN (2000) Mineralogy of Fe-rich sandstone and sources of clastic material of cu-bearing sedimentary rocks of the Udokan Group. Zap Vseros Miner Ob-va129 (2):44–53 (in Russian)

Zhu XK, Guo Y, Williams RJP, O'Nions RK, Matthews A, Belshaw NS, Canters GW, de Waal EC, Weser U, Burgess BK, Salvato B (2002) Mass fractionation processes of transition metal isotopes. Earth Planet Sci Lett 200:47–62

Satellite Sandstone-Hosted Cu-Ag-Fe Deposits in Rocks of the Udokan Supergroup

4

Abstract

In addition to the giant Udokan deposit, the rocks of the Udokan Supergroup contains several satellite Cu-Ag-Fe deposits. They are located in different formations at different stratigraphic levels. They differ strongly in mineral and chemical composition from the Udokan ores, often displaying much higher silver grades. The hypogene copper mineralization in these deposits consists of chalcopyrite, bornite, and, to a much lesser extent, chalcocite. Many rare minerals were found here. The deposits are of complex genesis, and for some, the hydrothermal origin was proven.

The rocks of the Udokan Supergroup contains not only the giant Udokan deposit, but there are also several satellite Cu-Ag-Fe deposits (Unkur, Krasnoe, Burpala, Saku; Fig. 3.2). They are located in different formations at different stratigraphic levels (Bogdanov et al. 1966a, b; Narkelyun et al. 1968, 1977, 1983; Krendelev et al. 1983; Tombasov and Sinitsa 1990; Chechetkin et al. 1997; Yurgenson 2008; Abramov 2008a, b, 2011). They are described in different works (Bogdanov et al. 1966a, 1973; Smirnov et al. 1971; Narkelyun and Trubachev 1978; Narkelyun et al. 1983; Trubachev et al. 2014). In mineral and chemical composition, they differ strongly from the Udokan ores, often displaying much higher silver grades. The hypogene copper mineralization in these deposits consists of chalcopyrite, bornite, and, to a much lesser extent, chalcocite (Bogdanov et al. 1966b). Pyrite is frequently present in large amounts. Pyrrhotite is reported throughout the mineralized sequence, but it is most common in underlying and overlying rocks. The cupriferous horizons in these deposits are widely varying in thickness from ones to tens of meters and a maximum length of 1 or 2 km. The mineralization spreads over 7–8 km of sedimentary sequence. If to consider the oldest to the youngest strata, they might have been accumulated during millions of years. The description of deposits is given below according to their stratigraphic position in the Udokan Supergroup (from the bottom to the top). It is based on the data by Narkelyun et al. (1983), supplemented by our data. At the Pravoingamakitskoe deposit, a new copper-precious metal mineralization of hydrothermal genesis was described (Gongalsky et al. 2007; Gongalsky and Krivolutskaya 2009). Most of these ore occurrences are still underexplored.

4.1 The Ikabiya Formation

The Ikabiya Formation comprises a horizon of quartzite-marble rocks enriched in (wt%) Cu (0.03), Zn (0.07), Pb (0.05), and Mo (0.005). The concentrations of Cu, Zn, and Pb reach up to 0.n wt%. The thickness of the ore-bearing strata in the Kodar subzone is 5–120 m, and it increases to the north to 250 m.

The main ore minerals are chalcopyrite, pyrite, sphalerite, galena, magnetite, and molybdenite. They form disseminated and veinlet mineralization. Galena and sphalerite dominate in carbonate rocks, while chalcopyrite and pyrite are found in quartzites. These two types of mineralization constitute two main zones: (i) Cu ores are located in the Kodar subzone and (ii) Pb-Zn ores are more often found in the Sulban subzone (Narkelyun et al. 1983). The mineralization in this formation remains underexplored.

4.2 The Chitkanda Formation

This formation comprises two deposits (Pravoingamakitskoe and Krasnoe) and some prospects (Khani-Sakukan, Luna, Kamustakh, Skalistyi, Ozero Mednoye).

4.2.1 The Pravoingamakitskoe Deposit

The Pravoingamakitskoe deposit is located in the south of the Katugin syncline (Fig. 4.1), comprising several forma-

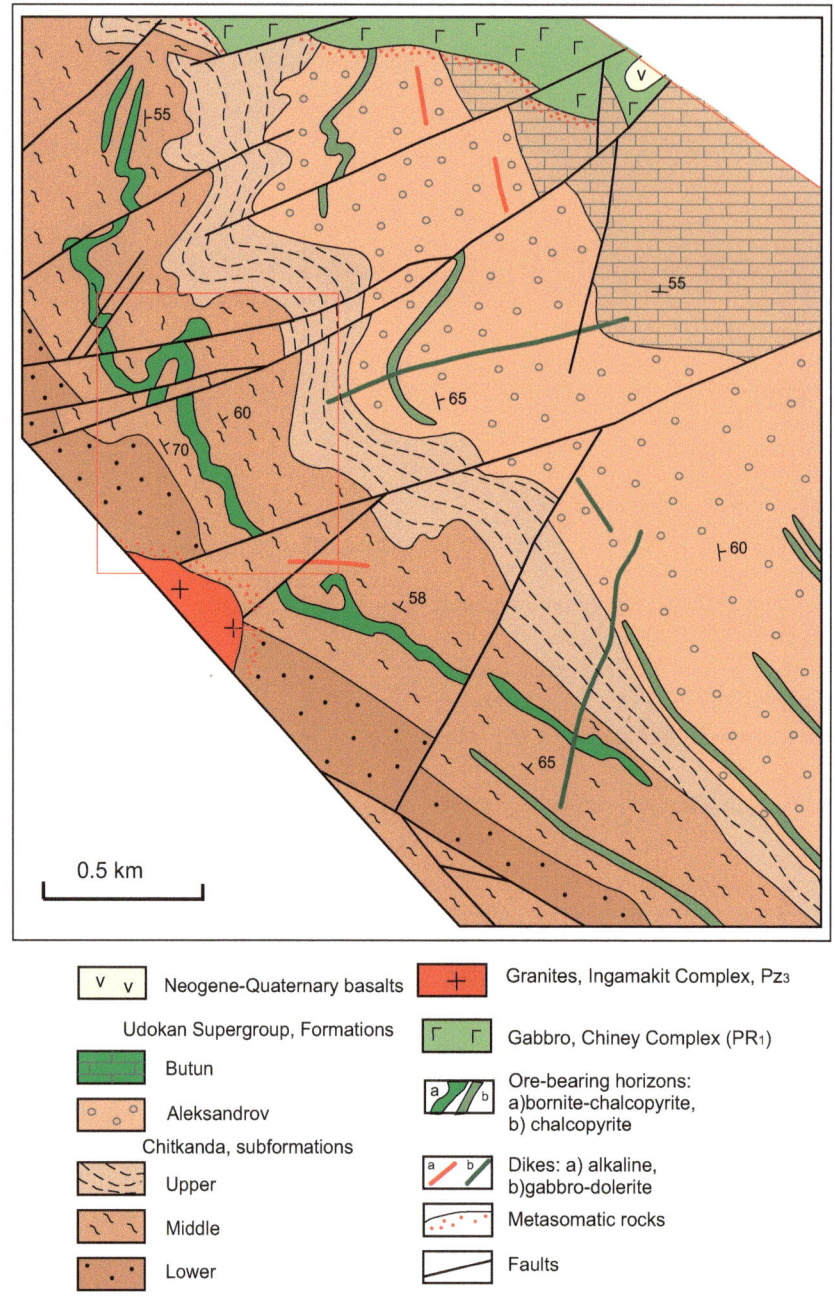

Fig. 4.1 Schematic geological map of the Pravoingamakitskoe deposit. (After Narkelyun et al. 1983) Red square shows the area in Fig. 4.3

Table 4.1 Au and Ag contents in channel samples from the Pravoingamakitskoe deposit

No	Cu, wt%	Ag, ppm	No	Cu, wt%	Ag, ppm
1	0.52	4.4	22	0.41	7.6
2	0.56	8.4	23	0.35	9
3	0.56	7.6	24	0.84	10
4	0.28	8.4	25	0.42	4.6
5	1	38.8	26	0.26	4.2
6	0.44	9.9	27	0.14	1.8
7	0.58	17	28	0.2	3
8	0.28	6.6	29	0.27	6.6
9	0.32	3.5	30	0.24	5.8
10	0.24	2.4	31	0.24	5.2
11	0.16	2	32	0.09	2
12	0.26	1.4	33	0.19	4.6
13	0.2	2.6	34	0.26	4.6
14	0.24	4	35	0.11	1.8
15	0.78	25.4	36	0.08	2.4
16	1.06	33.2	37	0.08	1.8
17	3.2	39.4	38	0.07	1.4
18	0.38	11.4	39	0.12	2.6
19	0.42	12.6	40	0.21	2.6
20	0.35	14	41	0.09	2.6
21	1.12	17.2	42	0.08	2.8

Note. Data of Udokan Expedition (analyses were carried out in Central Chemical Laboratory of Chitageologiya PGO)

tions: Chitkanda (lower, middle, and upper subformations), Alexandrov, and Butun. The first two formations consist of sandstone, while the third comprises mainly carbonate rocks. The Pravoingamakitskoe deposit occurs 4 km to the south of the Chiney pluton (Fig. 2.4), in immediate proximity to the Skvoznoe deposit related to this pluton. It is represented by disseminated ore in the exocontact zone of the pluton. The Pravoingamakitskoe deposit was studied by the geologists of the Udokan Expedition in 1966–1969 and 1988–1992. The deposit was considered to be an analogue of the Udokan deposit, although it markedly differs from the latter in mineralogy and chemical characteristics of the ore. It is enriched in silver in comparison with the Udokan deposit (Table 4.1). The chemical composition of major ore-forming minerals is also specific.

The Cu-bearing unit of the Pravoingamakitskoe deposit is traced at the surface for almost 10 km along the strike. Along approximately 4.5 km, it is considered to contain a potentially economic mineralization, which was traced for 400–500 m downdip. The host sandstone-siltstone-carbonate rocks are related to the middle portion of the Chitkanda Formation. In trenches, the orebodies make up en echelon lenses, up to 300–440 m along strike, 1.3–4.0 to 15–38 m in thickness, with an average grade of 0.47–2.5 wt% Cu. Disseminated ores are more widespread than massive ore. The boundaries of the orebodies were defined by assays. The second ore horizon relates to the rocks of the Alexandrov Formation. The deposit is estimated to contain 608,000 t Cu (Sekisov et al. 2014).

The deposit is characterized by complex fragmentation of the orebodies. Figure 4.1 shows a very simple structure of the deposit, but the deposit is subdivided into the Skvoznoe, Valunnoe, Bazaltovyi, and Pravoingamakitskoe mineral sites. The orebodies, exposed at the surface, differ in composition and morphology, and it is not easy to correlate them. This does not allow to reproduce their outlines in different interpretations. In many regards, fragmentation is controlled by the regional fault zone, separating the mineralization into blocks, displaced for hundreds of meters. The additional investigation of the regional geology, particular orebodies, mineralogy, and geochemistry constrained the evolution of this deposit in somewhat different mode. First, its Skvoznoe site strongly differs from other sites in composition and structure, revealing doubtless similarity to the exocontact ores of the Rudnoe deposit that is spatially related to the Chiney pluton.

At the same time, mineralization of the Valunnoe, Bazaltovyi, and Pravoingamakitskoe sites is obviously epigenetic hydrothermal in origin. The Bazaltovyi site is best studied (Figs. 4.2 and 4.3). Its structure is shown in Fig. 4.3 (Gongalsky et al. 2004). Orebodies at this site are represented by quartz veins and lenses, up to 1 m thick (Fig. 4.2b, c), and a few tens of meters in extent, with sulfide veinlets and pockets. The ore consists of pyrite-chalcopyrite (Fig. 4.4a, b) and chalcopyrite-pyrite (Fig. 4.4c, d) varieties, with breccia, massive, and disseminated structures (Fig. 4.5a, b). The quartz veins with breccia and massive texture (Figs. 4.6 and 4.7) and chalcopyrite-bornite (Figs. 4.6a, 4.7, and 4.8a), cubanite-chalcopyrite (Fig. 4.6b), chalcocite-bornite (Figs. 4.8b), and pyrite-chalcopyrite (Fig. 4.8c, d) mineralization and sandstone-hosted disseminated ore are less abundant. They are characterized by high and variable Cu/Ni ratio equal to 10–700. The highest enrichment in Ni is inherent to the quartz veins with elevated sulfide contents of up to 20% of rock volume (Fig. 4.6). The high Ni concentrations are due to enrichment of ore in millerite and pentlandite.

Gongalsky et al. (2007) described a new type of mineralization. It reveals elevated concentrations of noble metals, reaching 2.2 ppm Pt, 6.2 ppm Pd, and 0.4 ppm Au. Small (up to 10 μm) grains of clausthalite $Pb_{100}(Se_{0.78}S_{0.22})_{1.00}$ (Fig. 4.9), bravoite $(Ni_{0.73}Fe_{0.30})_{1.03}S_{1.97}$, and bogdanovichite $AgBiSe_2$ have been identified along with intermetallic Pd compounds that we failed to accurately determine because of too small (a few micrometers) grain sizes.

The high Cu grades (12.6–12.7 wt%) and low Ni and Co contents are typical of the pyrite-chalcopyrite ore from the Bazaltovyi site (Table 4.2). The quartz-pyrite-chalcopyrite ore, containing 5–7 wt% Cu and 0.14–0.76 wt% Ni, differs from mineralization at the Udokan deposit. The main ore minerals from the Pravoingamakitskoe deposit are distinguished by high Ni and Co concentrations, particularly in

Fig. 4.2 Bazaltovyi site of
the Pravoingamakitskoe
deposit. (**a**) General view; (**b**,
c) sulfide veins

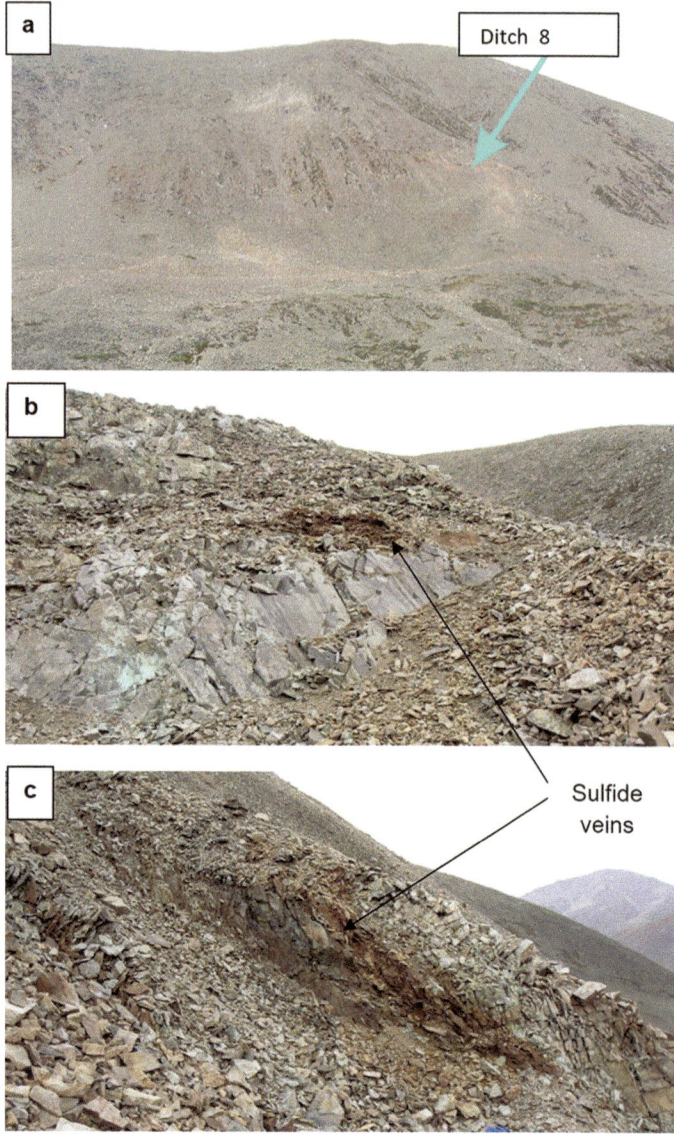

pyrite, where concentrations of these elements reach
1.75 wt% Ni and 1.48 wt% Co (Table. 4.2). Pentlandite is
represented by a low Co variety (0.22 wt% Co), and this feature differs from pentlandites in copper deposits from the
Chiney pluton, where Co contents vary from 2 to 18 wt%.
The disseminated pyrite-chalcopyrite ore from the Bazaltovyi
site resembles the exocontact ore at the Rudnoe and
Kontaktovyi deposits. In the bodies of massive sulfide highly
enriched in copper, the noble metal concentrations are much
lower (0.04–1.0 ppm Pt, 0.6–1.0 ppm Pd, and 0.1–0.4 ppm
Au) than in the veins, except for Ag (Table 4.1). The high
content of silver is typical for this type of mineralization,
reaching anomalous values up to 371 ppm Ag. It is noteworthy that millerite and pyrite contain up to 0.19 wt% Ag. The

zone of oxidation is widespread at the deposit as well
(Fig. 4.10). The ore of the Pravoingamakitskoe deposit is
essentially different from the Udokan ore. It reveals features
of typical hydrothermal deposits.

The fluid composition was studied in fluid inclusions
(Fig. 4.11). Primary, pseudo-secondary, and secondary two-
phase fluid inclusions in the form of a negative crystal or
irregularly shaped vacuole, 1–10 μm in size, have been found
in quartz veins. The heating and cooling results of the primary and secondary fluid inclusions in quartz are given in
Table 4.3. The two-phase fluid inclusions are homogenized
into the liquid phase at a temperature of 222–192 °C and
contain an aqueous solution with salinity of 2.6–2.7 wt%
NaCl. Equivalent Na and Mg chlorides dominate in the solu-

Fig. 4.3 Schematic geological map of the Bazaltovyi site, Pravoingamakitskoe deposit. (Based on data by Chitageologia)

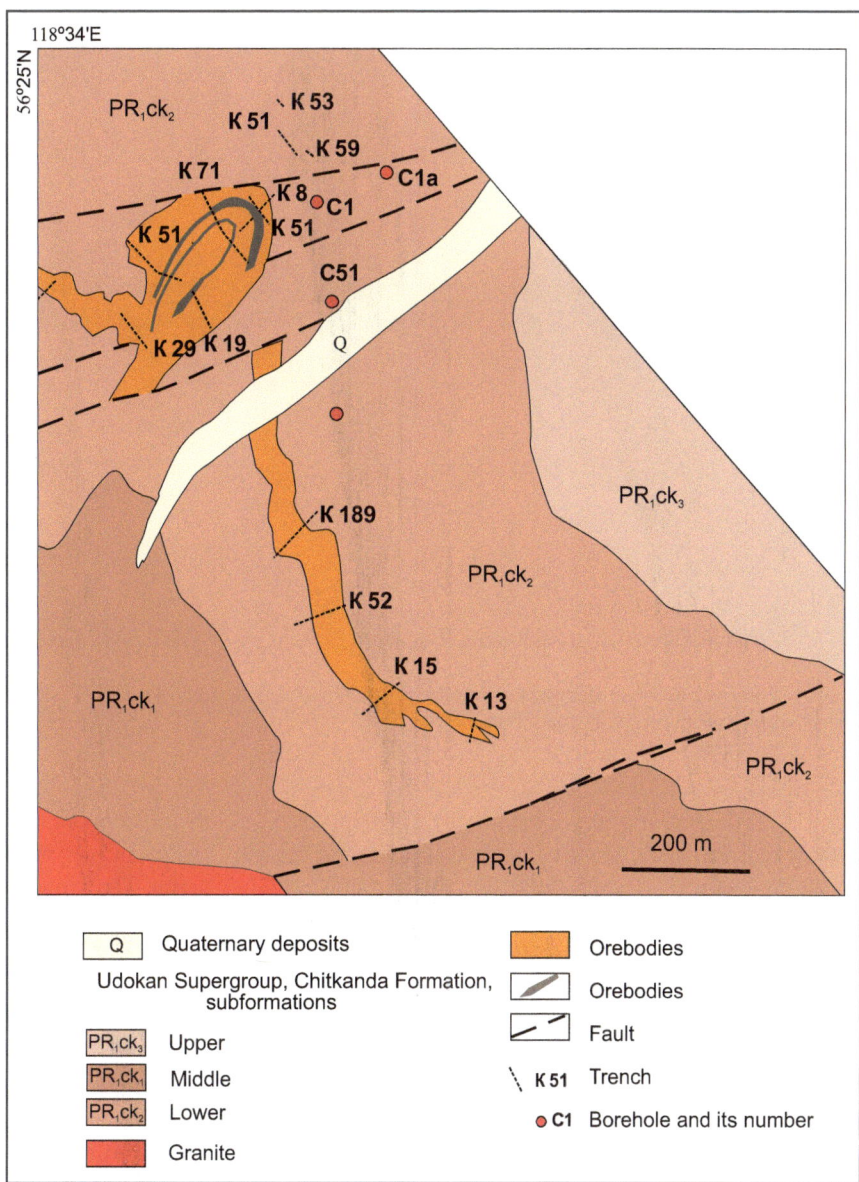

tion of these inclusions (eutectic temperature is −43 to −48 °C). The fluid density is 0.86–0.90 g/cm³ (analyst VYu Prokofiev, IGEM RAS). Similar values of the listed parameters are characteristic for the exocontact ores near the Chiney pluton. The magmatic nature of the sulfur in the sulfide minerals is indicated by its isotopic composition. The concentration of the heavy isotope $\delta^{34}S$ varies from −2 to +7‰ (analyst LP Nosik, IGEM RAS).

4.2.2 The Krasnoe Deposit

The Krasnoe deposit is localized in the northeastern part of the Kodar-Udokan Basin (Fig. 2.4) in the Chitkanda Formation, part of the lowermost group of the Udokan Supergroup, rather than in the Kemen Group that is a host

stratigraphic unit of the above-described copper deposits. The principal hosts are Alexandrov and Butun Formations. Seventeen en echelon units are recognized in the Cu-bearing sequence, related to the middle part of the Chitkanda Formation, which consists of fine-grained polymictic sandstone, siltstone, and shale (Bogdanov et al. 1966a, b). The thickness and strike of individual orebodies reach several meters and hundreds of meters, respectively. The dimensions of the Cu-bearing units at the other prospects are larger, varying from a few tens of meters in thickness to a few kilometers of strike (Table 4.4).

The highest Cu concentrations are commonly related to the lower part of the ore units. They are located in fine-grained shale and sandstone (Fig. 4.12). The shale consists of amphibole, pyroxene, plagioclase, titanomagnetite, and magnetite. Their compositions are shown in Table 4.5 and

Fig. 4.4 Samples of breccia ore from the Bazaltovyi site, Pravoingamakitskoe deposit
(**a**, **b**) Pyrite-chalcopyrite, (**c**, **d**) chalcopyrite-pyrite

Fig. 3.12. The ore-forming minerals are chalcopyrite and pyrrhotite (Fig. 4.13). The rare minerals are titanite (based on two analyses, wt%: SiO_2 = 32.62 and 31.21, CaO = 29.14 and 27.68, TiO_2 = 39.01 and 38.06, FeO = 0.85 and 3.20, Pr_2O_3 = 0.07 and 0.00, Nd_2O_3 = 0.04 and 0.07, respectively), apatite, and zircon.

Occasionally, the amount of sulfides decreases upward in the sequence. Pyrrhotite is noted throughout the Cu-bearing sequence, albeit in much lower quantities. This mineral is mostly abundant in the rocks that underlie and overlie the Cu-bearing sequence. In addition to ore-bearing units occurring conformably to host rocks, there are crosscutting sulfide bodies. They are lenticular in shape and consist of pyrrhotite in association with chalcopyrite, marcasite, pyrite, and gangue minerals. Their thickness reaches 0.5–1.0 m, and they are a few meters in extent.

Lateral and vertical zoning is outlined in the spatial localization of sulfide mineralization in the Cu-bearing sequences as a whole and in separate ore units. Each zonal series has a three- member structure: bornite-chalcopyrite-pyrrhotite (pyrite). Lateral zoning is expressed in consecutive northward replacement of the bornite zone into predominantly chalcopyrite and then pyrrhotite zones within each Cu-bearing unit. The zones gradually shift to the north and upward in the sequence, determining en echelon arrangement of Cu-bearing units and displaying a vertical zoning (Table 4.6).

As was mentioned above, the copper mineralization in the small Udokan-type deposits of the Ikabiya-Chitkanda dis-

trict is commonly represented by chalcopyrite, bornite, and chalcocite. Pyrrhotite and pyrite are subordinate in abundance. The ore of the Krasnoe deposit appears surprisingly different in mineralogy. In contrast to the other deposits and occurrences of this district and Kodar-Udokan Basin as a whole, the ore from the Krasnoe deposit bears no chalcocite at all and hosts widespread pyrrhotite (Bogdanov et al. 1966a). In addition, it is characterized by more complex mineralogy than is commonly recorded in the sandstone-hosted copper deposits (Figs. 4.13 and 4.14). Absolutely atypical minerals, such as arsenopyrite, sphalerite, pentlandite, tennantite, linnaeite, cobaltite, Ti-magnetite, ilmenite, molybdenite, native silver and gold, hematite, and millerite (Table 4.7), are known here. We have found some rare minerals of uraninite and monazite-(Ce) with high Th concentrations (Fig. 4.14e, f). The composition of minerals is shown in Fig. 4.14 and Tables 4.8, 4.9, 4.10, and 4.11.

Chalcopyrite, pyrite-chalcopyrite, pyrite, arsenopyrite, and pyrrhotite ore types are distinguished by the proportion of ore minerals, by variable concentration of copper, and, more distinctly, by gold and, especially, silver contents (Table 4.12). The samples with tennantite-tetrahedrite contain up to 240 ppm Ag and 0.32 ppm Au. The bornite-chalcopyrite ore bears 72 ppm Ag and 0.27 ppm Au. Bi, Sb, Mo, Co, Pb, Zn, Se, and Te are very rare in the ore. This ore is characterized by high As (up to 3 wt%).

The ore of the Krasnoe deposit is almost unaltered. A zone of oxidation is developed only at the surface. It consists of malachite, azurite, supergene chalcocite, tenorite, and cuprite.

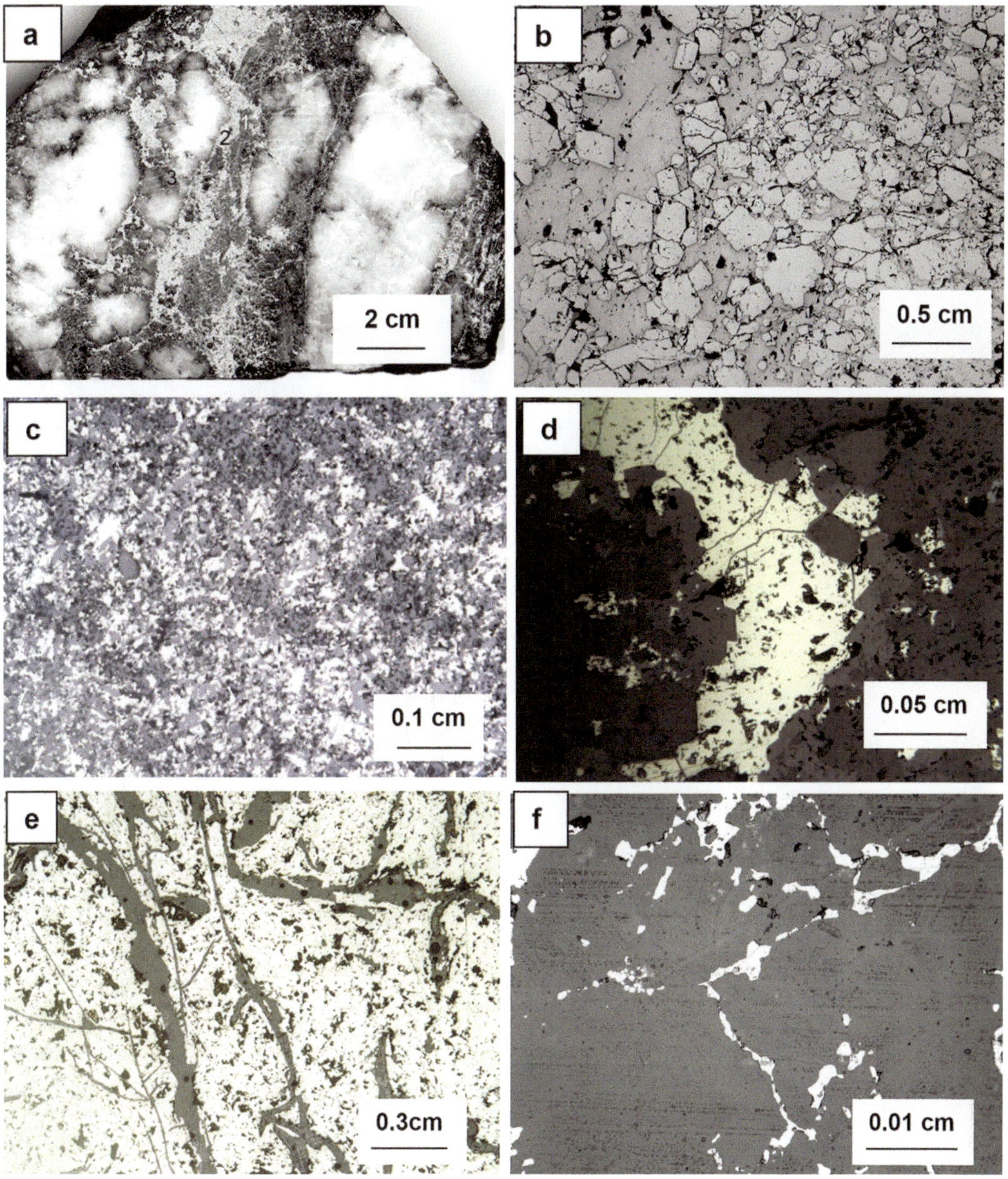

Fig. 4.5 Photomicrographs of pyrite and chalcopyrite breccia ore from the Bazaltovyi mineral site, Pravoingamakitsky deposit (reflected light). (**a**) Polished breccia ore (quartz cemented by sulfide+hydrothermal minerals), (**b**) pyritecrystals, (**c–f**) chalcopyrite

Fig. 4.6 Samples of quartz veins with breccia texture in bornite-chalcopyrite (**a**) bornite-chalcopyrite in pyrite-chalcopyrite (**b**) mineralization

Fig. 4.7 Quartz vein from the Bazaltovyi site

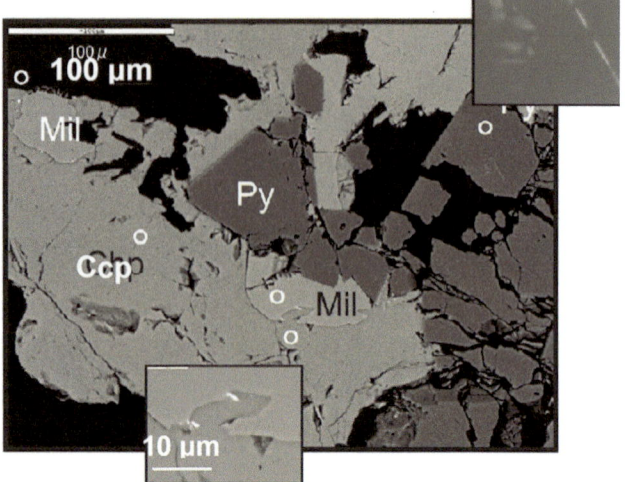

Fig. 4.9 Rare minerals in millerite-pyrite-chalcopyrite ore at the Pravoingamakitskoe deposit. Sample 45–3, bright phases in insets: clausthalite, bravoite and bogdanovichite. *Ccp* chalcopyrite, *Mill* millerite, *Py* pyrite. BSE images were taken at IGEM RAS (analyst EV Kovalchuk)

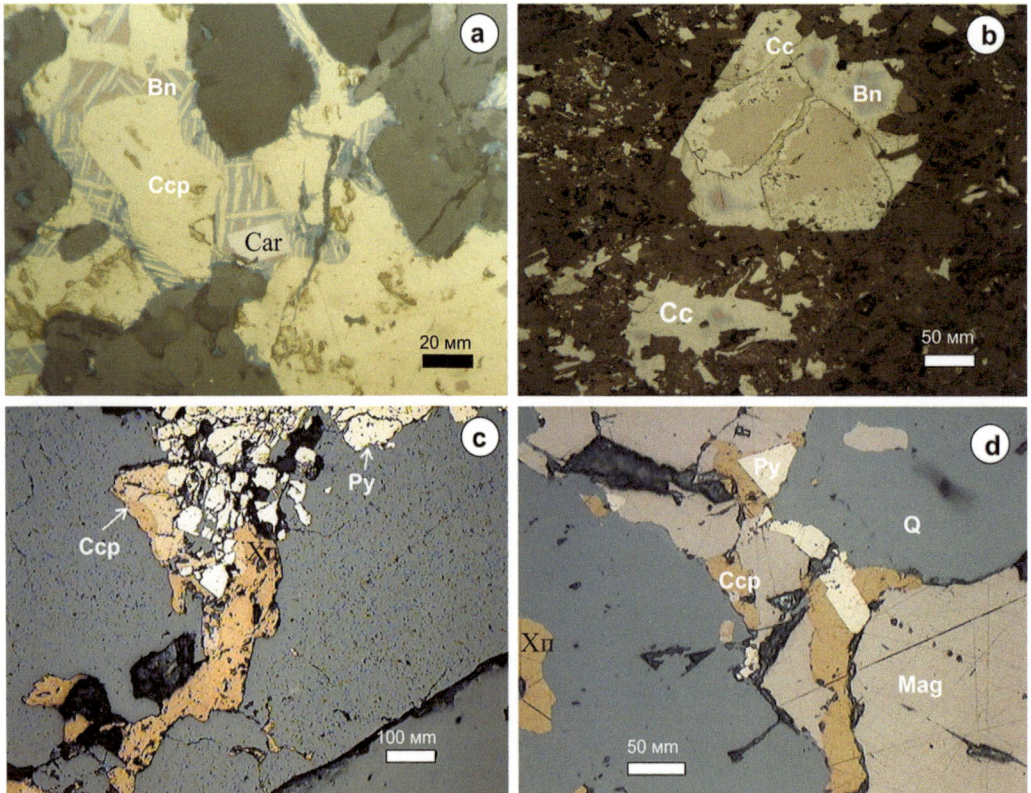

Fig. 4.8 Photomicrographs of sulfide minerals from quartz veins (reflected light). (**a**) Bornite-chalcopyrite, (**b**) chalcocite-bornite, (**c**) pyrite-chalcopyrite, (**d**) pyrite-chalcopyrite-magnetite. *Bn* bornite, *Ccp* chalcopyrite, *Py* pyrite, *Cc* chalcocite, *Car* carrollite, *Mag* magnetite, *Q* quartz

Table 4.2 Chemical compositions of minerals from the Pravoingamakitskoe deposit, wt%

No	As	S	Fe	Ni	Co	Se	Sb	Cu	Zn	Total
1	0.017	34.00	30.50	0.013	–	0.07	–	34.27	–	98.87
2	0.026	53.01	44.25	0.01	2.66	0.073	0.001	0.034	–	100.07
3	0.034	52.84	46.34	0.53	0.053	0.062	0.001	0.042	0.004	99.90
4	0	52.97	46.33	0.165	0.091	0.07	–	0.04	–	99.67
5	0.008	52.87	45.40	0.02	1.27	0.06	–	0.002	–	99.63
6	0.031	34.51	30.76	0.008	–	0.073	–	33.18	–	98.57
7	0.003	53.29	47.03	0.06	0.024	0.074	0.008	0.034	–	100.53
8	0.009	53.18	46.90	0.137	0.022	0.087	0.01	0.016	0.003	100.36
9	0.012	52.67	46.25	0.37	0.115	0.051	–	0.037	0.001	99.52
10	0.029	52.82	46.53	0.17	0.02	0.062	–	0.014	–	99.64
11	0.05	52.79	45.63	0.817	0.151	0.058	0.016	0.023	0.014	99.55
12	0.015	34.36	1.86	62.47	0.466	0.049	–	0.383	–	99.60
13	0.067	41.68	6.38	36.48	14.51	0.132	–	0.642	0.008	99.90
14	0.004	36.08	1.99	59.56	2.207	0.085	–	0.217	0.011	100.15
15	0.004	34.93	1.15	63.04	0.431	0.09	0.014	0.046	–	99.71
16	0.01	34.02	30.35	0.20	0.061	0.09	–	34.34	–	99.08
17	0.024	33.95	30.56	0.15	–	0.078	0.007	34.46	–	99.24
18	0.006	33.84	30.82	–	–	0.029	–	34.58	–	99.29
19	0.006	33.91	30.88	0.009	–	0.078	0.052	34.37	–	99.31
20	0.048	34.44	30.38	0.076	–	0.042	0.009	34.31	–	99.31
21	0	34.73	30.96	0.129	–	0.057	0.008	34.29	–	100.18
22	0	41.24	16.71	38.61	0.997	0.108	–	0.557	–	98.23
23	0.014	34.22	30.29	–	–	0.048	–	34.31	–	98.89
24	0.023	52.93	47.09	–	–	0.055	–	0.02	–	100.12
25	0.018	52.91	46.98	0.505	–	0.036	0.006	0.073	–	100.53
26	0.036	53.11	47.29	0.005	–	0.031	0.025	0.039	0.01	100.54
27	0	34.21	30.45	–	–	0.102	0.003	34.64	–	99.41
28	0.033	35.06	0.53	63.67	0.025	0.062	0.005	0.001	–	99.39
29	0.019	35.07	0.81	63.22	0.028	0.073	0.023	0.043	–	99.29
30	0.037	34.55	29.76	2.90	–	0.014	0.012	32.46	–	99.74
31	0.001	34.98	0.79	63.65	0.028	0.076	–	0.031	–	99.56
32	0	34.85	0.47	63.59	0.025	0.089	–	–	0.004	99.04
33	0.036	36.03	2.25	60.84	0.02	0.07	0.014	0.039	0.009	99.31
34	0.054	35.54	0.56	62.63	0.024	0.051	–	0.083	–	98.94
35	0.018	34.76	0.85	63.28	0.034	0.072	0.002	0.068	–	99.09
36	0.013	34.91	0.85	63.41	0.016	0.1	–	0.084	0.009	99.40
37	0	34.92	0.57	63.28	0.02	0.054	0.016	0.002	–	98.85
38	0	35.49	0.83	62.71	0.028	0.069	–	0.102	–	99.24
39	0.013	36.36	2.98	59.90	0.026	0.067	–	0.35	0.004	99.70
40	0.005	35.95	1.70	61.58	0.039	0.046	–	0.208	–	99.53
41	0.017	35.55	1.36	62.08	0.023	0.065	0.021	0.16	0.001	99.29
42	0.03	34.64	30.55	0.058	–	0.053	–	34.65	–	99.98
43	0.014	53.27	46.71	0.033	0.318	0.045	0.005	0.011	0.001	100.40
44	0.034	52.98	45.81	1.33	–	0.076	0.007	0.009	0.016	100.26
45	0.039	52.99	46.89	0.117	–	0.053	–	0.004	–	100.09
46	0.038	53.41	45.36	0.012	1.541	0.06	–	0.039	–	100.46
47	0.027	53.21	44.74	0.008	2.127	0.056	–	–	0.004	100.18
48	0.033	53.12	45.58	0.006	1.347	0.037	–	0.011	–	100.13
49	0.01	34.63	30.52	–	–	0.053	–	34.69	–	99.92
50	0.001	41.30	14.39	41.78	–	0.074	–	0.965	–	98.52
51	0	38.81	16.97	34.81	0.012	0.086	–	6.68	0.008	97.37
52	0.016	34.13	30.24	0.006	–	0.064	–	34.72	–	99.18
53	0	34.55	30.58	0.028	–	0.073	–	34.57	–	99.80
54	0.035	53.52	46.86	0.017	0.101	0.052	0.006	–	–	100.60
55	0.03	52.99	47.11	0.012	–	0.072	0.014	0.017	–	100.25
56	0.01	34.64	30.42	–	–	0.079	0.002	34.78	–	99.94
57	0	53.30	46.94	0.108	0.037	0.03	0.02	0.079	–	100.52

(continued)

Table 4.2 (continued)

No	As	S	Fe	Ni	Co	Se	Sb	Cu	Zn	Total
58	0.013	52.85	47.41	0.003	–	0.034	–	0.02	–	100.34
59	0.033	52.90	47.26	0.096	–	0.049	–	0.054	0.024	100.42
60	0.02	53.02	45.64	0.008	1.58	0.035	–	0.021	0.013	100.35
61	0.013	53.11	45.32	0.007	2.13	0.029	0.004	0.041	–	100.65
62	0.033	53.32	46.75	0.035	0.326	0.04	–	0.006	0.002	100.51
63	0.002	52.96	45.96	0.015	1.09	0.023	–	0.048	–	100.10

Note. 1, 16–21, 23, 27, 42, 49, 52, 53, 56, chalcopyrite; 2–11, 24–26, 43–48, 54, 55, 57–63, pyrite; 13, 22, 51, pentlandite; 12, 14, 15, 28, 29, 31–41, millerite; 12, pyrrhotite. Dash denotes concentrations below detection limit. Analyses were carried out in IGEM RAS, analyst EN Kovalchuk

Fig. 4.10 Zone of oxidation at the Pravoingamakitskoe deposit (**a**) Sulfate assemblage, (**b**) chrysocolla

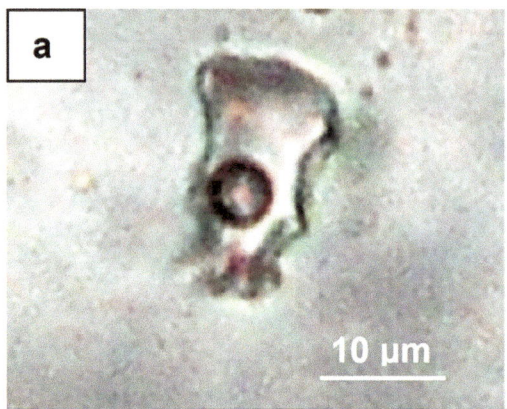
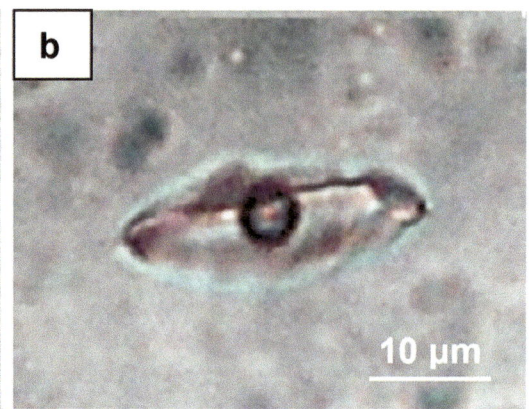

Fig. 4.11 Two examples of two-phase fluid inclusions in quartz at the Pravoingamakitskoe deposit
Temperature of homogenization is 222–192 °C; salinity is 2.7–2.6 wt% NaCl equiv

Table 4.3 Results of microthermometric study of individual fluid inclusions in quartz of the Pravoingamakitskoe deposit

Sample	Mineral	n	T_{hom}, °C	T_{eut}, °C	$T_{ice\ melt}$, °C	Salinity, wt% NaCl equiv.	D, g/cm³
5045–12	Quartz	4	221	−41	−1.6	2.7	0.86
	”	11	192	−38	−1.5	2.6	0.90
	”	3	222	−43	−1.6	2.7	0.86

Notes: Micrometric studies were carried out using measurement instrumentation created at IGEM RAS on the basis of THMSG-600 (Linkam, Great Britain) and Amplival microscope (Germany) equipped with a set of long-focal-length lenses, including Olympus 80ˣ lens (Japan), video camera, and controlling computer. The complex allows to measure phase transition temperature on-line and phase transition temperature within vacuoles (−196 to 600 °C), to observe them at a high magnification, and to take electron microphotographs. Solution salinity was determined from eutectic temperature (Borisenko 1977). Salt concentrations in inclusions were estimated from temperature of ice melting in NaCl-H₂O system (Bodnar and Vityk 1994). FLINCOR program (Brown 1989) was used to estimate salt concentrations and density of fluid

Table 4.4 Cu, Au, and Ag concentrations in disseminated ore at the Krasnoe deposit

No	Ore type	Number of samples	Average mean concentrations		
			Cu, wt%	Ag, ppm	Au, ppm
1	Chalcopyrite-bornite	7	3.82	238.9	0.314
2	Chalcopyrite	1	3.82	72.1	0.273
3	Pyrrhotite	56	1.88	11.2	0.186
4	Pyrite	49	0.27	4.8	0.037
5	Arsenopyrite	70	0.04	1.2	0.014

Note. Data after Bogdanov et al. (1966a, b)

Fig. 4.12 Samples (**a**, **b**) and polished sections (**c**, **d**, **e**) of disseminated ore from the Krasnoe deposit

Table 4.5 Rook-forming mineral compositions from the Krasnoe deposit, wt%

N°	SiO$_2$	FeO	Al$_2$O$_3$	CaO	Na2O	NiO	MnO	TiO$_2$	MgO	Total
1	37.92	15.73	19.12	23.04	0.00	0.25	0.02	0.00	0.00	96.08
2	53.52	22.06	0.56	11.97	0.28	1.14	0.29	0.02	9.38	99.23
3	30.66	2.04	1.17	28.38	0.00	0.13	0.04	36.38	0.01	98.80
4	55.19	15.60	0.66	12.63	0.17	0.51	0.26	0.03	13.65	98.70
5	38.45	7.08	26.54	23.93	0.01	0.00	0.03	0.05	0.00	96.10
6	52.72	16.55	1.37	12.23	0.31	0.69	0.23	0.03	12.72	96.84
7	55.09	14.74	0.47	12.65	0.15	0.65	0.21	0.03	14.01	97.99

Note. Here and in Tables 4.6 and 4.7, analyses were carried out in IGEM RAS, analyst EV Kovalchuk

Table 4.6 Composition of oxides from the Krasnoe deposit, wt%

No	FeO	NiO	MnO	TiO2	Cr$_2$O$_3$	V$_2$O$_3$	Total	Mineral
1	93.04	0.21	0.03	0.08	0.14	1.13	94.62	Mag
2	92.86	0.24	0.03	0.19	0.32	0.99	94.64	Mag
3	91.91	0.23	0.01	0.30	0.26	1.05	93.75	Mag
4	47.45	0.12	2.66	49.97	0.04	0.35	100.59	Ilm

Note. *Mag* magnetite, *Ilm* ilmenite

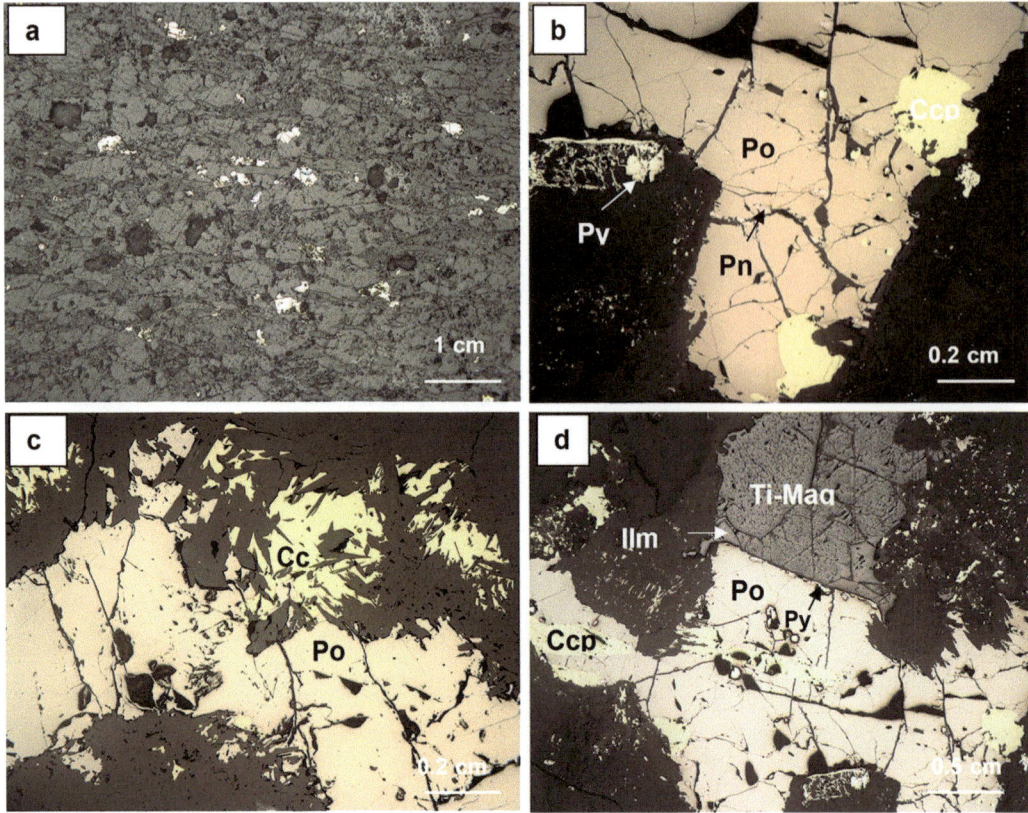

Fig. 4.13 Photomicrographs of disseminated chalcopyrite-pyrrhotite ore from the Krasnoe deposit (reflected light). (**a**) Disseminated sulfides in sandstone; (**b**) pyrrhotite with chalcopyrite, *pentlandite, and pyrite*; (**c**) chalcopyrite in silicate minerals; (**d**) association of sulfides with Fe-oxides. *Ccp* chalcopyrite, *Po* pyrrhotite, *Py* pyrite, *Pn* pentlandite, *Ti-Mag* titanomagnetite, *Ilm* ilmenite

4.3   The Alexandrov Formation

This formation is widespread in the Kodar-Udokan Basin. The most important ore is concentrated in the Kodar subzone, hosting the Sulban deposit. Only small occurrences were found in the Udokan subzone.

4.3.1   The Sulban Deposit

The Sulban deposit is in the valley of the River Sulban (Fig. 4.15). Four copper horizons were discovered in the rocks of the Alexandrov Formation. The mineralization is conformable to the host rocks, repeating all structures in deformed sandstone. The deposit is faulted (Fig. 4.15). The copper grade varies from 0.0n wt% to n%. The main ore minerals (50–90% of the volume) are chalcopyrite, pyrrhotite, and pyrite. The minor minerals (10–25 vol %) are bornite, ilmenite, magnetite, hematite, and graphite. The rare minerals (less than 10%) are chalcocite, arseno-

pyrite, molybdenite, marcasite, sphalerite, galena, millerite, carrollite, tennantite, vittechite, cubanite, and cobaltite.

The oxide mineralization includes blue chalcocite, covellite, malachite, brochantite, and tenorite. They form crusts and films on the surface of the hypogene minerals of copper and iron. It is important to note the presence of a large number of low-water sulfates. This distinguishes the zone of oxidation of the Sulban deposit from oxidation at Udokan (Narkelyun et al. 1983). Zn, Pb, Mo, Co, Ni, Ge, and As should be mentioned in the ore, but only Mo, Co, and Ni are of potential practical interest.

4.3.2   Occurences Near the Udokan Deposit

In the Udokan subzone, two copper-bearing horizons are established in the Alexandrov Formation (Ikabiya-Chitkanda district) (Bogdanov et al. 1966a, b). The upper horizon stretches for several tens of kilometers. Its thickness changes

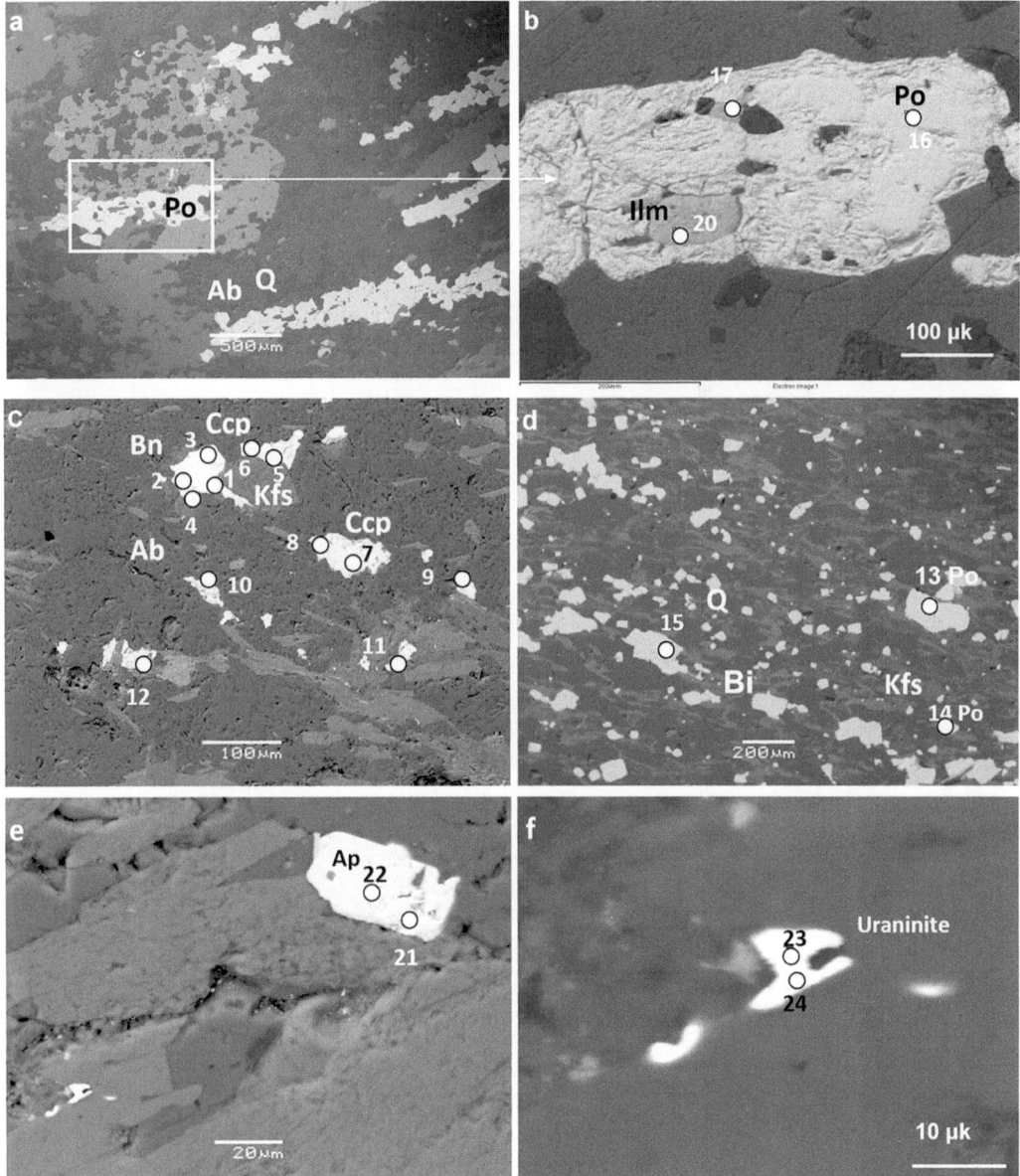

Fig. 4.14 BSM images of disseminated sulfide ore from the Krasnoe deposit. Images were taken at IGEM, RAS (analyst EV Kovalchuk). (**a**) Pyrrhotite and ilmenite grains in magnatite; (**b**) detail of Fig. 4.14a; (**c**) chalcopyrite grains in silicate rock; (**d**) pyrrhotite grains in silicate rock; (**e**) apatite; (**f**) uranium minerals. *Ccp* chalcopyrite, *Ap* apatite, *Ab* albite, *Po* pyrrhotite, *Bn* bornite, *Ilm* ilmenite, *Kfs* K-feldspar, *Q* quartz. N points = N analyses in Tables 4.8, 4.9, 4.10, and 4.11

from first centimeters to 1.5 m. It is composed of undulating horizontally laminated siltstone with sandstone beds. Mineralization is represented by pyrite, chalcopyrite, rarely bornite, and chalcocite. Pyrrhotite was noted as well. Chalcopyrite with pyrite dominates in the lower part of the horizon, and chalcocite with bornite occurs in the upper part. The copper grade reaches 1.5 wt%.

The lower ore-bearing horizon is composed of the rocks that are similar to the rocks of the upper horizon. Its thickness varies from 0.4 to 1.5 m. The mineralization is poor, and the copper content fluctuates within 0.3–0.5 wt%. The main ore minerals are chalcopyrite and pyrite.

Narkelyun et al. (1983) studied the mineralogy of the upper horizon. The ore of the Rossypnoi, Samotsvetnyi, and Glubokiy mineral sites consists of pyrite and chalcopyrite. There is zoned distribution of ore minerals. Pyrite is located at the bottom of this horizon, whereas chalcopyrite dominates in the upper part. The thickness of the pyrite-bearing rocks is much greater than the thickness of the chalcopyrite-bearing rocks (9–10:1).

Table 4.7 Composition of sulfide minerals from ore of the Krasnoe deposit, wt%

No	Cu	S	Fe	Ni	Co	Total	Formula
1	0.10	36.05	0.53	64.10	0.00	100.78	NiS
2	0.04	36.09	0.50	63.95	0.00	100.57	NiS
3	34.37	35.44	29.96	0.82	0.05	100.65	CuFeS$_2$
4	0.10	35.80	0.63	63.80	0.00	100.33	NiS
5	34.30	35.60	30.15	0.53	0.04	100.62	CuFeS$_2$
6	0.00	53.72	46.90	0.15	0.17	100.94	FeS$_2$
7	0.00	54.11	46.38	0.12	0.33	100.94	FeS$_2$
8	0.00	54.36	47.15	0.26	0.36	102.12	FeS$_2$
9	0.06	36.19	0.52	63.61	0.04	100.43	NiS
10	0.14	36.01	0.69	64.23	0.04	101.10	NiS
11	0.07	35.87	0.47	63.76	0.02	100.19	NiS

Note. FeS$_2$, pyrite. CuFeS$_2$, chalcopyrite. NiS, millerite

Table 4.8 Composition of sulfide minerals at the Krasnoe deposit, wt%

N°	S	Fe	Cu	Ag	Total
1	28.95	12.29	55.89	1.85	98.98
2	28.95	12.03	56.75	1.54	99.27
3	27.85	11.84	57.14	2.11	98.94
7	35.07	28.78	35.91	0.36	100.12
8	30.91	16.23	49.65	2.49	99.28
9	32.78	13.7	51.22	2.64	100.34
10	32.36	14.03	51.05	3.06	100.5
11	30.95	9.82	53.92	4.05	98.74
12	35.84	30.06	34.63	0.63	101.16
13	53.34	45.02			98.36
14	54.49	46.19			100.68
15	38.98	61.14			100.12
16	39.01	61.74			100.75

Note. N° here and in Tables 4.9–4.11 means point on Fig. 4.14

Table 4.9 Ilmenite composition from ore at the Krasnoe deposit, wt%

No	TiO$_2$	MnO	FeO	Total
17	52.13	4.45	43.02	99.6
20	52.26	6.59	40.62	99.47

Table 4.10 Monazite composition from ore at the Krasnoe deposit, wt%

No	P$_2$O$_5$	CaO	La$_2$O$_3$	Ce$_2$O$_3$	Pr$_2$O$_3$	Nd$_2$O$_3$	Sm$_2$O$_3$	Eu$_2$O$_3$	Gd$_2$O$_3$	ThO$_2$	Total
21	31.99	0.95	17.3	29.39		11.85		0.34	0.88	5.61	98.31
22	30.96	0.91	15.45	28.04	3.23	12.03	2.26	0.3	1.68	5.12	99.98

Table 4.11 Uraninite composition from ore at Krasnoe deposit, wt%

No	FeO	PbO	UO$_2$	Total
23	1.68	18.84	79.39	99.91
24	1.55	18.11	77.45	97.11

Table 4.12 Cu and Ag in ore of the Unkur deposit

Hole ID	Thickness, m	Interval in hole, beginning, m	Interval in hole, end, m	Ag, ppm	Cu, wt%
AM-001 (IS 1)	40	82.5	122.5	65.9	0.74
AM-001 subsection	22	82.5	103.5	111.9	1.13
AM-001 subsection	7	85.5	92.5	244.1	1.95
AM-001 (IS 2)	18	314.5	322.5	70.0	0.81
AM-001 subsection	7	320.5	327.5	137.7	1.51
AM-001 (IS 3)	5	339.5	344.5	49.2	0.32
AM-003	16	56.5	72.5	84.0	0.79
AM-003	5	59.5	64.5	200.4	1.39
AM-007	10	49.0	59.0	20.2	0.29
AM-011	13.7	140.2	153.9	38.6	0.5
AM-011 subsection	5.9	148.0	153.9	100.7	1.35

Note. Data from boreholes drilled in 2016 (CNW Group/Azarga Metals Corporation) with changes

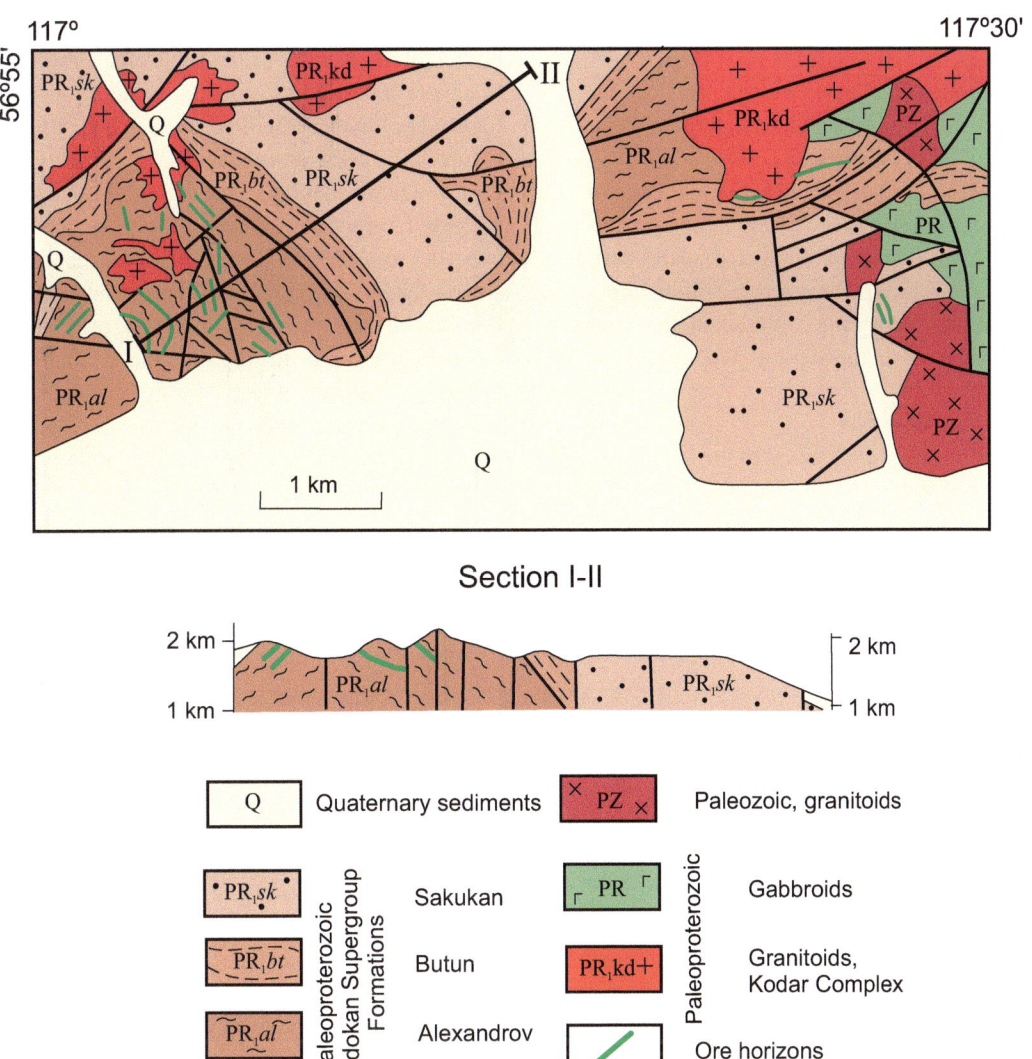

Fig. 4.15 Schematic geological map of the Sulban deposit. (After Narkelyun et al. 1983 based on Chitageologia data)

Cu-bearing rocks were discovered in other areas of the Ikabiya-Chitkanda district. They reveal significant areal extent, but the Cu grades are low. These areas remain under explored.

4.4   The Sakukan Formation

There are deposits and prospects located in both the Kodar and Udokan subzones of the Kodar-Udokan zone. They are related to the lower, middle, and upper Sakukan subformations. The last one contains the Udokan deposit described in Chap. 3 and the Saku deposit. Here, we characterize the deposits and prospects located in the rocks of the Sakukan Formation.

4.4.1   Occurrences of the Kodar Subzone

In the Kodar subzone, copper mineralization was discovered during geological mapping in the valley of the rivers Khilgando and Kukugunda. Chechetkin, Ermakov, and Kazanov recorded some closely spaced horizons with copper mineralization at the "Vershina Khadaktanda" site. The Cu-bearing rocks are approximately 130 m thick and consist of polymictic horizontally layered sandstone with intercalations of quartzite. The upper and lower copper horizons were traced along the strike. Mineralization consists of chalcopyrite, pyrite, and less often bornite. It was traced mainly in quartzite. The Cu contents in individual samples reach 0.5–1.5 wt%. In general, these horizons have not been sufficiently explored.

4.4.2 Deposits of the Udokan Subzone

The Udokan subzone contains two deposits (Unkur and Saku) in the lower and middle Sakukan subformations and one deposit (Burpala) in the upper Sakukan subformation. In addition to the sediment-hosted copper mineralization, the Ikabiya-Chitkanda ore district hosts numerous prospects of sulfide mineralization related to quartz and quartz-carbonate veins, as well as crush and fracture zones. Sulfide minerals in the veins and crush zones were formed in the course of hydrothermal activity of unknown age.

4.4.2.1 The Unkur Deposit

Host rocks. The Unkur deposit occurs to the southeast of the Luktur gabbronorite pluton (Figs. 2.4 and 4.16), not far from the border of the Chara trough (Fig. 4.17). The host rocks form a broad Unkur syncline, with an approximately vertical axial plane striking approximately 12 km from the northwest to the southeast (Narkelyun et al. 1983). The Alexandrov, Butun, and Sakukan Formations have been identified within the Unkur area. The rocks of the Alexandrov Formation are exposed in the southwestern limb of the syncline, comprising a package of interstratified siltstone and argillite, with quartzite beds, approximately 1 m thick, occurring every 25–30 m. The rocks correspond to magnetic lows. Based on geophysical data, the thickness of formation in the project area is approximately 450–600 m.

The upper part of the Butun Formation is exposed in the canyon of the River Unkur as a package of alternating siltstone and fine-grained sandstone. This formation corresponds to magnetic highs. Based on the geophysical data, its thickness is 500–600 m.

The Sakukan Formation is a principal host of copper mineralization. It occupies most of the Unkur area. In the east and northeast, this formation is intruded by the granite of the Kodar Complex, part of the Kemen granite layered mafic-ultramafic complex. The Sakukan Formation consists mainly of medium-grained gray sandstone. The middle and lower subformations of the Sakukan Formation have been identified in the area. The lower subformation comprises gray and pinkish-gray sandstone, alternating with gray and black siltstone. It is 1000–1200 m thick. The middle subformation mainly consists of gray and pinkish-gray sandstone, interlayered with calcareous sedimentary rocks. Rough cross-bedding is characteristic of the sandstone. The overall thickness of the middle subformation is approximately 1000 m.

Structure. The southwestern limb of the syncline dips to the northeast at 40–60° and is complicated by parasitic folding (Fig. 4.16). The Butun and Sakukan Formations outcrop in the northeast limb of the fold. They dip 15–30° to the southwest, increasing to 35–60° closer to the axial plane. To the southeast, the syncline gradually flattens. In the north-

west, geophysical evidence implies the syncline is cut by a branch of the Kemen Fault. The Kemen Fault is one of three main northwest-striking faults, in addition to the Burunga Fault. The displacement in the vertical direction on these major faults does not exceed 300 m (Narkelyun et al. 1983).

The Unkur syncline is also cut by the Khar northeast-striking fault system. Displacements on these faults do not exceed 150–200 m. All the faults were reactivated at various stages. There is no reliable information on the crosscutting relationships between the faults. The formations are intruded by the gabbro-diorite dikes of the Chiney Complex. The Luktur gabbronorite-layered mafic-ultramafic complex, also belonging to the Chiney Complex, is located in the north of the area. Borehole № 32 intercepted 2 km of intrusive rocks and never penetrated into sedimentary country rocks. The dikes range from meters to tens of meters thick, with observed strike lengths of 200–1000 m. The dikes strike northeast and northwest, corresponding to the strikes of the two main fault systems.

Mineralization. Two Cu-bearing horizons were discovered in the sandstone of the lower Sakukan subformation. The deposit was discovered and explored in the 1960s (Chechetkin and Kharitonov 2002; Chechetkin and Trubachev 2013). It is covered by Quaternary sediments. Several orebodies have been intercepted by drilling. The host rocks are medium- to fine-grained carbonate, sparse carbonate sandstone, and siltstone-argillite rock. The ratio of the fine-grained sandstone to siltstone is approximately 3:1. The rock-forming minerals are quartz, albite, mica (muscovite and biotite), calcite, and traces of chlorite. The accessory minerals are zircon, apatite, barite, and titanium oxides. Rhythmic structure in these rocks is common.

The fine-grained sandstone constitutes approximately 75% of the sequence. It contains abundant quartz and plagioclase, and the inter-grain cement is micaceous, also containing carbonate minerals. The structure is fine-grained; the micro-texture is massive and porous. These sandstones contain poorly disseminated iron oxides and rinds, films, and impregnations of copper oxides, such as malachite, azurite, and copper silicate. The iron oxides are hematite, magnetite, and martite. Zircon is also present. The fine-grained sandstones are rich in copper hydrocarbonates and contain 5–10% of malachite by volume.

The siltstone makes up 25% of the sequence. It is fine-grained, and the texture is massive and weakly layered. The siltstone contains disseminated impregnations of iron oxides (from individual grains to 2% of the rock volume). The siltstone contains rinds, films, and fine veinlets of copper minerals, usually malachite (from individual grains to 1–2% of the volume). The primary iron oxide is hematite, with small amounts of magnetite. The iron oxide grain size varies from 1 to 30 μm and occasionally up to 0.2 mm. The hematite and magnetite both contain 2–30 μm inclusions of the sulfide

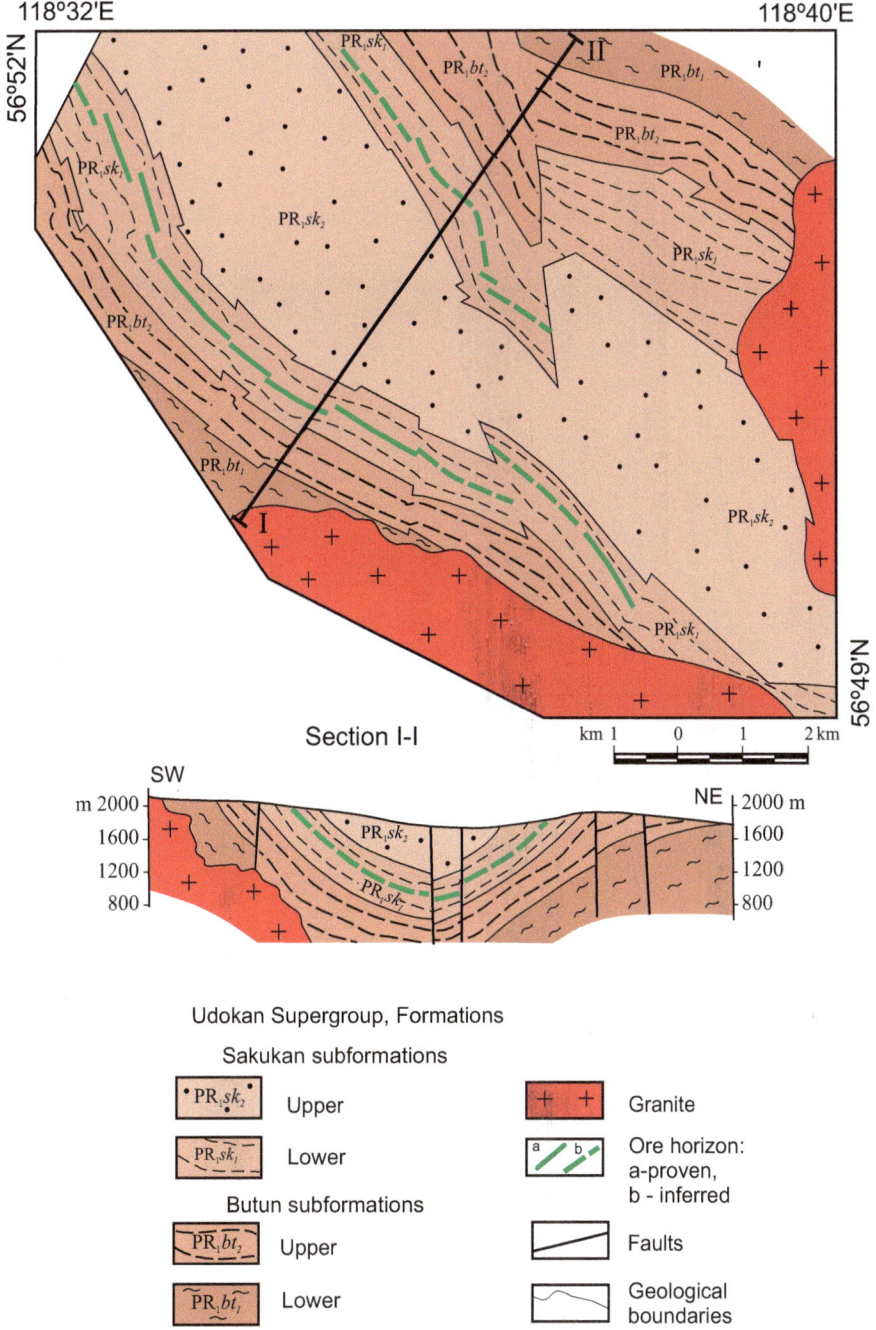

Fig. 4.16 Geological map of the Unkur deposit. (After Narkelyun et al. 1983 based on Chitageologia data)

Fig. 4.17 General view of the Unkur copper deposit area (looking west)

minerals present. The intergrowths are generally complex. Hematite is partially rimmed by copper hydrocarbonate films. The hematite grain size varies from 5–10 μm to 70–100 μm. The magnetite crystal size is 50 μm–0.15 mm and up to 1 mm in intergrowths with gangue minerals. Pyrite is disseminated and present as individual free grains in the gangue minerals.

The Cu-bearing layers are underlain by pyrite-bearing sandstone and siltstone. Orebodies, recognized by the results of sampling, occur in the central parts of the cupriferous layers and have the shape of layered lodes of 4200 × 12–50 × 300 m in size, locally divided by barren rock. The thickness of the ore-bearing units is 50–70 m. The stratiform and steeply dipping (45–50 °) orebodies are conformable to the host sandstone and limestone making up the northern limb of the Unkur syncline (Fig. 4.16). Intrusive rocks in this area are represented by the Kemen layered mafic-ultramafic complex and the dikes of gabbro-diabase.

Ore minerals. The primary ore minerals are chalcopyrite, bornite, pyrite, rare chalcocite, magnetite, and hematite. The disseminated ore dominates, and lenses, nests, and veinlets occur along the cleavage. The distribution of sulfides demonstrates a symmetrical vertical zonation, where chalcopyrite and pyrite occur in the bottom of the horizon and bornite is in the top. This zonation is also recorded at the southeastern

edge of the limb. The pyrite-chalcopyrite association forms the center of the ore body, whereas bornite and chalcopyrite are located on the northwestern edge. Economic mineralization is associated with the chalcopyrite-bornite and chalcopyrite zones.

We have found many rare minerals in ore. They are pyrrhotite, ilmenite, native gold, and the tiny palladium phases. The latter have remained unidentified because of small grain size (fractions of a micrometer). The rims around the chalcopyrite grains are composed of uranium oxide, and U-Th minerals are noteworthy (Fig. 4.18a, b, f). Grains and veinlets of native silver were found (Fig. 4.18c–e). The late barite occurs in veinlets and individual grains among the silicate minerals (Fig. 4.18g). Many zircons (Fig. 4.18h) with silicate minerals were recorded in the sulfide. The spectra of rare minerals are demonstrated in Fig. 4.19.

The mineralogy of one sample with disseminated ore has been studied by Azarga Metals Corporation (https://www. azargametals.com/). The mineralogical examination of the +1 mm fraction of the crushed sample indicates (Table 4.8) that the ore is a mixture of fine-grained sandstone and siltstone with disseminated impregnation of iron oxide and copper oxide minerals. Small amounts of sulfide minerals (bornite, chalcocite, covellite, chalcopyrite, and pyrite) are present together with small amounts of silver and silver sul-

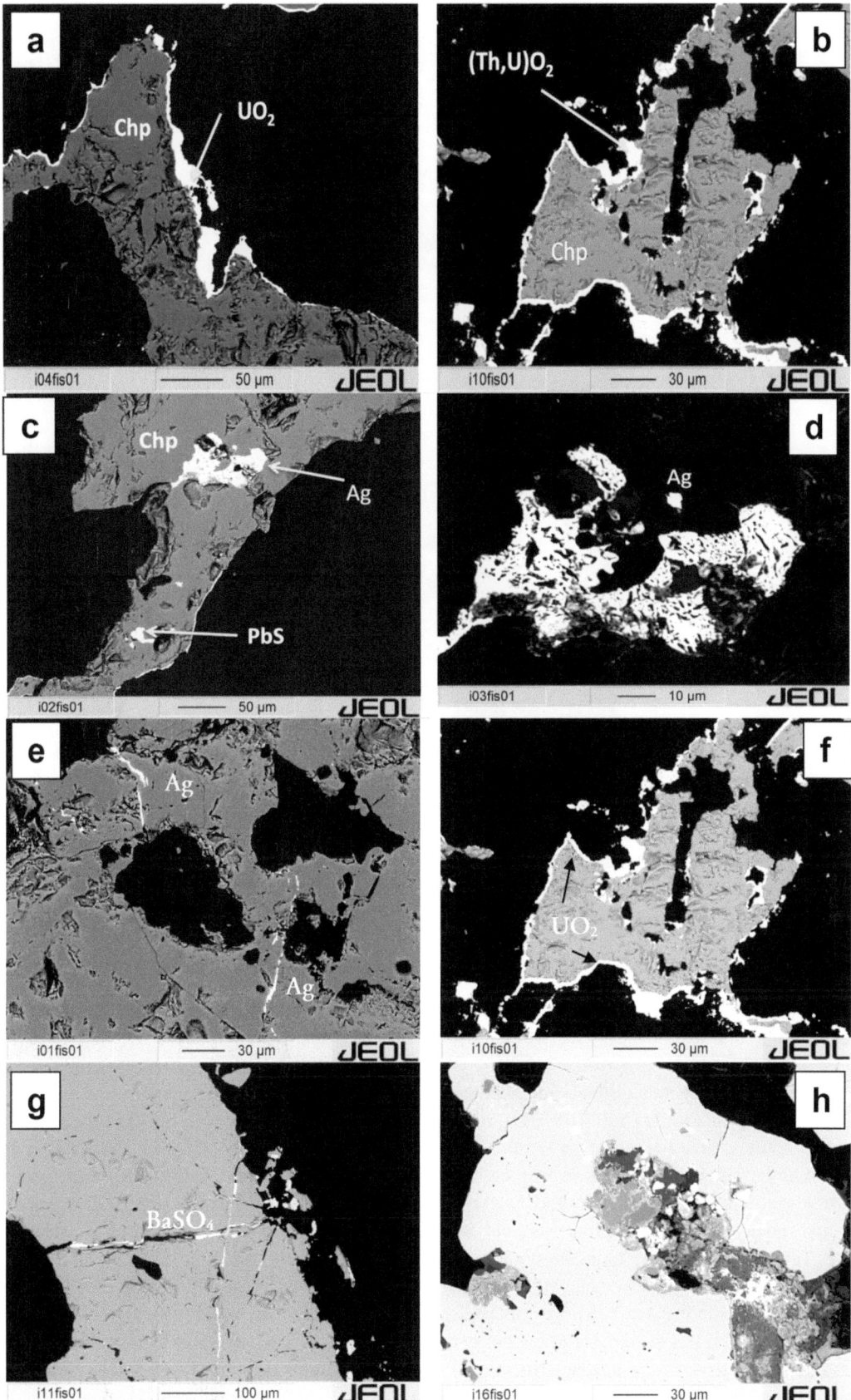

Fig. 4.18 BSE images of rare minerals from the Unkur deposit.
(**a, b**) – Uranium minerals around chalcopyrite grains (spectrum a on Fig. 4.19); (**c–e**) – veinlets and grains of native silver (spectrum b on Fig. 4.19); (**f**) – uraninite rims around sulfides; (**g**) – veinlet of barite (spectrum c on Fig. 4.19); (**h**) – zircon grains. BSE images were taken at Max-Planck-Institute of Geochemistry, Mainz, Germany (analyst DV Kuzmin)

Fig. 4.19 Spectra of rare
minerals from the Unkur
deposit. Spectrum of uranium
mineral (**a**), native silver (Ag)
(**b**) and barite (**c**), IGEM,
RAS (analyst LO Magazina)

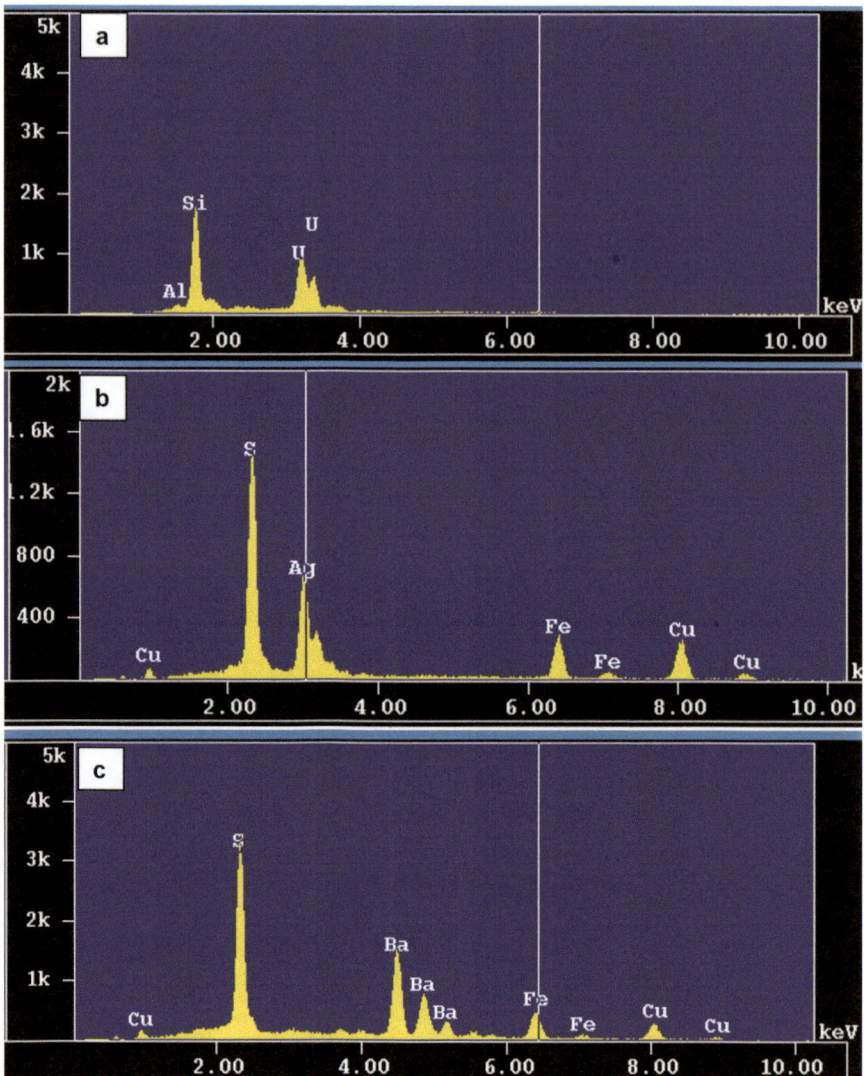

fide. The main copper oxide minerals occur in the form of
carbonates, malachite, and azurite. Chrysocolla is also
present.

The main copper-bearing minerals of the oxide zone are
malachite $Cu_2(CO_3)(OH)_2$ and azurite $Cu_3(CO_3)_2(OH)_2$. They
occur as rinds, thin films, and impregnation in the fine-
grained sandstone and siltstone, sometimes as veinlets and
stringers in siltstone and sometimes as rims around the grains
of magnetite and hematite. The malachite veinlets are
10–50 μm thick. The azurite is often intergrown with malachite. Some of the malachite aggregates contain an admixture of zinc and lead. In some instances, azurite contains a
small amount of lead. Sometimes in samples of copper and
lead arsenates, rare earth elements are present (Nd and Y).
Silver is found as silver sulfide and native silver too. Silver
sulfide Ag_2S occurs as inclusions and emulsions in malachite
or with fine native silver inclusions. Native silver occurs as

fine (0.5 μm) inclusions in silver sulfide and malachite. Their
concentrations in boreholes drilled by the Azarga Metals
Corporation in 2016 (Fig. 4.20) are summarized in Table 4.12
(https://www.azargametals.com/).

The Azarga Metals reported that the textural and structural features and mineralogical composition of the Unkur
ores are similar to the Udokan ores. For metallurgical pur-
poses, they are characterized as mixed carbonate-sulfide.

Chemical composition of ore. In the sample, containing
1.31 wt% Cu and 28.2 ppm Ag, Azarga Metals reported an
oxide copper mineralization, representing over 95% of the
contained copper with relatively low sulfur values. Both S_{Total}
and $S_{Sulfide}$ confirm the relatively low amount of copper in the
sulfide minerals compared with the minerals of the zone of
oxidation. The CaO and MgO contents together with the
relatively high LOI values are significant since they indicate
the presence of carbonate in the sample.

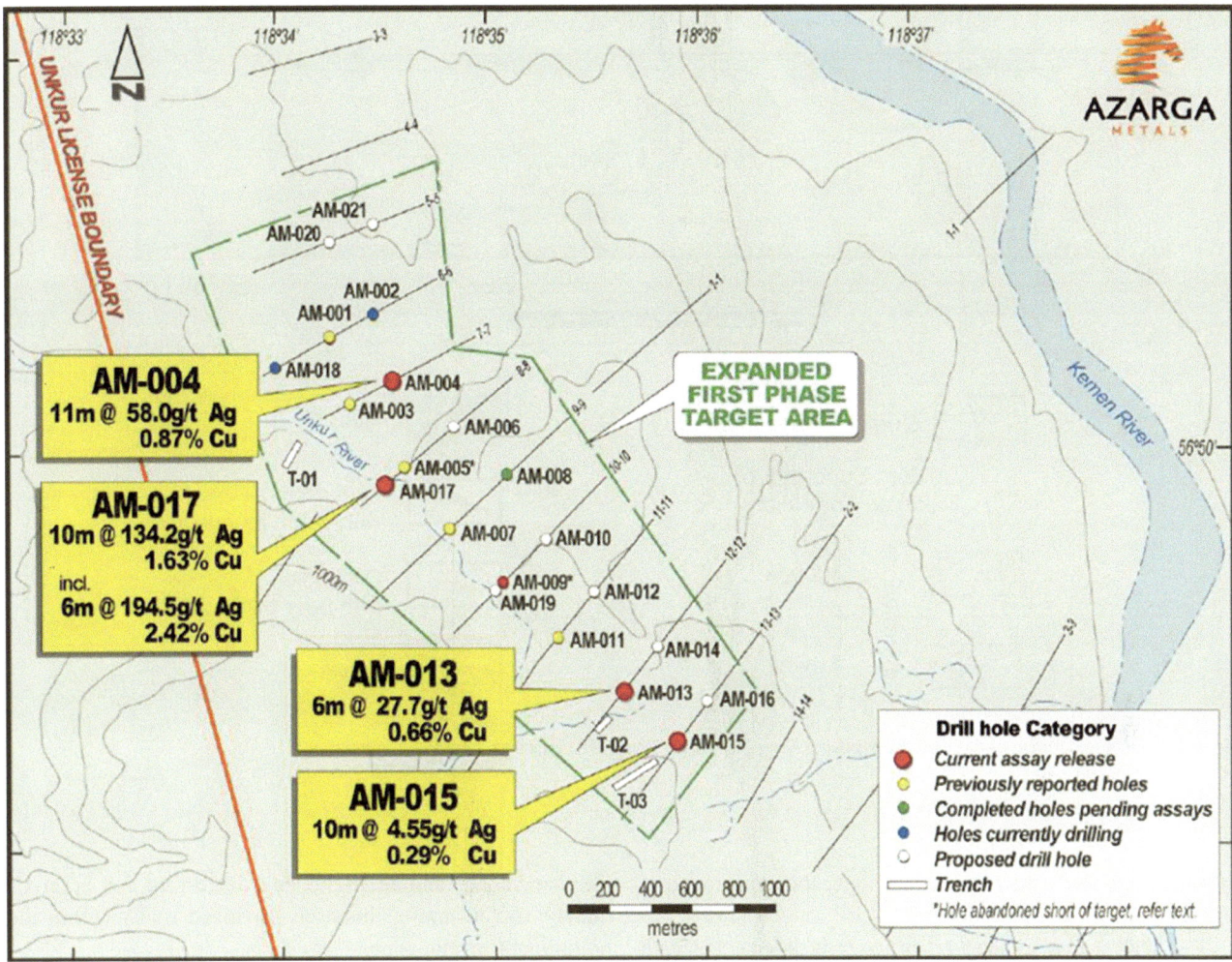

Fig. 4.20 Map of the Unkur deposit. (After Azarga Metals. https://www.azargametals.com/news/2016/azarga-metals-reports-new-drill-hole-assays-including-seven-meters-of-2441gt-silver-and-195-copper)

4.4.2.2 The Burpala Deposit

The deposit occurs in carbonate-clastic rocks of the lower Sakukan Formation (Figs. 2.4 and 4.21) that are cut through by sheetlike gabbro intrusions of the Doros Complex. The formation is a monotonous sequence of medium-grained oligomictic sandstone with sericite-quartz cement and by siltstone, fine-grained sandstone, carbonate, and quartzite interbeds. The sill-like gabbro bodies gently dip to the northwest, conformably with host rocks of the Sakukan Formation, or at an acute angle to the strike of these rocks. The thickness of the dikes varies from 20 to 200 m and extends for 2 kilometers. The endo-contact zones of the intrusions are fine-grained rocks. The slightly metamorphosed exocontact hornfels zone reaches a few meters in thickness. The gabbro rocks contain sporadic magnetite and pyrite disseminations. The deposit has a block structure caused by widespread faults. The NE-trending and gently dipping faults control the gabbro intrusions. The NE- and W-E-trending steeply dip-

ping faults (the latter ones are characterized by zones of intense pyritization), and numerous thin, variable in strike foliation zones can be traced (Bogdanov et al. 1966a, b).

The zone of mineralization is 9 km long and 150 m wide. The sulfide copper mineralization is concentrated in two major ore layers, while six-layered and lens-like orebodies are distinguished. The major ore layer extends for 3 km with thickness variable from 3 to 42 m (8–11 m on average). It is confined to the unit of medium-grained oligomictic sandstone with interbeds and to layers of limestone, sandy limestone, and calcareous sandstone.

Horizontal and less often wavy layering is typical for this rock. Ripple marks, along with lithologic features, are an evidence of coastal marine and lagoon formations. The comparatively weak degree of roundness and sorted clastogenic grains indicates the proximity of the denudation area. The rocks were subjected to the regional metamorphism at the greenschist, sometimes amphibolite facies. The sulfides are

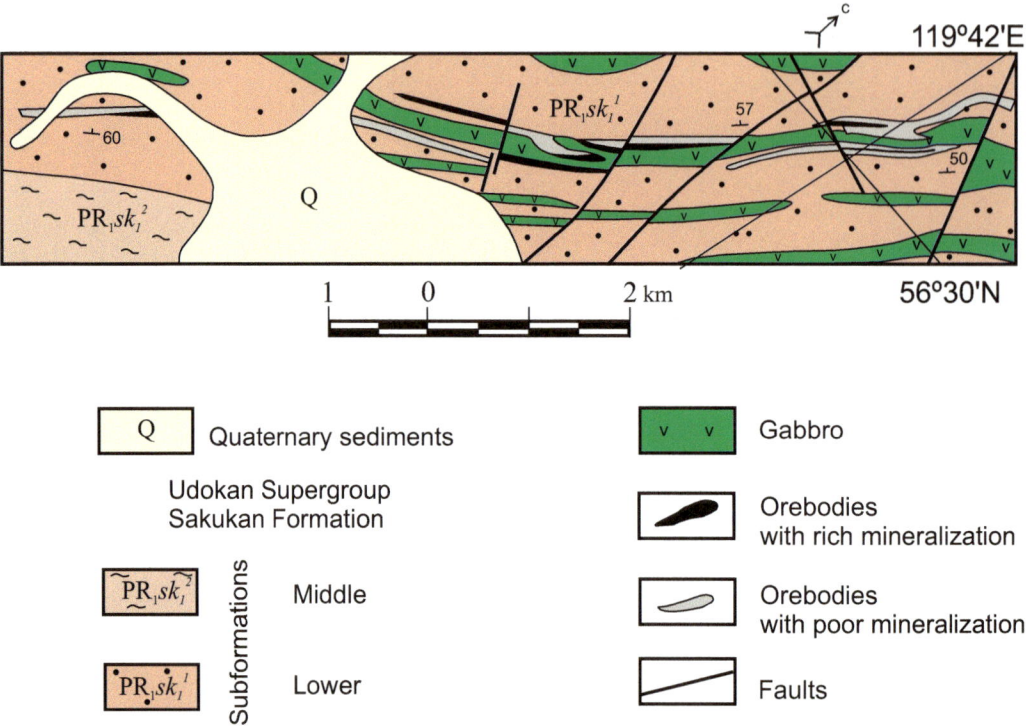

Fig. 4.21 Geological map of the Burpala deposit. (After Narkelyun et al. 1983 based on Chitageologia data)

extremely unevenly distributed, leading to the appearance of en echelon ribbon-shaped bodies.

The central part of the *first orebody* is intersected data acute angle to its strike by a gabbro dike, approximately 20 m thick, without a visible offset of the torn ore-bearing unit. As a result, the ore body is split into several small bodies included in the intrusive rocks together with the surrounding rocks. The copper mineralization is zoned from bottom upward, and the chalcocite-bornite ores are replaced by bornite, then by chalcopyrite, and finally by pyrite, with gradual transitions between them. The mineralized host rocks are altered (sericiticized, scapoliticized, and albitized). The mineralization consists of 8% chalcocite, 49% bornite, and 43% chalcopyrite. The *second ore body* in the center of the deposits extends for 5 km, with thickness varying between 2 and 70 m. Its main sulfide is chalcopyrite.

The orebodies differ from one another in their size, copper grade (the first is richer), and mineral composition. The main sulfides of the first body are bornite and chalcocite, and less often chalcopyrite, which dominates in the second one.

On the southwestern edge is the series of thin zones of poorly disseminated sulfides, mainly pyrite. The mineralized rocks are represented by massive and foliated calcareous sandstone, white crystalline limestone, massive quartzite-like sandstone, and quartzite that are traced for approximately 150 m among intensely pyritized rocks within the latitudinal tectonic zone.

Aksenova (1969, unpublished) studied the artificial heavy fractions. Ilmenite, titanite, rutile, magnetite, zircon, and galena, cerussite, sphalerite, scheelite, barite, and allanite occur in addition to the rock-forming and aforementioned ore minerals. The gold grade reaches 0.08–0.15 ppm in the bornite-chalcocite ore and 0.08 ppm in the bornite-chalcopyrite ore. The mean Se concentration is 646 ppm.

The highest silver concentrations are contained in the chalcocite-bornite ore (up to 125.2 ppm Ag at 3.01 wt% Cu on average of 23 analyses), the chalcopyrite-bornite ore (up to 113.3 ppm Ag at 2.06 wt% Cu on average of 34 analyses), and the pyrite-chalcopyrite ore (up to 26.7 ppm Ag at 1.63 wt% Cu on average of 37 analyses). In ore with a lower copper grade, the silver concentration abruptly drops down to 7.98 ppm Ag at 0.37 wt% Cu on average of 55 analyses and to 3.5 ppm Ag at <0.01% Cu on average of 12 analyses. At the mean Cu grade (0.6 wt%) of ore at the Burpala deposit, the mean Ag grade is 88.2 ppm, and this value is unique. In the Ag-richest copper ore at the Mansfeld deposit in Germany, which is similar to Burpala in composition and structure, the Ag content does not exceed 51 ppm.

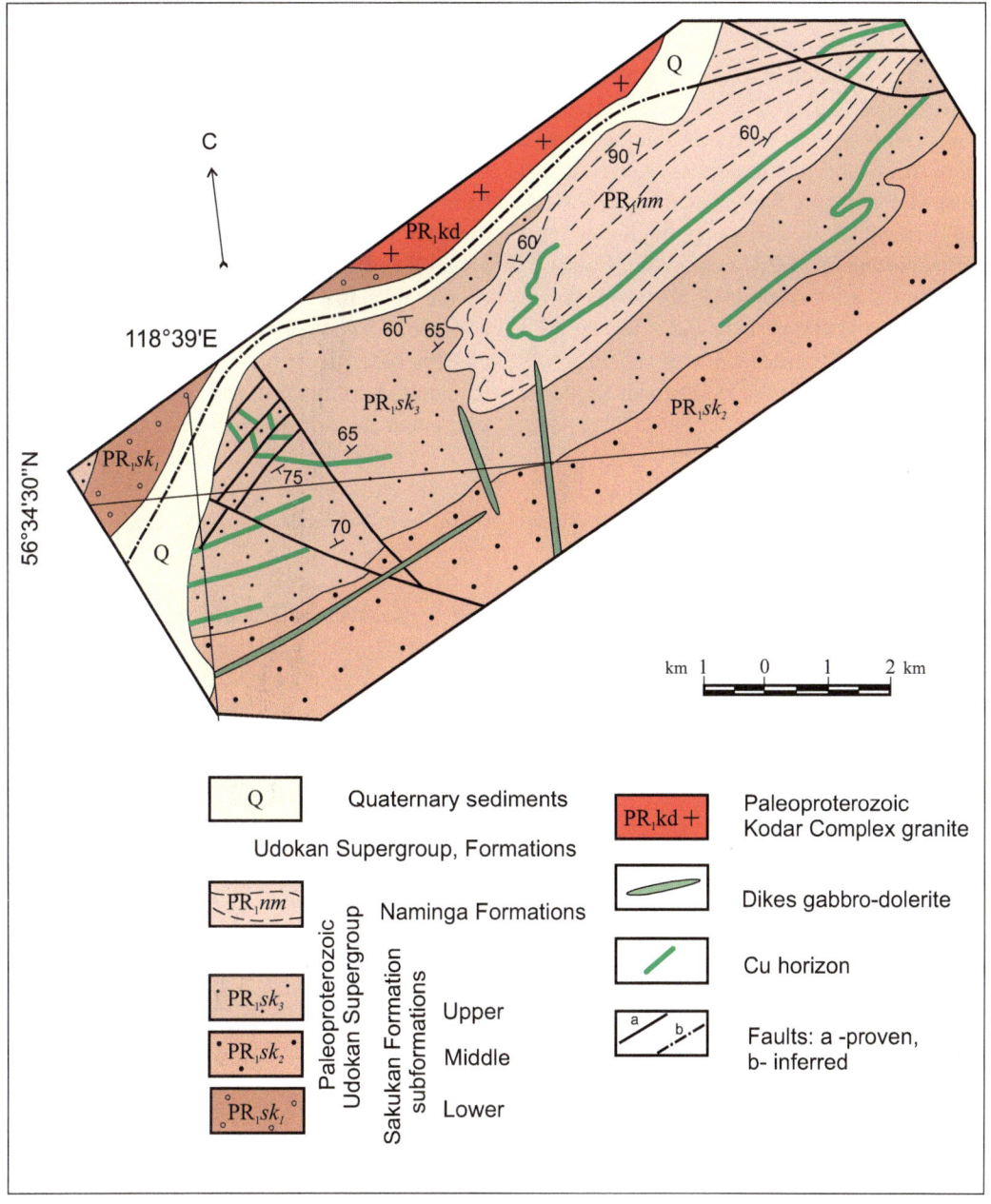

Fig. 4.22 Geological map of the Saku deposit. (Based on Chitageologia data, Narkelyun et al. 1983)

4.4.2.3 The Saku Deposit

The deposit is located in the Creek Saku, a tributary to the River Kemen (Narkelyun et al. 1983). The deposit occurs in the rocks of the lower, middle, and upper Sakukan and Naminga Formations (Fig. 4.22). They form a Saku syncline of northeastern strike that extends approximately for 15 km and is 2–3 km wide (Fig. 4.20). The sedimentary rocks are intruded by granite of the Kodar Complex. The southwestern periclinal of the Saku syncline has a very complex structure. It consists of many isoclinal folds and is broken by the many faults.

The Saku deposit occurs within a hinge at the southeastern edge of the Saku syncline. It is hosted in the middle part of the upper Sakukan subformation (650 m) and the lowermost part of the Naminga Formation (250 m). The formations consist of variegated sandstone, siltstone, and argillite formed in coastal marine, flood lagoon and on-land deltaic facies (Chechetkin and Kharitonov 2002).

The lower Sakukan subformation is represented by rhythmically interbedded fine-grained and pinkish-gray sandstone, siltstone, and argillite with characteristic wavy horizontal bedding textures. The middle Sakukan subformation is composed of magnetite-bearing cross-bedded quartzite sandstone, calcareous sandstone, siltstone, and conglobreccia. The upper Sakukan subformation comprises pinkish-gray fine-grained sandstone with rare interlayers of gray fine-grained sandstone and siltstone. An important role in the structure of the subformation plays rhythms, consisting of alternating pink and gray fine-grained massive sandstone and fine-grained wavy sandstone, including rare interlayers of siltstone and magnetite-bearing sandstone.

The Naminga Formation is subdivided into two packages in this area. A lower one is 400 m thick and is formed by a dark-gray wavy sandstone, interbedded with siltstone with ripple marks and cracks. The sand-siltstone sequence reaches up to 150 m. Typically, the rhythms have three parts: pinkish-gray fine-grained sandstone in the bottom (0.1–0.3 m), gray fine-grained sandstone in the middle part (0.8–1.2 m), and siltstone-argillaceous rocks in the upper part (0.1–0.2 m).

There are six copper-bearing horizons in the cross section, traced to different distances with copper grade up to 1.05%. The Cu mineralization changes from low grade to economic through the entire cross section. The orebodies are steeply dipping (50–80°), layered, and lenticular. The dimensions of the mineralization zones are 12,000 × 1–3 m. The ore horizons include sequences of barren rocks, 100–300 m thick.

The mineralization is hosted in quartzitic sandstone. The major ore minerals are bornite, chalcocite, chalcopyrite, and pyrite. The rare minerals are magnetite, hematite, molybdenite, and tennantite. They form disseminated ore. On the surface, the sulfides were oxidized into malachite, azurite, brochantite, antlerite, chrysocolla, and limonite.

4.5   Neoproterozoic to Cambrian Formations

The Vendian-Cambrian Formations (undivided sediments) were also identified in the Kodar-Udokan Basin (Fig. 2.4). A typical representative object of this type of mineralization is the Kilcheris deposit (Narkelyun et al. 1983), which is located on the water divide of the Kilcheris Creek and River Kemen (Fig. 2.4). The main area is occupied by sediments of the Pestrotsvetnaya Formation, to which the copper mineralization is also confined. The total thickness of the formation varies from 300 to 500 m. It is represented by a rhythmic interlayering of sedimentary conglobreccia, sandstone of variable granularity, siltstone, mudstone, dolomite, and limestone, formed in a marine lagoon. According to the unpublished data of Melnichenko, the shape of the mineralization is lenticular, with an average copper grade of 1.32% and a thickness of 5.5 m. To the northeast and southwest, the body wedges out. The style of mineralization along the XX m

strike and down dip does not change. The thickness of the individual ore lenses is 1.5–2 m, with copper grade of 0.8–1.2%. The mineralization occurs in quartzite, sandstone, and dolomite.

Ore minerals are noted in almost all types of rocks, but their bulk is confined to brecciated quartzite, where they form dissemination, nests, veins, and massive clusters. The main copper sulfide is chalcopyrite. The subordinate minerals are bornite, marcasite, and pyrite, followed by rare pyrrhotite, sphalerite, arsenopyrite, hematite, and enargite. They are subdivided into hypogene and supergene. The main associations are chalcopyrite-tennantite, chalcopyrite-pyrite, and tennantite-bornite. The rare associations are chalcopyrite-pyrite-marcasite-arsenopyrite and tennantite-enargite.

The chemical composition of mineralization is very simple, and only copper has potential economic value. Zinc is recorded in low concentrations here.

4.6   General Characteristics of Deposits

Narkelyun (Narkelyun 1962; Narkelyun et al. 1977, 1983), who studied the sediment-hosted Cu deposits in both Kazakhstan and Siberia, came to the following conclusions on their geology and genesis:

1. There is stratigraphic control of mineralization. Cu mineralization is located in the Chitkanda and Alexandrov-Sakukan Formations.
2. Mineralization is confined predominantly to coastal marine, lush lagoon, and delta lithofacies.
3. The Cu mineralization of different deposits is zonal due to the history of the basin.
4. The main mineral associations are chalcocite-bornite, bornite-chalcopyrite, pyrite-chalcopyrite, chalcopyrite-tennantite, and chalcopyrite-pyrite-pyrrhotite.
5. Additional metals are Ag, Ni, Mo, Zn, and Pb.
6. Magmatic processes are not proven to significantly affect the ore formation, being restricted to contact metamorphism.
7. The genesis of these deposits is uniquely sedimentary. Some geologists consider them formed purely by sedimentary process, the others believe in a sedimentary-diagenetic formation, the third group regards them as a product of catagenetic changes, and the fourth group suggests a metamorphic origin for the ore. Most likely, all these processes contributed to the ore formation;
8. The main source of the clastic material and ore was in the Archean rocks, including both the metamorphic and intrusive complexes of the Aldan Shield. Some researchers (Popov, Slivinskiy, and others) believe that some copper was introduced into the sediments during volcanic eruptions from the Muya zone.

The other geologists studied Dzhezkazgan depost (Satpaev 1977; Satpaeva 1985, 1995, 2007, 2008) emphasized the contribution of magmatic processes in ore formation.

The abovementioned features of the Cu stratiform deposits do not provide an unconditional evidence of sedimentary origin for Cu mineralization (Gongalsky et al. 2017). This problem is discussed in Chaps. 3 and 12.

In conclusion, numerous pyrrhotite-bearing stratigraphic units with disseminated chalcopyrite should be noted in the Ikabiya-Chitkanda mineral district as well. They have significant thickness and extent but remain underexplored, although they may be of interest because of anomalous concentrations of Co, Ni, and Ag in some of them.

Furthermore, the Cu horizons stretch from the Kodar-Udokan Basin to the northeast and west of the Aldan Shield (Fig. 3.1). There are two mineral clusters in the Olekma-Tokki and East Aldan areas, situated between the Aldan Shield and the Uchur-Maya terrane. The Nuya-Berezovo zone with copper occurrences was discovered in the left tributaries of the River Lena (Narkelyun et al. 1983).

4.7   Conclusions

1. Near the Udokan copper deposit, there are numerous satellite occurrences of sediment-hosted stratiform copper mineralization. They are located at different stratigraphic levels of the Udokan Supergroup. There are three main ore-bearing horizons: Chitkanda-Alexandrov, lower Sakukan, and upper Sakukan. Elsewhere, other prominent examples of this deposit type are the Dzhezkazgan copper deposits in Kazakhstan, Zambian-DRC copper belts, and Kupferschiefer in Poland and Germany.

2. Despite the commonality, each deposit has its own specific features, expressed mainly in variable composition of ore and distribution of ore associations in space. They are similar to the Dzhezkazgan deposits, but they have many differences in mineral composition.

3. In contrast to the Udokan deposit, chalcopyrite dominates in association with bornite or pyrrhotite and more rarely with pyrite in the satellite deposits. The deposits also differ in rare minerals.

4. The presence of potential by-products is also specific for each deposit. In addition to copper, some of them contain a significant amount of silver (Pravoingamakitskoe, Unkur) and lead-zinc.

5. In the satellite deposits, the hydrothermal (metamorphogenic) processes occurred to a different extent (from Pravoingamakitskoe to Krasnoe).

6. Despite localization in sedimentary rocks, the purely sedimentary origin of the ore is highly doubtful. For instance, it contradicts the isotopic composition of the ore.

References

Abramov BN (2008a) Petrochemistry of the Paleoproterozoic Udokan copper-bearing sedimentary complex. Lithol Miner Resour 43:37–43

Abramov BN (2008b) Specific features of formation of cupriferous sandstones of the Kemen and Chiney Groups in the Kodar-Udokan zone. Dokl Earth Sci 419:197–199 (in Russian)

Abramov BN (2011) Formation conditions and ore potential of black shales of the Udokan Group, East Siberia. Lithol Miner Resour 46(4):353–362

Aksenova SA, Rudakov VE, Sumatokhin VA (1969) Distribution of sulfides in copper-bearing sequences of the Burpala deposit. In: Sizikov AI (ed) Problems of geology of Transbaikalia. Chita, Transbaikalian Geographic Society 6:9–12 (in Russian)

Bodnar RJ, Vityk MO (1994) Interpretation of microthermometric data for H₂O–NaCl fluid inclusions. In: de Vivo B, Frezzotti ML (eds) Fluid inclusions in minerals: methods and applications. Siena, Pontignano, pp 117–130

Bogdanov YV, Kochin GG, Kutyrev EI, Feoktistov VP (1966a) Cupriferous sediments of the Olekma–Vitim highlands. Nedra, Leningrad, p 386 (in Russian)

Bogdanov YV, Kochin GG, Kutyrev EI, Travin LV, Feoktistov VP (1966b) Geology, formation conditions, and distribution of cupriferous sandstones in northeastern Olekma-Vitim mountain province. Int Geol Rev 8(11):1305–1315

Bogdanov YV, Buryanova EZ, Kutyrev EI (1973) Stratiform copper deposits of the USSR. Nedra, Moscow, p 312 (in Russian)

Borisenko AS (1977) Study of salt composition of gas-liquid inclusions in minerals by method of cryometry. Sov Geol Geophys 8:16–27 (in Russian)

Brown P (1989) FLINCOR: a computer program for the reduction and investigation of fluid inclusion data. Am Mineral 74:1390–1393

Chechetkin VS, Kharitonov YuF (2002) Geology and mineral deposits of the Chita segment of BAM. Chita, ZabNii, p 63 (in Russian)

Chechetkin VS, Kharitonov YuF (2009) Mineral resources of the Transbaikalia. Chita, ZabNII, p 23 (in Russian)

Chechetkin VS, Trubachev AI (2013) Mineral resources of Transbaikalia. Chita, ZabGU, p 231 (in Russian)

Chechetkin VS, Asoskov VM, Voronova LI, Chaban NN (1997) Mineral resources of Chita region. Chita, Chitageologia, p 124 (in Russian)

Gongalsky BI, Krivolutskaya NA (2009) Udokan-Chiney ore-magmatic system. Russ Northwest Geol 42:180–184

Gongalsky BI, Izokh AE, Krivenko AP, Tolstykh ND (2004) Giant copper concentrations in the deposits of the Kodar-Udokan area (Northern Transbaikalia). In: Rundkvist DV (ed) Large and extra-large deposits: formation and distribution. IGEM, Moscow, pp 206–218 (in Russian)

Gongalsky BI, Safonov YG, Krivolutskaya NA, Prokof'ev VY, Yushin AA (2007) A new type of copper–noble metal mineralization in Northern Transbaikalia. Dokl Earth Sci 414(5):645–648

Gongalsky BI, Krinov DV, Magazina LO, Krivolutskaya NA, Kovalchuk EN (2017). Mineralogy and geochemistry of the Cu-sandstone deposits from the Kodar-Udokan area, South Siberia, Russia. Ore deposits of Asia: China and beyond. Inc. SEG 2017 Conference P010

Krendelev FP, Bakun NN, Volodin RN (1983) Udokan copper sandstone. Nauka, Moscow, p 248 (in Russian)

Narkelyun LF (1962) Geology and ore of the Dzhezkazgan deposit. IGEM, Moscow, p 128 (in Russian)

Narkelyun LF, Trubachev AI (1978) Ore-bearing sedimentary formations of the Siberian platform. In: Narkelyun LF, Krasinets SS (eds) Sedimentary formations of the Siberian platform and Transbaikalia. Irkutsk, pp 5–67 (in Russian)

Narkelyun LF, Bezrodnykh YuP, Trubachev AI, Yurgenson GA (1968) Geology and genesis of the Udokan copper sandstone deposit. In: Geology of some deposits in the Transbaikalian region. Chita, ZabNII, pp 70–90 (in Russian)

Narkelyun LF, Bezrodnykh YP, Trubachev AI, Salikhov VS (1977) Copper sandstone and shale in the southern Siberian Platform. Nedra, Moscow, p 223 (in Russian)

Narkelyun LF, Salikhov VS, Trubachev AI (1983) Cu-sandstone and shale of the world. Nedra, Moscow, p 414 (in Russian)

Satpaev KI (1977) Dzhezkazgan copper region. Alma-Ata, Izd AN KazSSR, 112 (in Russian)

Satpaeva MK (1985) Dzhezkazgan ores and conditions of their formation. Nauka, Almaty, p 206 (in Russian)

Satpaeva MK (1995) Mineralogy and zoning of Dzhezkazgan lodes based on material from Annensky ore district. Bylym, Almaty, p 122 (in Russian)

Satpaeva MK (2007) Mercury-arsenic-silver mineralization on the lower horizons of Zhezkazgan. Izv National Academy of Sciences of the Republic of Kazakhstan, Ser Geol 5:17–31 (in Russian)

Satpaeva MK (2008) On mantle plumes. Izv NAS of the RK, Ser Geol 1:15–24 (in Russian)

Sekisov AG, Chechetkin VS, Trubachev AI (2014) New geotechnology for mineral raw materials development (non-ferrous and precious metals) of Transbaikalia. Vestnik Zab GU 110(7):28–38 (in Russian)

Smirnov VI, Popov VM, Domarev VS (1971) Stratiform deposits of metals. Ulan-Ude, Izd Zab Geogr Ob-va, 127 (in Russian)

Tombasov IA, Sinitsa SM (1990) Stratigraphy of the Udokan complex in the Ikabiya-Chitkanda region. In: Rozhnov SV (ed) Lower Precambrian stratigraphy of the Far East. Vladivostok, Dal'nevost Otd Akad Nauk SSSR, pp 56–61 (in Russian)

Trubachev AI, Chechetkin VS, Sekisov AG, Salikhov VS, Lavrov AY, Manzirev DV (2014) Stratiform deposits of BAM zone and problems of their exploration. Vestnik Zab GU 12:33–44 (in Russian)

Yurgenson GA (2008) Mineral raw materials of Transbaikalia. Chita, Poisk 1(3): 256 (in Russian)

The Chiney Layered Pluton: Structure and Mineral Composition

Abstract

The Chiney pluton is unique in distinct layering and rhythmic structure in Russia. It contains enormous vanadium resources (up to 30 Gt of ore) and PGE-Cu-Ni. The ores related to the Chiney intrusion differ from the ores of other Cu-Ni deposits in their high Cu/Ni ratio (~100). Its age is estimated as 1880–1850 Ma. It consists of the Western, Central, Eastern, and Southeastern blocks. Four main groups of intrusive rocks were recognized within the Chiney pluton, with two types of layering:(1) sharply bound isomodal layers, forming sandwiched leucocratic and melanocratic rocks (usually anorthosite and titanomagnetite gabbro and melagabbro), and (2) density-graded layers with titanomagnetite and/or pyroxene near the base. Maximum thickness of gabbro rocks has been estimated as 2.5 km. Internal structure and distribution of major components have been studied in the vertical sequence of the pluton. Composition of rock-forming minerals varies in separate rhythms but not in sequence.

5.1 General Characteristics

The Chiney pluton is located in the Kalar Ridge, where it merges with the Udokan Ridge (Fig. 5.1). It is the best-layered mafic-ultramafic intrusion in Russia. This mafic-ultramafic intrusion is unique in distinct layering and rhythmic structure. It is comparable in dimensions to the famous Skaergaard pluton and to the upper zone of Bushveld in petrography. Some portions of this text were previously published in the paper by Gongalsky et al. (2016).

The Chiney pluton contains enormous vanadium resources (30 Gt ore). The large Magnitnyi and Etyrko Fe-Ti-V deposits are related to gabbro of the Chiney pluton (http://www.metalbulletin.ru/analytics/color/165/). The gabbro-anorthosite comprises Ni-Cu-PGE mineralization. The sulfide ores display very high Cu/Ni ~100, differing from other magmatic sulfide deposits. Only some deposits have similar Cu/Ni ratio, such as Volkov pluton in the Urals (Murzin et al. 1988; Volchenko et al. 1998; Zaccarinni et al. 2004), Kevitsa intrusion in Finland (Yang et al. 2013), and Okiep in South Africa (Maier et al. 2013).

The Chiney pluton is a constituent of the transregional Yenisei-Aldan metallogenic belt, which was defined by Dodin (2002) on the western and southern periphery of the Siberian craton, including the Taimyr Peninsula in the north. He included unique Ni-Cu-PGE deposits of Noril'sk and Kodar-Udokan mineral districts (Permian-Triassic and Paleoproterozoic, respectively) in this belt. The Chiney magmatic complex comprises Chiney, Luktur, and Mylovskiy plutons, as well as Main Udokan Dike and other smaller intrusive bodies.

Mafic composition of the Chiney pluton is its specific feature. In contrast to the most layered mafic-ultramafic plutons, the Chiney pluton is dominated by gabbro, enriched in titanomagnetite, widespread monomineralic rocks (clinopyroxenite, orthopyroxenite, titanomagnetite, anorthosite), and combination of oxide and sulfide mineralization.

This attracted attention of geologists to the Chiney pluton over many years since its discovery in 1938 (Petrusevich 1946). Many publications contain comprehensive information about this unusual geological phenomenon (Lebedev 1962; Fedotova et al. 1977; Belova 1980; Shabalin and Sharapov 1981; Konnikov 1986; Gongalsky and Krivolutskaya 1993; Tatarinov et al. 1998; Chechetkin and Kharitonov 2000; Tolstykh et al. 2008; Gongalsky 2015; Gongalsky et al. 2016). Additional data can be found in unpublished reports of geological teams who conducted geological surveying, prospecting, and exploration under the guidance of Chechetkin, Golev, Kazanov, Devi, Goleva, and other specialists.

© Springer Nature Switzerland AG 2019
B. Gongalsky, N. Krivolutskaya, *World-Class Mineral Deposits of Northeastern Transbaikalia, Siberia, Russia*,
Modern Approaches in Solid Earth Sciences 17, https://doi.org/10.1007/978-3-030-03559-4_5

Fig. 5.1 General view of the Kalar and Udokan ridges

5.2   Structural Position and Age

The Chiney pluton intruded the deformed and metamorphosed Paleoproterozoic carbonate-clastic rocks, overlying unconformably the Archean basement of the Chara-Olekma block of the Aldan Shield (~3.0 Ga; Glebovitsky et al. 2008). The mapped metasedimentary successions represent remnants of a much larger basin (Arkhangelskaya et al. 2004). The carbonate-clastic sequence was distinguished as the Udokan Supergroup, deformed into a series of brachysynclines and anticlines. A combined 12 km thickness of clastic rocks is recorded in the Kodar-Udokan basin.

There are three largest structures in the southern part of the basin (Fig. 2.4): Naminga and Katugin synclines and Chiney anticline (Chechetkin et al. 2000). The 80 × 10 km Naminga syncline has a steeper northern limb and an overturned southern limb. It hosts the cupriferous sandstones of the giant Udokan deposit (see Chap. 3). The strata of the Katugin syncline have been almost completely replaced by Paleoproterozoic granite of the Kodar Complex, whose U-Pb zircon age is 1876–1873 Ma (Larin et al. 2000). The Chiney pluton is located in the center of the Chiney anticline (10–12 km in width). The marginal facies rocks of the pluton were dated as 1880 ± 16 Ma (Ar-Ar age; Polyakov et al. 2006) and 1867 ± 3 Ma (U-Pb zircon age; Popov et al. 2009), whereas the central portion was dated as 1850 ± 90 Ma (Sm-Nd; Gongalsky et al. 2008a, b). SHRIMP-II zircon dating

yielded ages of 1858 ± 17 Ma for the titanomagnetite-bearing gabbro (known as Group 2), but a much younger age of 1811 ± 27 Ma for norite (Group 3; Gongalsky 2012).

The position of the Chiney pluton is controlled by the intersection of the near-latitudinal and northwest-trending fault systems, which can be seen on satellite images and geophysical data. The northwest-trending fault zone bounds the Kodar-Udokan basin and consists of the Katugin-Ingamakit and Chukchudu branches. No due attention was attached earlier to the northwest-trending faults in the area adjoining the Chiney pluton, because they are mostly older in age. Nevertheless, they are clearly expressed at the bottom of this pluton. The Ingamakit and Chiney faults control localization of minor gabbro intrusions of the Neoproterozoic Doros Complex in the east of the basin, as well as Mesozoic and Neogene-Quaternary dikes. The Ingamakit Fault Zone (Fig. 5.2) bounds the Chiney pluton in the north and locally reaches a kilometer in width, being represented by hydrothermally reworked folded and foliated rocks, with abundant chlorite-actinolite, carbonate, and zeolite veinlets. The zone is perfectly expressed in topography and partly coincides with the Lower Ingamakit river valley (Fig. 5.3).

The important role of this fault in the origin of the Chiney intrusion was emphasized by many geologists. For example, Mel'nikova and Belova (1979) considered the Ingamakit Fault to be a conduit for mafic melt and as a natural barrier that limited northward propagation of magma. Konnikov

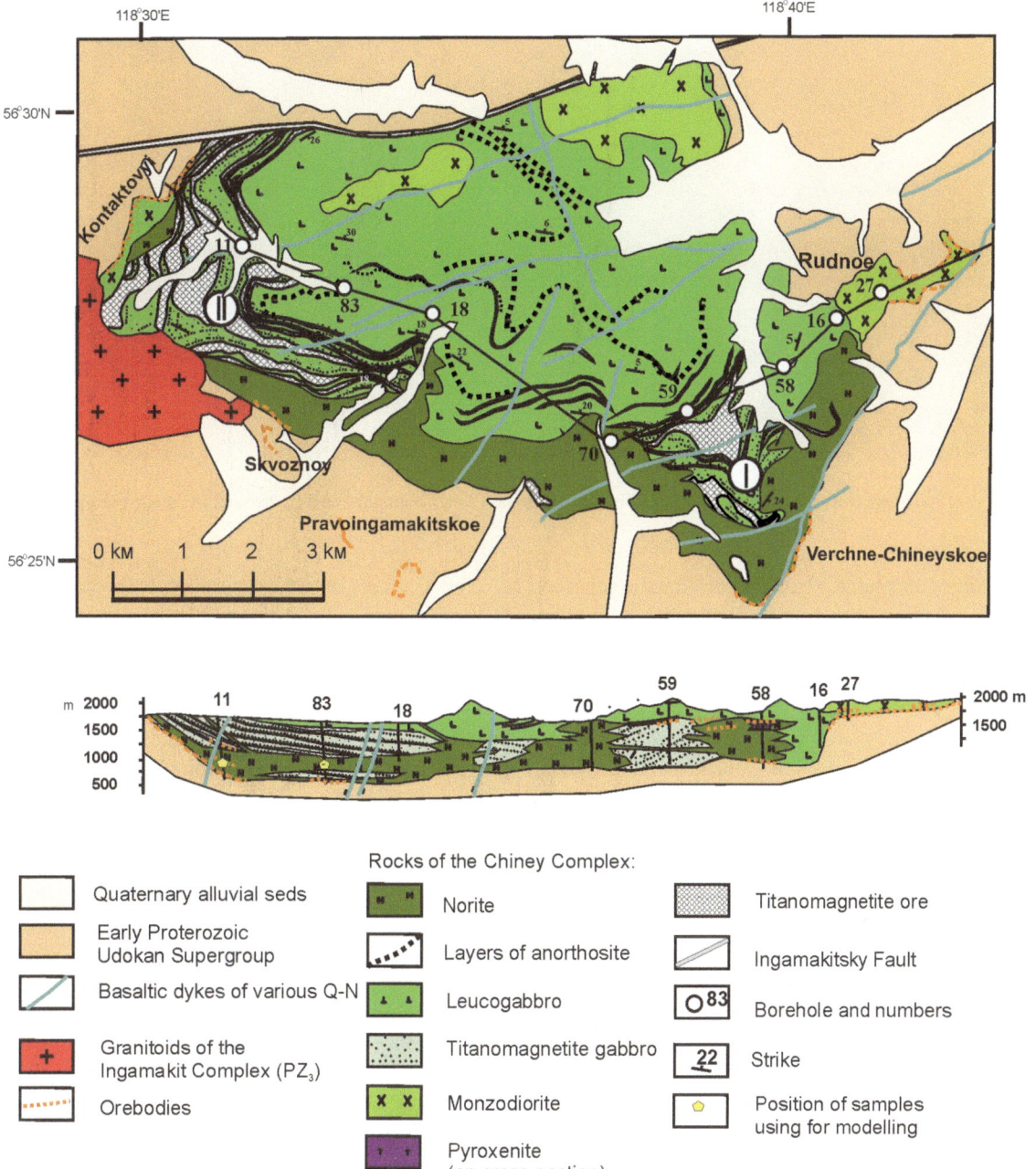

Fig. 5.2 Schematic geological map of the Chiney pluton. Compiled by Gongalsky and Krivolutskaya (1993) using the data of VV Golev and KS Kazanov from the Udokan Expedition

(1986) suggested that the northern part of the pluton has been downfaulted for 3–5 km and is now located beneath the Udokan deposit.

5.3 Morphology

The intrusion forms an asymmetric lopolithic body whose basal contact dips toward the center at 10–25° in the western part, but it is nearly horizontal in its eastern part. The north-

ern contact of the intrusion is faulted (Fig. 5.2). The Chiney pluton is close to the oval in plain view. The pluton extends for 18 km in the near-latitudinal direction and has a maximum width of about 11 km. The exposed portion of this pluton occupies around 120 km² (Fig.5.2). The maximum thickness of gabbro rocks has been estimated as 2.5–3.0 km. This was proved by deep drilling performed in the 1960–1980s and can be seen in its gravity response (Trofimovich and Chechetkin 1969). The western contact of the Chiney pluton with country rocks dips at 25–30° to the northeast.

Fig. 5.3 A view of the Ingamakit Fault (Pravyi Ingamakit Branch) from the Chiney pluton (**a**) and on the satellite image (**b**)

The eastern part of the pluton, i.e., its offset (Rudnoe deposit), is almost horizontal. Nevertheless, its gently dipping bottom is complicated by basins and swells. Its western part is more uniform. The deeps in the bottom coincide with the near-latitudinal Chiney Fault Zone.

In the southwest, the pluton is intruded by late Paleozoic granite and is separated from the Udokan deposit by the Ingamakit Fault in the north.

The present pluton's morphology is determined by the combination of structural elements formed during pre-, syn-, and post-intrusive stages of Phanerozoic reactivation (Mel'nikova and Belova 1979).

5.4 Internal Structure

5.4.1 Block Structure and Rock Groups

As it was mentioned above, the Chiney layered intrusion is primarily composed of mafic rocks, predominantly gabbronorite, norite, and gabbro. Pyroxenite and anorthosite are subordinate. Quartz diorite and monzodiorite are limited in abundance. Our results are based on the study of rocks and ore, outcropping on the surface and extracted from drill core (Fig. 5.4).

The pluton has complex and irregular inner structure. Some geologists (Belova 1980; Mel'nikova 1981; Konnikov 1986) assumed that pluton was formed as a result of single episode of magma emplacement. The intrusive consists of several rocks varieties different in their composition, structure, and texture. These features allow us to recognize the several rock groups related to four episodes of magma emplacement.

The Chiney pluton consists of four distinct blocks—Western, Central, Eastern, and Southeastern (Fig. 5.2). The Western and Southeastern blocks host layered rocks with high concentration of titanomagnetite gabbro. The Eastern block consists largely of gabbro, diorite, and monzodiorite, containing numerous carbonate xenoliths derived from the host rocks of the pluton. This complex internal architecture has made reconstruction of the original structure of the pluton a challenging task. Four main groups of intrusive rocks were recognized within the pluton.

Rocks of the first group have been retained as xenoliths (Fig. 5.5) and remnants that occupy about 5% of pluton's volume mainly in its eastern and northern parts, represented by massive anorthosite and leucogabbro. High-Ti gabbro (Group 2) and low-Ti gabbro and norite (Group 3) occupy ~90 vol % of the intrusion and are characterized by distinct magmatic layering, including thick packages of rhythmic units of variable mineral compositions, and small-scale rhythmic layering. They comprise leucocratic to melanocratic gabbro, gabbronorite, anorthosite, pyroxenite, and gabbro enriched in titanomagnetite. The location of both rock groups coincides in strike and dip; however, numerous crosscutting relationships are also observed. They are emphasized by the arrangement of plagioclase laths in norite at an angle to the trachytoid structure of titanomagnetite gabbronorite. Similar facts were noted by Lebedev (1962). Variations in dip and strike of rocks were also recorded by Mel'nikova and Belova (1979). These authors have described intrusive relationships between gabbroic rocks and interpreted them as repeated local injections. The xenoliths, enriched in titanomagnetite, were mentioned by Shabalin and Sharapov (1981) in barren gabbro.

We documented evidence for the unconformable relationships between high- and low-Ti gabbro, especially in

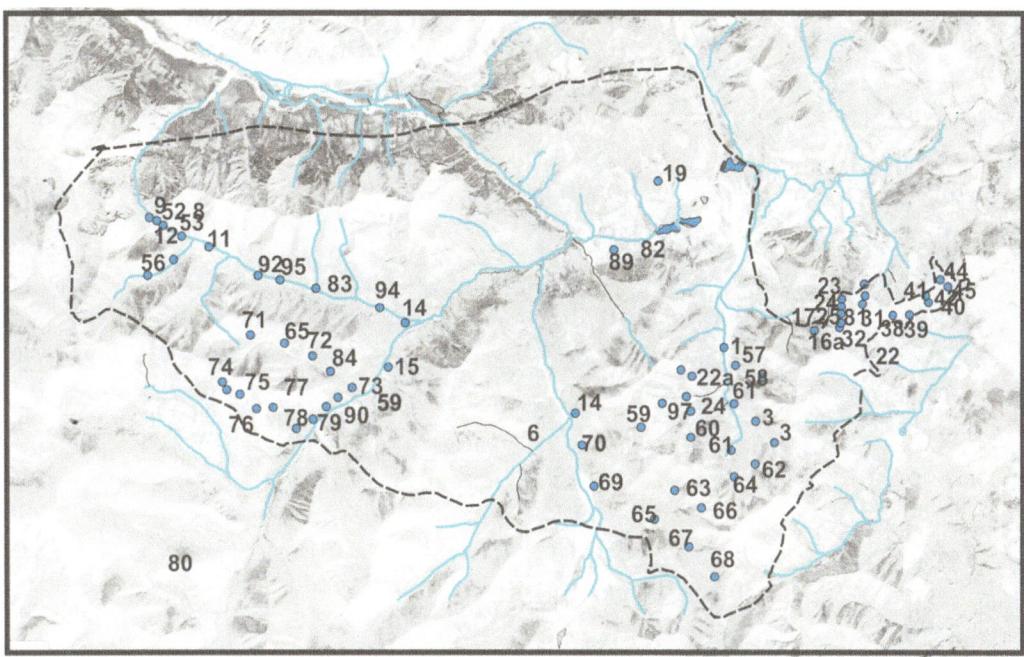

Fig. 5.4 Location of boreholes in the Chiney pluton plotted on satellite image (after Chitageologia) Dashed line outlines the extent of the pluton. 501–522—line of sampling in leucogabbro series

Fig. 5.5 Altered anorthosite xenoliths (rock Group 1) among titanomagnetite gabbro (rock Group 2) (**a**) xenolith in the southwestern part of the Chiney pluton, (**b**) xenolith in the southern part of the pluton

the southern part of the pluton. Large blocks of titano-magnetite gabbro, irregular in shape, are incorporated here into the massive gabbronorite with leucogabbro interlayers. Stratification in blocks is oriented almost perpendicular to the trachytoid structure of country rocks. The contacts between barren and mineralized gabbro rocks are sharp, often rectilinear; however, in some cases, the norite offsets penetrate the titanomagnetite gabbro along the layering (Fig. 5.6). The regenerated lenticular segregations of massive titanomagnetite ore (1.5–2.0 m thick) occur almost everywhere in the contact zone, irrespective of contact morphology. The thickness of segregations varies from a few centimeters to 1–2 m. Taking into account the intrusive relationships between low- and high-titanomagnetite gabbro rocks, abrupt variations in their thickness, irregular position in section, and discordant strike and dip, we concluded that these groups of rocks were formed as a result of multiple injections of different portions of melt, separated in time. The available factual material does not allow us to consider the formation of these rocks as a result of crystallization of the magma chamber as a whole.

Fig. 5.6 Relationships between low-titanomagnetite (pale) and high-titanomagnetite (black) rocks at the Verkchnechineyskoe deposit
(**a**) Norite offsets in titanomagnetites with titanomagnetite and anorthosite xenoliths in norite, (**b**) norite offsets in titanomagnetite
Thin-polished sections, samples 11/652 and 11/647 (sample number = borehole number/depth in m)

The marginal (lower, upper, lateral) facies are primarily composed of fine-grained gabbro and gabbronorite, up to 10–15 m thick in the south of pluton, as well as quartz diorite and monzodiorite of higher alkalinity (the fourth group of rocks up to 100 m thick in the southeast and south of the pluton).

5.4.2   Layering and Rhythmicity

The Chiney pluton is a layered intrusive body (Lebedev 1962; Konnikov 1986), characterized by two types of layering (Gongalsky and Krivolutskaya 1993; Gongalsky 2015) formed by (1) alternating leucocratic and melanocratic rocks (usually anorthosite and titanomagnetite gabbro and melanogabbro, gabbronorite; Fig. 5.7) and (2) gravitational distribution of titanomagnetite (Fig. 5.8) and/or pyroxene near the bottom. The thickness of layers varies from 4–5 cm to 1.5–2 m. Layering is typical for titanomagnetite gabbro, occupying ~60% of the total volume of the pluton, and it is rare in norite.

The Fe_{total}/Fe_2O_3 ratio in rocks marks the boundaries between macrorhythms (Gongalsky et al. 2008a) in rocks enriched in magnetite. The titanomagnetite gabbro series consists of four macrorhythmic units (Group 2). The rhythmic layering in the norite series can be recorded in changes in clinopyroxene and orthopyroxene contents.

The complete sequence of the Western block was studied in drill core of hole 83 (lower part) and outcrops (upper part). Lateral variations of rock rhythmicity in the lower zone of the pluton were studied in drill core from hole 11 (Fig. 5.2).

5.5   Petrography

As it was mentioned above, the Chiney intrusive complex mainly consists of mafic rocks. Ultramafic varieties were distinguished in very rare cases (Lebedev 1962; Konnikov 1986; Gongalsky and Krivolutskaya 1993). Very specific rocks were documented in the southern part of intrusion, so-called "leopard" gabbro (where pyroxenes form spots of 1–8 cm in diameter) (Fig. 5.9). The main rocks from the Chiney pluton are gabbro-diorite, gabbro, titanomagnetite gabbronorite and gabbro, norite, anorthosite, and pyroxenite. They are described below from top to bottom based on the study of cross section of the pluton from hole 11 (penetrated the thickest part of the sequence) and outcrops (Fig. 5.10).

The photomicrographs of *gabbro-diorite* are shown in Fig. 5.10a, b. This is a silicified, feldspathized, and then regressed massive crystalline rock. Dark-colored minerals (25–30 vol %) are represented by relatively small euhedral crystals of amphibolitized or chloritized ortho- and clinopyroxenes that occur in equal amounts (up to 10 vol %), reddish brown biotite (5–7%), and secondary pseudomorphs of greenish brown biotite and chlorite. The light-colored minerals (up to 65 vol %) are represented by inequigranular (up to porphyry-like) aggregates of resorbed tabular crystals of partly albitized zonal plagioclase (An_{50-55}) (50 vol %) with myrmekite quartz ingrowths and dactyl type intergrowths of quartz and K-feldspar in interstices (up to 10 vol %). Secondary calcite, prehnite, zoisite, clinozoisite, hornblende, and actinolite are noted. Ore minerals (up to 7 vol %) are represented by magnetite, titanomagnetite, ilmenite, and hematite. Apatite and titanite are accessory minerals.

The *anorthosite* (Fig. 5.10 h, i, aa) is a leucocratic massive coarse-grained rock with panidiomorphic-granular and sporadic sideronite structure. This rock forms separate layers of varying thickness, 1–2 m on average. Light-colored minerals (up to 80 vol %) are represented by euhedral cumulate crystals (76 vol %) of polysynthetically twinned labrador (An_{66-58}) and interstitial quartz grains (4 vol %). Dark-colored intercumulus grains of clinopyroxene (4–5 vol %) and orthopyroxene (4 vol %) are combined with biotite (2 vol %) and hornblende (2–3 vol %). The latter mineral occurs as reaction rims at the pyroxene-plagioclase contacts. Ore minerals (5 vol %) are represented by single grains and schlieren of titanomagnetite and sporadic ilmenite grains. This rock was recorded at different horizons of the sequence in hole 11 (Fig. 5.10h, i, o, cc, dd).

As was emphasized above, a distinguishing feature of the titanomagnetite gabbro series is the presence of titanomag-

Fig. 5.7 Layering in intrusive rocks of the Chiney pluton
(**a**) Anorthosite layer with titanomagnetite ore, (**b**) anorthosite layers in melagabbro, (**c**) interlayering of anorthosites and melagabbro (Anorthosite Creek), (**d**) rough layering in rocks of leucogabbro series consisting of anorthosite layers in massive gabbronorite, (**e**) anorthosite layers in gabbro, (**f**) anorthosite layers inside massive norite, (**g**) thin layering in titanomagnetite gabbro series, (**h**) thin interlayering of leucogabbro and melagabbro

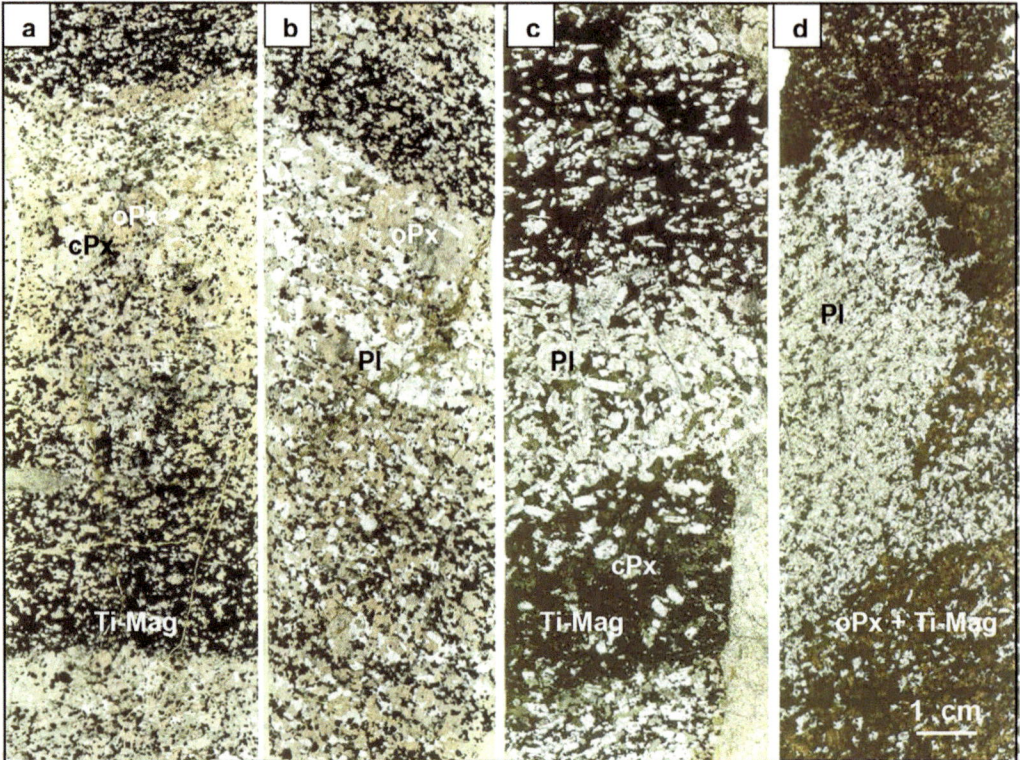

Fig. 5.8 Thin layering in rocks of Group 2 in the Chiney pluton
(**a**) Sample 11/642, (**b**) sample 11/1057 (from the bottom to the top: titanomagnetite norite-norite), (**c**) sample 11/485 of Group 2 (titanomagnetite gabbro-anorthosite), (**d**) sample 11/285 of Group 2 (titanomagnetite gabbronorite—anorthosite)

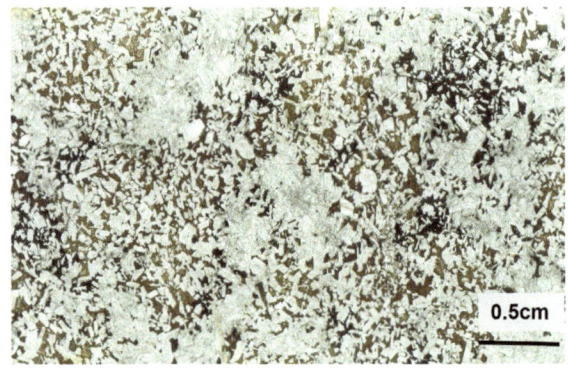

Fig. 5.9 Photomicrograph of "leopard" gabbro (sample ch-12, River Pravyi Ingamakit, polished thin section)

netite as a major mineral. If a rock consists of titanomagne-tite and plagioclase in widely variable proportions, it was called *ore anorthosite* (Konnikov 1986; Lebedev 1962; Shabalin and Sharapov 1981), *magnetite anorthosite*, or *fer-roanorthosite* (Wager and Brown, 1968; Molyneux 1970), but not all these terms have a clear meaning. These rocks, which are widespread in the Chiney pluton, were called *chineyite* after the type locality (Gongalsky 1992, 1993).

According to the recommendation of the Terminological Commission of the Petrographic Committee, USSR Academy of Sciences (1981), chineyite can be subdivided into leuco-, meso-, and melanocratic varieties, with 10–35%, 35–65%, and 65–90% titanomagnetite, respectively, and less than 10% of other minerals. The total amount of dark-colored minerals should be <10%. If the clinopyroxene content exceeds 10%, the rock may be called gabbrochineyite. For the titanomagnetite-clinopyroxenite rock, which is also abundant in the layered units of the pluton, the term kosvite (after Mount Kos'va in the northern Urals) should be kept. Massive titanomagnetite rock passes into kosvite at the base of microrhythms and in the lower parts of macrorhythms (members), whereas chineyite occurs in the upper parts of layered micro- and macrounits. This type of rock is very common in the middle and upper part of the pluton (Fig. 5.10h, i, aa). Position of these rocks in relation to tit-anomagnetite gabbro is shown on diagram (Fig. 5.11).

Plagioclase and titanomagnetite are major minerals of chineyite. Clinopyroxene, quartz, K-feldspar, biotite, apatite, and sulfides may occur as minor minerals. The structure of rocks is equigranular and medium-grained, subophitic, with elements of corona structure. Plagioclase occurs as mainly

Fig. 5.10 Photomicrograph of rocks from the Chiney pluton

Number under photo means number of borehole 11 and depth (m); oPx, orthopyroxene; cPx, clinopyroxene; Pl, plagioclase; Ti-Mag, titanomagnetite; Bi, biotite; Q, quartz. See text for explanations. (**a, b**) – Gabbro-diorite; (**c, j, l, s, t, u, uu, dd**) – gabbronorite; (**d, e, f, g, z**) – titanomagnetite gabbro; (**h, i, aa**) – anorthosite; (**k, x, y, bb, cc, ee**) – norite; (**o, q, r**) – chineyite; (**n, p**) – lamprophyre

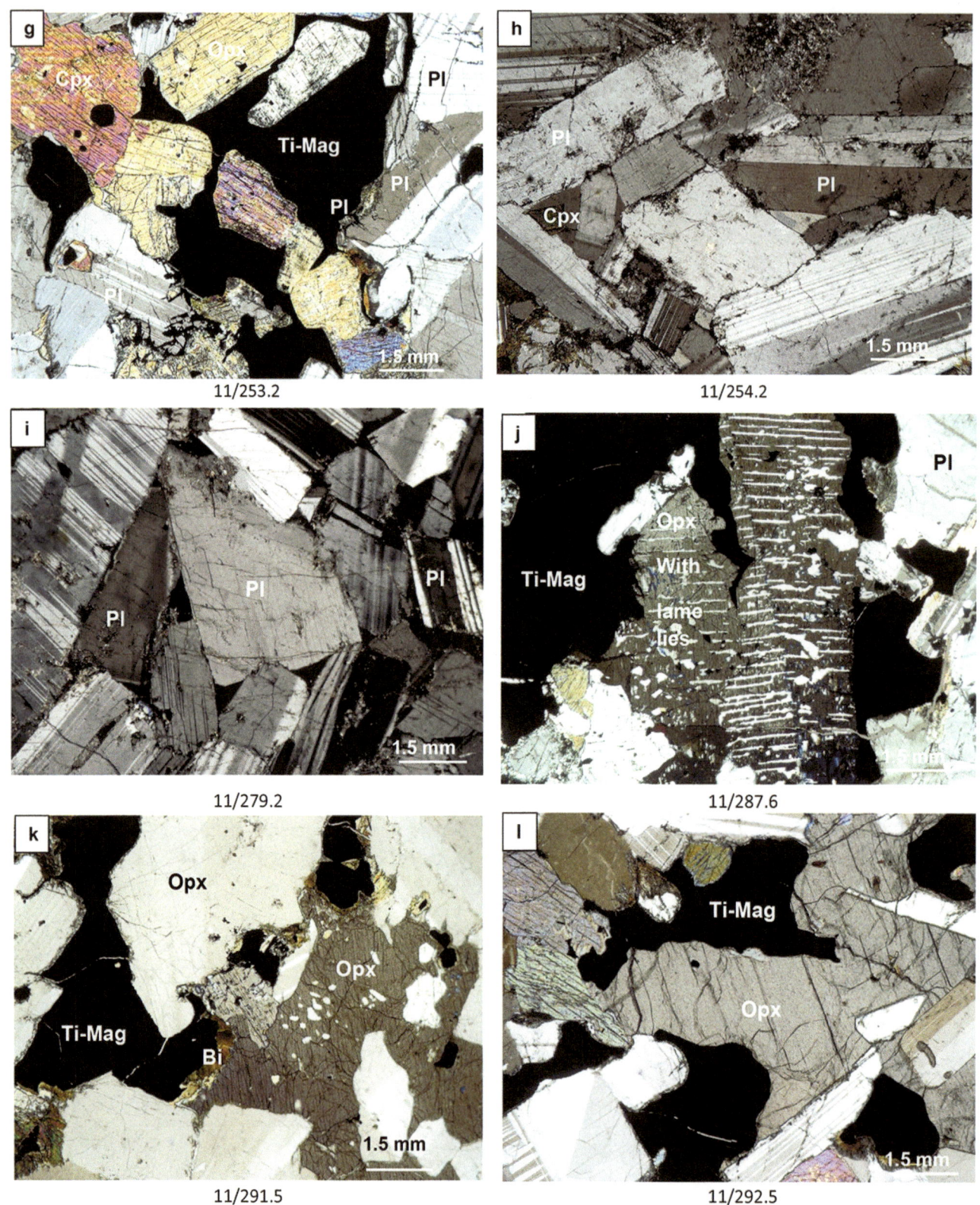

11/253.2

11/254.2

11/279.2

11/287.6

11/291.5

11/292.5

Fig. 5.10   (continued)

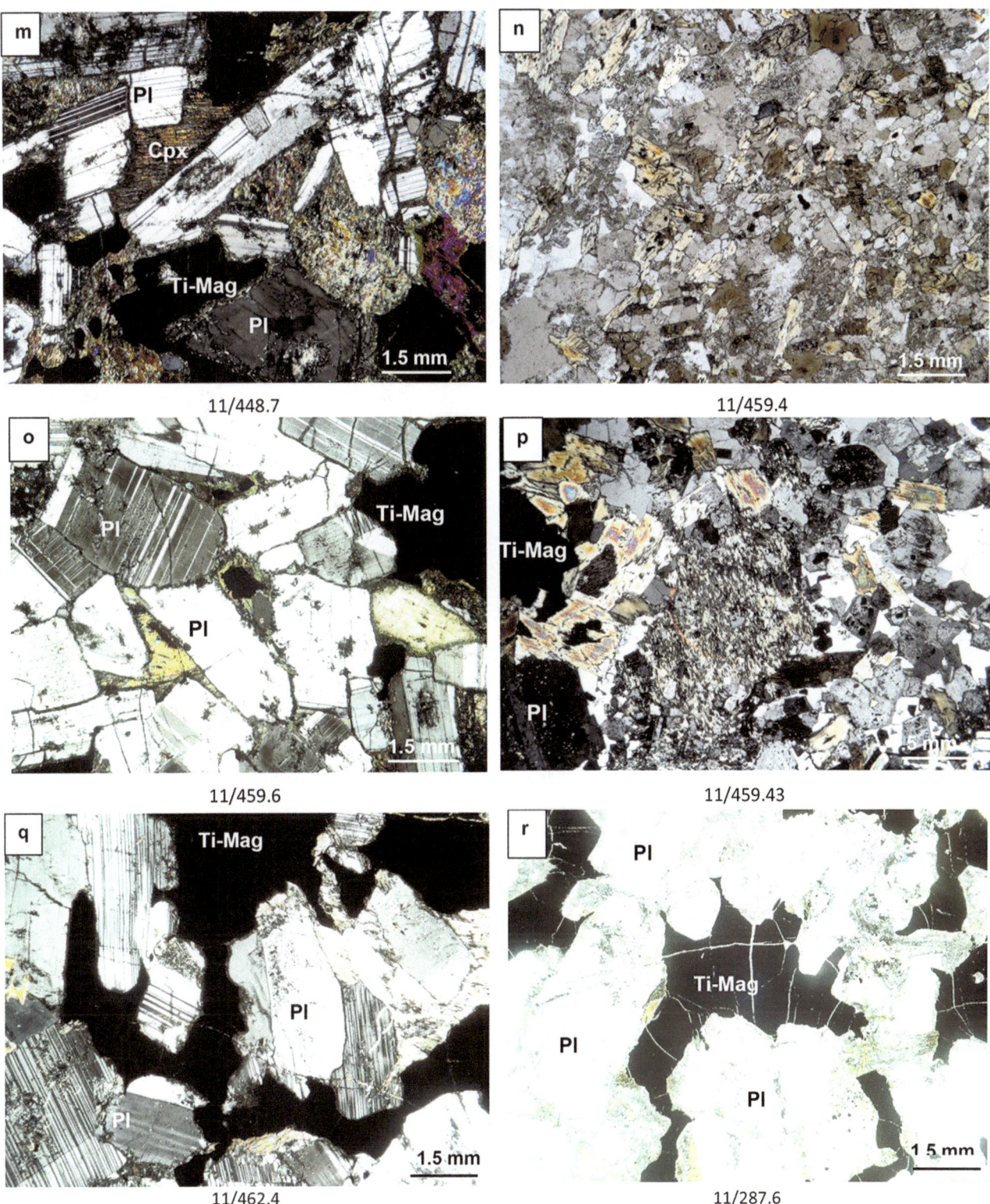

Fig. 5.10 (continued)

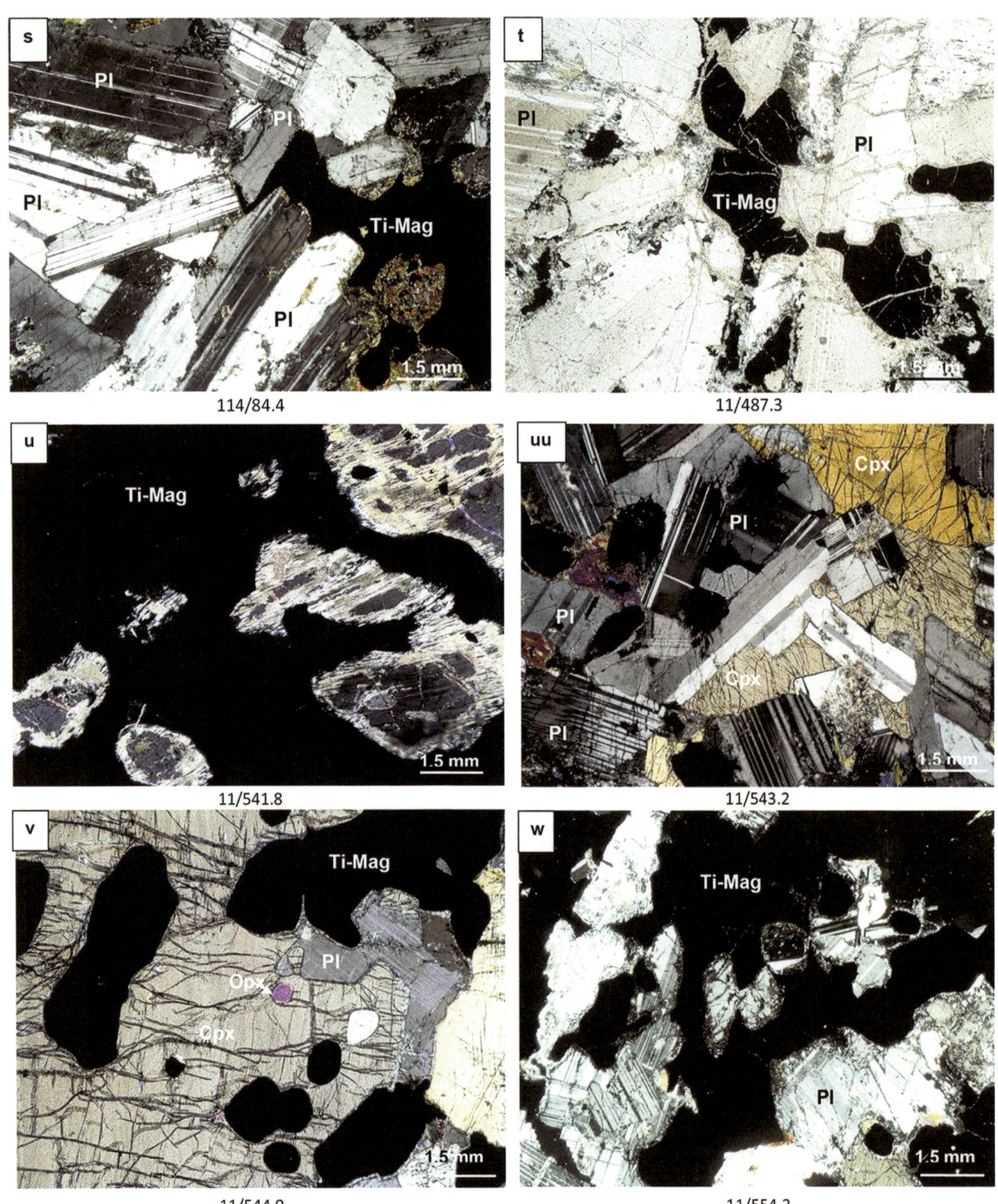

Fig. 5.10 (continued)

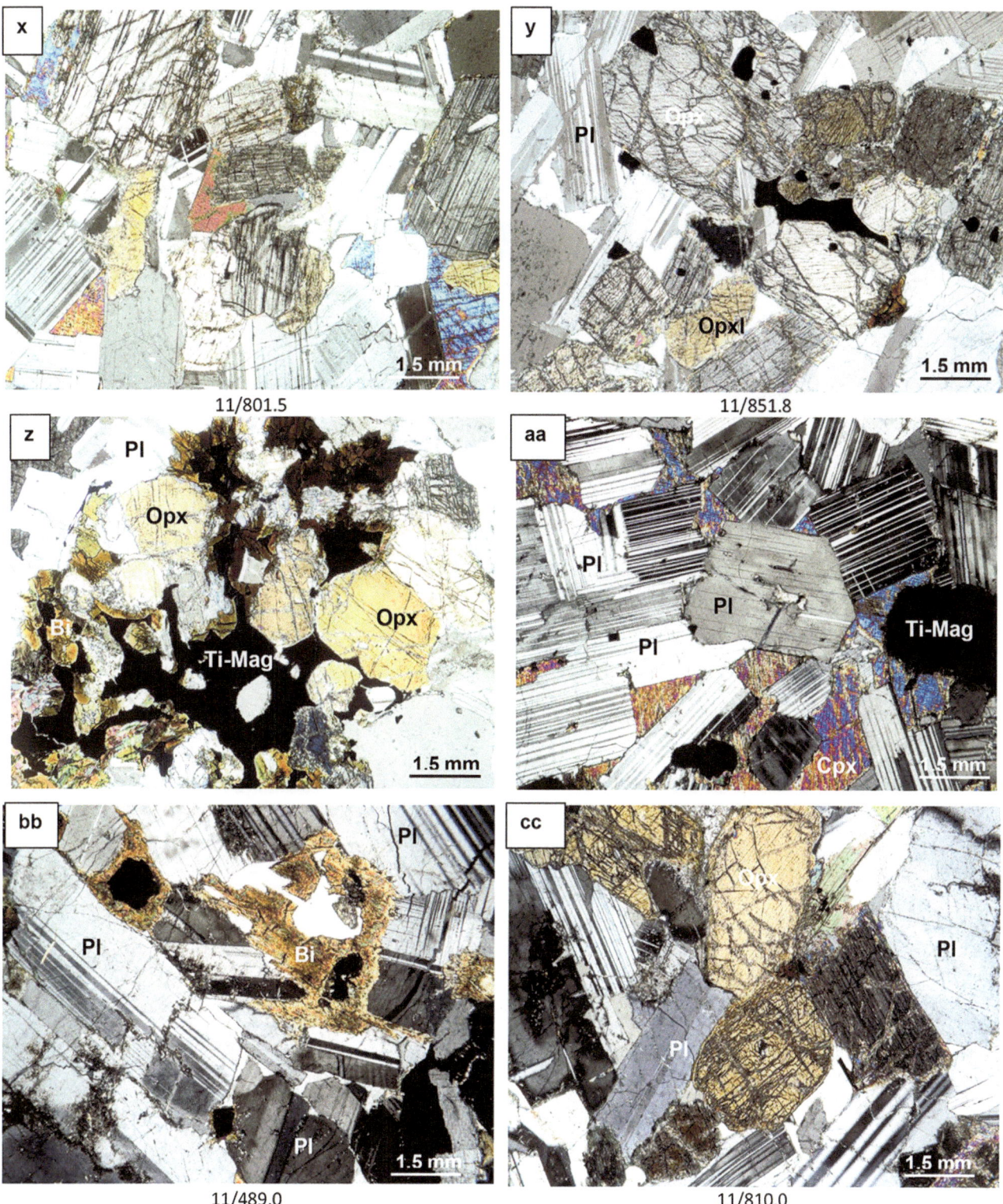

Fig. 5.10 (continued)

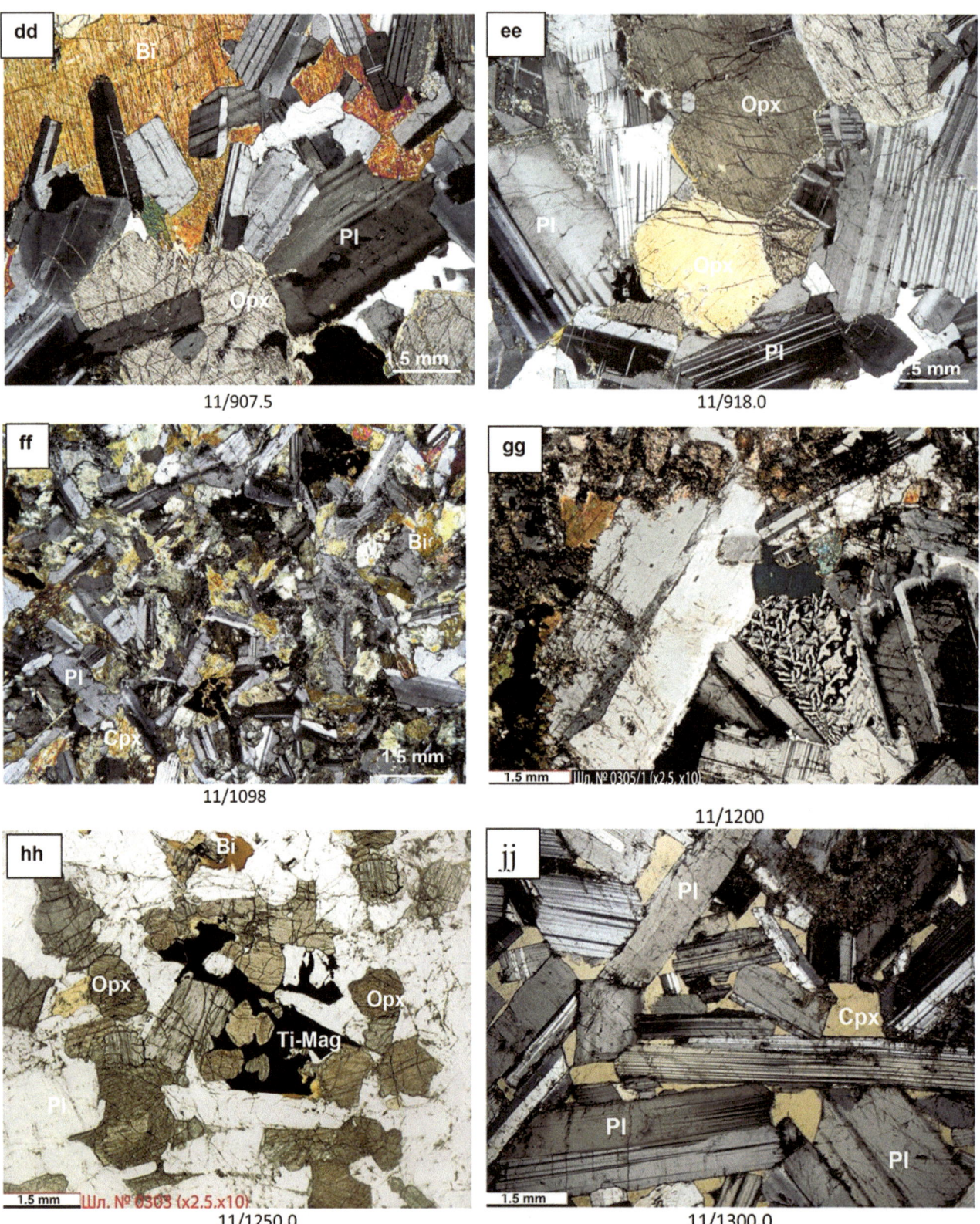

Fig. 5.10 (continued)

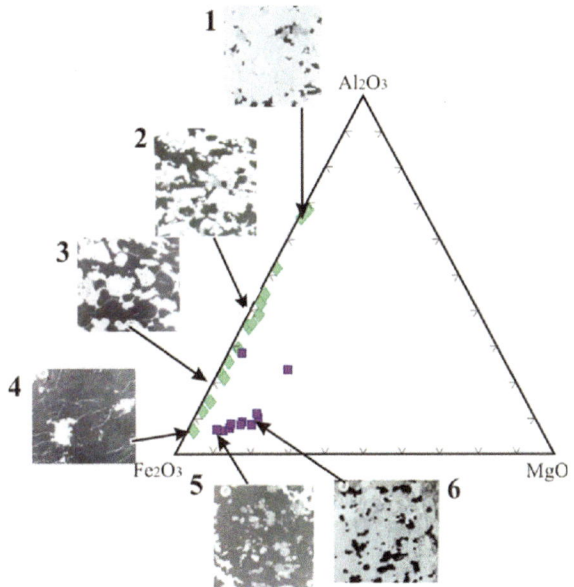

Fig. 5.11 Diagram Al$_2$O$_3$-Fe$_2$O$_3$-MgO for anorthosite-titanomagnetite rocks
Marks (squares): green, chineyite; blue, titanomagnetite gabbro

up to 50% of rock volume. Dark-colored minerals (30 vol %) comprise amphibolized clino- and orthopyroxene (15 and 10 vol %, respectively) and biotite as rims around elongated titanomagnetite and ilmenite grains. The rims of Fe-rich hornblende surround grains of actinolitized pyroxenes at their contacts with labradorite. Accessory apatite is noted.

The biotite-bearing norite (Fig. 5.10k, x, y, bb, cc, ee) and gabbronorite (Fig. 5.10c, j, l, s, t, u, uu, dd) are massive equigranular melanocratic rocks with gabbro structure. Dark-colored minerals (40 vol %) are represented by euhedral orthopyroxene crystals (25 vol %), sporadic clinopyroxene grains (5 vol %), and reddish brown biotite flakes (up to 5–6 vol %). Light minerals (60 vol %) are represented by tabular crystals of sericitized zonal bytownite-labrador (An$_{62-70}$) and interstitial quartz (<5 vol %). Zoisite and clinozoisite are secondary minerals. Chlorite occurs as thin rims around resorbed pyroxene grains. Ore minerals (up to 5 vol %) are represented by titanomagnetite, ilmenite, and small flakes of secondary hematite. Apatite and baddeleyite are accessory minerals.

5.6 Main Rock Groups

5.6.1 Group 1: Coarse-Grained Anorthosite and Monzodiorite

Rocks of the first group occur mostly in the eastern part of the pluton and are observed as large sheet-like blocks and xenoliths variable in size (meters to dozens of meters), incorporated into low-Ti and high-Ti gabbro.

In the eastern apophysis (Fig. 5.1), these rocks are represented by quartz-biotite and hornblende gabbro, gabbro-diorite, and monzodiorite (Fig. 5.10b), up to lenticular segregations of potassic granite. All these rocks are characterized by inequigranular porphyry-like structure and non-uniformly spotted texture. The zonal plagioclase in these rocks is occasionally overgrown by K-feldspar rim (monzonitic structure) or corroded by micropegmatitic quartz-K-feldspar aggregate. Brown hornblende and biotite dominate among dark-colored minerals. The minerals, containing volatile components (biotite, amphibole, apatite) and direct determinations of gas phase, show that melts were fluid-rich.

The large xenoliths are composed of amphibolized gabbro and gabbro-diorite, gabbro- and diorite-pegmatites, monzodiorite, albitized and saussuritized anorthosite, and leucogabbro. At the headwaters of the Katugin Creek and at the junction of the eastern apophasis with pluton, a xenolith, up to 50 m across, composed of coarse-

elongated, perfectly faceted tablets, varying from 0.3 to 5 mm in size and twinned according to the albite and less frequent pericline law or merely remaining poorly zoned (Fig. 5.10). Plagioclase corresponds to An$_{60-65}$ in composition and is almost always saussuritized. Titanomagnetite crystallizes in the form of angular grains, as a rule, and xenomorphic in respect to plagioclase tablets or as inclusions in clinopyroxene. Titanomagnetite is always surrounded overall by a thin (0.02–0.1 mm) rim, consisting of biotite flakes, creating elements of a corona structure. The sharply xenomorphic clinopyroxene (augite) grains fill interstices between plagioclase and titanomagnetite and are partly replaced with amphibole and biotite. Xenomorphic quartz grains occur sporadically in interstices as well. Accessory apatite is represented by short (<0.1 mm) and long (up to 2 mm) prismatic crystals with clearly expressed transverse jointing. Sulfide minerals—chalcopyrite, pyrrhotite, and less abundant pyrite—are identified in polished or thin sections.

On the plagioclase (Pl)-titanomagnetite (Ti-Mag)-clinopyroxene (Cpx) triangular diagram, the gabbro clusters along the Pl-Cpx side of the triangle, and the chineyite plots along the Pl-Ti-Mag side and kosvite along the Ti-Mag-Cpx side (Fig. 5.11). The area within the triangle is occupied by rocks of mixed composition.

Figure 5.10d, e, f, g, z show photomicrographs of titanomagnetite gabbronorite. This is a massive medium-grained rock with relict gabbro structure, nearly parallel ataxitic orientation of tabular labrador (An$_{60-55}$) grains, occupying

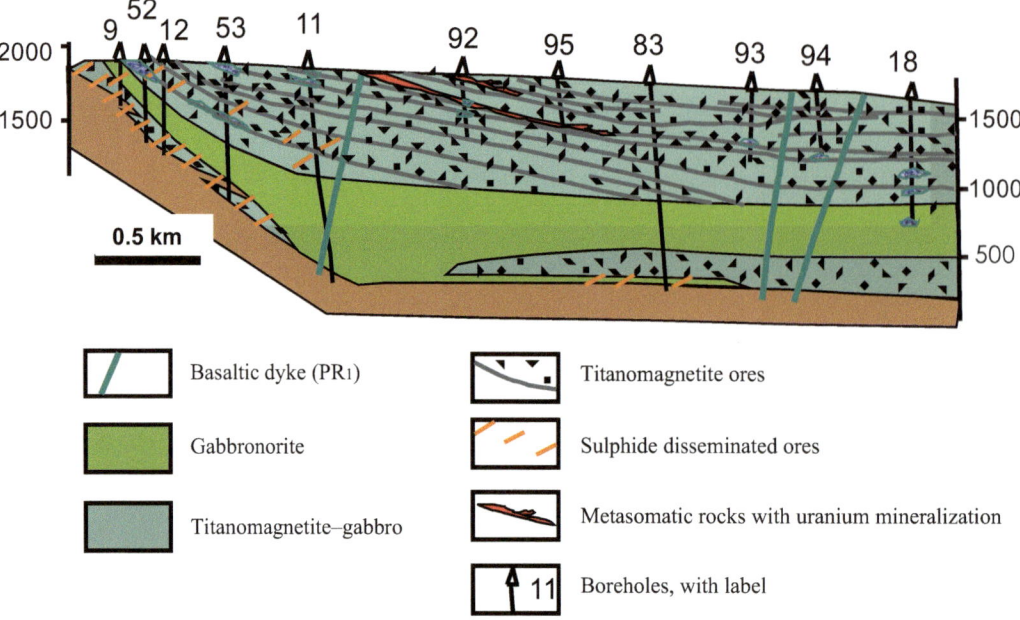

Fig. 5.12 Cross section along the boreholes 9–18 at the Kontaktovyi and Etyrko deposits

grained monzodiorite with a granite pegmatite vein, 0.8 m thick, occurs among fine-grained trachytoid gabbronorites from the group of high-Ti gabbroids (Gongalsky & Krivolutskaya 1993). Pegmatite is traced only in monzodiorite and is abruptly cut off by the xenolith's boundary.

The numerous small (~1 m) rounded (Fig. 5.5) or slightly elongated xenoliths, called "altered anorthosite" by Lebedev (1962), occur in the central part of the pluton (Peak Magnitnyi, Etyrko Creek). They are characterized by a coarse-grained panidiomorphic-granular structure, owing to which they sharply differ from anorthosites of the layered units. The xenoliths consist of tabular plagioclase crystals, up to 3–5 cm in size (70–80 vol %). Interstices between them are filled with amphibole, chlorite, and coarse-crystalline epidote.

The origin of small xenoliths remained unclear for a long time. They were considered to be younger rocks relative to the host gabbroids (anorthosite bubbles of Konnikov 1986) formed from a residual leucocratic melt. Indeed, it is not easy to establish their nature because the contact zone of anorthosite with the host rock is commonly destroyed due to contrasting resistance to weathering (Fig. 5.5b). At the same time, the finds of anorthosite in the polished alluvial blocks do not leave a doubt as to their older age. Figure 5.5a clearly shows fused margins of xenolith and the fine-grained chilled zone, 2–3 cm thick, on the side of the host gabbro. It is also evident that large plagioclase crystals are cut off at the boundary with gabbro.

5.6.2   Group 2: High-Ti Gabbro Rocks

This is the most important rock group due to economic value of Fe-Ti-V mineralization. The high-Ti gabbro rocks occupy ~60% of the total pluton's volume (~85% of its eroded surface). They possess the most distinct layering as compared with low-Ti varieties. Subdivision of the layered rocks into high-Ti and low-Ti varieties seems to be most natural because they can be readily distinguished in the field by content of titanomagnetite. These rocks were earlier grouped into two (Belova 1980; Golev et al. 1987) to six (Konnikov 1986) zones. In general, the rocks lie parallel to the bottom of the intrusive body and are traced from west to east along the strike up to the River China valley for almost 10 km. Thus, layering of gabbro rocks is almost parallel to the pluton's bottom, and this tendency is distorted in its central part (Fig. 5.12).

5.6.2.1   Titanomagnetite Gabbro Series

Due to the style of layering and amount of titanomagnetite, the rocks of this group are subdivided into (i) titanomagnetite gabbro series in the lower position in the sequence and (ii) leucogabbro series in the upper position. The contact between these series is traced along the extended unit of spotty leucogabbro-anorthosite associated with the lenticular and extended bodies of titanomagnetite ore.

The *titanomagnetite gabbro series* is characterized by fine layering caused by fluctuations of contents of major minerals within separate layers (gravity layering, after

Buddington et al. 1936; Wager and Brown 1968), whereas the leucogabbro series is distinguished by rough layering made up of anorthosite and leucogabbro layers, 2–3 m thick, against the background of massive gabbronorite and gabbro. The first type of layering is rarely established in rocks exposed at the surface, but it is commonly observed in drill hole core. This layering has been described by Golev and other authors (Kulikov et al. 1980, 1981; Mel'nikova and Belova 1979; Konnikov 1986; Gongalsky and Krivolutskaya 1987, 1993). The rough layering is typical of the leucogabbro series perfectly exposed in the upper part of the pluton (Fig. 5.7d, e). That is why Lebedev (1962) characterized the Chiney pluton as a "moderately differentiated intrusion with mainly rough type of layering and almost without large-scale banding." The titanomagnetite gabbro series reaches a maximum thickness (~1000 m) in the western part of the pluton. This series consists of various melanocratic (pyroxenite, melanorite), mesocratic (gabbro, gabbronorite), and subsidiary leucocratic gabbro rocks (leucogabbro, anorthosite) with gradual transitions between them. Titanomagnetite occurs in all rocks as a major mineral (10–80 vol %).

Norite and gabbronorite (especially in thin layers) with high-titanomagnetite contents are dominant in the titanomagnetite gabbro series. Composition of plagioclase widely varies from An_{35} to An_{72}. The overwhelming majority corresponds to An_{48-62}. Andesine and oligoclase are commonly associated with interstitial quartz and K-feldspar. The altered plagioclase is saussuritized in core and albitized in outer rims. Orthopyroxene occurs as prismatic crystals, frequently resorbed and rimmed by clinopyroxene. Orthopyroxene and plagioclase occur as inclusions in large clinopyroxene grains, imparting poikilophitic structure to the rock. Brown hornblende and sheet-like biotite inclusions develop along fractures and grain boundaries of clinopyroxene. Biotite with strong pleochroism also occurs as rims around titanomagnetite (Fig. 5.10z; and as isolated irregular segregations up to 0.5 mm in size (Fig. 5.10bb, dd, hh). Quartz fills interstices between major minerals. Apatite occurs as thickened prismatic or elongated columnar crystals. The average size of titanomagnetite is 1–2 mm; its grains make up elongated segregations, consisting of 5–6 items to form discontinuous chains. Plagioclase and some clinopyroxene crystals are oriented in the same direction; however, clinopyroxene grains are mostly equant. Titanomagnetite-pyroxene rocks dominate at the bottom of members. Pyroxene is occasionally replaced with biotite or serpentinite but most frequently with hornblende. Titanomagnetite occurs as small (up to 1 mm) inclusions in ortho- and clinopyroxene or as larger (up to 3 mm) aggregates with formation of sideronite patterns.

The structure of titanomagnetite gabbro series markedly varies in the lateral direction. From west to east, at the interfluve of the Rivers Pravy and Lower Ingamakit, the section becomes more leucocratic, with markedly decreased amount of titanomagnetite. This series either rests on low-Ti gabbro rocks (western and central parts of the pluton) or immediately contacts the sedimentary rocks of the Udokan Supergroup.

The titanomagnetite gabbro series is characterized by the most distinct small-scale layering (Figs. 5.8 and 5.13). In general, fluctuations of titanomagnetite contents in rocks of

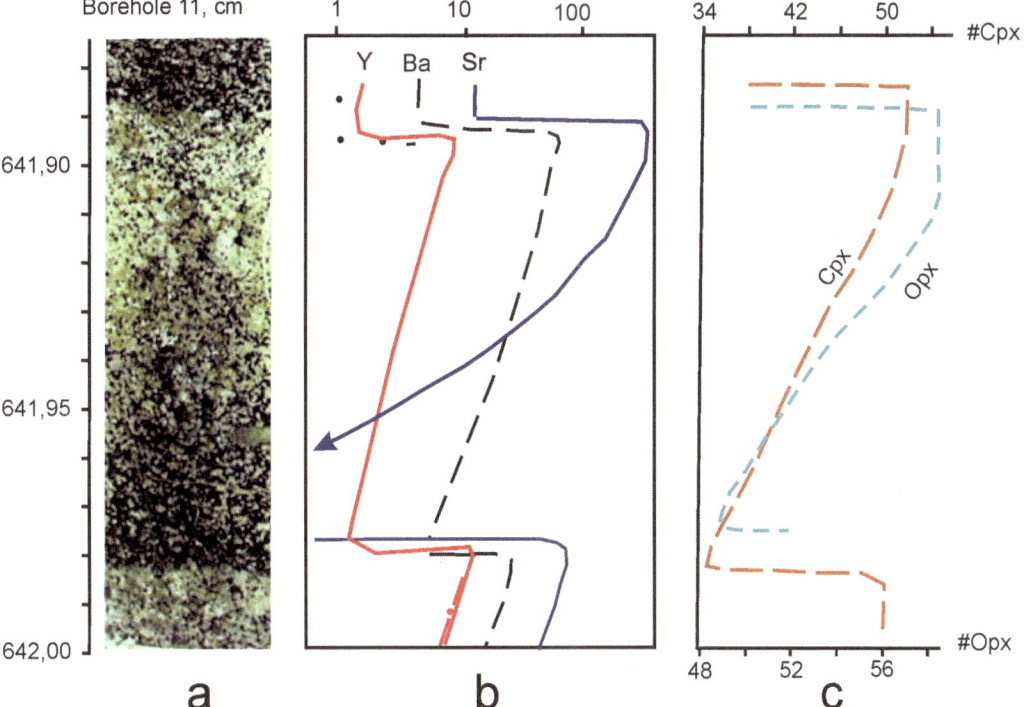

Fig. 5.13 Trace element distribution and variation of pyroxene compositions in microrhythm 11/642
Cpx clinopyroxene, *Opx* orthopyroxene

this series, with the formation of microrhythms in its lower part, mainly occur in pyroxene and plagioclase rocks of the upper part. In the latter case, the ore mineral contents change more contrastingly: the massive and high-grade titanomagnetite ore is associated with barren anorthosite.

The microrhythms vary in thickness from 2–3 cm to 1.5 m; no regular trends of such variation have been established. The boundaries of microrhythms are commonly distinct and sharp. The central part of the series is generally characterized by fine banding and correspondingly by thin rhythms. The layers differentiate and are sharply distinguished from one another by their content of titanomagnetite, which is either distributed more or less uniformly within a rhythm or accumulates near the layer bottom. The gravity accumulation of ore mineral in each microrhythm advances incoherently. The contrast of rhythms in this respect decreases with reducing titanomagnetite content in rocks, when their boundaries become diffuse. In the upper parts of microrhythms, rocks, as a rule, are leucocratic, coarse-grained, and often pegmatoidal. Chalcopyrite and pyrrhotite mineralization are most frequently hosted just in these leucocratic segregations. The characteristics of microrhythms are presented below.

1. *Microrhythm 11/642* (borehole 11, depth 642.0 m), 9 cm in thickness, was an object of thorough study (Figs. 5.8a and 5.13) (Gongalsky and Krivolutskaya 1987). This microrhythm reveals the most distinct gravity stratification and is completely devoid of xenoliths or other complications, such as epidote-chlorite veinlets and secondary alteration of rock-forming minerals represented by titanomagnetite, ortho- and clinopyroxene, and plagioclase.

   The concentration of titanomagnetite gradually decreases from bottom (73%) to roof (18%); the concentration of pyroxenes first increases to the middle of rhythm from 27% to 59% and then decreases to 37% in its upper part. Plagioclase appears as a major mineral only in the middle of the rhythm, and its percentage gradually increases up to 40 vol % in the upper part of the rhythm. Biotite, chalcopyrite, and pyrrhotite are the subordinate minerals in abundance. Their total content does not exceed 5 vol %, and they are mostly contained in the leucocratic rocks of the upper part of the rhythm.

   Morphology of minerals and their composition also change from bottom to top. For example, titanomagnetite occurs at the bottom of the rhythm as large (3–4 mm) cumulate grains, multiangular in shape. Orthopyroxene is characterized here by xenomorphic outlines and cements titanomagnetite crystals. Upwards in the sequence, its grains obtain an idiomorphic habit, and the size increases

up to 1.5 cm, while titanomagnetite is clustered in the form of small (< 1 mm) grains and chains (up to 5 mm long) between pyroxene and plagioclase grains. Single xenomorphic plagioclase grains, up to 1 mm in size, sporadically occur in the lower third of the rhythm.

The brownish-green reaction rims, composed of fine-grained aggregates and reaching up to 0.1 mm in thickness, develop at the titanomagnetite-plagioclase contact. In the central parts of rhythm, plagioclase crystallizes as grains, up to 2 mm in size. The amount of them and degree of idiomorphism gradually increase approaching the roof, where lenticular segregations composed of large tabular crystals occur, and small xenomorphic pyroxene grains are localized between them.

Along with the visible variations in mineralogy and chemistry of rocks within the described microrhythm, the variable composition and properties of rock-forming minerals reflect the so-called cryptic layering. Pyroxene is the most indicative in this respect. As can be seen from Fig. 5.13, the content of the ferrosilite component in ortho- and clinopyroxenes increases toward the upper boundary of microrhythm 1 from 47.6 mol % to 59.0 mol % and from 34.8 mol % to 46.9 mol %, respectively. Orthopyroxenes can be subdivided by size and morphology of exsolution clinopyroxene lamellae into three groups, containing (1) relatively large (0.2–0.3 mm) euhedral augite crystals (Fig. 5.14), (2) wide (0.05 mm) lamellae parallel to (100), and (3) very thin (<0.01 mm) lamellae parallel to (100). Plagioclases in this microrhythm are classified into three discrete groups, containing 28–37 mol %, 47–56 mol %, and 58–63 mol % anorthite end member.

The anorthite content in plagioclase changes from 56–47 mol % at the rhythm bottom, via 52–55 mol % in the middle, to 50–52 mol % in the upper part of rhythm. The sodic plagioclase (An_{28-32}) occurs in interstices between more calcic plagioclase crystals, and An-rich plagioclase (An_{58-63}) has been detected only in one grain with a block structure and 1.2 cm in size. Titanomagnetite is characterized by insignificant upward depletion in MgO and Al_2O_3 along with enrichment in TiO_2 and V_2O_5. The MgO concentration is maximal in the middle part of the rhythm and reflects a higher concentration of pyroxene here. The trace element contents sharply grow toward the rhythm's roof.

2. *Microrhythm 11/620* (borehole 11, depth 620 m). A more complex structure is exemplified in microrhythm 2, which reaches 20 cm in thickness. Gravity accumulation of titanomagnetite and dark-colored rock-forming minerals develops here more feebly; however, the trends revealed

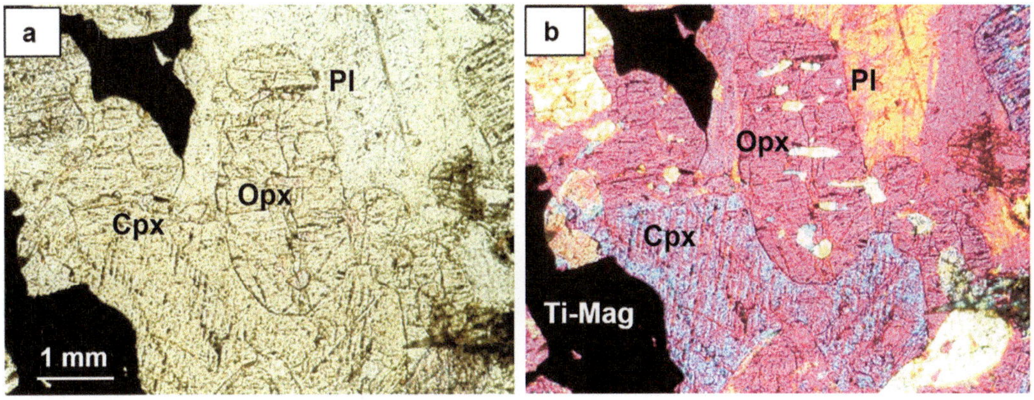

Fig. 5.14 Photomicrographs of gabbronorite. Clinopyroxene lamellae in orthopyroxene are in the center of the photo: (**a**) plain light, (**b**) crossed polars

in microrhythm 1 are confirmed by microrhythm 2. The contents of major and minor elements in rocks, iron mole fraction of pyroxene, composition of plagioclase, and titanomagnetite vary up sequence as in the former microrhythm but less contrastingly.

The trends in cryptic layering in microrhythms 1 and 2 are also similar. The iron mole fraction of ortho- and clinopyroxenes first increases from bottom to top very gradually and more sharply only in the upper third of the rhythm, in general, from 51.7 to 61.8% in orthopyroxene and from 44.9 to 52.2% in clinopyroxene. The latter is also distinguished by upward enrichment in MnO. The most striking cryptic layering in microrhythm 2 is established for plagioclase: the An content diminishes upwards from 66.8 to 59.0 mol %. Leucogabbronorite in the upper part of the rhythm is distinctly separated from underlying melanocratic rocks. A variable composition of minerals is illustrated in Table 5.2.

3. *Microrhythm 11/638* (borehole 11, depth 620 m). The increase in concentrations of certain oxides and decrease in others established for microrhythms 1 and 2 are disturbed in microrhythm 3 (Gongalsky and Krivolutskaya 1993) due to occurrence of a thin kosvite interlayer in the upper third. This interlayer is probably a part of the overlying rhythm. Disappearance of leucocratic rocks at the roof is apparently explained by ascent of the corresponding melt after solidification of the bottom in the overlying rhythm.

All aforementioned rhythms reveal a similar tendency: the data points plotted on the Fe_2O_3-Al_2O_3-MgO diagram are being displaced upward in line with the crystallization of titanomagnetite and then pyroxene.

The change of melanocratic varieties of rocks for leucocratic ones within particular layers also develops at higher levels with the formation of rhythmicity of various ranks:

macrorhythms (members), tens to hundreds of meters in thickness, rhythms (a few meters), and microrhythms (separate layers measured in centimeters and up to a few meters). In microrhythms, the upward decrease in contents of dark-colored mineral is clearly expressed (see above), and then in the macrorhythms (members), this tendency is not seen visually being smoothed over by microlayering but is always expressed in depletion in MgO, TiO_2, and Fe_2O_3 along with enrichment in SiO_2, Al_2O_3, and Na_2O up the section, as will be demonstrated below for reference sections of the Chiney pluton. If rhythmicity of high-Ti gabbro rocks is manifested mainly in sharp variations of titanomagnetite contents, then in the low-Ti rocks variations of clinopyroxene, contents become crucial.

5.6.2.2 Leucogabbro Series

In the leucogabbro series, the mesocratic gabbronorite passing into leucocratic varieties becomes dominant. A great amount of anorthosite and leucocratic varieties as distinct layers and making up rough layering is a typical feature (Fig. 5.7b–f). The thickness of rocks reaches 1500 m. The lower (500 m), middle (800 m), and upper (200 m) members differ in composition and structure.

The lower member is characterized by anorthosite layers in association with massive titanomagnetite ore (Fig. 5.7a). They are distinctly recognized in a sequence of massive mesocratic gabbronorite and gabbro. One of the titanomagnetite layers is exposed at the bank of the River Lower Ingamakit, where the two-member section crops out. The indicated titanomagnetite layer and the second one, fragmentarily exposed below, are bases of two rhythms. Light gray compact leucogabbro and anorthosite lenses occur upsection. The thickness of titanomagnetite layers in the lower and upper rhythms are 1.5 and 3.0 m, respectively. The upper parts of the rhythms are composed of massive gabbronorite.

The titanomagnetite layers are strongly fragmented. Being the most brittle, they undergo intense bedding-plane deformations.

Leucogabbro, spotty gabbro rocks, and anorthosite occasionally make up a lateral series with gradual transitions between these varieties. The contacts with country rocks are commonly sharp. The mean thickness is 1–2 m, locally increasing to 5–6 m. The boundaries of leucocratic layers most frequently are nearly parallel to the layering as a whole; the angular unconformities and crosscutting relationships are less abundant.

The middle part of leucogabbro series is characterized by abundant leucocratic interlayers, frequently occurring closely to one another and hosted in barren gabbro. The thickness of anorthosite layers is commonly less than a meter, locally reaching 2–3 m; their boundaries are rather distinct. In the upper part of the section, the combinations of the rocks, contrasting in composition, are not observed, as this takes place in the lower part. Certain anorthosite layers extend for 1–2 km. However, most of them are traced for tens or hundreds of meters, being en echelon arranged. Insignificant variations in strike and dip as well as the swells of layers are noted.

The leucocratic rocks occur not only as extended separate layers, persistent along the strike, but also as lenses or bodies, irregular in shape; their contacts with country rocks are distinct. Ordinary anorthosite closely associates with spotty varieties. The light-colored matrix is similar to white anorthosite, while the dark spots are closer to gabbronorite. Augite is more idiomorphic in spots and frequently even more idiomorphic than plagioclase. Quartz, which is xenomorphic to major minerals, occurs in both dark spots and light matrix. Leucogabbronorite consists of plagioclase (An_{54-64}) (60–70 vol %) and hypersthene and augite (15–25 vol % in total). Pyroxene crystals are commonly larger than plagioclase tablets. Biotite occurs as wide (up to 0.3 mm) sheets; apatite is identified as elongated grains at boundaries of rock-forming minerals.

In the uppermost part of the leucogabbro series, the amount of quartz increases, and K- feldspar appears, so that rocks pass into the category of quartz diorite of limited abundance. Many authors (Lebedev 1962; Kulikov et al. 1981) refer to the formation of quartz diorite due to the assimilation of country sandstone (Udokan Supergroup) by gabbro rocks with enrichment in alkali metals and silica. This interpretation is, however, unequivocal, because the clastic country rocks themselves are not enriched in alkali and alkali-earth metals (Konnikov et al. 1981). In the opinion of Mel'nikova et al. (1983), the rocks of anomalous alkalinity are products of fractionation of initial mafic melt.

5.6.3   Group 3: Low-Ti Gabbro Rocks

The Group 3 rocks in the Chiney pluton are represented by norite, gabbronorite, and much less abundant melanonorite and leucogabbro. The content of titanomagnetite, as a rule, does not exeedes 5 vol %. The intervals, a few meters thick and enriched in titanomagnetite up to 10–20 vol %, are extremely rare. The rocks of this group mainly occur in the lower part of the pluton under the high-Ti gabbro rocks. They are also observed as thin sills accommodated in gabbro rocks of the second group (boreholes 83, 59, 63, 66, etc.).

A series of repeated macro-layered units (members) and more numerous micro-layered units (rhythms) are traced in the section of low-Ti gabbro rocks. Three rock members—1 M, 2 M, and 3 M—are recognized. Their thickness decreases in a vertical direction, from the first to the third members, and in a lateral direction, from the west to the east. Not only visible but also cryptic layering, expressed in variation of orthopyroxene composition from hypersthene to bronzite, is established in these members. Plagioclase does not reveal such distinct trends because of a narrow range of their compositions (50–58 mol % An).

The lower parts of members are characterized by the most distinct rhythmic structure of the second order. The mean thickness of microrhythms coinciding with separate layers is about 1.5–2.0 m. The boundaries between them are distinct, with abrupt change of leucocratic into melanocratic rocks. Gravity layering is controlled by primary accumulation of pyroxene at the base of rhythm.

Norite is a massive rock, with grain size of minerals varying from 0.5–3.0 to 4–5 mm. The major minerals are plagioclase and orthopyroxene; subsidiary minerals are clinopyroxene, titanomagnetite, hornblende, biotite, quartz, and K-feldspar; apatite, zircon, and monazite are accessory minerals; and secondary minerals are sericite, epidote, zoisite, carbonate, talc, serpentine, actinolite, and chlorite. Plagioclase occurs as perfectly faceted laths, 0.5–4.0 mm in size; the mineral is characterized by polysynthetic twinning and occasionally by vague zoning. Its composition corresponds to labradorite (An_{70-75}). Orthopyroxene is crystallized as prismatic crystals 3–6 mm in size which are idiomorphic relative to plagioclase, although inverse relationships are also noted. Melanocratic rocks in the lower parts of rhythms contain bronzite, whereas leucocratic rocks in the upper parts contain hypersthene. Talc and serpentine develop along fractures.

Clinopyroxene (augite) is represented by small (0.3–0.5 mm) grains, xenomorphic relative to major minerals. In some cases clinopyroxene grows over orthopyroxene. Biotite occurs as separate flakes and replaces hornblende. Symplectic

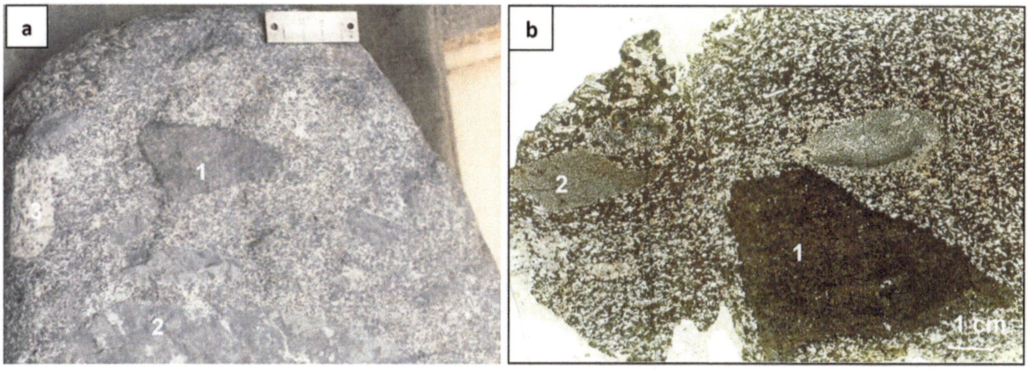

Fig. 5.15 Magmatic breccia from western part of the Chiney pluton: (**a**) sample, (**b**) thin section
Xenoliths: 1, pyroxenite; 2, sandstone; 3, anorthosite; 4, gabbro; 5, skarn; 6, gabbro-diorite; 7, lamprophyre

biotite-quartz intergrowths locally surround pyroxene grains. Quartz and K-feldspar make up micropegmatitic intergrowths in sporadic interstices. Scarce titanomagnetite grains are spider-shaped; biotite, zircon, and apatite are associated with this mineral.

With an increasing amount of clinopyroxene in the rock, a poikilitic structure appears (Fig. 5.10v). In gabbronorite, where clinopyroxene prevails over orthopyroxene, this structure is predominant. Large clinopyroxene grains in gabbronorite differ in composition (diallage) from sporadic equant augite grains in norite.

The low-Ti gabbroids (the second group of rocks) are markedly distinguished from the high-Ti gabbro rocks (the third group), monzodiorites (the first group), and lamprophyres (the fourth group). The so-called leopard gabbro (Fig. 5.9) occurs among low-Ti gabbro rocks in the southern part of the pluton close to the Bazaltovyi Creek. This rock is so named owing to its spotty texture caused by round segregations of dark-colored minerals against the light-colored feldspathic matrix. The size of these segregations varies from 1–2 to 8–10 cm. The gradual increase in spot size is traced from the southern contact inward the pluton for a distance of 2.5–3.0 km. The contacts between leopard gabbro rocks with gabbronorite and norite are rather sharp.

5.6.4   Group 4: Fluid-Magmatic Breccia with Lamprophyre Cement

The rocks of this group are limited in abundance. They are primarily confined to the contact zone between gabbro rocks of Group 2 and 3 with country sedimentary rocks of the Udokan Supergroup. Their thickness varies from a few meters to tens of meters, and its localization is controlled mainly by near-horizontal faults at the bottom of the pluton. Certain dike-shaped breccia bodies are vertical. The abrupt increase in thickness is observed at fault intersections. These breccias are the most widespread in the eastern part of the

pluton (Rudnoe deposit), where they occur as sheet-like bodies, varying in thickness from 50–60 m to few meters and then to complete pinch-out. Breccias with lamprophyre cement have been found in the southern part of the pluton in the form of almost vertical thin (10–15 m) offsets, cutting through the low-Ti and high-Ti gabbro rocks. Some boreholes intercepted up to 20 such offsets. Furthermore, they are traced into country rocks for a distance of 300 m from the intrusive contact. In the west, they are noted at a distance of 1 kilometer from the contact. The boundaries of lamprophyre with country rocks are distinct and sharp. Gabbro rocks in a contact zone with lamprophyre are enriched in biotite. The contacts of lamprophyres with sedimentary rocks are often characterized by a migmatitic appearance.

The magmatic breccia was formed in crush zones, commonly at the contact between gabbro rocks and sandstone, where lamprophyre has been injected (Fig. 5.10n, p). The amount of clastic material varies from a few percent to 60–70% in the central part of breccia bodies. Gabbro rocks, skarn, quartz crystals, and numerous schlieren of dark-colored minerals (pyroxene and hornblende replaced with actinolite, talc, and chlorite) have been documented as fragments (Fig. 5.15). In many cases, lamprophyre contains numerous curved and angular fused fragments of recrystallized hornfels, developing after sedimentary rocks. Cement is composed of inequigranular fine-crystalline rocks, mainly consisting of sodic and intermediate plagioclase (An_{35-60}), biotite (up to 40 vol %), actinolite, and chlorite; quartz and K-feldspar occur occasionally.

5.7   Detailed Sections of the Chiney Layered Complex

We studied two vertical sections across the pluton in detail, where the above-listed rocks, excluding the first group, have been identified: (i) the section in the central part of the pluton (borehole 83 and outcrops upsection) and (ii) the section in

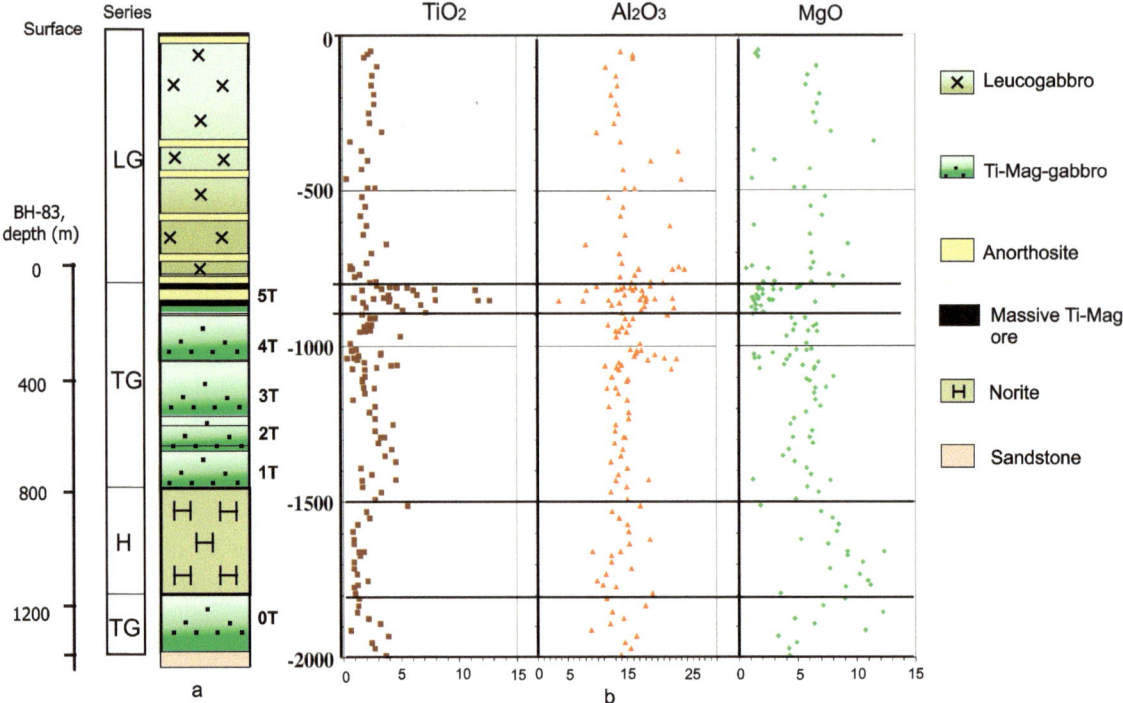

Fig. 5.16 Cross section (**a**) of the Chiney pluton (constructed on borehole 83 and outcrops) and distribution of TiO$_2$, Al$_2$O$_3$, and MgO (**b**)
1 T, number of macrorhythm
Series: TG, titanomagnetite gabbro; LG, leucogabbro; H, norite. Triangular symbols part of borehole 11 shown in Fig. 5.17

the Etyrko river valley (borehole 11). The first one is the most representative for the Chiney intrusion.

5.7.1 Central Part of Pluton (Borehole 83 and Outcrops)

Since the complete magmatic sequence of the pluton cannot be studied in outcrops, it was made up of two parts (Figs. 5.4 and 5.16, section 1). The lower portion of the Chiney pluton was constructed based on drill core 83, whereas the upper portion was constructed based on the results of sampling of surface outcrops in the Western block of the pluton, where samples were collected at distances of 20 and 50 m from one another, respectively (samples No 501–522). Large gaps between samples make it possible to establish only tendencies in distribution of major oxides. For example, the 80 m interval of rocks stands out, consisting only of plagioclase and titanomagnetite with a minor admixture of clinopyroxene.

Borehole 83 penetrated the following succession of rocks (from bottom to top): (1) high-Ti gabbro (Group 2), (2) low-Ti gabbro (Group 3), and (3) high-Ti gabbro (Group 2). The upper part of the intrusion contains many anorthosite layers, i.e., it is represented by rocks of leucogabbro series. The

uppermost 400 m of this leucogabbro series could not be sampled as it was completely eroded in the Western block. The top of the section consists of monzodiorite.

The lower zone (200 m thick) consists of gabbroids and anorthosites with titanomagnetite. Its cumulus minerals are titanomagnetite (Ti-Mag) and plagioclase (Pl). The rhythmic layering of this zone is pronounced weakly. The zone is dominated by coarse-grained gabbro, leucogabbro, and anorthosites, which consist mostly of plagioclase (An$_{50-55}$) and titanomagnetite with minor admixtures of orthopyroxene. Titanomagnetite typically occurs as euhedral grains among large tabular non-zonal plagioclase crystals. This zone is overlain by norites with subordinate amounts of gabbronorites and leucogabbro (~400 m thick norite series). The composition of the cumulus minerals (orthopyroxene± plagioclase) varies within En$_{59-61}$ Fs$_{36-38}$ Wo$_3$ and An$_{50-60}$ (Tables 5.3, 5.4), composition of other minerals are in Tables A3.1–A3.4. The rhythmic layering of the rocks is pronounced clearly enough and is accentuated by variations in the chemical composition of the rocks, because compositionally homogenous rocks systematically alternate in rhythmic units in the vertical section: the pyroxenites grade into norites and then gabbronorites. The thicknesses of the individual units are 1.5–2 m.

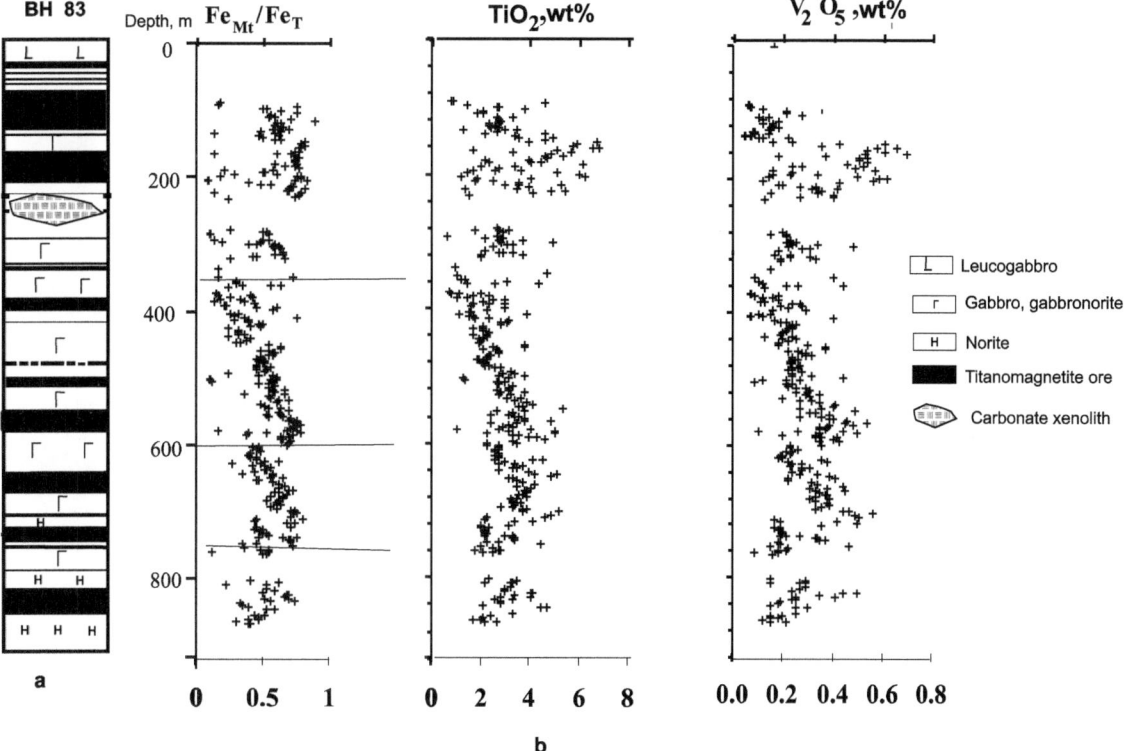

Fig. 5.17 Lithology of the Chiney pluton in borehole 83 (**a**) and distribution of Fe_{Mt}/Fe_t, TiO_2 and V_2O_5 in the sequence (**b**) Fe_{Mt}, iron in magnetite; Fe_t, total iron in rock

The central part of the vertical section consists of the rocks of the titanomagnetite gabbro series (700 m thick), which are characterized by clearly pronounced layering. The cumulus mineral is clinopyroxene En_{37-42} Fs_{15-20} Wo_{38-42} (or clinopyroxene + titanomagnetite), in the lower parts of the rhythmic units, and plagioclase An_{52-57}, in their upper parts. The rocks are dominated by clinopyroxene, although many rock units also contain, along with this mineral, orthopyroxene En_{53-68} Fs_{29-45} Wo_{2-3} and inverted pigeonite. The rocks of the titanomagnetite gabbro series often have trachytoid structures and sometimes contain xenoliths of the first-group pyroxenite.

Farther up the vertical sequence, these rocks give way to the leucogabbro series (800 m), which consists of predominant anorthosites and leucogabbro. These rocks are enriched in titanomagnetite in the lower part (these varieties are referred to as *chineyites*, with beds of massive titanomagnetite). The main cumulus minerals are An_{47-57} and, in rocks rich in titanomagnetite, also titanomagnetite (Ti-Mag ± Pl). The rocks contain subordinate concentrations (no more than 10 vol%) of clinopyroxene En_{37-38} Fs_{19-20} Wo_{40-42}. The apical part of the pluton is made up of monzodiorites and quartz diorites, which often have a granophyric texture. The composition of cumulus in discrete zones in the vertical section varies unsystematically: An_+ Ti-Mag–Opx-Pl – Cpx-Mag–An +Ti-Mag.

Proportions of total and magnetite iron in ore intervals of continuous core sampling in Borehole 83 (data of the Udokan Expedition) corroborate the occurrence of macrorhythms (Fig. 5.17) in the section of the titanomagnetite gabbro series (Gongalsky et al. 2009, 2016).

If the titanomagnetite gabbro series is characterized by steadily high-Ti contents, then the Ti distribution in the leucogabbro series acquires a sawtooth pattern, reflecting alternation of anorthosite and gabbro rocks enriched in titanomagnetite. The selected five members in the titanomagnetite gabbro series in boreholes 83 and 11 gradually pinch-out toward the western end of the pluton and in the eastern direction. The lower (IT-3 T) members have been studied thoroughly.

As was mentioned above, the recognized series consist of a variable number of rock units. For example, the titanomagnetite gabbro series consists of five units, while norite comprises only three. This is clearly manifested in the ratio of "magnetite" iron to the total iron concentration in the rocks (FeO_{Mt}/FeO_t), with the latter determined separately by conventional chemical techniques at the Central Laboratory of Chitageologia, and in the variations of TiO_2 and V_2O_5, which are more typical of the titanomagnetite gabbro series (Fig. 5.17). The same series is characterized by clearly pronounced layering, which could be caused by the gravitational accumulation of titanomagnetite near the bottom of layers.

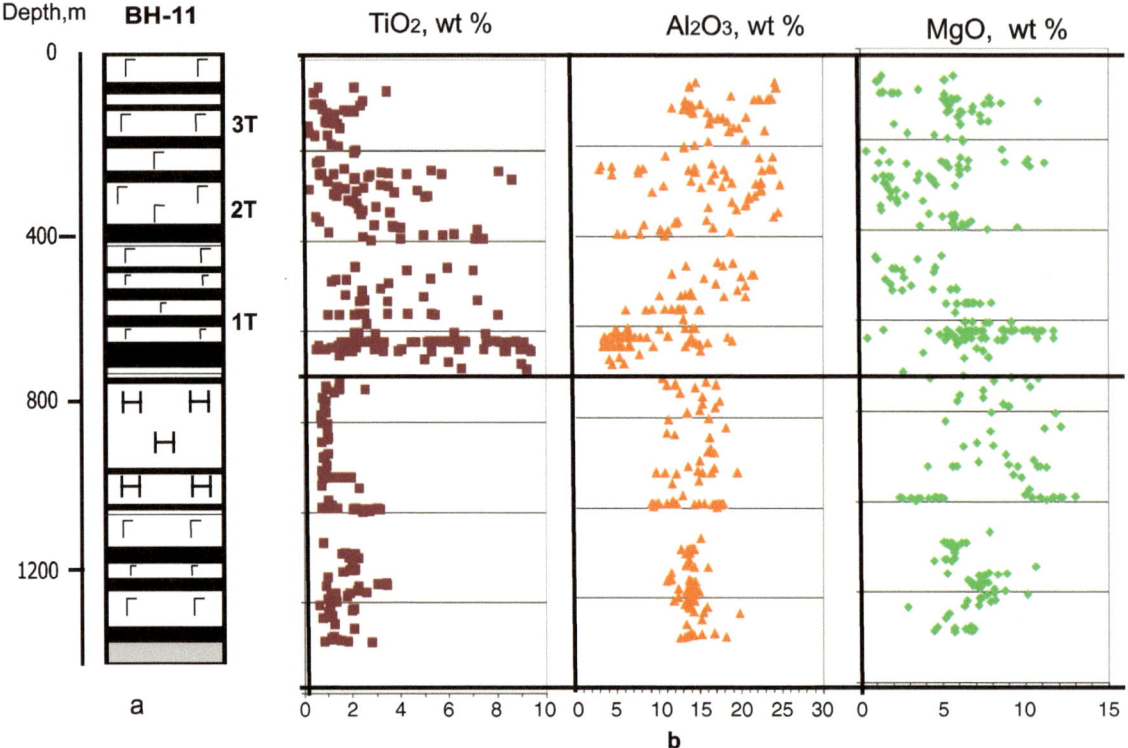

Fig. 5.18 Lithology of the Chiney pluton in borehole 11 (**a**) and distribution of TiO₂, Al₂O₃, and MgO (**b**). Legend is in Fig. 5.17

Boundaries between microrthythms are usually sharp and clear (Fig. 5.8). From the lower to the upper contact, interstitial anhedral plagioclase appears, with the size of its grains gradually increasing in the same direction. These grains compose lenses and thin layers, in which plagioclase acquires a euhedral habit. Rare sulfides are usually restricted to leucocratic segregations in the upper parts of the rhythms, although they were also often found in the lower parts of the rhythms. The data presented above show that the titanomagnetite gabbro series is characterized by clearly pronounced rhythmicity of variable scale: microrthythms, rhythms, and units with a systematic gradation of titanomagnetite gabbro into anorthosites enriched in titanomagnetite (chineyites).

The composition of rock-forming minerals also varies in vertical direction. The results of microprobe analysis of major and minor elements in plagioclase, ortho-, and clinopyroxenes from the rocks intercepted in Borehole 83 are given in Table 5.3 and A3.1–A3.4. In addition to major elements, the elevated Fe concentrations varying from 0.22 to 0.57 wt% in different grains are established in plagioclase. Ti and Na are the leading minor elements in pyroxenes. Their contents in clinopyroxene (0.22–0.76 wt% TiO₂, 0.11–0.78 wt% Na₂O) are generally higher than in orthopyroxene (0.20–0.30 wt% TiO₂, up to 0.11 wt% Na₂O), whereas MnO

concentrations, to the contrary, are higher in orthopyroxene (0.45–0.65 wt%) than in clinopyroxene (0.20–0.39 wt%).

5.7.2   Western Part of Pluton (Borehole 11)

The drill core from borehole 11 is another source of valuable information on the structure of the lower zones of the Chiney pluton (Fig. 5.18, section 2). It helps to understand the variations along the strike.

Borehole 11 penetrated the following rock groups (Gongalsky et al. 2008) from bottom to top (Fig. 5.18): (1) titanomagnetite gabbro (Group 2), (2) low-Ti gabbro rocks (Group 3, three rock members), and (3) titanomagnetite gabbro series (Group 2, three rock members). In general, the structure of this lower part of intrusion is similar to the structure described for borehole 83. There are some changes in thickness of rocks belonging to different series.

The lower zone (230 m thick) consists of rocks of the titanomagnetite gabbro series. Cumulus minerals are titanomagnetite (Ti-Mag), pyroxene (Cpx), and plagioclase (Pl). The rhythmicity of this zone is pronounced weakly, but nevertheless three mesorhytms were diagnosed based on chemical compositions. Their thicknesses are 80, 95, and 55 m (from bottom to top). Rocks of norite series (400 m) compose the central part of cross section, where rhythms were formed mainly by variations of pyroxenes and plagioclase. The lower parts of the microrhythms consist of pyroxenite, while the upper one is formed by gabbronorite or norite. Their thickness changes from 0.5 m up to 2–3 m.

The upper part of the borehole 11 consists of the rocks of the titanomagnetite gabbro series (700 m thick), which are characterized by clearly pronounced layering. In contrast to the analogous part of borehole 83, only three macrorhythms were recognized. They correlate with three lower macrorhythms of borehole 83. The upper two, noted in borehole 83, are absent in borehole 11 due to its lower location in the sequence. Microrhythmicity is very typical of these rocks.

The textural features of the rocks, penetrated by borehole 11, are very similar to the texture of the rocks from the drill core of borehole 83. Titanomagnetite gabbro and gabbronorites of the Group 2 represent middle- to coarse-grained rocks with massive or trachytoid texture. Sometimes rocks have poikilitic texture (Fig. 5.10m, t, v), where clinopyroxene contains oval grains of titanomagnetite and orthopyroxene. These rocks alternate with leucogabbro or anorthosite, consisting of large tabular crystals of plagioclase (Fig. 5.10jj), with twins formed by Carlsbad and albite laws.

Norite occurs in the central part of borehole 11. They consist of idiomorphic crystals of orthopyroxene and plagioclase with small amount of titanomagnetite (Fig. 5.10y, ee, cc). The latter has rims of biotite at the contact with plagioclase (Fig. 5.10bb).

The upper part of this cross section consists of titanomagnetite gabbro (Fig. 5.10d), gabbronorite (Fig. 5.10g), and sometimes norite (Fig. 5.10j, k, l). Orthopyroxene contains (Fig. 5.10j, k) lamellae of clinopyroxenes (up to 40 vol %). The amount of anorthosite growths in the upper part of the sequence in comparison with the lower one (Fig. 5.10h, i). Their characteristic feature is enrichment in titanomagnetite. Thus, these rocks are chineyite (Fig. 5.10q–t).

5.8 Chemistry of Rock-Forming Minerals

The main rock-forming minerals of the Chiney pluton are ortho- and clinopyroxene, plagioclase, and titanomagnetite. The minor minerals include olivine and ilmenite. The rare minerals are apatite, zircon, and titanite.

All studied samples from the main groups of rocks are summarized in Table 5.1. They characterize different groups and series of rocks (see "Geochemical features of rocks" in Chap. 6): Group 2, titanomagnetite gabbro, sample 307

Table 5.1 Main rock types of the Chiney pluton

Group	Rock type	Sample, N	Place of sampling
Group 1	Monzodiorite	321, 39/104, 39/103	Upper and eastern part of massif
Group 2, titanomagnetite gabbro series	Titanomagnetite gabbro	306, 307	Bh 11/280, 11/253
Group 2, leucogabbro series	Titanomagnetite gabbronorite,	501–518	Section upward to top from Bh 83
Group 3, norite series	Norite	Bh 83/843, 83/940, 83/947, 83/936, 83/1036, 83/1101 83/1100; 303, 11/980	West of massif, Bh 83 and 11, east Bh 58/68
Group 3	Pyroxenite	301, 302	Bh 58/24, 58/36
Group 4	Breccia magmatic, cement	Br-cem 1, Br-cem 1	Rudnoe deposit (east of massif)

(11/280, i.e., borehole 11/depth 280 m), 83/100–700, leucogabbro 501–518; Group 3, norite 301,302 (pyroxenite from borehole 58: 301–58/24, 302–58/36), norite 303 (58/68), 83/800–1000; and Group 4 (magmatic breccia-brem 1, monzodiorite 321, hole 39/103–104).

5.8.1 Clinopyroxene

Clinopyroxene was found almost in all types of rocks—gabbro, gabbronorites, pyroxenites, anorthosites—where its quantity varies considerably. However, the mineral composition does not fluctuate very much (Table 5.2). The variations of the major components (Ca and Mg) in the main types of rocks are reflected in the CaO-Mg# diagram (Mg# = MgO/40.3/(MgO/40.3 + FeO/71.8) * 100)). Figure 5.19 shows that the compositions of minerals from different groups (main components) differ insignificantly. Pyroxenes from more magnesian rocks (norite series) are magnesium and calcium rich, while minerals from monzodiorite contain higher Fe and less calcium. The distribution of trace elements in clinopyroxene—titanium, aluminum, manganese, and sodium—indicates their similar behavior in minerals from different types of rocks. Thus, the contents of TiO_2, Al_2O_3, and Na_2O in clinopyroxenes (Fig. 5.20a, b, d)

Table 5.2 Chemical composition of rock-forming minerals (wt%) from the rocks of the Chiney pluton (Borehole 11)

No.	Depth, m	SiO$_2$	TiO$_2$	Al$_2$O$_3$	FeO	MnO	MgO	CaO	Na$_2$O	Total
Clinopyroxene										
1	104.2	50.06	0.52	2.31	11.60	0.30	12.64	20.29	0.22	97.98
2	104.2	50.94	0.49	1.67	11.06	0.31	13.30	21.06	0.20	99.03
3	174.2	50.41	0.76	2.06	14.71	0.37	13.64	17.50	0.27	99.73
4	174.2	50.85	0.60	1.84	11.78	0.32	12.98	20.78	0.11	99.27
5	174.2	50.57	0.61	1.90	11.57	0.32	12.77	20.80	0.19	98.74
6	174.2	50.92	0.52	1.55	11.09	0.28	13.14	21.41	0.19	99.12
7	200	51.69	0.53	2.06	10.57	0.23	13.25	20.29	0.16	98.80
8	200	51.22	0.48	1.97	10.37	0.27	13.63	21.05	0.28	99.31
9	200	51.18	0.49	2.00	10.42	0.28	13.54	21.34	0.21	99.46
10	200	51.25	0.56	2.09	9.86	0.25	13.53	21.87	0.14	99.54
11	240	50.74	0.51	2.49	13.40	0.35	14.19	17.86	0.32	99.88
12	240	50.61	0.54	2.43	10.36	0.27	13.45	21.21	0.20	99.10
13	240	51.07	0.54	1.75	10.33	0.27	13.81	21.37	0.18	99.32
14	320.2	53.78	0.03	1.60	16.95	0.09	13.14	12.73	0.08	98.54
15	320.2	53.68	0.22	2.11	12.19	0.17	14.95	16.31	0.34	100.00
16	320.2	51.25	0.53	2.27	11.04	0.28	13.07	21.71	0.26	100.42
17	320.2	51.79	0.43	1.96	11.20	0.27	13.06	21.74	0.24	100.69
18	320.2	51.39	0.51	2.45	11.07	0.28	12.91	21.74	0.19	100.55
19	380.3	59.16	0.14	2.25	10.45	0.18	15.58	11.06	0.23	99.10
20	380.3	54.06	0.48	2.37	10.06	0.24	13.58	20.59	0.21	101.63
21	380.3	48.61	0.56	2.70	11.06	0.28	12.37	20.73	0.39	97.16
22	380.3	50.25	0.65	3.20	10.73	0.26	12.95	21.62	0.26	99.94
23	400	51.05	0.49	2.26	10.22	0.22	13.77	21.49	0.21	99.73
24	400	55.40	0.51	2.24	10.43	0.24	15.42	22.05	0.20	106.49
25	460	52.06	0.23	3.16	13.13	0.21	15.14	13.08	0.36	97.44
26	460	51.28	0.57	2.52	11.65	0.28	15.06	18.42	0.27	100.08
27	460	51.01	0.57	2.64	10.01	0.26	14.53	20.17	0.30	99.53
28	460	51.75	0.67	2.93	9.72	0.24	14.41	21.29	0.37	101.44
29	460	51.61	0.56	2.66	9.60	0.27	14.05	21.76	0.25	100.78
30	500	51.09	0.25	1.93	12.58	0.29	12.99	20.33	0.34	99.84
31	500	51.13	0.45	2.10	11.76	0.31	12.82	21.28	0.26	100.12
32	500	51.82	0.26	1.80	11.84	0.30	13.06	21.31	0.28	100.70
33	500	50.53	0.54	2.73	10.38	0.26	13.20	21.45	0.41	99.51
34	500	51.29	0.33	1.92	11.55	0.29	12.88	21.52	0.30	100.09
35	760.1	51.37	0.38	2.36	16.63	0.37	15.55	13.95	0.06	100.70
36	760.1	50.83	0.48	2.65	11.87	0.30	13.79	19.05	0.20	99.19
37	760.1	51.02	0.41	2.48	10.26	0.27	13.40	21.57	0.24	99.68
38	760.1	50.66	0.46	2.58	10.26	0.27	12.91	21.70	0.27	99.14
39	843.7	50.89	0.45	2.40	12.36	0.29	13.32	19.70	0.33	99.77
40	912.3	53.27	0.12	2.49	14.05	0.28	15.47	11.82	0.16	97.74
41	927.9	50.63	0.63	3.24	10.12	0.23	13.13	21.26	0.41	99.70
42	940.6	51.02	0.55	3.09	9.86	0.24	12.96	21.74	0.33	99.82
43	1036.3	52.08	0.16	1.67	8.53	0.25	14.24	22.48	0.19	99.64
44	1101.2	49.94	0.63	3.51	9.37	0.21	13.07	21.97	0.32	99.04
45	1101.2	50.44	0.51	3.08	9.43	0.23	13.17	22.01	0.21	99.14
46	1200.4	50.96	0.39	3.61	12.77	0.29	13.71	16.92	0.78	99.46
47	1200.4	51.15	0.32	2.09	11.75	0.33	12.50	21.15	0.37	99.67
48	1200.4	51.04	0.41	2.16	10.86	0.29	12.67	21.59	0.31	99.34
49	1200.4	51.17	0.13	1.65	11.16	0.31	12.90	21.69	0.24	99.36
Ort	1285.7	52.05	0.20	1.83	8.48	0.29	14.43	21.89	0.12	99.28
Orthopyroxene										
51	174.2	50.61	0.33	0.90	24.84	0.57	19.21	1.36	0.07	98.28
52	174.2	51.46	0.37	0.91	25.79	0.59	19.54	1.45	0.00	100.16

(continued)

Table 5.2 (continued)

No.	Depth, m	SiO$_2$	TiO$_2$	Al$_2$O$_3$	FeO	MnO	MgO	CaO	Na$_2$O	Total
53	174.2	51.47	0.30	0.80	25.77	0.56	19.57	1.19	0.00	99.68
54	174.2	51.21	0.36	0.88	25.59	0.57	19.61	1.40	0.10	99.75
55	174.2	51.79	0.29	0.95	25.79	0.56	19.64	1.32	0.00	100.39
56	174.2	51.43	0.35	0.97	25.31	0.58	19.79	1.26	0.03	99.83
57	240	51.38	0.31	1.62	22.85	0.51	21.14	1.88	0.11	99.82
58	240	51.98	0.32	1.59	23.06	0.51	21.38	1.71	0.17	100.75
59	240	51.89	0.31	0.95	23.16	0.52	21.62	1.29	0.03	99.80
60	240	51.82	0.32	1.55	23.22	0.51	21.71	1.18	0.00	100.32
61	400	52.63	0.23	1.92	23.03	0.45	19.39	3.02	0.00	100.70
62	400	52.05	0.29	1.70	24.04	0.49	21.32	1.20	0.00	101.12
63	400	51.69	0.27	1.58	23.62	0.46	21.58	0.99	0.07	100.29
64	400	51.65	0.26	1.63	23.81	0.48	21.59	0.88	0.00	100.35
65	500	51.05	0.27	1.42	28.51	0.64	17.96	1.28	0.00	101.18
66	843.7	51.18	0.37	1.65	25.44	0.53	19.92	1.29	0.04	100.43
67	843.7	51.64	0.33	1.66	24.86	0.50	19.98	1.85	0.00	100.82
68	843.7	51.45	0.31	2.03	24.09	0.44	20.78	1.88	0.00	101.02
69	912.3	51.37	0.26	2.19	23.60	0.48	20.67	2.16	0.06	100.84
70	912.3	52.32	0.20	1.50	23.04	0.49	21.27	1.86	0.00	100.71
71	927.9	52.11	0.31	1.93	22.84	0.47	21.78	1.42	0.00	100.87
72	927.9	51.65	0.22	1.59	22.25	0.45	21.89	1.71	0.19	100.05
73	940.6	52.26	0.28	1.79	24.24	0.46	21.62	0.96	0.00	101.63
74	940.6	52.86	0.15	1.67	22.53	0.42	22.80	1.34	0.02	101.83
75	1036.3	49.97	1.97	2.20	17.04	0.35	21.11	2.16	0.09	94.99
76	1036.3	52.10	0.21	1.58	21.70	0.49	23.24	0.77	0.00	100.10
77	1100	52.17	0.29	2.18	21.15	0.43	22.34	1.82	0.18	100.64
78	1100	51.58	0.26	2.04	20.67	0.41	22.53	2.10	0.00	99.64
79	1100	51.58	0.30	2.14	21.27	0.41	23.27	0.79	0.04	99.81
80	1101.2	51.34	0.24	2.04	22.87	0.45	21.33	1.96	0.00	100.24
81	1101.2	51.64	0.26	2.14	21.59	0.42	21.91	2.24	0.08	100.33
82	1101.2	51.84	0.26	2.21	20.85	0.40	22.65	2.07	0.01	100.33
83	1200.4	50.64	0.28	1.60	27.59	0.65	18.15	1.52	0.00	100.46
84	1200.4	51.17	0.29	1.53	28.37	0.65	18.36	1.02	0.14	101.55
85	1200.4	51.07	0.27	1.68	26.53	0.60	19.21	1.16	0.00	100.55

Note: Here and in Table 5.3, 5.4, and 5.6, analyses were carried out at MPI, Mainz, Germany, analysts DV Kuzmin and NA Krivolutskaya

Fig. 5.19 Diagram
CaO-Mg# for clinopyroxene
of the main rock types
Here and in Figs. 5.20, 5.26, and
5.28, rock types are from
Table 5.1 and from notes to
Table 5.2

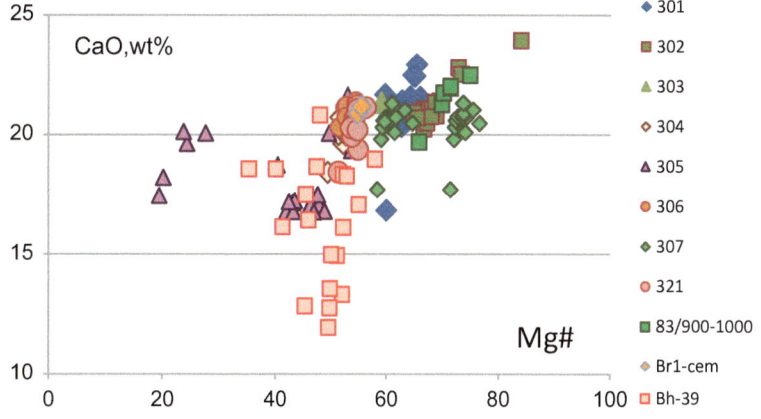

Fig. 5.20 Diagrams
Mg#—TiO$_2$ (**a**), Al$_2$O$_3$ (**b**),
MnO (**c**), and Na$_2$O (**d**) for
clinopyroxene of the main
rock types

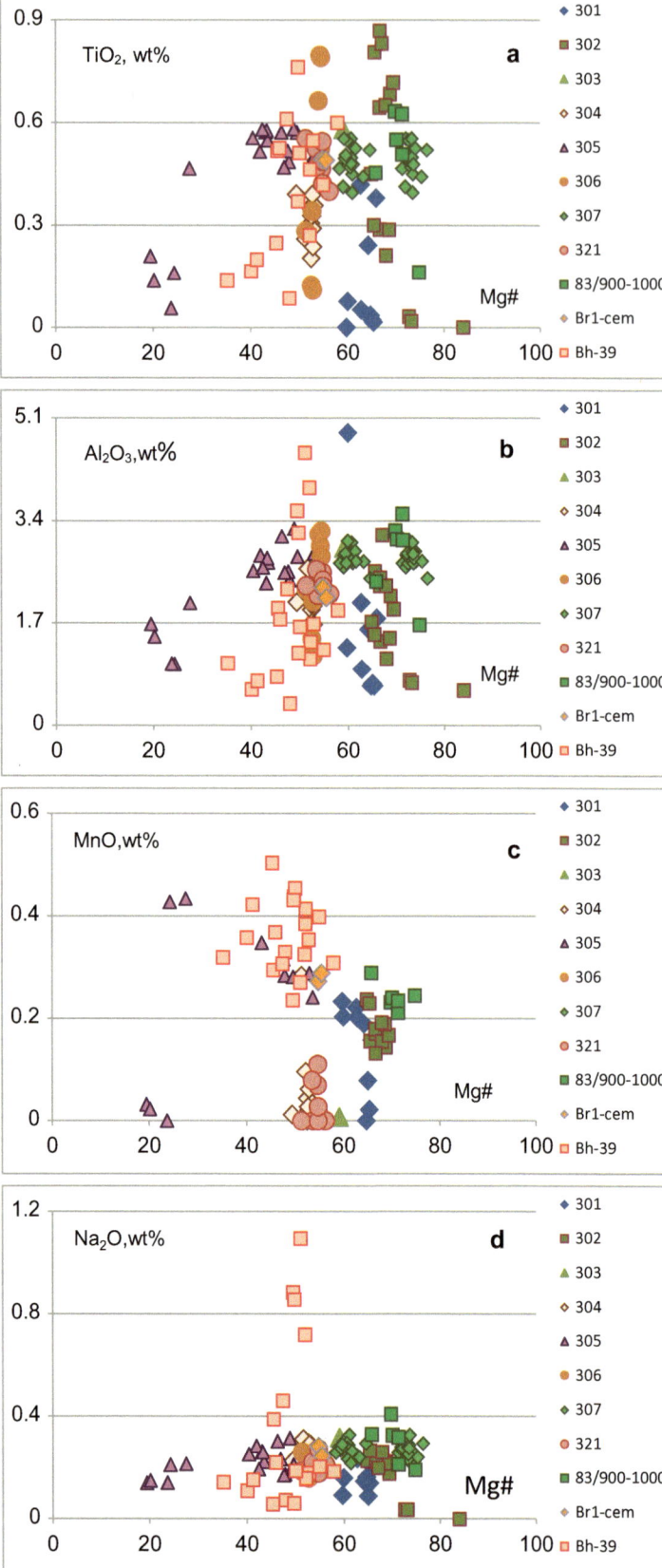

are practically the same in all species of rock (excluding several elevated concentrations of sodium in the pyroxenes from monzodiorites, Fig. 5.20d). Significant differences are recorded only for MnO, whose elevated concentrations are established in the pyroxene of monzodiorites (Fig. 5.20c, samples 305 and hole 39).

Despite the consistency in the mineral composition of samples in sections 1 and 2 (Figs. 5.21, 5.22, and 5.23), the clinopyroxene composition inside individual microrhythms varies considerably from bottom to top (Fig. 5.13), but they are not significant in the sequence (Fig. 5.24). Similar variations were recorded in the rhythms of the titanomagnetite gabbro series (Gongalsky & Krivolutskaya 1993).

5.8.2 Orthopyroxene

Orthopyroxene is less common than clinopyroxene and dominates in the rocks of the Group 3. As a rule, it forms idiomorphic grains, without the zoning in them (as well as in clinopyroxene) (Fig. 5.10). Its composition is close to clinopyroxene, but it differs in the trace element: in orthopyroxene there is no sodium and chromium appears (their contents are lower than the detection limit of EPMA method) (Table 5.3). The composition of mineral changes more significantly than the composition of clinopyroxene. Orthopyroxene contains high U and Th (Fig. 5.25). First of all, orthopyroxene from norite is more magnesian and less calcic in comparison with orthopyroxene from leucogabbro series (Fig. 5.26a). Secondly, it is enriched in TiO_2 and depleted in MgO (Fig. 5.26b, c). Practically, there is no difference in Al concentrations between pyroxenes (Fig. 5.26d). Orthopyroxene from titanomagnetite series of rocks was studied in detail in borehole 11 (Fig. 5.27). CaO and TiO_2 contents do not depend on Mg# of mineral (Fig. 5.27a, b).

But other data show good correlation between the main characteristics of mineral (Mg#) and trace elements (Fig. 5.27c, d). Thus, Al_2O_3 has direct correlation with Mg# (Fig. 5.27d), while MnO demonstrates opposite trendline (Fig. 5.27c). R^2 for sample 11/876 is 0.68 is the best correlation found in measured samples (Fig. 5.28).

5.8.3 Plagioclase

Plagioclase was found in all rock types inside the Chiney pluton. It dominates in anorthosite and leucogabbro, especially in the upper part of the Chiney sequence. Its composition changes significantly in one sample and in cross section (Table 5.4) and does not depend directly on its morphology (Fig. 5.29a–d). The main trace element in plagioclase is iron. But it is not possible to recognize different rock types based on plagioclase composition. The plagioclase grains consist of twins. Twinning in plagioclase follows the albite and Carlsbad laws. The results of microprobe studies of major and some minor elements in plagioclase are summarized in Table 5.4.

5.8.4 Olivine

Olivine is a rare mineral in the rocks of the Chiney pluton. It was found as fresh crystals only in high-Mg rocks, in pyroxenite from borehole 11 (at depth 992–994 m). Olivine forms rounded grains located in clinopyroxene or plagioclase. More often, it can be diagnosed as pseudomorph of serpentine (samples 83/840 and 83/1100; Fig. 5.29e,f). The composition of olivine, measured in sample 11/992 (11 grains), varies from $Fo_{72.9}$ to $Fo_{76.5}$, i.e., range of composition in one sample is around 4 mol %.

Fig. 5.21 Diagrams
Mg#—TiO$_2$ (**a**), Al$_2$O$_3$ (**b**),
MnO (**c**), and Na$_2$O (**d**) for
clinopyroxene of section 1
(borehole 83 + 500)
Legend shows sample number

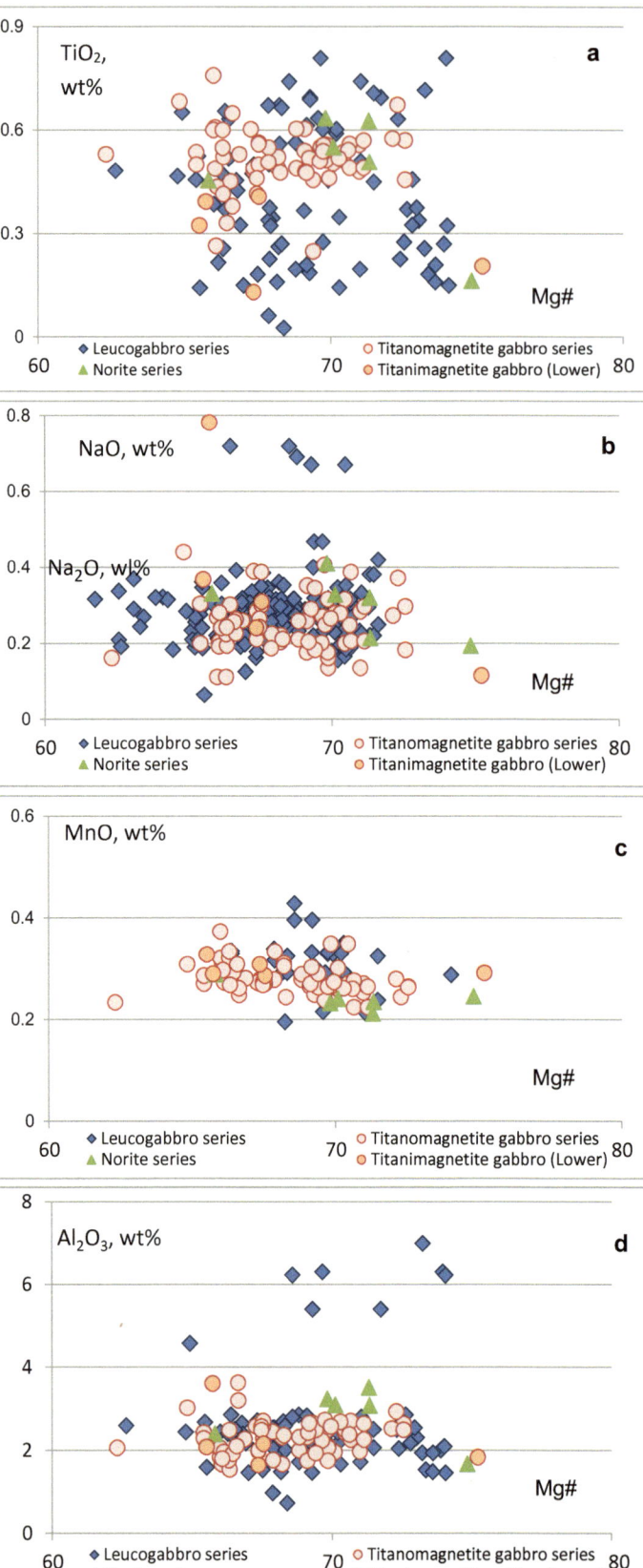

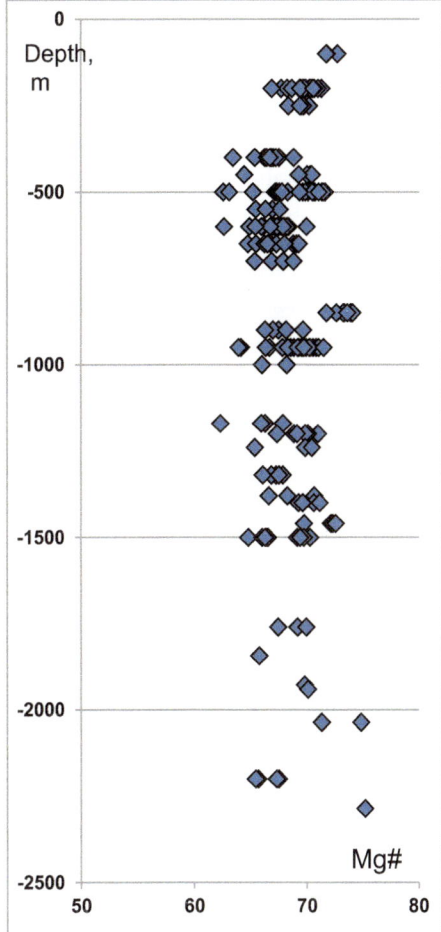

Fig. 5.22 Variations of Mg# in clinopyroxenes from borehole 83

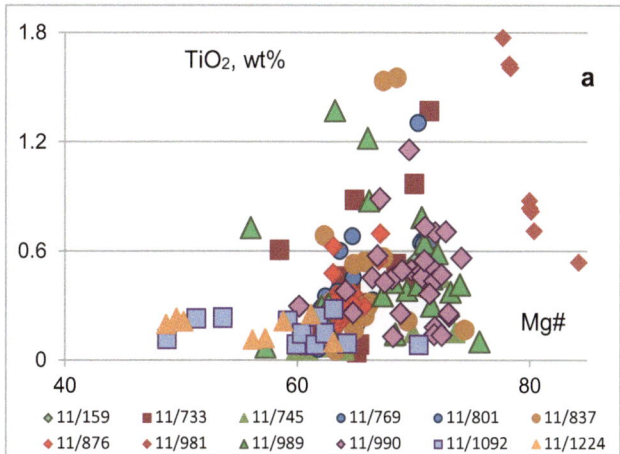

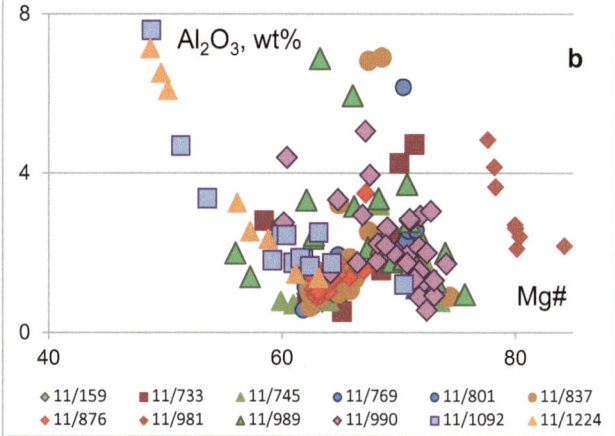

Fig. 5.23 Diagrams Mg# vs TiO₂ (**a**) and Al₂O₃ (**b**) for clinopyroxene from borehole 11

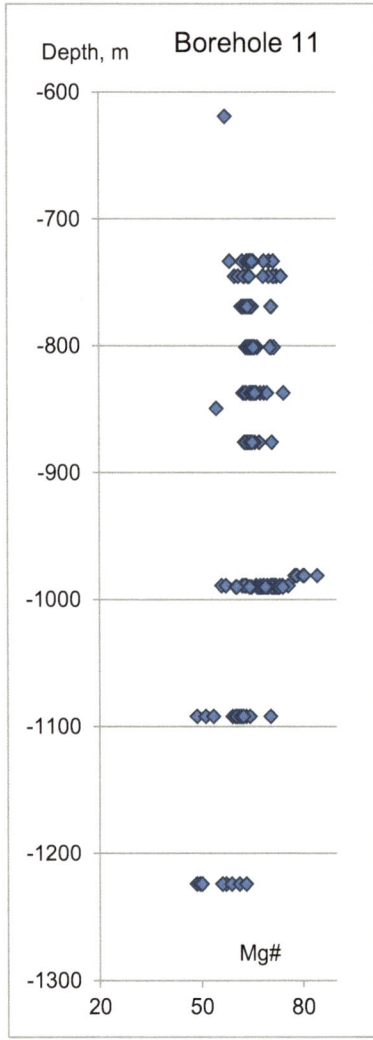

Fig. 5.24 Variations of Mg# in clinopyroxene from borehole 11

Olivine contains two trace elements—Ni and Mn (Table 5.5). The NiO concentrations are 0.11–0.12 wt% and do not correlate with forsterite in olivine. MnO contents change more significantly, from 0.39 to 0.43 wt%, and have negative correlation with forsterite mineral in olivine (olivine $Fo_{76.46}$ contains the highest MnO concentration equal to 0.39 wt%, while olivine $Fo_{42.92}$ contains 0.43 wt% MnO) (Fig. 5.30).

5.8.5 Titanomagnetite

Titanomagnetite content varies essentially in rocks from 0 vol % to 100% when it forms massive titanomagnetite (Fig. 5.7a). As a rule, titanomagnetite forms intergrowths with ilmenite in rocks and ore. Titanomagnetite has very complicated texture, which will be described in Chap. 7. This mineral occurs as interstitial grains between orthopyroxene, clinopyroxene, and plagioclase crystals (Table 5.6).

Composition of some grains of titanomagnetite is given in Table 5.6. Besides major components (Fe and Ti), it contains many trace elements that are typical of spinel group minerals. All iron, measured by electron microprobe, was calculated as FeO. Thus, FeO concentration in mineral changes from 68.81 to 87.8 wt%. FeO correlates with TiO_2, $R^2 = 0.56$ due to Ti presence in magnetite not only as isomorphic impurity but as ilmenite phase as well. Vanadium is the most important element in titanomagnetite (it is concentrated in magnetite matrix) (Table 5.6, analyses N 5–30 and 1–3, respectively) due to its economic value. Its content in titanomagnetite reaches up to 1.64 wt%. Zinc is distributed very irregularly in the mineral; ZnO concentrations in titanomag-

Table 5.3 Composition of orthopyroxene from the rocks of the Chiney pluton, wt%

No	Sample. N	Mg#	SiO$_2$	TiO$_2$	Al$_2$O$_3$	FeO	MnO	MgO	CaO	Cr$_2$O$_3$	Total
1	11/159-35	63.01	52.43	0.322	0.88	22.86	0.572	21.83	1.60	0.023	100.52
2	11/159-36	62.64	52.67	0.283	0.92	23.20	0.569	21.82	1.43	0.037	100.93
3	11/159-37	63.04	51.84	0.306	1.01	22.41	0.567	21.44	2.20	0.043	99.82
4	11/159-38	63.27	52.39	0.305	0.92	22.83	0.567	22.06	1.42	0.038	100.54
5	11/159-39	62.65	52.14	0.299	0.96	23.03	0.592	21.66	1.63	0.054	100.37
6	11/159-40	62.47	52.23	0.268	0.79	23.21	0.575	21.68	1.33	0.051	100.14
7	11/159-41	62.74	52.25	0.293	1.02	22.13	0.550	20.90	3.05	0.048	100.25
8	11/159-43	62.93	52.36	0.286	0.83	22.85	0.562	21.75	1.58	0.027	100.24
9	11/159-44	62.50	52.39	0.276	0.82	23.32	0.582	21.81	1.36	0.030	100.60
10	11/159-45	62.77	52.24	0.262	0.84	23.02	0.580	21.77	1.31	0.033	100.05
11	11/159-47	63.49	52.36	0.234	0.76	22.70	0.569	22.14	1.16	0.036	99.96
12	11/159-49	63.07	52.67	0.242	0.73	22.93	0.593	21.97	1.33	0.035	100.49
13	11/159-50	62.74	52.00	0.274	0.94	23.07	0.569	21.79	1.35	0.028	100.01
14	11/159-52	62.10	52.29	0.294	0.87	23.55	0.592	21.64	1.48	0.033	100.75
15	11/159-53	62.10	52.01	0.284	0.97	23.56	0.573	21.65	1.49	0.050	100.58
17	11/733-168	64.19	52.64	0.310	1.43	22.26	0.456	22.38	1.50	0.019	100.98
18	11/733-170	62.15	51.85	0.881	1.25	22.28	0.453	20.51	2.92	0.014	100.16
19	11/733-171	63.48	52.73	0.280	1.10	22.53	0.475	21.96	2.04	0.010	101.12
20	11/733-172	58.46	52.88	0.529	1.57	21.36	0.618	16.86	2.83	0.021	96.68
21	11/733-175	65.38	55.87	0.142	0.75	19.02	0.568	20.15	0.74	0.012	97.25
22	11/733-176	63.77	52.55	0.411	1.18	22.28	0.461	22.00	2.32	0.016	101.22
23	11/733-177	64.93	52.36	0.371	1.43	20.97	0.432	21.78	3.10	0.015	100.46
24	11/733-178	64.43	52.85	0.316	1.11	22.23	0.462	22.59	1.29	0.017	100.86
25	11/733-180	65.02	52.92	0.241	0.98	21.80	0.451	22.73	1.76	0.015	100.90
26	11/733-181	65.36	52.46	0.272	1.47	21.39	0.440	22.64	1.91	0.019	100.61
27	11/733-182	63.80	52.64	0.460	1.18	22.54	0.458	22.29	1.62	0.007	101.20
28	11/733-183	65.15	55.77	0.086	1.26	19.21	0.523	20.13	0.72	0.017	97.71
29	11/733-185	62.20	55.59	0.045	0.52	21.02	0.553	19.40	0.94	0.011	98.08
30	11/733-186	64.60	53.07	0.228	1.10	21.34	0.460	21.84	2.68	0.007	100.73
31	11/745-144	59.89	54.50	0.053	0.79	22.18	0.623	18.58	0.91	0.013	97.64
32	11/745-146	62.51	55.26	0.071	0.87	20.86	0.583	19.51	0.76	0.044	97.97
33	11/745-148	64.06	55.63	0.057	1.02	19.75	0.554	19.74	0.57	0.018	97.34
34	11/745-149	64.07	55.76	0.054	0.79	19.83	0.523	19.83	0.54	0.003	97.33
35	11/745-150	60.96	55.06	0.052	0.71	21.33	0.606	18.69	1.22	0.013	97.68
36	11/745-151	63.83	55.46	0.053	1.01	18.82	0.458	18.62	3.99	0.009	98.42
37	11/745-154	64.23	52.59	0.328	1.21	22.09	0.433	22.25	1.89	0.016	100.80
38	11/745-155	62.74	55.30	0.077	1.12	20.13	0.567	19.01	1.20	0.017	97.42
39	11/745-157	64.04	55.49	0.076	1.25	19.87	0.531	19.85	0.57	0.008	97.65
40	11/745-165	61.84	55.55	0.060	0.57	21.07	0.615	19.15	1.05	0.010	98.07
41	11/769-125	63.50	52.20	0.378	1.15	22.25	0.446	21.70	2.62	0.017	100.76
42	11/769-126	62.99	52.18	0.327	1.10	22.19	0.465	21.18	2.95	0.009	100.40
43	11/769-127	63.78	52.17	0.349	1.42	21.88	0.438	21.61	2.65	0.027	100.54
44	11/769-128	64.13	52.68	0.298	1.16	22.10	0.458	22.16	1.98	0.017	100.84
45	11/769-129	63.52	52.48	0.228	1.19	22.60	0.491	22.07	1.52	0.016	100.59
46	11/769-132	62.16	52.12	0.291	0.94	23.53	0.484	21.68	1.39	0.007	100.43
47	11/769-133	64.42	52.36	0.254	1.06	21.73	0.456	22.07	2.31	0.018	100.27
48	11/769-134	63.67	52.34	0.601	1.20	22.78	0.467	22.39	1.24	0.009	101.03
49	11/769-135	64.02	52.29	0.306	1.54	21.82	0.424	21.78	2.37	0.030	100.56
50	11/769-136	61.99	51.98	0.277	1.14	23.26	0.501	21.28	1.68	0.007	100.14
51	11/769-137	65.00	52.47	0.271	1.16	20.29	0.430	21.13	4.30	0.011	100.05
52	11/769-138	63.25	52.19	0.276	1.19	22.93	0.470	22.13	1.06	0.010	100.26
53	11/769-139	64.65	52.16	0.321	1.33	20.87	0.439	21.41	3.62	0.009	100.16
54	11/769-140	62.93	52.83	0.249	0.86	22.39	0.486	21.32	2.70	0.002	100.83
55	11/769-141	64.66	52.69	0.282	1.07	21.75	0.442	22.32	1.93	0.013	100.50

(continued)

Table 5.3 (continued)

No	Sample. N	Mg#	SiO$_2$	TiO$_2$	Al$_2$O$_3$	FeO	MnO	MgO	CaO	Cr$_2$O$_3$	Total
56	11/769-142	62.45	52.37	0.352	1.04	23.39	0.478	21.82	1.12	0.006	100.58
57	11/769-143	63.78	52.46	0.247	1.23	21.57	0.450	21.31	3.79	0.013	101.07
58	11/801-100	64.91	52.33	0.348	1.58	21.15	0.429	21.95	2.57	0.024	100.36
59	11/801-101	64.55	52.89	0.260	1.15	22.15	0.466	22.62	1.26	0.018	100.81
60	11/801-102	64.07	52.62	0.316	1.19	22.30	0.462	22.31	1.37	0.013	100.58
61	11/801-104	64.86	53.05	0.277	0.98	21.30	0.449	22.05	2.58	0.006	100.70
62	11/801-105	64.40	52.75	0.264	1.35	21.90	0.462	22.21	1.67	0.020	100.62
63	11/801-107	66.54	52.55	0.330	1.65	19.64	0.394	21.90	4.01	0.028	100.50
64	11/801-108	63.22	52.47	0.296	1.16	22.95	0.481	22.12	1.12	0.014	100.61
65	11/801-109	64.95	53.06	0.267	1.13	21.81	0.464	22.67	1.59	0.018	101.02
66	11/801-110	64.96	52.49	0.290	1.22	21.61	0.451	22.47	1.77	0.019	100.31
67	11/801-111	63.74	52.38	0.274	1.30	22.46	0.469	22.15	1.34	0.017	100.40
68	11/801-112	65.46	52.16	0.302	1.58	20.81	0.425	22.12	2.75	0.020	100.17
69	11/801-113	64.73	52.20	0.319	1.36	21.78	0.452	22.42	1.73	0.019	100.27
70	11/801-115	64.77	52.58	0.263	1.43	20.09	0.420	20.72	4.91	0.013	100.43
71	11/801-117	65.16	52.54	0.270	0.96	21.63	0.458	22.69	1.74	0.011	100.31
72	11/801-118	65.46	52.55	0.315	1.38	20.94	0.425	22.25	2.47	0.014	100.34
73	11/801-119	64.62	52.57	0.296	0.97	22.22	0.463	22.76	1.15	0.010	100.44
74	11/801-120	64.76	52.10	0.682	1.12	21.47	0.450	22.13	2.24	0.019	100.21
75	11/801-121	64.19	52.42	0.260	1.32	22.04	0.455	22.16	1.77	0.018	100.44
76	11-801/122	65.81	52.63	0.268	1.33	20.63	0.418	22.28	2.78	0.024	100.36
77	11/801-123	64.81	52.50	0.324	1.23	22.02	0.471	22.75	1.05	0.017	100.36
78	11/801-124	63.96	52.11	0.348	1.43	21.81	0.452	21.70	2.69	0.019	100.56
79	11/801-96	65.43	52.72	0.254	1.26	21.76	0.432	23.10	1.07	0.021	100.61
80	11/801-97	64.72	52.91	0.294	1.31	21.37	0.427	21.99	2.67	0.017	100.99
81	11/801-98	64.86	52.93	0.277	1.46	22.06	0.446	22.84	1.30	0.027	101.35
82	11/801-99	65.15	53.00	0.196	1.19	21.63	0.451	22.67	1.79	0.021	100.95
83	11/837-67	62.33	54.80	0.685	1.03	21.07	0.562	19.55	0.72	0.010	98.43
84	11/837-68	65.31	52.82	0.257	1.44	21.75	0.435	22.97	1.08	0.010	100.75
85	11/837-69	69.53	54.15	0.216	2.06	15.32	0.403	19.61	6.99	0.012	98.77
86	11/837-70	65.30	52.45	0.284	1.41	21.83	0.431	23.04	0.98	0.018	100.44
87	11/837-73	65.85	52.67	0.242	1.07	21.54	0.432	23.29	0.84	0.018	100.09
88	11/837-74	65.77	52.84	0.298	1.32	20.78	0.413	22.40	2.72	0.015	100.79
89	11/837-76	65.19	52.82	0.269	1.27	21.00	0.429	22.05	2.55	0.014	100.40
90	11/837-78	64.95	52.28	0.288	1.32	21.89	0.430	22.75	1.07	0.011	100.03
91	11/837-80	63.23	54.86	0.115	1.52	19.99	0.549	19.29	1.44	0.020	97.78
92	11/837-81	64.50	52.39	0.302	1.43	22.16	0.457	22.58	0.99	0.015	100.32
93	11/837-82	63.23	52.93	0.191	0.86	23.10	0.489	22.27	0.92	0.016	100.78
94	11/837-83	66.28	52.72	0.316	1.35	20.32	0.407	22.40	3.03	0.021	100.57
95	11/837-84	63.30	55.14	0.055	0.96	19.00	0.594	18.38	4.01	0.003	98.14
96	11/837-85	66.08	52.87	0.293	1.57	21.29	0.417	23.27	1.22	0.022	100.95
97	11/837-86	63.07	52.54	0.327	1.04	23.04	0.480	22.06	1.15	0.019	100.64
98	11/837-88	64.91	52.89	0.166	0.99	22.20	0.436	23.03	0.86	0.011	100.58
99	11/837-89	65.69	52.95	0.297	1.21	21.69	0.429	23.29	1.17	0.015	101.04
100	11/837-90	65.54	53.03	0.253	1.07	21.19	0.426	22.61	2.21	0.018	100.81
101	11/837-91	64.76	53.00	0.224	1.23	21.43	0.426	22.09	2.33	0.012	100.73
102	11/837-92	64.89	52.84	0.528	3.23	19.29	0.358	20.00	4.59	0.016	100.85
103	11/837-93	64.29	52.46	0.290	1.34	21.53	0.457	21.74	2.22	0.012	100.06
104	11/837-94	62.41	55.32	0.139	0.64	20.89	0.682	19.46	0.55	0.003	97.69
105	11/837-95	65.81	54.33	0.542	1.86	17.59	0.521	18.99	4.12	0.025	97.98
106	11/876-31	63.18	52.37	0.318	1.37	21.94	0.449	21.12	2.83	0.019	100.42
107	11/876-32	63.78	52.52	0.264	1.09	21.67	0.471	21.41	3.02	0.016	100.47
108	11/876-33	64.86	52.50	0.366	1.77	19.55	0.398	20.25	5.30	0.028	100.16
109	11/876-34	63.30	52.02	0.328	1.42	22.80	0.471	22.06	1.01	0.025	100.13

(continued)

Table 5.3 (continued)

No	Sample. N	Mg#	SiO$_2$	TiO$_2$	Al$_2$O$_3$	FeO	MnO	MgO	CaO	Cr$_2$O$_3$	Total
110	11/876-36	62.71	51.96	0.284	1.08	22.60	0.480	21.33	2.19	0.010	99.93
111	11/876-37	63.29	52.47	0.310	0.94	22.41	0.455	21.66	2.19	0.008	100.44
112	11/876-38	64.62	52.35	0.295	1.38	21.38	0.423	21.90	2.44	0.023	100.18
113	11/876-39	64.10	52.29	0.315	1.55	21.67	0.423	21.69	2.48	0.028	100.45
114	11/876-40	64.40	52.42	0.291	1.41	21.41	0.431	21.72	2.66	0.032	100.37
115	11/876-41	63.15	52.70	0.172	1.03	23.09	0.468	22.19	0.90	0.013	100.56
116	11/876-42	63.14	52.32	0.243	1.10	22.34	0.455	21.46	2.32	0.011	100.24
117	11/876-43	64.09	51.77	0.302	1.40	21.92	0.439	21.94	2.04	0.014	99.82
118	11/876-44	65.13	52.55	0.318	1.31	21.53	0.435	22.56	1.79	0.021	100.51
119	11/876-45	63.29	52.36	0.324	1.12	22.28	0.454	21.54	2.23	0.008	100.31
120	11/876-46	63.78	52.25	0.357	1.44	21.90	0.418	21.64	2.54	0.022	100.57
121	11/876-47	62.81	52.26	0.324	1.37	22.65	0.472	21.45	2.11	0.024	100.65
122	11/876-48	64.45	52.03	0.343	1.68	21.28	0.424	21.63	2.71	0.031	100.11
123	11/876-49	64.87	52.05	0.312	1.46	21.20	0.416	21.96	2.17	0.025	99.59
124	11/876-50	63.15	52.44	0.626	1.07	22.61	0.459	21.73	2.22	0.020	101.17
125	11/876-51	62.79	52.40	0.272	0.95	22.63	0.460	21.42	1.86	0.007	99.99
126	11/876-52	64.00	52.65	0.303	1.02	21.78	0.466	21.72	2.58	0.013	100.53
127	11/876-54	63.41	52.72	0.270	1.06	22.77	0.472	22.13	1.24	0.015	100.68
128	11/876-55	63.40	52.37	0.362	1.22	21.73	0.449	21.12	3.07	0.020	100.34
129	11/876-56	64.32	52.51	0.279	1.33	22.00	0.432	22.24	1.62	0.025	100.43
130	11/876-58	63.81	52.10	0.217	1.46	21.30	0.434	21.06	3.68	0.016	100.27
131	11/876-59	63.01	52.33	0.275	0.91	22.58	0.471	21.58	2.19	0.011	100.34
132	11/876-60	63.48	52.84	0.227	1.00	22.11	0.451	21.56	2.43	0.012	100.64
133	11/876-61	65.79	52.47	0.298	1.38	20.32	0.408	21.91	3.23	0.017	100.03
134	11/876-62	63.13	52.30	0.479	1.06	22.44	0.460	21.54	2.06	0.016	100.36
135	11/876-63	63.91	52.39	0.292	1.57	21.71	0.438	21.56	2.54	0.027	100.52
136	11a6p3	0.41	41.70	54.53	0.10	29.03	0.50	0.07	0.20	0.000	101.81
137	11a7p5s	0.08	18.92	55.11	0.16	28.49	0.41	0.01	0.05	0.067	101.42
138	11a1	0.16	39.64	55.13	0.11	28.18	0.36	0.03	0.13	0.000	101.11
139	11a7p3	1.11	36.86	68.56	0.54	19.93	0.57	0.13	0.19	0.034	103.08
140	83/174,2	57.46	51.46	0.37	0.91	25.79	0.59	19.54	1.45	0.025	100.11
141	83/174,2	57.96	50.61	0.33	0.90	24.84	0.57	19.21	1.36	0.377	97.82
142	83/174,2	58.23	51.43	0.35	0.97	25.31	0.58	19.79	1.26	0.099	99.68
143	83/174,2	57.74	51.21	0.36	0.88	25.59	0.57	19.61	1.40	0.015	99.62
144	83/174,2	57.52	51.47	0.30	0.80	25.77	0.56	19.57	1.19	0.006	99.66
145	83/174,2	57.58	51.79	0.29	0.95	25.79	0.56	19.64	1.32	0.044	100.34
146	83/240	62.47	51.89	0.31	0.95	23.16	0.52	21.62	1.29	0.016	99.74
147	83/240	62.50	51.82	0.32	1.55	23.22	0.51	21.71	1.18	0.000	100.30
148	83/240	62.25	51.38	0.31	1.62	22.85	0.51	21.14	1.88	0.015	99.69
149	83/240	62.30	51.98	0.32	1.59	23.06	0.51	21.38	1.71	0.000	100.56
150	83/843,7	58.27	51.18	0.37	1.65	25.44	0.53	19.92	1.29	0.009	100.37
151	83/843,7	58.90	51.64	0.33	1.66	24.86	0.50	19.98	1.85	0.003	100.81
152	83/843,7	60.60	51.45	0.31	2.03	24.09	0.44	20.78	1.88	0.020	100.99
153	83/927,9	63.69	51.65	0.22	1.59	22.25	0.45	21.89	1.71	0.093	99.76
154	83/927,9	62.97	52.11	0.31	1.93	22.84	0.47	21.78	1.42	0.004	100.85
155	83/940,6	64.34	52.86	0.15	1.67	22.53	0.42	22.80	1.34	0.008	101.77
156	83/940,6	64.51	52.75	0.26	1.91	21.83	0.42	22.26	2.28	0.016	101.70
157	83/940,6	61.38	52.26	0.28	1.79	24.24	0.46	21.62	0.96	0.001	101.61
158	83/940,6	63.33	52.52	0.29	1.85	23.00	0.44	22.28	1.07	0.002	101.46
159	83/1036,3	68.84	49.97	1.97	2.20	17.04	0.35	21.11	2.16	0.064	94.79
160	83/1036,3	65.63	52.10	0.21	1.58	21.70	0.49	23.24	0.77	0.017	100.07
161	83/1100	66.26	52.27	0.26	2.00	20.91	0.41	23.03	2.12	0.015	101.00
162	83/1100	65.31	52.17	0.29	2.18	21.15	0.43	22.34	1.82	0.016	100.37
163	83/1100	65.57	51.93	0.26	2.18	21.33	0.42	22.78	1.31	0.020	100.20

(continued)

Table 5.3 (continued)

No	Sample. N	Mg#	SiO$_2$	TiO$_2$	Al$_2$O$_3$	FeO	MnO	MgO	CaO	Cr$_2$O$_3$	Total
164	83/1100	66.02	51.58	0.26	2.04	20.67	0.41	22.53	2.10	0.034	99.58
165	83/1100	66.11	51.58	0.30	2.14	21.27	0.41	23.27	0.79	0.001	99.76
166	83/1101,2	65.95	51.84	0.26	2.21	20.85	0.40	22.65	2.07	0.000	100.29
167	83/1101,2	64.40	51.64	0.26	2.14	21.59	0.42	21.91	2.24	0.016	100.21
168	83/1101,2	62.44	51.34	0.24	2.04	22.87	0.45	21.33	1.96	0.004	100.23
169	83/1200,4	53.97	50.64	0.28	1.60	27.59	0.65	18.15	1.52	0.005	100.44
170	83/1200,4	53.58	51.17	0.29	1.53	28.37	0.65	18.36	1.02	0.002	101.39
171	83/1200,4	56.35	51.07	0.27	1.68	26.53	0.60	19.21	1.16	0.003	100.52
172	516-17	60.99	51.92	0.247	0.77	24.27	0.504	21.28	1.43	0.016	100.44
173	516-16	59.33	52.13	0.208	0.82	25.20	0.484	20.62	1.46	0.000	100.92
174	516-19	59.10	51.68	0.237	0.83	25.03	0.474	20.29	1.49	0.001	100.03
175	516-17a	61.79	51.70	0.288	0.95	23.53	0.468	21.34	1.63	0.000	99.91
176	516-26	59.75	52.89	0.280	1.14	21.93	0.475	18.26	5.71	0.001	100.69
177	508-54	58.34	51.74	0.278	1.10	25.50	0.580	20.03	1.28		100.50
178	508-64	59.32	51.48	0.313	1.19	24.70	0.579	20.21	1.73		100.20
179	508-61	61.03	51.64	0.376	1.35	24.01	0.542	21.09	1.44		100.46
180	503-16	67.71	52.85	0.357	1.46	20.25	0.563	23.81	1.39		100.67
181	503-17	68.91	53.08	0.412	1.72	19.56	0.531	24.32	1.43		101.06
182	307	70.56	53.56	0.266	1.71	19.04		25.59	1.04	0.082	101.29
183	307	68.27	52.23	0.293	1.75	20.09	0.376	24.25	1.25	0.000	100.24
184	307	71.38	53.17	0.269	1.83	18.63	0.385	26.06	1.27	0.000	101.61
185	307	70.05	53.54	0.293	1.82	19.04		24.98	1.32	0.000	100.99
186	307	70.17	53.14	0.248	1.62	19.07	0.364	25.16	1.33	0.000	100.93
187	307	70.01	53.50	0.288	1.67	19.01	0.393	24.89	1.37	0.000	101.12
188	307	67.53	52.19	0.280	1.85	20.44	0.405	23.84	1.39	0.000	100.40
189	307	69.17	52.62	0.258	1.76	19.64	0.405	24.72	1.40	0.000	100.80
190	307	69.99	53.22	0.297	1.75	19.03	0.392	24.89	1.52	0.000	101.10
191	307	70.62	52.88	0.219	1.73	18.47	0.336	24.90	1.62	0.000	100.16
192	307	69.94	53.22	0.286	1.86	19.00		24.79	1.74	0.008	100.90
193	307	68.86	52.91	0.251	1.58	19.62	0.389	24.33	1.75	0.000	100.83
194	307	67.41	52.77	0.282	2.16	20.09	0.419	23.31	1.95	0.000	100.98
195	307	70.96	53.37	0.302	1.75	18.26	0.345	25.03	2.20	0.000	101.26
196	307	68.81	51.70	0.305	1.77	19.65	0.397	24.32	2.24	0.000	100.38
197	307	67.81	52.94	0.305	1.83	19.70	0.361	23.28	2.64	0.000	101.06
198	307	69.71	51.38	0.338	1.81	18.79	0.397	24.25	2.75	0.000	99.72
199	307	69.37	53.72	0.224	1.97	18.78		23.86	2.86	0.092	101.51
200	307	69.47	51.61	0.296	1.76	18.46	0.343	23.56	3.05	0.000	99.08
201	307	69.19	52.27	0.271	1.77	18.33	0.352	23.09	4.26	0.000	100.34
202	307	70.30	53.92	0.140	2.41	16.21		21.52	5.09	0.000	99.29
203	307	67.17	53.03	0.357	2.08	18.88		21.67	5.29	0.058	101.37
204	307	69.35	53.37	0.206	2.13	16.51	0.466	20.95	6.54	0.000	100.17
205	39/103.9-11	43.02	51.03	0.174	0.47	33.89	0.638	14.35	0.88	0.000	101.43
206	39/103.9-13	45.51	51.43	0.229	0.62	32.53	0.614	15.24	1.00	0.000	101.66
207	39/103.9-14	40.33	50.33	0.157	0.68	34.74	0.609	13.17	1.12	0.000	100.81
208	39/103.9-17	46.54	51.61	0.185	0.53	31.80	0.589	15.53	1.26	0.002	101.1
209	39/103.9-8	29.96	50.03	0.093	0.28	39.73	0.553	9.53	1.37	0.000	101.59
210	39/103.9-2	39.71	50.95	0.103	0.70	34.84	0.630	12.87	1.43	0.026	101.55
211	39/103.9-1	34.89	50.90	0.105	0.23	37.59	0.665	11.30	1.57	0.007	102.37
212	39/103.9-7	41.77	50.11	0.146	0.32	33.38	0.675	13.43	2.65	0.000	100.71
213	39/103.9-6	40.12	50.29	0.204	0.57	32.92	0.583	12.37	4.45	0.000	101.39
214	39/103.9-18	46.70	52.39	0.178	0.51	29.26	0.525	14.38	5.50	0.011	102.75
215	39/103.9-19	43.53	51.20	0.239	1.01	27.59	0.474	11.93	9.30	0.035	101.78

Note: Rock types seen in Table 5.1. Here and in Tables 5.4 and 5.5, Sample N = 39/103.2-1 = borehole number/depth (m)—point of measurement

Fig. 5.25 Spider-diagram for orthopyroxenes from titanomagnetite series Samples: borehole 11, depth 260 m. Normalized to Primitive Mantle (after Hofmann 1988)

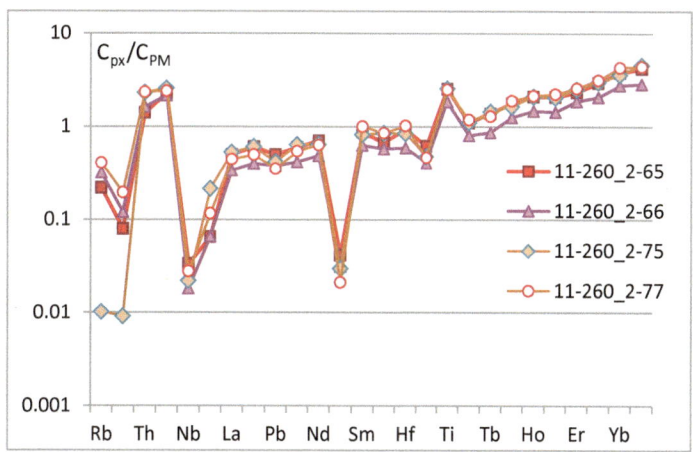

netite range from 0 to 0.55 wt%, while in ilmenite this element is absent. Titanomagnetite contains Ni (up to 0.44 wt%), Al_2O_3 (maximum 5.5 wt%), and SiO_2 (0.37 wt%) as well. Besides ilmenite, titanomagnetite contains spinel (Table 5.6).

Fig. 5.26 Diagram Mg# vs
CaO (**a**), TiO₂ (**b**), MnO (**c**),
and Al₂O₃ (**d**) for
orthopyroxene from rocks of
the Chiney pluton

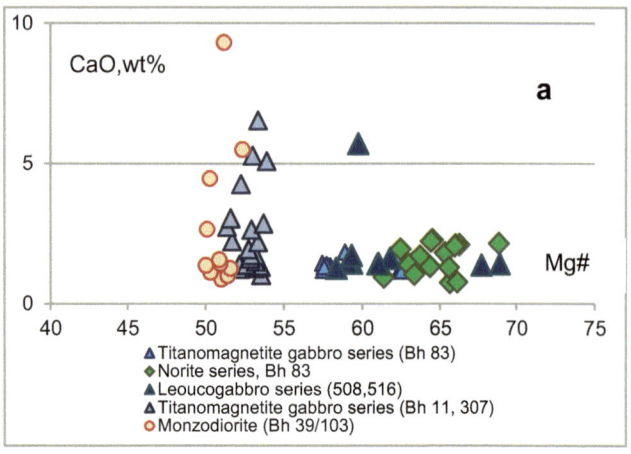

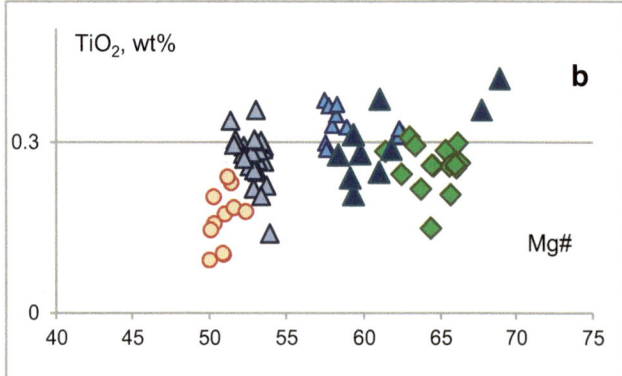

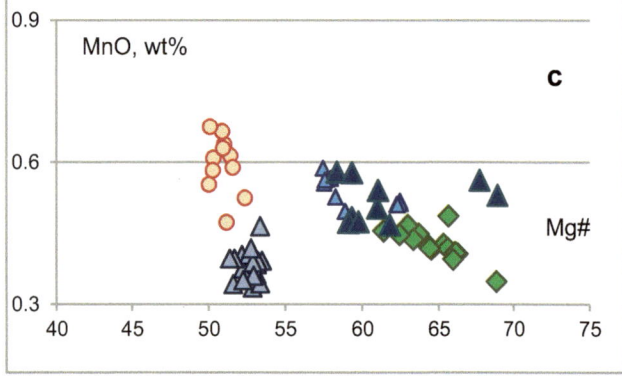

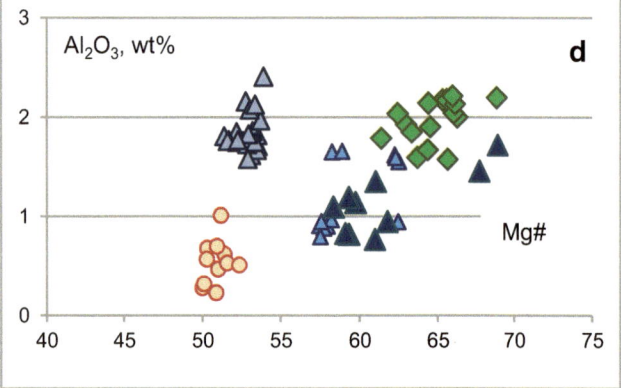

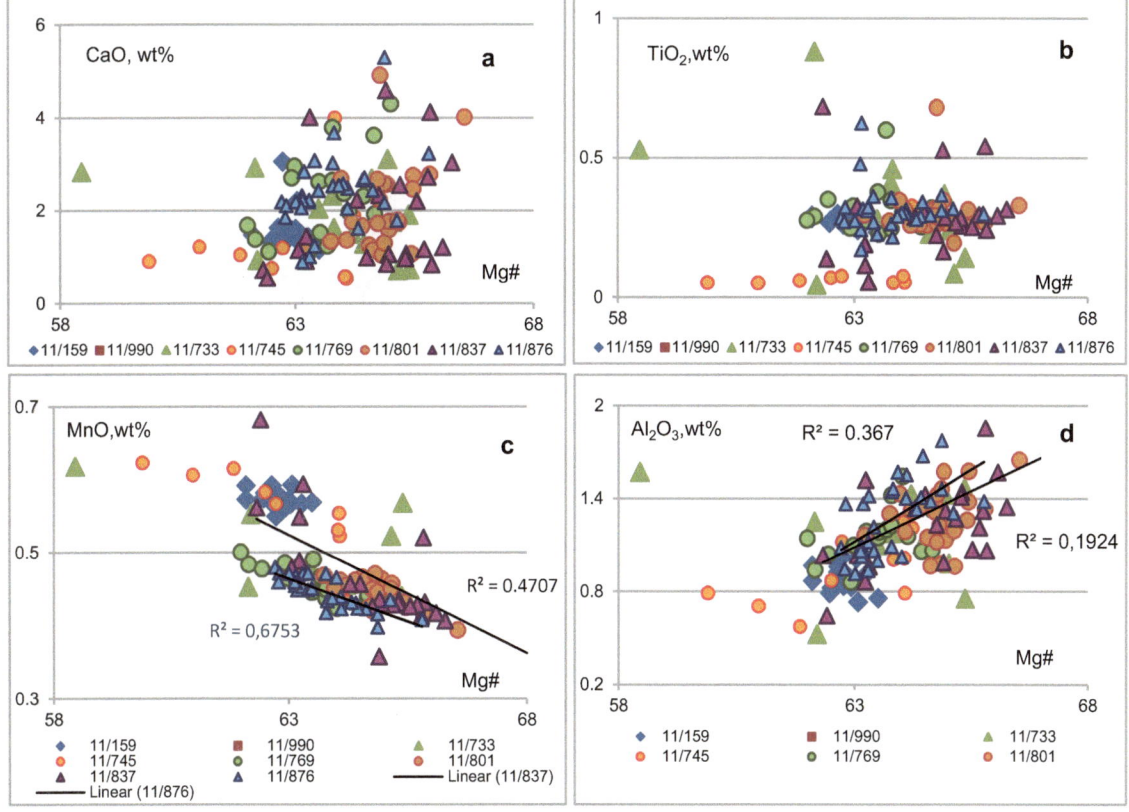

Fig. 5.27 Diagram Mg#vsCaO (**a**), TiO₂ (**b**), Al₂O₃ (**c**), and MnO (**d**) for orthopyroxene from borehole 11

Fig. 5.28 Diagram
An-FeO for plagioclase
from different rock types
of the Chiney pluton

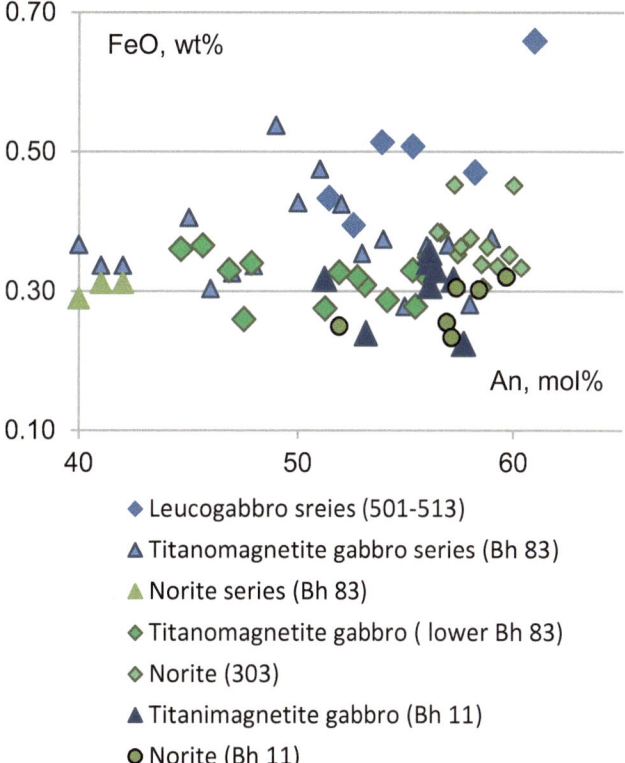

Table 5.4 Plagioclase composition from the rocks of the Chineysky pluton, wt%

N	Comment	An, mol%	SiO$_2$	TiO$_2$	Al$_2$O$_3$	FeO	MnO	CaO	Na$_2$O	K$_2$O	Total
1	501-9	55.35	54.98	0.08	29.05	0.51	0.01	10.92	4.87	0.36	100.78
2	506	51.43	56.49	0.03	26.94	0.43	0.04	9.82	5.13	0.55	99.43
3	507	52.59	55.63	0.04	27.53	0.40	0.04	10.18	5.08	0.36	99.27
4	509	53.94	55.29	0.09	27.69	0.51	0.00	10.41	4.92	0.38	99.30
5	513	58.22	54.49	0.05	28.39	0.47	0.02	11.25	4.47	0.38	99.52
6	513	60.94	53.45	0.03	28.64	0.66	0.00	11.81	4.19	0.29	99.08
7	83-104,2	54.61	53.68	0.06	30.86	0.37	0.00	10.94	5.03	0.32	101.28
8	83-104,2	55.74	53.71	0.06	30.64	0.43	0.00	11.03	4.85	0.47	101.20
9	83-104,2	54.23	53.61	0.05	30.59	0.38	0.00	11.02	5.15	0.48	101.29
10	83-104,2	56.29	52.90	0.06	30.23	0.40	0.00	10.86	4.67	0.47	99.59
11	83-121,6	52.84	53.80	0.07	31.36	0.26	0.00	11.01	5.44	0.16	102.09
12	83-121,6	32.45	59.02	0.03	26.65	0.22	0.00	6.58	7.58	0.27	100.35
13	83-174,2	52.54	54.18	0.08	30.41	0.36	0.00	10.40	5.20	0.46	101.09
14	83-174,2	53.11	54.26	0.08	30.93	0.28	0.01	10.67	5.21	0.38	101.83
15	83-174,2	49.98	54.81	0.02	30.46	0.30	0.00	10.25	5.68	0.47	102.00
16	83-174,2	52.15	54.93	0.08	30.41	0.32	0.01	10.36	5.26	0.48	101.85
17	83-174,2	55.80	53.67	0.08	30.75	0.28	0.00	10.87	4.77	0.43	100.84
18	83-174,2	55.06	51.46	0.08	30.05	0.32	0.00	10.85	4.90	0.42	98.09
19	83-200	52.17	53.94	0.06	30.61	0.40	0.01	10.60	5.38	0.48	101.49
20	83-200	53.00	54.15	0.09	29.96	0.32	0.00	10.59	5.20	0.40	100.71
21	83-200	52.53	54.22	0.07	30.22	0.39	0.00	10.47	5.24	0.44	101.05
22	83-200	51.70	54.43	0.07	30.62	0.31	0.00	10.40	5.38	0.49	101.70
23	83-200	52.86	54.47	0.07	30.17	0.41	0.00	10.48	5.17	0.40	101.17
24	83-200	55.04	53.51	0.07	30.95	0.38	0.01	11.17	5.05	0.37	101.50
25	83-200	54.28	53.59	0.06	30.85	0.36	0.00	11.03	5.14	0.39	101.43
26	83-240	54.86	53.83	0.07	28.63	0.36	0.00	11.02	5.02	0.42	99.36
27	83-240	55.35	53.85	0.07	28.24	0.39	0.01	10.87	4.85	0.46	98.74
28	83-240	53.63	54.38	0.07	27.64	0.35	0.00	10.44	5.00	0.39	98.28
29	83-240	48.38	54.70	0.07	27.70	0.39	0.00	10.01	5.91	0.24	99.02
30	83-320,2	54.65	54.39	0.06	27.80	0.35	0.00	10.76	4.94	0.48	98.78
31	83-320,2	53.83	54.43	0.06	27.78	0.33	0.01	10.65	5.05	0.49	98.79
32	83-320,2	56.88	53.66	0.07	28.21	0.35	0.00	11.33	4.76	0.40	98.79
33	83-380,3	54.89	53.84	0.06	28.00	0.40	0.00	10.97	4.99	0.42	98.69
34	83-380,3	54.97	53.95	0.06	27.67	0.43	0.00	10.85	4.92	0.40	98.30
35	83-380,3	53.85	53.97	0.07	27.70	0.42	0.00	10.87	5.16	0.45	98.64
36	83-380,3	40.84	55.65	0.05	25.95	0.34	0.00	8.00	6.41	0.40	96.82
37	83-460	53.34	54.21	0.07	27.80	0.39	0.00	10.73	5.20	0.51	98.92
38	83-460	54.46	54.70	0.05	27.86	0.54	0.01	10.93	5.06	0.36	99.52
39	83-460	58.04	53.34	0.07	28.12	0.52	0.00	11.58	4.63	0.47	98.73
40	83-460	52.80	53.43	0.04	27.32	0.57	0.00	10.51	5.20	0.52	97.59
41	83-460	56.47	53.65	0.08	28.36	0.39	0.01	10.95	4.67	0.57	98.68
42	83-500	52.92	53.70	0.06	28.27	0.38	0.00	10.84	5.34	0.45	99.04
43	83-500	53.25	54.33	0.08	27.69	0.28	0.00	10.33	5.02	0.57	98.29
44	83-500	53.70	54.52	0.06	27.72	0.31	0.00	10.64	5.08	0.39	98.72
45	83-500	55.53	53.19	0.06	28.81	0.42	0.01	11.41	5.06	0.45	99.42
46	83-500	53.57	53.35	0.06	27.96	0.37	0.00	10.76	5.16	0.50	98.17
47	83-500	53.89	53.44	0.06	28.33	0.34	0.00	10.91	5.17	0.49	98.74
48	83-500	54.56	53.64	0.07	27.84	0.34	0.00	10.84	5.00	0.46	98.18
49	83-500	57.32	52.68	0.07	28.56	0.41	0.00	11.35	4.68	0.47	98.21
50	83-600	54.02	53.70	0.06	28.00	0.30	0.00	10.90	5.14	0.37	98.47
51	83-600	54.23	53.87	0.06	27.60	0.33	0.00	10.66	4.98	0.24	97.74
52	83-760	53.87	53.74	0.05	27.91	0.34	0.00	10.88	5.16	0.42	98.50
53	83-760	57.03	53.19	0.05	28.70	0.54	0.00	11.62	4.85	0.35	99.29
54	83-760	55.89	53.38	0.05	28.27	0.43	0.00	11.24	4.91	0.39	98.67

(continued)

Table 5.4 (continued)

N	Comment	An, mol%	SiO_2	TiO_2	Al_2O_3	FeO	MnO	CaO	Na_2O	K_2O	Total
55	83-760	56.48	53.40	0.06	28.25	0.48	0.00	11.29	4.82	0.35	98.63
56	83-760	55.75	53.42	0.08	27.95	0.43	0.01	11.12	4.89	0.42	98.30
57	83-760	54.62	53.62	0.05	28.11	0.35	0.01	10.98	5.05	0.42	98.58
58	83-760	56.27	53.05	0.05	28.14	0.38	0.00	11.27	4.85	0.38	98.11
59	83-760,1	51.43	53.84	0.06	27.53	0.28	0.00	10.49	5.48	0.08	97.76
60	83-760,1	53.58	53.84	0.06	27.86	0.37	0.01	10.75	5.15	0.46	98.50
61	83-760,1	55.55	53.99	0.05	27.66	0.37	0.00	10.73	4.75	0.35	97.91
62	83-760,1	53.51	53.33	0.04	27.93	0.28	0.01	10.88	5.23	0.35	98.06
63	83-760,1	55.69	53.60	0.07	28.23	0.38	0.00	11.10	4.89	0.46	98.72
64	83-843,7	47.76	55.51	0.06	27.09	0.24	0.01	9.60	5.81	0.31	98.62
65	83-843,7	56.46	53.29	0.06	28.25	0.32	0.01	11.34	4.84	0.36	98.48
66	83-843,7	47.66	53.42	0.05	26.81	0.27	0.01	9.60	5.84	0.30	96.29
67	83-843,7	55.42	53.50	0.08	28.22	0.39	0.00	11.24	5.00	0.34	98.77
68	83-843,7	59.07	53.15	0.07	28.59	0.31	0.00	11.74	4.50	0.36	98.72
69	83-843,7	57.05	52.66	0.06	28.37	0.31	0.00	11.49	4.79	0.30	97.98
70	83-912,3	56.79	53.31	0.06	28.14	0.48	0.01	11.45	4.82	0.31	98.58
71	83-912,3	56.06	53.40	0.06	28.69	0.28	0.00	11.43	4.96	0.34	99.16
72	83-912,3	56.87	52.88	0.06	28.56	0.31	0.01	11.43	4.80	0.36	98.41
73	83-912,3	57.83	52.44	0.08	28.75	0.36	0.00	11.89	4.80	0.32	98.63
74	83-912,3	57.45	52.69	0.07	27.72	0.38	0.00	11.25	4.61	0.32	97.04
75	83-927,9	52.13	54.15	0.07	27.74	0.30	0.01	10.50	5.34	0.35	98.46
76	83-927,9	51.71	54.16	0.05	28.13	0.26	0.00	10.63	5.50	0.30	99.02
77	83-927,9	55.11	53.20	0.06	28.85	0.32	0.00	11.39	5.13	0.27	99.22
78	83-927,9	54.59	52.15	0.05	28.96	0.62	0.02	11.37	5.23	0.15	98.54
79	83-927,9	58.95	52.57	0.07	28.74	0.33	0.00	11.70	4.51	0.26	98.18
80	83-927,9	55.28	52.70	0.06	28.60	0.33	0.01	11.52	5.16	0.31	98.68
81	83-927,9	57.66	52.75	0.06	28.64	0.32	0.01	11.48	4.67	0.29	98.23
82	83-940,6	53.19	53.86	0.05	28.04	0.49	0.01	10.96	5.34	0.21	98.96
83	83-940,6	53.05	53.93	0.05	27.32	0.33	0.02	10.69	5.24	0.23	97.79
84	83-940,6	50.95	54.79	0.05	27.37	0.34	0.01	10.36	5.52	0.18	98.62
85	83-940,6	44.90	55.92	0.03	26.76	0.31	0.00	9.23	6.27	0.24	98.76
86	83-1036,3	53.16	53.59	0.06	28.28	0.30	0.00	10.89	5.31	0.21	98.64
87	83-1036,3	57.54	52.54	0.05	29.28	0.27	0.00	11.85	4.84	0.19	99.03
88	83-1036,3	55.50	52.80	0.08	28.70	0.30	0.00	11.31	5.02	0.23	98.43
89	83-1100	50.00	54.00	0.03	27.58	0.22	0.00	10.23	5.66	0.34	98.06
90	83-1100	41.90	56.91	0.03	25.90	0.25	0.00	8.26	6.34	0.47	98.15
91	83-1100	54.39	53.33	0.07	27.59	0.31	0.00	10.82	5.02	0.37	97.50
92	83-1100	58.95	53.53	0.02	27.92	0.28	0.01	11.58	4.47	0.22	98.03
93	83-1100	54.15	53.14	0.07	28.00	0.30	0.00	10.91	5.11	0.32	97.85
94	83-1101,2	54.27	53.68	0.07	27.51	0.28	0.00	10.69	4.99	0.37	97.59
95	83-1101,2	58.06	52.26	0.06	28.91	0.30	0.01	11.93	4.77	0.31	98.55
96	83-1101,2	59.44	52.13	0.06	28.13	0.30	0.01	11.77	4.45	0.33	97.17
97	83-1101,2	60.23	51.65	0.06	28.64	0.35	0.00	12.20	4.46	0.29	97.65
98	83-1101,2	57.48	52.41	0.07	28.09	0.30	0.01	11.50	4.71	0.32	97.40
99	83-1101,2	58.51	52.53	0.06	28.72	0.22	0.00	11.89	4.67	0.28	98.37
100	83-1101,2	58.15	52.59	0.06	28.46	0.25	0.00	11.74	4.68	0.31	98.08
101	83-1101,2	56.44	52.72	0.07	28.01	0.29	0.01	11.37	4.86	0.33	97.64
102	83-1101,2	58.13	51.75	0.06	27.90	0.31	0.00	11.72	4.67	0.30	96.71
103	83-1101,2	60.02	51.61	0.05	28.39	0.31	0.00	11.95	4.40	0.32	97.04
104	83-1200,4	51.26	54.14	0.07	27.30	0.28	0.01	10.32	5.43	0.31	97.86
105	83-1200,4	55.50	53.11	0.07	27.95	0.28	0.01	11.35	5.04	0.25	98.07
106	83-1285,7	53.15	53.88	0.04	27.27	0.31	0.00	10.69	5.21	0.35	97.75
107	83-1285,7	51.92	53.93	0.06	27.66	0.33	0.01	10.58	5.42	0.25	98.23
108	83-1285,7	47.89	54.94	0.06	26.32	0.34	0.00	9.62	5.79	0.39	97.46

(continued)

Table 5.4 (continued)

N	Comment	An, mol%	SiO$_2$	TiO$_2$	Al$_2$O$_3$	FeO	MnO	CaO	Na$_2$O	K$_2$O	Total
109	83-1285,7	47.55	55.04	0.07	26.68	0.26	0.00	9.66	5.90	0.33	97.94
110	83-1285,7	46.87	55.16	0.04	26.83	0.33	0.01	9.61	6.03	0.28	98.29
111	83-1285,7	45.63	55.82	0.07	26.61	0.37	0.00	9.40	6.20	0.17	98.62
112	83-1285,7	44.65	55.90	0.06	26.19	0.36	0.00	9.01	6.19	0.39	98.10
113	83-1285,7	54.20	53.57	0.07	27.27	0.29	0.00	10.83	5.07	0.27	97.37
114	83-1285,7	56.09	52.92	0.05	27.92	0.32	0.01	11.43	4.95	0.29	97.90
115	83-1285,7	52.79	52.95	0.07	26.90	0.32	0.00	10.74	5.32	0.32	96.62
116	83-1285,7	55.36	52.84	0.06	27.36	0.33	0.00	10.93	4.88	0.31	96.70
117	0303-1	56.65	54.31	0.05	29.10	0.38	0.02	11.16	4.73	0.34	100.11
118	0303-5	58.02	53.65	0.04	29.43	0.38	0.03	11.26	4.51	0.34	99.62
119	303	59.81	53.55	0.03	29.72	0.35	0.00	11.84	4.40	0.28	100.18
120	303	60.01	53.79	0.06	29.65	0.45	0.01	11.84	4.37	0.32	100.49
121	303	58.61	54.35	0.07	29.40	0.31	0.00	11.56	4.52	0.35	100.55
122	303	59.27	54.46	0.01	29.48	0.34	0.00	11.77	4.48	0.30	100.84
123	303	60.36	52.22	0.01	29.66	0.33	0.00	11.64	4.23	0.27	98.37
124	303	57.28	51.86	0.04	29.14	0.45	0.01	10.99	4.54	0.30	97.33
125	303	58.54	53.09	0.04	29.64	0.34	0.00	11.39	4.47	0.31	99.28
126	303	58.81	51.83	0.02	29.31	0.36	0.00	11.42	4.43	0.28	97.66
127	303	55.97	54.44	0.05	28.70	0.35	0.01	10.89	4.74	0.37	99.55
128	303	57.44	51.90	0.03	29.34	0.35	0.01	11.25	4.61	0.32	97.83
129	303	57.59	53.64	0.05	29.15	0.36	0.01	11.29	4.60	0.35	99.45
130	303	56.51	53.82	0.07	29.25	0.39	0.01	11.18	4.76	0.34	99.81
131	0307a-4	56.10	54.43	0.10	29.30	0.34	0.04	11.00	4.76	0.30	100.27
132	Ch2_11-1	57.25	54.70	0.05	29.34	0.32	0.00	11.56	4.78	0.20	100.96
133	Ch2_11-2	56.15	53.97	0.07	29.27	0.31	0.00	11.27	4.87	0.15	99.90
134	Ch2_11-3	53.19	54.76	0.04	28.79	0.24	0.01	10.51	5.12	0.24	99.71
135	Ch2_11-4	51.22	56.93	0.02	28.42	0.32	0.02	10.19	5.37	0.28	101.55
136	Ch2_11-5	56.41	53.79	0.03	29.25	0.33	0.03	11.18	4.78	0.27	99.67
137	Ch2_11-6	57.72	53.96	0.06	29.39	0.23	0.01	11.38	4.62	0.23	99.89
138	Ch2_11-7	56.14	54.43	0.04	29.27	0.36	0.00	11.27	4.88	0.20	100.44
139	11_679.5-1	56.94	53.64	0.04	29.36	0.26	0.00	11.44	4.79	0.27	99.78
140	11_849.5-11	57.39	53.45	0.06	30.94	0.30	0.00	11.27	4.63	0.18	100.84
141	11_849.5-12	58.43	54.26	0.07	30.73	0.30	0.00	11.37	4.48	0.38	101.59
142	11_849.5-13	59.69	53.64	0.04	31.05	0.32	0.00	11.68	4.37	0.35	101.45
143	11_849.5-14	51.93	55.85	0.03	29.33	0.25	0.00	10.01	5.13	0.48	101.08
144	11_849.5-15	57.17	53.95	0.02	30.43	0.23	0.00	11.33	4.70	0.29	100.96

Fig. 5.29 Photomicrographs of rocks from the Chiney pluton
(**a**) upper leucogabbro, sample 83/44, (**b**) chineyite (sample 83/97), (**c** and **d**) lower titano-magnetite gabbro: c, 83/460; d, 83/500; (**e** and **f**) norite, (e) 83/840, and f, 83/1101
Relic Ol, relics of olivine (**e, f**)

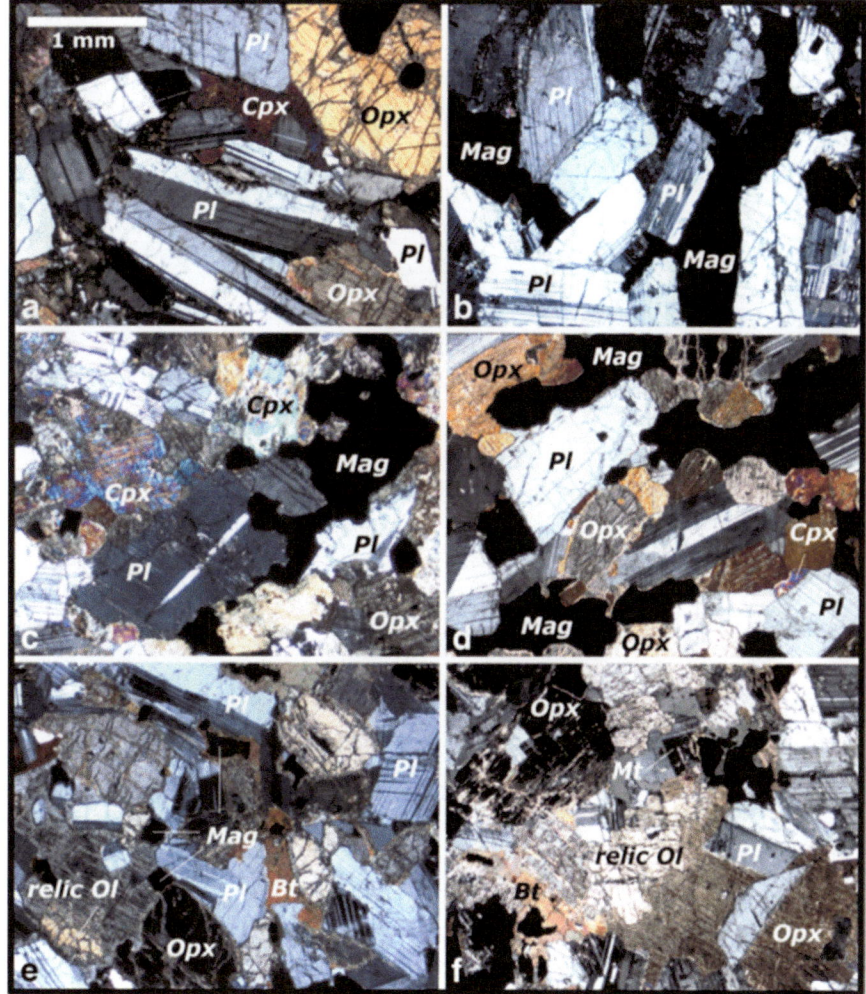

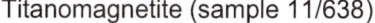

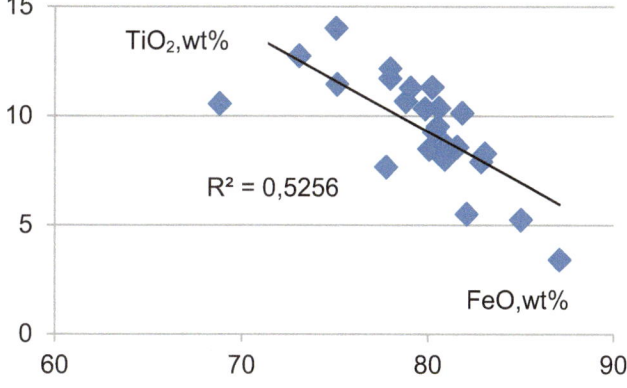

Fig. 5.30 Diagram TiO$_2$-FeO for titanomagnetite

Table 5.5 Olivine composition from the rocks of the Chiney pluton, wt%

No	SiO$_2$	FeO	CaO	MnO	MgO	NiO	Total	Fo, mol %
1	38.46	21.74	0.09	0.39	39.64	0.11	100.44	76.46
2	38.19	23.71	0.07	0.41	38.04	0.12	100.54	74.08
3	37.91	24.24	0.08	0.42	37.83	0.12	100.59	73.55
4	37.66	24.63	0.05	0.43	37.22	0.12	100.10	72.92
5	38.14	23.23	0.09	0.41	38.60	0.12	100.58	74.75
6	38.08	22.69	0.10	0.40	38.72	0.12	100.10	75.25
7	38.23	22.13	0.09	0.40	39.25	0.11	100.21	75.97
8	37.94	23.61	0.09	0.41	37.95	0.12	100.11	74.12
9	37.78	24.21	0.07	0.42	37.48	0.12	100.08	73.39
10	38.21	23.17	0.09	0.40	38.42	0.11	100.41	74.71
11	38.00	23.66	0.09	0.41	38.01	0.12	100.28	74.11

Note. Olivine from sample 11/992 (borehole 11, depth 992 m). Analyses were carried out at IGEM RAS, analyst EV Kovalchuk

Table 5.6 Composition of oxide minerals from the rocks of the Chiney pluton, wt%

	SiO$_2$	TiO2	Al$_2$O$_3$	FeO	MnO	MgO	CaO	Cr$_2$O$_3$	NiO	V$_2$O$_5$	ZnO	Total
1	0.17	49.96	0.14	48.87	0.63	0.14	0.00	0.03	0.00	0.33	0.00	100.35
2	0.11	49.71	0.05	48.66	0.64	0.30	0.04	0.00	0.00	0.27	0.00	100.06
3	0.30	53.10	0.11	43.89	1.09	0.00	0.07	0.00	0.19	0.44	0.00	99.35
4	0.17	25.12	0.24	67.60	0.32	0.07	0.08	0.11	0.48	1.16	0.13	95.76
5	0.21	11.45	4.02	75.11	0.12	1.79	0.00	0.09	0.11	1.64	0.00	94.70
6	0.24	11.31	2.00	80.21	0.10	0.30	0.01	0.11	0.31	1.38	0.24	96.27
7	0.15	10.37	1.84	80.56	0.26	0.26	0.02	0.05	0.06	1.50	0.00	95.67
8	0.16	14.02	2.17	75.06	0.35	1.49	0.03	0.00	0.16	1.65	0.00	95.35
9	0.16	12.16	1.93	77.96	0.32	0.34	0.03	0.06	0.21	1.28	0.05	94.54
10	0.20	10.67	3.24	78.81	0.05	0.48	0.05	0.00	0.29	1.40	0.05	95.52
11	0.15	11.72	1.68	77.97	0.17	0.27	0.07	0.00	0.26	1.28	0.06	94.05
12	0.16	10.33	1.68	79.83	0.05	0.37	0.08	0.06	0.15	1.40	0.00	94.45
13	0.07	8.47	3.10	80.05	0.26	0.49	0.00	0.06	0.25	1.20	0.07	94.14
14	0.15	8.01	1.78	80.86	0.24	0.19	0.00	0.11	0.21	1.11	0.00	92.85
15	0.22	7.89	1.07	82.81	0.00	0.00	0.00	0.12	0.31	1.49	0.00	94.22
16	0.24	9.52	0.87	80.50	0.06	0.25	0.00	0.00	0.44	1.35	0.40	93.74
17	0.25	8.51	1.16	81.00	0.14	0.25	0.00	0.15	0.23	1.57	0.17	93.61
18	0.26	8.58	1.55	81.53	0.00	0.18	0.00	0.03	0.08	1.32	0.00	93.68
19	0.37	9.26	1.93	80.31	0.15	0.34	0.01	0.09	0.27	1.27	0.00	94.24
20	0.20	7.65	6.40	77.76	0.10	0.54	0.02	0.10	0.46	1.23	0.23	94.91
21	0.31	8.27	1.48	83.01	0.02	0.00	0.03	0.14	0.27	1.77	0.00	95.67
22	0.35	8.61	1.46	80.95	0.10	0.23	0.04	0.05	0.37	1.73	0.27	94.56
23	0.15	5.25	1.61	84.95	0.07	0.23	0.05	0.08	0.31	1.51	0.31	94.59
24	0.11	5.50	4.36	82.04	0.00	0.61	0.07	0.09	0.32	1.43	0.24	94.93
25	0.20	3.41	0.73	87.08	0.08	0.06	0.08	0.04	0.26	1.72	0.24	94.03
26	0.20	6.54	31.04	49.98	0.23	10.70	0.00	0.11	0.17	1.10	0.55	101.32
27	0.52	10.56	1.02	68.81	0.12	0.17	0.10	4.37	0.17	1.00	0.14	87.55
28	0.25	10.14	1.19	81.79	0.17	0.00	0.11	0.00	0.46	1.35	0.02	95.62
29	0.28	11.28	3.18	79.07	0.16	0.34	0.11	0.25	0.19	1.27	0.36	96.79
30	0.22	12.75	5.50	73.07	0.16	0.70	0.12	0.16	0.26	1.21	0.28	94.55

Note: Sample 11/638, N 1–3, ilmenite; 26, spinel; other analyses, titanomagnetite

5.9 Conclusions

1. The Chiney pluton is the best-layered magmatic pluton in Russia. Contrary to most layered mafic-ultramafic plutons, the Chiney pluton is dominated by gabbro rocks, enriched in titanomagnetite. The Chiney pluton comprises Fe-Ti-V and PGE-Cu deposits.
2. The pluton has a complicated inner structure and consists of four blocks, the Western, Central, Eastern, and the Southeastern. Titanomagnetite gabbro is widespread in the Western and Southeastern blocks, characterized by distinct layering.
3. Rhythmic layering of several scales is typical for the Chiney pluton: microrhythms (centimeters-decimeters), mesorhythms (several meters), and macrorhythms (tens to hundreds of meters).
4. Four main groups of intrusive rocks were recognized within the Chiney pluton: xenoliths of anorthosite and leucogabbro (5 vol %), high-Ti gabbro, low-Ti gabbro, and norite (~90 vol %) and magmatic breccias.
5. Composition of rock-forming minerals does not change essentially from bottom to top of the pluton.

References

Arkhangelskaya VV, Bykhov Yu, Volodin RN, Narkelyun LF, Skursky VC, Trubachev AI, Chechetkin VS (2004) Udokan copper and Katuginskoe rare metal deposits in the Chita region, Russia. Administration of Chita, Chita, Russia, p 519 (in Russian)

Belova NB (1980) Structure of the Chiney intrusive massif. Dissertation Moscow, MGRI, 187 (in Russian)

Buddington AF, Faney T, Vlisidis A (1936) Gravity stratification as a criterion for the interpretation of the structure of certain intrusives of the Northwestern Adirondacks. In: Reports of the 16 IGC. Washington, DC, pp 27–29

Chechetkin VS, Kharitonov YF (2000) Geology and mineral deposits of the Chita segment of BAM. Razvedka I Okhrana Nedr 1:12–18 (in Russian)

Chechetkin VS, Yurgenson GA, Narkelyun LF, Trubachev AI, Salikhov VS (2000) Geology and ore of the Udokan copper deposit: a review. Russian Geol Geophys 41:710–722

Dodin DA (2002) Metallogeny of the Taimyr–Norilsk Region, Northern Central Siberia. St. Petersburg, Nauka, 822 (in Russian)

Fedotova VM, Chechetkin VS, Savchenko AA, Kuzmina LS (1977) Fe–Ti mineralization in the Chiney gabbronorite pluton, Transbaikalia. Soviet Geol 4:136–141 (in Russian)

Glebovitsky VA, Khiltova VY, Kazakov IK (2008) Tectonic structure of the Siberian craton: interpretation of geological, geophysical, geochronological and isotope-geochemical data. Geotectonics 1:12–26

Golev VK, Gongalsky BI, Davy MN, Zinoviev YI, Krivenko VA, Narkekyun LF, Pereyaslovsky IV, Rutshtein IG, Sunkinzyan VS, Trubachev AI, Chechetkin VS (1987) Excursion metallogeny of Siberia. In: Krendelev FP (ed) Guidebook of the 11th all-union metallogenic conference. Novosibirsk, IGIG, p 81 (in Russian)

Gongalsky BI (1992) Chineyite–plagioclase-titanomagnetite rock. Bull MOIP, Section Geol 67(1):145–147 (in Russian)

Gongalsky BI (1993) A role of chineyite (plagioclase–titanomagnetite rock) in formation of the Chiney layered pluton, the Northern Transbaikalia. Bull MOIP, Section Geol 68(2):83–88 (in Russian)

Gongalsky BI (2012) Proterozoic metallogeny of the Udokan-Chiney ore district (Northern Transbaikalia). Abstract Doctor Sci dissert, IGEM RAS, 48 (in Russian)

Gongalsky BI (2015) Deposits of the unique metallogenic province of Northern Transbaikalia. Moscow, VIMS, 248 (in Russian)

Gongalsky BI, Krivolutskaya NA (1987) Microrhythm 1106420 in the Chiney pluton. Dokl Akad Nauk USSR 296(5):1199–1203 in Russian

Gongalsky BI, Krivolutskaya NA (1993) The Chiney layered pluton. Novosibirsk, Nauka, 184 (in Russian)

Gongalsky BI, Krivolutskaya NA, Ariskin AA, Nikolaev GS (2008a) Internal structure, composition and genesis of the Chineysky anorthosite–gabbronorite pluton. North Transbaikalia Geochem Int 46:637–665

Gongalsky BI, Sukhanov MK, Holtzman YV (2008b) Sm-Nd isotope system of the Chiney anorthosite – gabbronorite pluton (Eastern Transbaikaliaia). In: Laverov NP (ed) Problems of geology of ore deposits, mineralogy, petrology and geochemistry. IGEM, Moscow, pp 57–60 (in Russian)

Gongalsky BI, Makariev LB, Voyakovsky SK (2009) Meso–Cenozoic magmatism of the Udokan–Chiney district and uranium mineralization. In: Gordeev EI (ed) Volcanism and geodynamics. Petropavlovsk-Kamchatsky, pp 321–323 (in Russian)

Gongalsky BI, Krivolutskaya NA, Ariskin AA, Nikolaev GS (2016) The Chiney gabbronorite-anorthosite layered massif (Northern Transbaikalia, Russia): its structure, Fe-Ti-V and Cu-PGE deposits, and parental magma composition. Mineral Deposita 51(8):113–1034

Hofmann AW (1988) Chemical differentiation of the earth: relationship between mantle, continental crust and oceanic crust. Earth Planet Sci Lett 90:297–314

Konnikov EG (1986) Precambrian differentiated mafic–ultramafic complexes in the Transbaikalia region. Novosibirsk, Nauka, 224 (in Russian)

Konnikov EG, Epelbaum MB, Chekhmir AS (1981) The causes of potassium concentration in endocontact zone of the Chiney gabbronorite pluton. Geochemia 2:257–263

Kulikov AI, Kryukov VK, Morozova NN, Grechishnikov DN (1980) Ore types of the Chiney titanomagnetite deposits and their compositions. Geol Ore Deposits 22(5):85–88 in Russian

Kulikov AI, Golev VK, Grigor'ev VM, Kryukov VK (1981) Geology and titanomagnetite ore of the Chiney gabbro pluton. In: Mitrofanov GL (ed) Geology, prospecting and exploration of ore deposits. Irkutsk, pp 26–35 (in Russian)

Larin AM, Kovalenko VI, Kotov AB, Sal'nikova EB, Kovach VP, Makar'ev LB, Timashkov AN, Berezhnaya NG, Yakovleva SZ (2000) New data on the age of granites of the Kodar and Tukuringra complexes. East Sib: Geodyn Constraints Petrol 8(3):238–248

Lebedev AP (1962) The Chiney gabbro–anorthosite pluton, eastern Siberia. Moscow, USSR Acad Sci, 100 (in Russian)

Maier WD, Andreoli MAG, Groves DI, Barnes S-J (2013) Petrogenesis of Cu-Ni sulfide ores from O'Okiep and Kliprand, Namaqualand, South Africa: constraints from chalcophile metal contents. South Afr J Geol 115:499–514

Mel'nikova KM (1981) Conditions of ore localization and resource potential of the Chiney layered basic pluton. In: Kuznetsov VA (ed) Igneous rock associations in fold regions of Siberia: their origin, ore resource potential and mapping. OIGGM, Novosibirsk, pp 203–205 (in Russian)

Mel'nikova KM, Belova NB (1979) Structure of the Chiney ore field, southern part of the Kodar–Udokan trough. Geologiya I Razvedka 3:46–53 (in Russian)

Mel'nikova KM, Kryukov VK, Belova NB (1983) Patterns of localization of mineralization in the Chineisky stratified massif of the main rocks (Udokan ore district). Endogenous processes and metallogeny in the area of BAM. Novosibirsk Nauka 2:25–30

Mel'nikova TG (1970) The geology of the area in the vicinity of magnetic heights. Eastern Transvaal. With special reference to the magnetic iron ore. Geol Soc S Afr Spec Publ 1:228–241

Murzin VV, Moloshag VP, Volchenko VV (1988) Mineralogical paragenesis of PGE from Cu-Fe-V ores in Volkovsky type in the Urals. Dokl Earth Sci 300:1200–1202 (in Russian)

Petrusevich MN (1946) The Chiney titanomagnetite deposit. Sovetskaya Geologiya 10:91–94 (in Russian)

Polyakov GV, Isokh AE, Krivenko AP (2006) Platiniferous ultramafic–mafic formations of mobile belts of central and southeastern Asia. Russian Geol Geophys 47(12):1227–1241

Popov NV, Kotov AB, Postnikov AA, Sal'nikova EB, Shaporina MN, Larin AM, Yakovleva SZ, Plotkina YV, Fedoseenko AM (2009) Age and tectonic position of the Chiney layered massif, Aldan Shield. Dokl Earth Sci 424(1):64–67

Shabalin LI, Sharapov VN (1981) Elements of differentiation dynamics of the Chiney gabbro pluton. In: Bannikov OL, Velinskiy VV, Polyakov GV (eds) Problems of genetic petrology. Nauka, Novosibirsk, pp 163–180 (in Russian)

Tatarinov AV, Yalovik LI, Chechetkin VS (1998) A dynamometamorphic model of the formation of basic layered plutons: a case of the Chiney Pluton in Northern Transbaikalia. Novosibirsk, Nauka, 120 (in Russian)

Tolstykh ND (2008) PGE mineralization in marginal sulfide ores of the Chinei layered intrusion. Russia Mineral Petrol 92:283–306

Tolstykh ND, Orsoev DA, Krivenko AP, Izokh AE (2008) Noble-metal mineralization in mafic–ultramafic layered plutons in the southern Siberian platform. Novosibirsk, Papallel, 194 (in Russian)

Trofimovich DV, Chechetkin VS (1969) The application of the gravity method for studying the basic massifs at the estimation stage in conditions of Alpine-type relief. In: Mokshantsev KB (ed) Geology and deposits of southern Yakutia, pp 82–93 (in Russian)

Volchenko YA, Zoloev IN, Koroteev VA (1998) New prospective types of PGE mineralization in the Urals. In: Koroteev VA (ed) Geology and metallogeny of the Urals, vol 1. OAO UGSE, Yekaterinburg, pp 238–255 (in Russian)

Wager L, Brown G (1968) Layered igneous rocks. Oliver and Boyd, Edinburg

Yang S, Maier WD, Hanski EJ, Lappalainen M, Santaguida F, Määttä S (2013) Origin of ultra-nickeliferous olivine in the Kevitsa Ni-Cu-PGE-mineralized intrusion, northern Finland. Contrib Mineral Petrol 166:81–95

Zaccarinni F, Anikina E, Pushkarev E, Rusin I, Garuti G (2004) Palladium and gold minerals from the Baronskoe-Kluevsky ore deposit (Volkovsky complex, Central Urals, Russia). Mineral Petrol 82:137–156

Chemistry and General Typification of Intrusive Rocks

6

Abstract

Chemical composition of the Chiney mafic-ultramafic layered intrusion was studied in two main sections and from other rock types in the eastern part. All rocks, except norite, are enriched in Fe and Ti. The MgO and TiO_2 variations demonstrate a rhythmic structure of the Chiney pluton. Distribution of rare elements in rocks is similar to the distribution in typical crustal rocks with negative Ta-Nb and positive Pb anomalies. Techniques of geochemical thermometry with the use of the COMAGMAT computer program allowed to estimate the phase and chemical composition of the parental magma. The Chiney magma was represented by differentiated suspension of olivine, plagioclase, and magnetite crystals in ferrobasaltic melt at a temperature of approximately 1130 °C. The gravitational separation of these phases in the melt before its emplacement into the chamber and during the subsequent emplacement of various portions of the initial magma into the modern chamber predetermined the heterogeneity of the pluton.

6.1 Geochemical Characteristics

The above-presented data on the Chiney mafic-ultramafic intrusion testify that its inner structure is notably heterogeneous, as is pronounced in the variations in the chemical and mineralogical composition of the rocks, often on a small scale (from a few meters to a few centimeters, Figs. 5.8 and 5.13).

All chemical data (847 analyses) demonstrate (Fig. 6.1) their compact plotting on the diagram SiO_2-Mg# (Mg#= MgO/(MgO+FeO)*100), where some rock varieties composing the intrusion were distinguished. Analyses of two major rock groups (the Group 2, including the titanomagnetite gabbro and leucogabbro series and gabbro of the Group 3) form one field due to their low MgO (5.7 wt%) and high

TiO_2 (2.6 wt%) contents. Another field is formed by norite and pyroxenite of Group 3 because of their elevated MgO concentrations (>9.3 wt% on average) and relatively low concentrations of Fe and Ti (1.1 wt%). The fourth group (magmatic breccia) occupies a special place on the diagram due to its higher concentrations of SiO_2 and alkalis (Appendix 2, Table A2.1). Some data on rock geochemistry were published in our previous work (Gongalsky and Krivolutskaya 2009).

6.1.1 Major Elements

To analyze the behavior of major elements, we choose the representative data from different rock groups (Table 6.1): titanomagnetite gabbro (boreholes 83, 11), norite (boreholes 83,11, 301–303), and leucogabbro (501–518). The SiO_2 content changes in the rocks from 38.82 to 54.63 wt%. The lower values for this (and other) oxide are due to high titanomagnetite concentration (and opposite high FeO and TiO_2) in many samples. Most samples are gabbro. Anomalous MgO (8–14 wt%) confirms attribution of samples to norite and pyroxenite varieties (Fig. 6.2). The K_2O concentration does not exceed 1 wt%, except one sample, while most points lie inside the range of 0–0.2 wt% P_2O_5 and 0–3 wt% TiO_2. All mentioned oxides (SiO_2, FeO, TiO_2, K_2O, P_2O_5 –Fig. 6.2a, b, d, g, h) do not reveal any correlation with MgO. But some elements show positive (MnO; Fig. 6.2e) and negative (CaO, Na_2O; Fig. 6.2c, f) correlations with MgO contents.

The contrasting structure of the detailed sequences is accentuated by the distribution of major oxides, first of all, MgO and TiO_2 (Figs. 5.16 and 5.18). The rocks of the norite series are characterized by an elevated content of MgO, while the titanomagnetite gabbro series displays high TiO_2 concentrations. The rocks of the same group in the leucogabbro series exhibit a saw-shaped TiO_2 distribution due to alternating anorthosites and gabbroids enriched in titanomagnetite (Table 6.2).

© Springer Nature Switzerland AG 2019
B. Gongalsky, N. Krivolutskaya, *World-Class Mineral Deposits of Northeastern Transbaikalia, Siberia, Russia*,
Modern Approaches in Solid Earth Sciences 17, https://doi.org/10.1007/978-3-030-03559-4_6

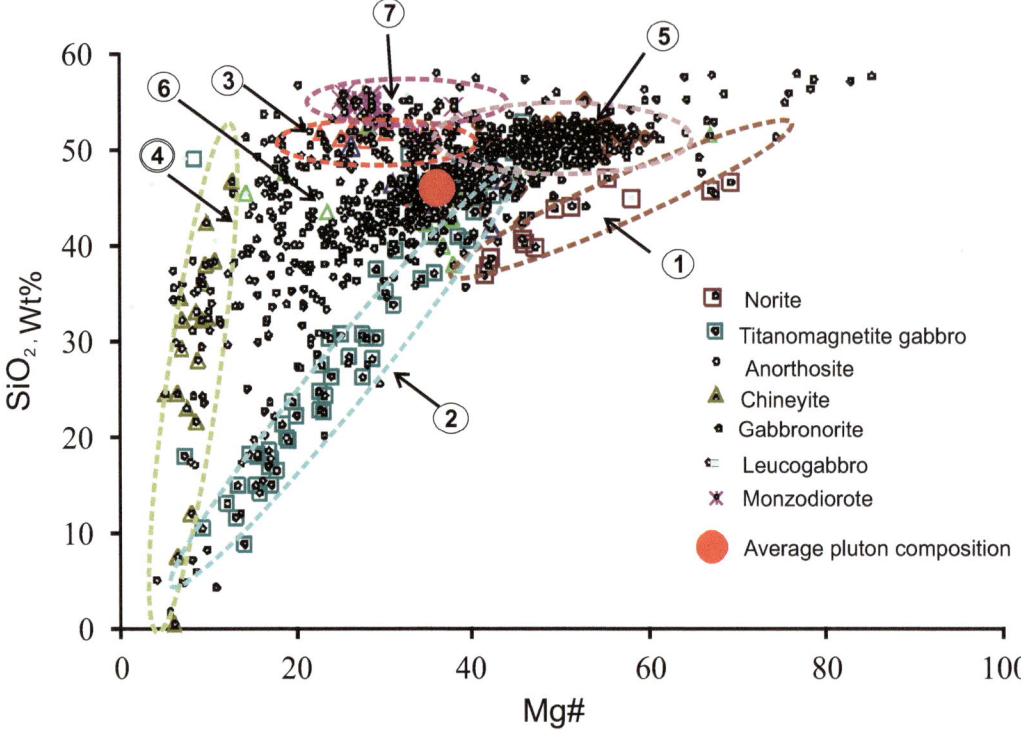

Fig. 6.1 SiO$_2$-MgO diagram for the Chiney rocks

The same series is characterized by clearly pronounced layering, which could be caused by the gravitational accumulation of titanomagnetite near the bottom of the layers. Boundaries between microrhythms are usually sharp and clear (Table 6.3, Fig. 6.3). From the lower to the upper contact, interstitial anhedral plagioclase appears, with the size of its grains gradually increasing in the same direction. From the bottom to the top, SiO$_2$, Na$_2$O, and K$_2$O increase, while Fe$_2$O$_3$ decreases. Thus, the chemical analyses of the Chiney rocks demonstrate their enrichment in Fe and Ti.

6.1.2 Trace and Rare Elements

Table 6.4 includes a set of representative analyses of rocks from the Chiney pluton for the isotopic data obtained. General characteristics of REE distributions in these samples, normalized to chondrite, are shown in Fig. 6.4. All these spectra have negative Ta and Nb anomalies combined with a positive Pb anomaly that are typical of the crust rocks. The presented spidergrams are subdivided into three types. The first type is the characteristic of Groups 2 and 3 (titanomagnetite gabbro, leucogabbro, and norite series) that occupy the main body of pluton. Both groups are distinguished by low Gd/Yb ratios (1.87–2.06 for Group 2 and <1.6 for Group 3) and an Eu minimum. The second type of

spidergrams is inherent to rocks of marginal facies (quartz diorite, monzodiorite), and the third type characterizes lamprophyre of Group 4.

Extended list of samples is given in Table 6.1. Patterns, including all spectrum of elements normalized to primitive mantle (Hofmann 1988), were constructed for these samples. Figure 6.5 demonstrates the characteristics of the rocks from the detailed Sects. 6.1 and 6.2 (from borehole 83 (Fig. 6.5a) and borehole 11 (Fig. 6.5b) and from outcrops of leucogabbro series (Fig. 6.5c)). Data for titanomagnetite gabbro from eastern part of the layered mafic-ultramafic complex mafic-ultramafic intrusion (boreholes 63 and 70) are summarized in Fig. 6.6. All these spectra have similar topology that indicate their cognation and belong to products of fractionation of the same primary magma. These patterns resemble spidergrams of the continental crust, which are characterized by high concentrations of incompatible elements (steep slope of patterns in their left parts, where (La/Sm)n varies from 0.9 to 6.0), negative Ta and Nb anomalies, positive Pb anomalies, and a pronounced Ti peak. The patterns of rocks pertaining to the norite series are distinguished by the absence of Ti anomaly owing to low concentrations of Fe-Ti oxides as solid phases. The above features of spidergrams sharply differ them from MORB, OIB, and other mantle-derived rocks (Joron and Treuil 1988; Hofmann 2003). The gentle slopes of the right parts of the patterns, characterized by a low (Gd/

Table 6.1 Composition of rocks from the Chiney pluton

N	1	2	3	4	5	6	7	8
Sample, N	**82/862**	**83/160**	**83/260**	**83/520**	**83/561**	**83/800**	**83/961**	**83/1020**
SiO_2	52.57	26.22	53.19	38.82	46.18	47.89	52.41	44.81
TiO_2	0.82	6.65	0.56	3.77	2.81	1.79	0.96	2.02
Al_2O_3	15.26	15.33	17.65	14.69	14.83	13.98	12.88	10.73
FeO	11.90	42.20	9.12	29.40	18.61	16.90	14.54	24.57
MnO	0.25	0.16	0.20	0.16	0.22	0.20	0.22	0.22
MgO	8.24	1.73	5.68	4.05	5.40	6.49	9.86	10.31
CaO	7.69	5.54	9.85	6.39	9.21	9.44	6.42	5.30
Na_2O	2.14	1.97	2.91	2.78	2.38	2.16	1.85	1.63
K_2O	0.68	0.25	0.54	0.36	0.35	0.87	0.72	0.44
P_2O_5	0.08	0.02	0.04	0.02	0.03	0.05	0.07	0.06
Total	99.64	100.1	99.75	100.4	100.00	99.77	99.93	100.10
LOW	0.85	0.64	0.18	0.37	0.98	0.74	1.86	2.03
Rb	21.3	4.8	14.0	9.9	7.5	28.5	30.7	19.7
Ba	266	118	209	128	163	308	251	172
Th	1.88	0.16	1.47	0.32	1.05	1.87	0.96	1.81
U	0.45	0.08	0.49	0.17	0.40	0.91	0.69	0.49
Nb	2.65	1.65	0.75	1.42	1.83	3.29	2.43	2.78
Ta	0.19	0.13	0.05	0.11	0.14	0.24	0.19	0.20
La	10.2	2.1	5.7	2.3	5.6	10.2	7.7	8.3
Ce	20.6	3.6	10.7	3.7	11.3	22.0	16.4	16.4
Pb	5.10	3.56	2.61	0.89	3.36	4.28	5.44	5.31
Pr	2.45	0.39	1.27	0.40	1.44	2.89	2.07	1.92
Nd	9.84	1.44	5.15	1.56	6.22	13.31	9.10	7.90
Sr	292	97	404	201	271	257	207	176
Sm	2.09	0.22	1.15	0.33	1.48	3.33	2.07	1.68
Zr	67.1	16.1	21.1	17.0	27.9	46.2	18.1	38.4
Hf	1.58	0.53	0.58	0.56	0.88	1.51	0.57	1.10
Eu	0.77	0.32	0.60	0.30	0.73	0.85	0.61	0.50
Ti	4697	42,7	3222	21,4	16,4	10,7	5012	12,6
Gd	2.00	0.21	1.18	0.47	1.57	3.38	2.16	1.64
Tb	0.32	0.03	0.19	0.09	0.25	0.59	0.34	0.26
Dy	2.08	0.16	1.26	0.61	1.71	3.61	2.26	1.70
Ho	0.44	0.03	0.25	0.12	0.34	0.72	0.46	0.35
Y	11.7	0.8	7.0	3.4	9.2	18.9	12.6	9.6
Er	1.29	0.09	0.79	0.40	0.99	1.97	1.40	1.01
Tm	0.20	0.02	0.11	0.06	0.14	0.29	0.19	0.15
Yb	1.36	0.08	0.79	0.34	0.99	1.85	1.44	1.06
Lu	0.21	0.02	0.12	0.06	0.14	0.24	0.21	0.17
Ni	92	382	17	121	72	123	123	229
Cu	82	1097	71	73	481	1032	337	152
Zn	113	228	46	165	138	61	78	137
Mn	1514	1339	1245	1139	1398	1396	1433	1982
Sc	35	22	36	35	56	43	41	40
Co	67	16	21	17	28	46	18	38

(continued)

Table 6.1 (continued)

N	9	10	11	12	13	14	15	16
Sample, N	83/1327	83/1240	63/309.2	63/342	63/41.5	63/521	11/121.2	11/260.2
SiO_2	40.50	45.01	52.79	49.95	54.63	51.18	46.99	44.22
TiO_2	2.81	2.68	0.74	1.72	0.64	0.97	1.96	1.86
Al_2O_3	9.67	14.83	13.99	15.82	13.37	16.98	13.50	7.75
FeO	25.33	20.79	12.38	13.06	11.55	11.23	18.96	21.41
MnO	0.14	0.17	0.21	0.16	0.18	0.19	0.17	0.29
MgO	4.46	4.04	8.21	5.15	8.92	5.78	6.83	9.90
CaO	16.64	8.67	8.02	10.13	7.16	9.89	8.01	12.88
Na_2O	0.35	2.50	2.44	2.56	2.30	2.70	2.32	1.06
K_2O	0.20	0.82	0.62	0.61	0.73	0.57	0.68	0.17
P_2O_5	0.05	0.09	0.06	0.07	0.22	0.03	0.10	0.01
Total	100.1	99.59	99.46	99.22	99.68	99.51	99.51	99.56
LOW	0.37	0.56	0.84	1.05	0.67	0.81	0.31	0.25
Rb	1.0	24.4	31.2	18.7	20.2	21.3	24.5	5.1
Ba	91	273	311	185	241	178	234	75
Th	2.38	1.01	2.94	1.45	2.04	2.39	2.50	0.92
U	0.99	0.51	0.90	0.69	0.44	0.51	0.64	0.29
Nb	2.59	1.92	4.02	2.91	2.25	2.89	3.66	0.86
Ta	0.19	0.14	0.30	0.20	0.16	0.23	0.25	0.07
La	45.4	9.9	12.9	8.5	14.4	10.7	11.9	4.4
Ce	74.9	19.5	26.5	18.1	29.8	20.7	25.1	10.5
Pb	0.40	5.71	5.37	4.19	4.34	3.87	5.44	2.62
Pr	8.30	2.24	3.17	2.27	3.48	2.44	3.18	1.50
Nd	32.80	8.73	13.08	9.79	13.74	9.85	13.87	7.17
Sr	600	357	236	214	246	301	245	191
Sm	5.57	1.63	2.69	2.16	2.77	2.08	2.98	1.93
Zr	53.9	20.9	56.6	48.1	17.2	33.9	56.2	26.4
Hf	1.41	0.54	1.60	1.45	0.56	0.99	1.63	0.86
Eu	1.38	0.76	0.86	0.66	0.79	0.75	0.80	0.57
Ti	17,2	3193	16,0	12	3709	5868	11,959	11,6
Gd	4.94	1.41	2.61	2.33	2.82	2.14	2.97	2.21
Tb	0.65	0.21	0.40	0.35	0.36	0.32	0.47	0.36
Dy	3.78	1.25	2.55	2.36	2.22	2.02	2.95	2.38
Ho	0.68	0.26	0.52	0.48	0.47	0.42	0.59	0.49
Y	19.7	6.9	13.9	12.9	13.0	11.2	15.8	13.2
Er	1.84	0.70	1.51	1.38	1.26	1.20	1.63	1.46
Tm	0.24	0.11	0.21	0.20	0.19	0.17	0.23	0.21
Yb	1.51	0.78	1.45	1.28	1.25	1.14	1.62	1.40
Lu	0.21	0.10	0.22	0.20	0.19	0.18	0.24	0.22
Ni	150	84	135	146	102	146	175	140
Cu	1397	224	122	141	240	324	320	323
Zn	56	72	131	109	80	81	109	162
Mn	1138	1192	1167	1318	1228	1190	1405	2166
Sc	41	18	40	59	34	38	45	88
Co	54	21	57	48	17	34	56	26

(continued)

Table 6.1 (continued)

N	17	18	19	20	21	22	23	24
Sample, N	11/295	11/379.8	11/507.3	11/561.9	11/603.43	11/837	11/989.4	11/991.3
SiO_2	50.05	49.95	44.32	47.97	46.00	51.85	52.48	43.74
TiO_2	1.11	1.03	2.23	1.91	2.28	0.97	0.84	2.41
Al_2O_3	23.47	16.63	21.24	12.92	15.12	12.53	11.61	15.21
FeO	9.36	11.86	18.39	18.47	16.66	14.58	13.93	21.02
MnO	0.05	0.17	0.07	0.22	0.21	0.21	0.25	0.18
MgO	1.42	6.27	1.22	7.87	5.52	10.80	11.83	4.67
CaO	10.09	11.16	8.30	7.31	10.63	6.15	6.27	9.70
Na_2O	3.56	2.51	3.30	2.28	2.51	1.82	1.73	2.38
K_2O	0.51	0.20	0.47	0.41	0.35	0.57	0.52	0.52
P_2O_5	0.08	0.03	0.07	0.03	0.11	0.06	0.05	0.04
Total	99.69	99.81	99.61	99.41	99.38	99.53	99.50	99.87
LOW	0.69	0.31	0.74	0.88	1.01	1.76	1.45	0.43
Rb	12.5	3.0	12.0	14.1	10.2	17.2	19.2	18.2
Ba	251	133	233	167	195	198	178	193
Th	1.51	0.12	1.72	1.75	0.87	2.48	2.44	1.99
U	0.46	0.06	0.49	0.51	0.35	0.64	0.61	0.92
Nb	1.65	0.28	1.91	1.90	2.16	2.35	2.16	2.29
Ta	0.12	0.03	0.14	0.14	0.16	0.16	0.16	0.17
La	7.8	2.6	8.2	7.9	8.0	11.5	8.1	7.9
Ce	15.0	4.9	15.9	16.0	16.0	22.6	15.9	16.1
Pb	4.97	1.73	5.06	3.99	6.64	5.46	3.14	1.79
Pr	1.77	0.67	1.85	1.91	2.16	2.56	1.85	1.99
Nd	7.23	3.29	7.18	8.03	9.26	9.83	7.48	8.20
Sr	466	316	397	241	281	208	229	276
Sm	1.45	1.01	1.36	1.71	1.90	1.80	1.53	1.78
Zr	28.5	7.4	28.8	32.7	34.7	31.4	50.9	41.6
Hf	0.79	0.35	0.85	1.02	0.99	0.90	1.31	1.12
Eu	0.71	0.68	0.68	0.62	0.74	0.55	0.54	0.62
Ti	6885	6002	13,043	12,162	13,879	5654	4963	14,0
Gd	1.42	1.28	1.31	1.75	1.99	1.68	1.49	1.85
Tb	0.22	0.23	0.19	0.27	0.29	0.25	0.24	0.29
Dy	1.32	1.52	1.14	1.78	1.39	1.69	1.58	1.95
Ho	0.26	0.29	0.23	0.37	0.40	0.34	0.35	0.40
Y	7.1	7.7	6.2	10.2	10.1	9.7	9.1	9.8
Er	0.78	0.83	0.64	1.07	1.05	1.04	1.02	1.17
Tm	0.10	0.11	0.09	0.16	0.16	0.16	0.16	0.16
Yb	0.70	0.81	0.63	1.09	1.13	1.11	1.05	1.09
Lu	0.10	0.12	0.09	0.17	0.15	0.17	0.17	0.14
Ni	111	181	187	143	115	159	115	144
Cu	311	198	865	745	639	106	143	592
Zn	59	67	77	125	91	126	115	66
Mn	532	1204	676	1797	1300	1703	1705	1138
Sc	13	34	14	43	44	41	42	42
Co	28	7	29	33	35	31	51	42

(continued)

Table 6.1 (continued)

N	25	26	27	28	29	30	31
Sample, N	11/1205	11/1289.2	70/177.7	70/455.3	70/558	301	302
SiO$_2$	52.49	49.75	52.64	47.77	53.80	48.74	43.63
TiO$_2$	0.67	1.70	0.78	1.91	0.94	1.28	1.91
Al$_2$O$_3$	12.44	14.09	15.29	14.73	14.24	5.48	1.78
FeO	13.79	16.28	11.71	14.52	11.93	10.50	27.21
MnO	0.22	0.19	0.20	0.15	0.20	0.20	0.30
MgO	9.75	5.54	8.10	3.56	7.24	11.15	14.21
CaO	6.60	8.42	7.70	13.56	7.38	22.50	11.27
Na$_2$O	1.54	2.37	2.26	2.32	2.55	0.06	0.10
K$_2$O	1.61	0.84	0.58	0.74	0.90	0.01	0.03
P$_2$O$_5$	0.07	0.09	0.06	0.07	0.11	0.04	0.02
Total	99.18	99.26	99.33	99.34	99.29	99.96	100.46
LOW	0.99	0.56	1.02	0.97	0.35	0.99	1.01
Rb	65.8	28.7	19.3	21.7	29.6	0.5	2.3
Ba	242	274	212	251	321	14	15
Th	2.88	3.13	2.02	2.54	2.50	0.69	0.38
U	0.85	0.85	0.54	0.65	0.72	0.19	0.12
Nb	3.35	3.75	2.76	2.71	4.06	0.54	2.63
Ta	0.24	0.28	0.19	0.22	0.30	0.05	0.19
La	11.8	12.5	9.1	10.3	15.9	4.1	2.5
Ce	23.5	24.9	18.9	21.7	32.0	11.4	7.5
Pb	4.57	4.40	4.23	2.69	6.63	2.38	1.84
Pr	2.81	2.96	2.30	2.66	3.75	1.80	1.21
Nd	11.54	12.08	9.37	11.26	15.09	9.03	6.11
Sr	165	242	265	326	270	17	21
Sm	2.40	2.47	1.99	2.42	3.00	2.46	1.68
Zr	55.2	71.4	29.3	55.0	24.7	59.4	20.0
Hf	1.52	1.94	0.90	1.79	0.82	2.52	0.68
Eu	0.70	0.83	0.59	0.64	0.91	0.44	0.27
Ti	3582	10,330	5151	11,781	5480	7608	11,437
Gd	2.41	2.49	1.99	2.27	2.93	2.85	1.79
Tb	0.40	0.40	0.30	0.33	0.45	0.44	0.29
Dy	2.44	2.52	2.02	2.01	2.85	2.89	1.79
Ho	0.51	0.52	0.41	0.39	0.58	0.55	0.35
Y	14.2	13.9	11.2	10.4	15.4	14.4	9.3
Er	1.51	1.52	1.22	1.12	1.68	1.52	1.00
Tm	0.23	0.21	0.18	0.16	0.24	0.21	0.14
Yb	1.54	1.45	1.21	1.07	1.66	1.42	0.95
Lu	0.23	0.23	0.19	0.18	0.25	0.21	0.15
Ni	422	86	80	44	101	132	134
Cu	1682	191	89	163	157	118	77
Zn	75	103	119	105	99	35	142
Mn	1409	1492	1451	987	1328	1440	2330
Sc	39	43	37	39	34	76	94
Co	55	71	29	55	25	59	20

(continued)

Table 6.1 (continued)

N	32	33	34	35	36	37
Sample, N	303	520	514	513	518	34/487
SiO_2	54.18	47.50	52.77	52.02	46.31	52.31
TiO_2	0.70	2.26	0.42	0.52	2.05	0.50
Al_2O_3	14.73	14.23	15.16	14.45	13.80	15.43
FeO	11.01	16.76	10.45	11.07	16.82	10.07
MnO	0.21	0.20	0.26	0.23	0.17	0.22
MgO	8.44	5.39	10.35	10.79	6.41	10.99
CaO	7.18	10.45	8.19	8.09	11.31	7.69
Na_2O	2.41	2.41	2.14	2.07	2.09	1.96
K_2O	0.70	0.59	0.26	0.25	0.31	0.38
P_2O_5	0.07	0.07	0.08	0.06	0.10	0.23
Total	99.64	99.86	100.08	99.55	99.37	99.76
LOW	0.66	0.65	0.99	0.93	0.47	0.97
Rb	21.4	16.0	3.4	3.3	8.3	6.2
Ba	239	212	197	190	154	256
Th	3.07	1.70	0.18	0.17	0.69	0.55
U	0.88	0.45	0.05	0.04	0.20	0.11
Nb	2.66	2.50	0.58	0.61	0.78	2.58
Ta	0.21	0.19	0.03	0.03	0.06	0.13
La	10.6	7.8	4.6	4.4	6.2	9.1
Ce	21.4	16.2	9.6	8.6	13.2	18.1
Pb	4.96	1.12	6.44	6.87	3.20	19.46
Pr	2.54	2.12	1.23	1.05	1.77	2.20
Nd	10.23	9.20	5.30	4.58	7.91	9.03
Sr	267	291	328	334	292	465
Sm	2.07	2.20	1.22	1.06	1.86	1.67
Zr	70.8	40.1	12.4	11.3	17.1	26.8
Hf	1.62	1.23	0.42	0.36	0.54	0.59
Eu	0.62	0.82	0.56	0.57	0.74	0.65
Ti	3954	15,084	3166	2849	10,764	3284
Gd	2.05	2.29	1.23	1.08	2.07	1.53
Tb	0.31	0.35	0.19	0.17	0.32	0.21
Dy	1.95	2.25	1.24	1.17	1.94	1.33
Ho	0.41	0.45	0.25	0.24	0.39	0.27
Y	11.0	12.0	7.1	6.5	10.4	7.4
Er	1.21	1.28	0.80	0.69	1.09	0.75
Tm	0.18	0.18	0.12	0.11	0.15	0.12
Yb	1.20	1.20	0.81	0.79	0.99	0.82
Lu	0.18	0.18	0.13	0.12	0.15	0.13
Ni	67	57	265	204	62	172
Cu	85	230	135	93	191	81
Zn	97	43	122	108	92	171
Mn	1377	1358	1579	1481	1322	1555
Sc	32	54	35	34	52	31
Co	71	40	12	11	17	27

Note: Oxides are given in %, elements in ppm. Major components were analyzed by XRF at IGEM (analyst AI Yakushev), trace elements by LA-ICP-MS at MPI (analyst DV Kuzmin). Sample, N = 83/160 = borehole number/depth, m

Fig. 6.2 Harker diagrams for rocks from the detailed sections of the pluton MgO-SiO₂ (**a**), FeO (**b**), CaO (**c**), TiO₂ (**d**), MnO (**e**), Na₂O (**f**), K₂O (**g**), and P₂O₅ (**h**)

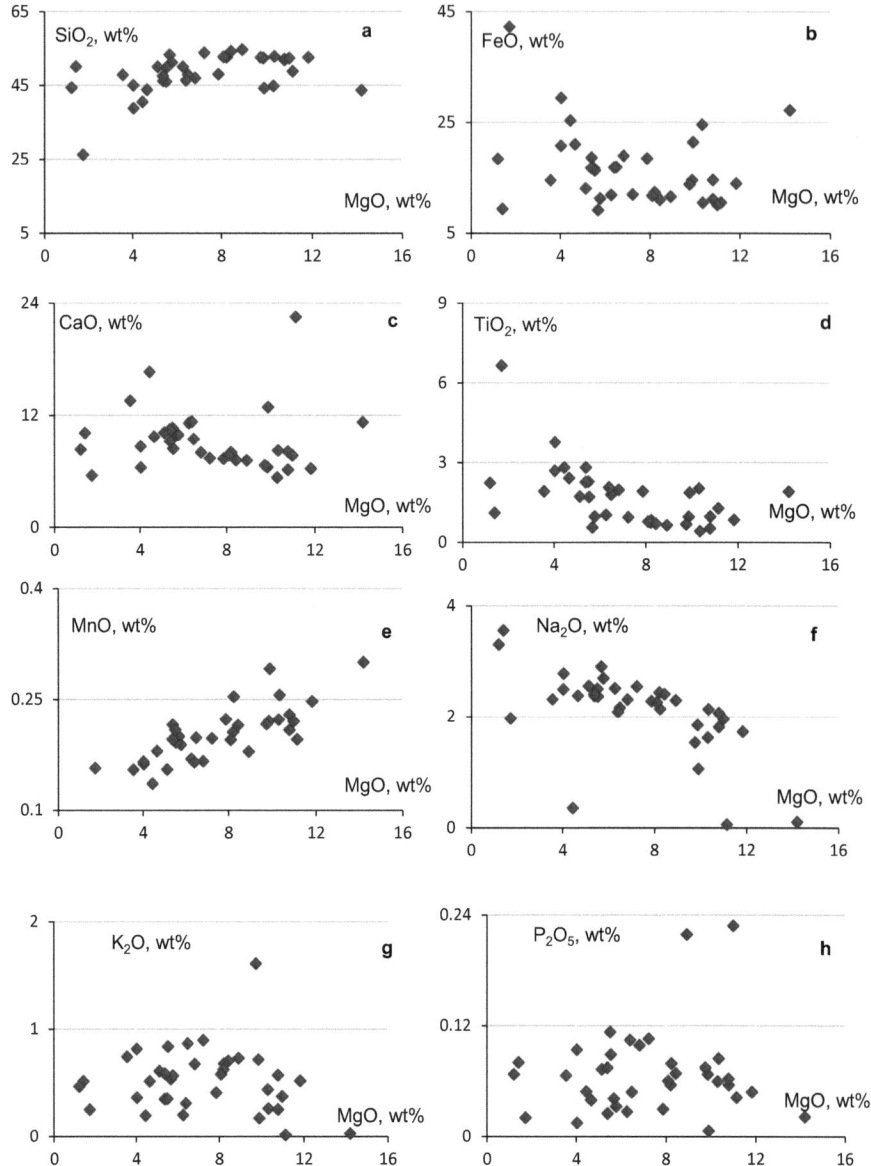

Yb)n = 1.0–2.7, emphasize common geochemical specification of the primary magma for the rocks of Groups 2 and 3 and provide evidence for lack of garnet in the magma source.

Nevertheless, some difference in patterns of rocks exists. It can be clearly seen in the right part of the spectra (HREE), especially for rocks of leucogabbro series (Fig. 6.6c). This difference is demonstrated in Fig. 6.7b, where (La/Yb)n ratio (2.2–9.4, two results are outside this interval for unknown reason) characterizes the slope of spectrum. At the same time, they have a wide range in (La/Sm)n ratio (Figs. 6.6 and 6.7a).

The higher degree of enrichment in incompatible elements (samples 0305, 0321/1) in marginal facies and lamprophyres is caused by fractionation of primary magma with

the formation of relatively enriched residual melts. The main distinction of the second-type pattern, without a positive Ti anomaly and even with a slight Ti minimum in spidergrams, is apparently related to the same cause. These relationships are typical of magma fractionation under conditions of magnetite crystallization.

The characteristics of patterns of pyroxenite from xenoliths of Group 1 differ markedly from patterns of the other rock groups. Pyroxenite is depleted in LREE, e.g., the (La/Sm)n ratio varies from 0.99 to 1.12, and the REE contents are close to those in a primitive mantle. They are also depleted in LILE (Cs, Rb, Ba, Sr, Rb, K) and enriched in HFSE (Th, U). In addition, these rocks are devoid of Ta and Nb minima and have a distinct negative Sr anomaly due to

Table 6.2 Composition of rocks in cross section of the Chiney massif, wt%

№.	SiO$_2$	TiO$_2$	Al$_2$O$_3$	Fe$_2$O$_3$	MnO	MgO	CaO	Na$_2$O	K$_2$O	P$_2$O$_5$	LOW	Total	№ sample
1	53.79	2.40	13.94	14.75	0.23	1.63	6.56	3.83	1.70	0.70	0.47	100.00	321/1
2	53.77	2.12	15.93	11.76	0.24	1.45	5.67	4.85	2.95	0.54	0.71	100.00	321/2
3	56.82	1.82	15.94	11.35	0.20	1.63	6.15	3.40	1.66	0.60	0.43	100.00	321/4
4	42.44	2.93	11.37	21.68	0.22	6.63	10.74	2.38	0.62	0.10	1.16	100.27	522
5	43.61	2.49	13.07	20.10	0.18	5.79	10.20	2.79	0.92	0.09	0.06	99.30	521
6	44.98	2.48	13.62	19.87	0.18	5.80	11.00	2.96	0.67	0.11	0.18	101.82	520
7	43.09	2.69	12.42	20.05	0.21	6.91	12.11	2.88	0.42	0.07	0.00	100.85	519
8	42.94	2.68	13.21	19.23	0.2	6.68	11.52	2.87	0.39	0.11	0.38	100.21	518
9	45.58	2.29	13.87	17.84	0.19	6.44	11.96	2.79	0.47	0.07	0.34	101.85	517
10	44.83	2.31	12.85	17.70	0.19	6.52	12.21	2.64	0.52	0.06	0.14	99.97	516
11	39.77	3.34	9.88	23.49	0.22	7.77	11.93	2.28	0.35	0.08	0.29	99.38	515
12	51.42	0.61	13.91	9.99	0.21	11.42	8.93	2.40	0.35	0.06	0.44	99.73	514
13	47.97	1.59	23.69	9.77	0.06	1.27	10.34	4.07	0.73	0.08	1.35	100.93	513
14	43.40	2.15	18.87	17.30	0.11	3.03	9.67	4.35	0.68	0.09	0.20	99.85	512
15	46.16	1.64	14.54	15.88	0.16	6.16	12.26	3.71	0.39	0.07	0.38	101.34	511
16	52.61	0.31	24.14	4.93	0.05	1.10	9.90	4.14	1.25	0.28	1.62	100.34	510
17	46.15	2.17	14.86	16.16	0.02	5.65	12.2	2.96	0.44	0.07	0.16	100.83	509
18	47.98	1.66	12.06	15.85	0.21	7.47	12.08	3.39	0.60	0.12	0	101.40	508
19	46.11	1.89	14.26	15.35	0.16	6.07	11.31	3.03	0.62	0.10	0.20	99.10	507
20	48.67	1.49	14.05	14.15	0.2	7.12	11.41	3.00	0.49	0.07	0.17	100.81	506
21	45.77	2.02	21.58	13.69	0.07	1.28	9.00	4.13	1.00	0.09	1.32	100.65	505
22	46.82	1.79	14.94	15.97	0.18	6.13	10.79	3.93	0.61	0.08	0.18	101.43	504
23	38.09	3.77	8.07	26.59	0.29	9.28	11.74	2.46	0.28	0.05		100.14	503
24	43.45	2.43	13.76	18.46	0.20	6.17	11.34	3.42	0.53	0.11	0	99.85	502
25	45.98	2.01	14.10	16.29	0.19	6.3	11.38	2.89	0.51	0.08	0.11	99.82	501
26	49.55	0.78	22.50	8.55	0.08	2.47	10.51	3.22	0.61	0.14	1.11	99.51	8300201
27	46.54	1.43	16.20	15.97	0.16	7.65	8.04	2.49	0.53	0.09	0.51	99.61	8300388
28	43.03	2.82	18.83	19.35	0.13	3.00	8.89	3.05	0.49	0.08		99.54	8300600
29	33.36	4.69	16.49	32.19	0.12	1.70	7.47	3.68	0.39	0.03	0.52	100.64	8300800
30	32.32	4.62	17.91	32.95	0.12	1.87	6.44	3.16	0.46	0.02	0.30	100.16	8301400
31	23.34	7.19	14.34	45.65	0.16	2.07	5.45	2.98	0.33	0.02		101.20	8301600
32	42.96	2.76	15.91	19.91	0.14	4.48	10.82	2.64	0.39	0.03		99.95	8301786
33	45.28	2.17	11.96	19.81	0.19	6.68	12.10	2.25	0.49	0.03	0.29	101.24	8302000
34	44.93	1.96	13.86	17.69	0.17	5.98	11.26	2.58	1.09	0.02	1.85	101.38	8302200
35	43.31	2.80	16.56	22.04	0.15	4.80	8.23	2.74	0.53	0.03	0.23	101.41	8302400
36	52.03	0.64	17.21	10.36	0.16	5.76	9.88	2.83	0.65	0.04	0.68	100.23	8302600
37	51.68	0.64	17.05	11.49	0.14	6.14	8.96	2.69	0.73	0.06	1.11	100.69	8302830
38	45.94	1.36	16.38	15.90	0.14	4.38	10.10	4.75	0.69	0.08	1.55	101.25	8303000
39	45.85	1.83	13.92	17.28	0.16	6.91	10.14	2.48	0.49	0.03	2.80	101.87	8303202
40	43.12	2.92	13.27	20.22	0.18	6.46	10.33	2.24	0.52	0.03	1.15	100.44	8303366
41	44.65	1.84	12.36	17.16	0.17	7.96	11.75	2.37	0.49	0.03	0.37	99.15	8303655
42	46.16	1.69	15.08	16.47	0.17	6.12	12.02	2.79	0.61	0.02	0.31	101.44	8303805
43	45.00	1.77	13.11	18.88	0.20	7.41	11.34	2.43	0.32	0.02		100.16	8304000
44	46.46	1.91	13.70	16.85	0.18	6.42	12.07	2.27	0.40	0.03	0.71	100.99	8304200
45	49.83	0.87	15.11	12.06	0.14	6.52	13.15	2.18	0.32	0.02	0.42	100.62	8304401
46	40.26	2.73	11.87	24.15	0.19	6.92	10.87	2.41	0.34	0.03	0.25	100.01	8304602
47	44.63	2.26	15.42	18.51	0.17	5.76	10.83	2.71	0.48	0.03	0.33	101.12	8304800
48	35.68	4.29	13.21	32.34	0.16	4.42	6.61	3.21	0.43	0.02	0.51	100.88	8305202
49	43.44	2.77	13.18	22.28	0.20	6.30	10.95	2.33	0.42	0.04		101.56	8305400
50	37.99	3.49	14.30	26.55	0.16	4.57	9.51	2.93	0.45	0.03		99.57	8305600
51	45.25	3.01	13.82	19.99	0.19	5.69	7.90	2.53	0.45	0.03	0.43	99.29	8305617
52	44.23	3.03	12.90	20.69	0.20	6.30	10.31	2.32	0.48	0.03		100.40	8305800
53	34.07	4.20	14.35	31.88	0.16	4.21	8.59	2.42	0.38	0.02	0	100.27	8306000

(continued)

Table 6.2 (continued)

№.	SiO₂	TiO₂	Al₂O₃	Fe₂O₃	MnO	MgO	CaO	Na₂O	K₂O	P₂O₅	LOW	Total	№ sample
54	37.71	3.63	14.08	27.67	0.16	3.74	9.35	3.50	0.66	0.04	0.28	100.82	8306200
55	34.32	4.54	12.37	32.12	0.14	4.77	8.28	3.31	0.34	0.02	0.84	101.06	8306400
56	47.79	1.50	15.20	15.35	0.17	5.85	12.14	2.97	0.59	0.06	0	101.62	8306600
57	43.64	2.50	13.60	20.75	0.19	6.18	10.06	2.78	0.68	0.09	0.22	100.69	8306803
58	45.33	1.60	13.27	19.22	0.22	7.89	10.74	2.50	0.39	0.03	0.42	101.60	8307000
59	44.97	1.68	14.96	15.46	0.16	5.85	9.55	2.62	0.87	0.07	4.43	100.62	8307197
60	40.29	3.29	12.28	24.38	0.21	6.78	10.25	2.65	0.45	0.05		100.17	8307380
61	40.44	2.70	14.86	23.77	0.15	4.08	9.57	2.48	0.48	0.04		99.07	8307600
62	35.00	5.48	16.97	27.81	0.14	1.77	7.63	3.36	0.56	0.03	0.31	99.06	8307815
63	45.95	1.97	12.37	18.39	0.19	6.92	9.71	2.39	1.05	0.07	0.37	99.37	8308000
64	44.62	2.23	13.58	20.43	0.20	7.89	7.39	2.50	0.54	0.04	0.23	99.65	8308200
65	48.93	1.22	14.97	14.56	0.20	8.43	7.68	2.66	0.81	0.08	0	99.54	8308400
66	50.99	0.78	15.14	12.61	0.19	8.26	7.26	2.58	0.78	0.07	0.61	99.26	8308628
67	50.57	0.91	18.96	10.45	0.13	5.33	8.95	2.58	1.02	0.10	1.96	100.94	8308875
68	51.25	0.90	15.54	12.46	0.17	7.68	8.08	2.28	1.06	0.09	1.84	101.34	8309032
69	49.04	1.33	13.98	15.92	0.19	9.32	6.97	2.21	0.99	0.09	1.34	101.37	8309270
70	47.41	1.76	9.20	21.69	0.27	12.4	5.67	1.79	0.71	0.08	0.07	101.05	8309286
71	48.29	1.42	12.28	18.80	0.20	9.18	6.37	2.56	0.77	0.06		99.83	8309406
72	49.21	0.93	12.45	16.04	0.20	10.59	6.84	2.87	0.92	0.09	0.68	100.82	8309610
73	49.13	0.91	15.47	13.16	0.16	7.64	8.15	2.44	0.95	0.08	0.95	99.04	8309812
74	49.72	1.20	11.61	16.73	0.23	10.41	6.88	3.64	0.92	0.09	0.29	101.70	8310000
75	42.45	2.10	9.93	26.72	0.23	10.93	5.48	2.39	0.57	0.06		100.25	8310208
76	49.94	1.22	10.88	16.55	0.22	11.18	6.67	2.73	0.85	0.07	0.43	100.73	8310365
77	46.58	0.87	12.92	17.55	0.18	8.96	7.12	2.30	0.73	0.06	1.85	99.11	8310410
78	56.26	1.01	19.02	9.62	0.07	3.46	1.21	3.45	4.28	0.11	0.82	99.29	8310620
79	50.63	1.36	11.73	16.64	0.21	9.19	7.55	3.68	0.86	0.08	0	101.92	8310800
80	53.51	1.28	18.24	8.44	0.18	7.23	9.18	2.48	0.82	0.03	0.29	101.67	8311012
81	51.03	1.16	12.37	13.60	0.23	12.17	6.80	1.80	0.66	0.04		99.69	8311226
82	46.88	2.13	14.31	18.76	0.16	4.75	9.50	2.42	0.91	0.09	0	99.91	8311435
83	41.29	3.17	12.18	26.56	0.24	6.38	5.95	2.35	1.41	0.12	0.36	100.02	8311600
84	50.18	0.67	9.07	10.09	0.16	10.91	17.88	1.76	0.37	0.07	0.65	101.79	8311800
85	40.67	3.94	16.83	25.36	0.16	3.34	7.83	2.91	0.99	0.09		101.73	8312004
86	46.65	2.44	14.79	18.19	0.17	4.94	9.14	3.18	1.15	0.12	0.34	101.10	8312200
87	43.52	2.67	15.44	21.89	0.14	4.16	8.03	2.57	0.98	0.09	0	99.49	8312400
88	37.80	3.66	13.99	28.33	0.15	4.26	6.85	3.29	1.68	0.08	0.34	100.44	8312610
89	51.44	1.24	13.94	14.59	0.17	6.66	7.62	2.24	1.24	0.090	1.03	100.26	1112750
90	53.81	0.81	18.10	9.76	0.12	4.42	8.58	2.45	0.95	0.120	0.94	100.06	1112845

Note: (1) Analyses were made by XRF (analysts NS Baluev, LM Bad'ina, TP Mikhailova; Chita Institute of Natural Resources, Siberian Division, Russian Academy of Sciences); (2) analysis numbers correspond to the following series: 321/1–321/3 and 501–522, leucogabbro; 8300201–8308875, titanomagnetite gabbro; 8309032, gabbronorite; 1112750–1112845, lamprophyres and hole 11; (3) samples for the analyses were taken at sites spaced 20 m apart in hole 83 and above it (from exposures). Seven-digit numbers: the first two digits correspond to the hole number, and the third through seventh digits correspond to the depths (in decimeters)

depletion in plagioclase. Only positive Ti anomaly and HREE pattern combine pyroxenites together with the mafic rocks, most abundant in the Chiney pluton.

6.2 Petrochemistry

The ferrobasaltic composition of the initial magma from the Chiney pluton poses a question on the nature of initial and derivative magmatic melts. The substantial internal heterogeneity of the Chiney pluton is emphasized by significant variations of mineral and chemical compositions of the

rocks, often documented at the microlevel (few centimeters). Nevertheless, the variation diagram (Fig. 6.1) displays that relatively stable groups (varieties of rocks) actually exist. They obey general petrologic laws of magma evolution and rock distribution throughout the pluton.

To reveal these groups and establish their spatial distribution, we used a method of petrochemical typification of igneous rocks, which was elaborated at the Department of Geochemistry of the Moscow State University. It was applied to describe internal structures of the Kivakka layered pluton in Karelia and Yoko-Dovyren pluton in the northern Baikal region (Yaroshevsky 2004; Yaroshevsky et al. 2006). This

Table 6.3 Compositions of rocks inside microrhythms, wt%

Depth, cm	Rhythm	SiO$_2$	TiO$_2$	Al$_2$O$_3$	Fe$_2$O$_3$	FeO	MnO	MgO	CaO	Na$_2$O	K$_2$O	P$_2$O$_5$
68	5	17.82	8.84	4.07	27.26	29.51	0.29	6.73	3.47	1.79	0.20	0.01
69		18.67	8.74	3.61	28.41	28.06	0.31	6.58	3.94	1.48	0.19	0.01
71	4	41.42	3.36	7.21	10.12	16.89	0.26	9.96	9.35	0.98	0.37	0.08
73		35.49	4.67	6.81	15.11	19.84	0.28	8.81	7.76	0.94	0.26	0.03
74		30.81	5.74	6.03	19.41	21.95	0.30	9.12	5.30	1.01	0.30	0.03
75		21.41	8.13	4.33	26.47	27.77	0.32	7.05	3.27	0.99	0.24	0.03
76		17.02	9.05	3.40	29.92	29.62	0.31	6.91	2.14	1.43	0.19	0.01
78	3	33.84	5.09	5.42	16.84	22.66	0.32	10.40	4.17	0.92	0.29	0.05
79		28.45	6.70	5.71	21.06	23.73	0.31	9.15	3.68	0.94	0.25	0.02
80		23.64	7.53	4.77	26.79	25.23	0.32	7.31	3.21	0.95	0.23	0.02
81		19.99	8.50	4.43	27.49	27.93	0.30	7.43	2.86	0.86	0.21	0.01
83	2	49.92	1.58	18.48	1.07	12.74	0.13	4.15	9.02	2.36	0.49	0.06
84		45.26	2.51	8.53	7.85	17.08	0.28	10.88	5.88	1.27	0.42	0.04
85		43.34	2.91	7.75	10.33	17.25	0.29	10.97	5.68	1.09	0.33	0.06
86		37.08	4.45	5.37	13.79	21.94	0.32	11.63	4.14	0.99	0.24	0.05
87		30.25	6.13	4.11	19.15	24.77	0.32	10.49	3.59	0.98	0.20	0.02
89	1	48.17	1.94	14.20	8.32	9.57	0.19	6.51	8.42	2.12	0.48	0.08

Note: Analyses were made by XRF (analysts NS Baluev, LM Bad'ina, TP Mikhailova; Chita Institute of Natural Resources, Siberian Branch, Russian Academy of Sciences); samples were taken from the core of hole 11, depth interval 622.68–622.89 m

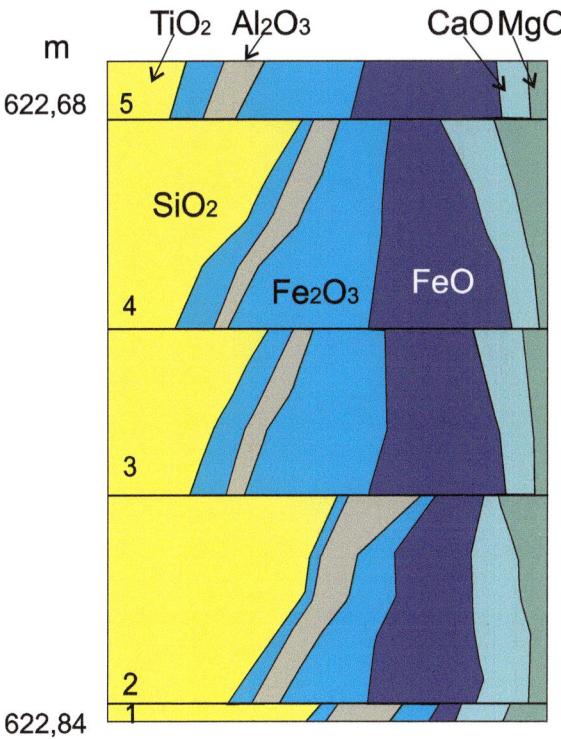

Fig. 6.3 Variations in major oxides in microrhythms of the lower part of the titanomagnetite gabbro series (rock Group 2). Numbers correspond to discrete microrhythms

approach allowed us to optimize the handling of the petrochemical data and to select the representative chemical compositions for using them at the next stage of research to estimate chemical composition of the initial magma and to determine physicochemical parameters of its crystallization. The typification of rocks and computer simulation of crystal-

lization conditions and phase sequences were carried out with assistance of Ariskin and Nikolaev (Gongalsky et al. 2008).

This method of petrochemical typification is based on the computer program that realized a principle of hierarchical clustering of objects by formal parameters using the Ward's algorithm (Ward 1963). Being applied to a great number of chemical characteristics, this principle suggests a consecutive combination of composition, each step of which is characterized by a minimal increment of average "chemical distance" between the compositions combined into groups (Yaroshevsky et al. 2006). The geochemical distances are calculated in the Euclidean metrics by means of relative concentrations normalized to dispersion for each component in the initial selection of compositions. This allows us to reduce the contents of various components to a common scale and makes comparable their contributions to the calculation of cluster compactness. Classification of rocks from the Chiney pluton was performed using a special version of the PETROTYPE software, which was developed at the Institute of Geochemistry and Analytical Chemistry, Russian Academy of Sciences for the petrochemical typification of volcanic rocks. To recognize stable petrochemical types of rocks, eight major components were used. They are SiO$_2$, TiO$_2$, Al$_2$O$_3$, FeOtot, MgO, CaO, Na$_2$O, and K$_2$O.

We took chemical compositions of rock from detailed Sect. 6.1 (Fig. 5.16, samples from borehole 83 and along the reference section in the Western block of pluton, collected at a distance of 20 m from each other, numbers 501–522). The composition of rocks from the eastern part of pluton, which does not occur in the Western block, was additionally included. The published chemical composition of core samples from borehole 11 (Gongalsky and Krivolutskaya 1993)

Table 6.4 Representative analyses of the main rock types from the Chiney Pluton

Sample	O301	2101/2	O302	O303	O304	O305	O306	O307	O321/1
Type	1	1	1	9	12	10	7	12	10
SiO_2	47.54	41.83	43.62	53.54	46.38	55.78	45.5	45.87	53.28
TiO_2	1.24	2.72	1.96	0.71	2.47	1.14	3.17	2.50	2.38
Al_2O_3	4.85	4.42	1.76	15.37	17.11	17.15	16.06	21.08	13.81
Fe_2O_3	10.62	26.99	24.35	10.88	17.27	9.35	17.28	14.40	14.61
MnO	0.17	0.32	0.29	0.18	0.17	0.13	0.16	0.11	0.23
MgO	11.62	11.22	15.46	7.76	3.76	2.07	3.68	1.74	1.61
CaO	21.02	12.17	11.27	7.15	8.26	7.80	9.62	9.58	6.50
Na_2O	0.08	0.26	0.16	2.63	3.40	3.75	3.16	3.54	3.79
K_2O	0.01	0.16	0.02	0.63	0.60	1.49	0.42	0.33	1.68
P_2O_5	0.012	0.07	0.006	0.08	0.10	0.13	0.08	0.033	0.69
Cr_2O_3	0.011	0	0.028	0.014	0.014	0.004	0.01	0.01	0.002
LOI	2.52	0.24	0.63	0.36	0.11	0.23	0.18	0.27	0.7
Total	99.69	100.40	99.55	99.30	99.64	99.02	99.32	99.46	99.32
Li	9.8		6.5	5.6	6.3	9.5	5.4	4.2	9.2
Sc	47.2	99.4	83.4	19.1	25.0	22.9	16.1	17.2	28.4
Ti	7476	14,7	10,5	4175	12,5	6228	17,0	13,7	11,1
V	725	989	2335	234	935	281	1675	1112	27.2
Cr	105	103	159	119	71.3	11.3	17.8	50.3	11.4
Mn	1339	2136	2145	1458	1109	888	1147	737	1525
Co	32.7	103	151	73.8	76.4	32.1	81.8	71.0	21.9
Ni	104	156	150	94.2	124	28.1	103	151	8.87
Cu	79.4	189	55.5	62.5	171	87.2	148	403	15.4
Zn	41.3	139	126	85.9	103	79.8	113	88.4	149
Ga	10.0	14.8	4.5	18.1	26.8	22.9	25.1	27.9	22.4
Rb	0.52	3.38	2.26	17.4	20.9	60.1	14.1	8.73	48.3
Sr	16.5	41.3	21.7	295	365	304	292	406	306
Y	13.6	22.6	8.97	10.6	10.6	23.1	11.1	5.01	33.6
Zr	55.1	32.3	18.3	52.9	51.5	143	44.9	24.0	115
Nb	1.85	1.94	2.89	2.66	3.03	6.60	2.76	1.58	9.03
Mo	0.31	0.69	0.41	0.38	0.44	1.26	0.57	0.32	1.75
Cs	0.04	0.20	0.12	0.66	0.64	1.33	0.51	0.29	1.52
Ba	14.2	43.0	16.5	255	298	524	220	200	611
La	3.7	7.35	2.29	10.1	13.6	24.7	9.02	6.01	31.8
Ce	10.7	17.1	7.2	20.7	22.6	50.2	18.6	10.8	76.6
Pr	1.63	2.42	1.15	2.40	2.65	5.83	2.19	1.21	8.94
Nd	8.08	11.6	5.55	9.62	10.2	22.0	8.79	4.74	36.0
Sm	2.15	3.09	1.49	1.91	2.00	4.31	1.83	0.94	7.31
Eu	0.43	0.91	0.26	0.66	0.87	1.25	0.72	0.65	2.54
Gd	2.36	3.39	1.60	1.79	1.85	3.82	1.78	0.82	6.69
Tb	0.39	0.55	0.25	0.29	0.29	0.60	0.28	0.12	0.98
Dy	2.41	3.48	1.57	1.78	1.68	3.57	1.71	0.81	5.74
Ho	0.47	0.72	0.33	0.38	0.35	0.74	0.36	0.16	1.16
Er	1.26	1.95	0.86	1.12	0.97	2.11	1.03	0.46	3.15
Tm	0.18	0.28	0.13	0.17	0.14	0.31	0.15	0.07	0.45
Yb	1.08	1.82	0.80	1.09	0.90	2.04	0.95	0.42	2.77
Lu	0.16	0.28	0.12	0.17	0.13	0.30	0.14	0.06	0.42
Hf	2.24	1.04	0.62	1.37	1.36	3.62	1.21	0.67	3.30
Ta	0.22	0.47	0.28	0.22	0.23	0.52	0.52	0.22	0.96
Pb	2.32	2.38	2.33	4.05	4.34	14.0	4.18	4.43	9.43
Bi	0.12	0.10	0.09	0.05	0.09	0.18	0.07	0.15	0.12
Th	0.54	1.85	0.36	2.57	2.25	7.89	2.24	1.12	6.50
U	0.17	0.27	0.11	0.75	0.73	2.12	0.51	0.31	2.17

Notes: Analyses were carried out with XRF (major components, wt%) at the Vernadsky Institute of Geochemistry and Analytical Chemistry, Russian Academy of Sciences, analysts IA Roshchina and TV Romashova and with ICP-MS (trace elements, ppm) at Institute of the Mineralogy, Geochemistry, and Crystal Chemistry of Rare Elements, analyst DZ Zhuravlev. Rocks: pyroxenite (samples 0301, 2101/2, 0302), norite (0303), leucogabbro (0304), monzodiorites (0305, 0321/1), total, titanomagnetite leucogabbro (0306), titanomagnetite gabbro (0307), and gabbrodiorite from the Luktur pluton (0308)

Fig. 6.4 Chondrite-normalized patterns of trace elements for rocks of the Chiney pluton. (After Gongalsky 2015)

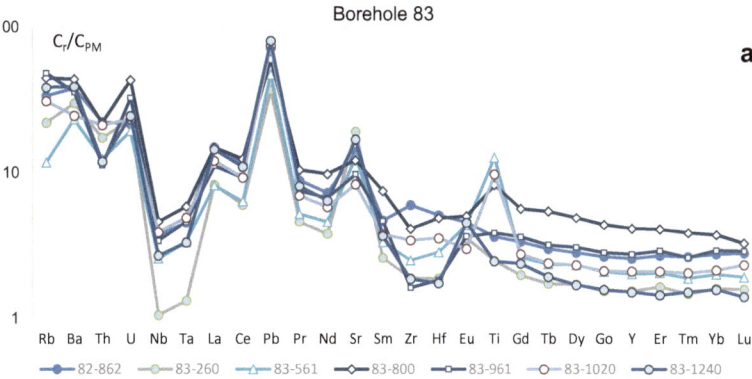

Fig. 6.5 Spider diagrams of rocks from Sects. 6.1 and 6.2 of the Chiney pluton
Here and in Figs. 6.6 and 6.7, C_r/C_{PM} content in rock/content in primitive mantle. (After Hofmann 1988). Rocks from (**a**) borehole 83, (**b**) borehole 11, (**c**) outcrops

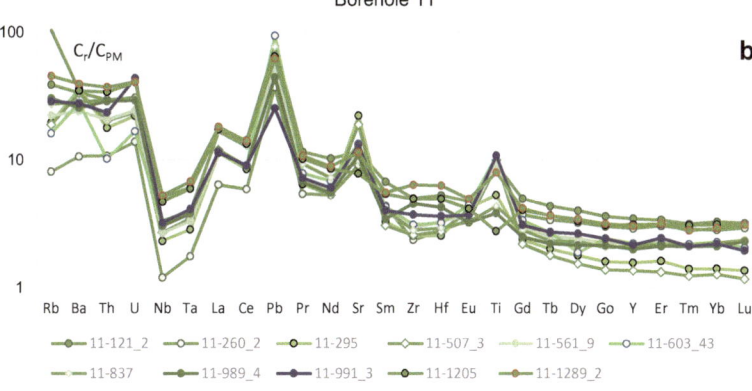

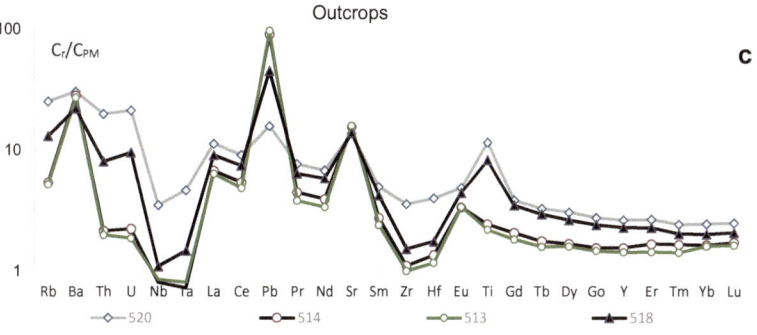

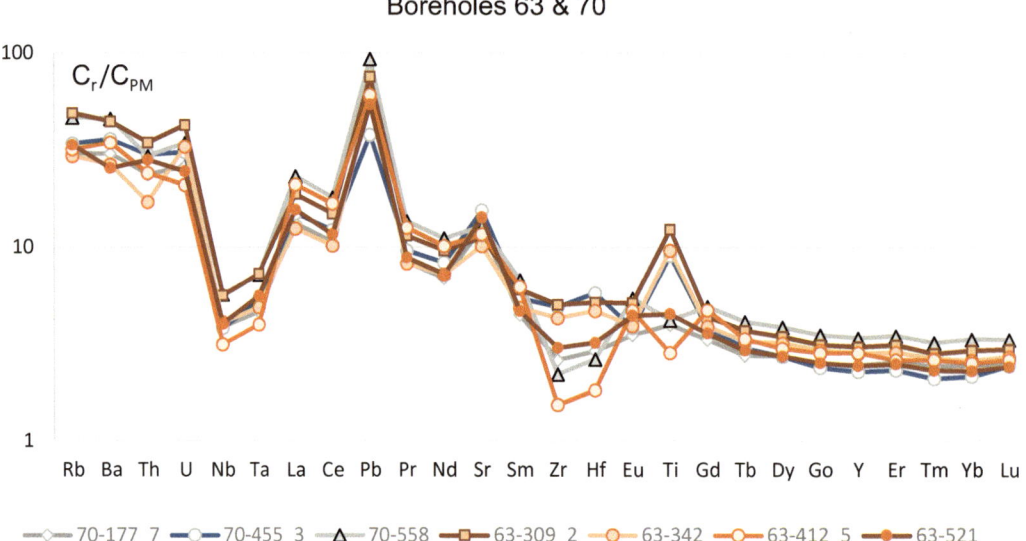

Fig. 6.6 Spider diagram for rocks from boreholes 63 and 83

Fig. 6.7 Diagrams (Gd/Yb)
n-La/Sm)n (**a**), (La/Ta)n-(La/
Yb)n (**b**) for rocks from the
reference sections of the
Chiney pluton

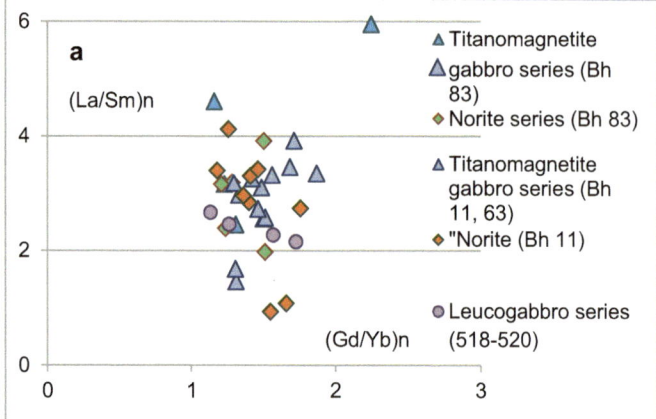

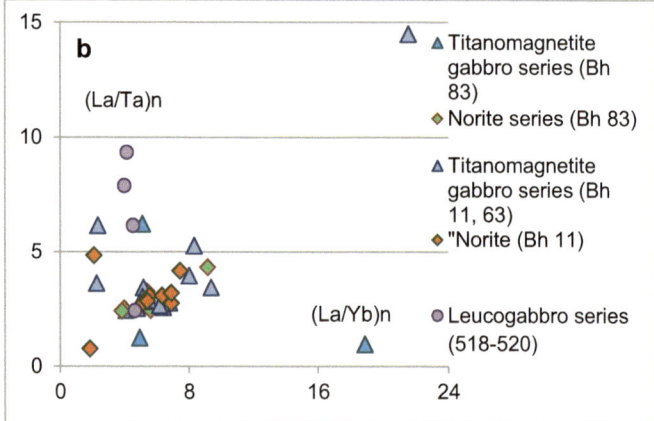

has been involved as well. The contents of ferric and ferrous ions were determined in these samples along with XRF determinations of total iron, which have been carried out for all other samples. The total number of rock compositions in the dataset is 216.

The results of consecutive hierarchical clustering of these chemical data are shown in Fig. 6.8. No formal criteria for the classification of the entire dataset of samples into types or groups were used in the procedure of clustering, i.e., a numerical level of unification is not set up. Clusterization is

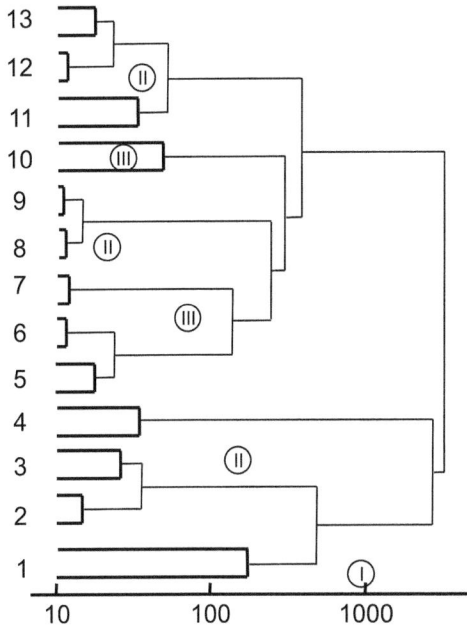

Fig. 6.8 Petrochemical rock types in the Chiney pluton based on the results of cluster analysis
The diagram shows a fragment of the general cluster diagram for the formal degree of geochemical differences of the rock collection >10. Roman numbers denote major rock types, and Arabic numerals show their petrochemical types (see text and Tables 6.5 and 6.6). Here and Figs. 6.9, 6.10, and 6.11. (After Gongalsky et al. 2008)

carried out on the basis of dendrogram and independent information, including structural, petrographic, and geological data (Yaroshevsky 2004). In the classification of rocks from the Chiney pluton, we were guided by the dendrogram of consecutive combinations of chemical compositions (Fig. 6.8), verifying correspondence of selected geochemical types to the aforementioned petrographic varieties. In this way, we selected 13 stable petrochemical types (Tables 6.5 and 6.6) and provided insight into the fine pattern of four rock groups.

The first type of rocks, represented by pyroxenite xenoliths (Group 1 from geological data), stands out in the dendrogram quite distinctly (Fig. 6.8). The same can be concluded about plagioclase-titanomagnetite rocks from leucogabbro series, called chineyite (the fourth type with mean FeO content at ~59 wt%). Types 11, 12, and 13 of leucogabbro, differing from one another in relative amounts of titanomagnetite, are generally similar in chemical composition. These rocks occur in both the leucogabbro series and the upper parts of the certain rhythms in the titanomagnetite gabbro series (Group 2). The types 5, 6, and 7 represented by gabbroids with limited variations in pyroxene, titanomagnetite, and plagioclase contents also reveal an appreciable similarity. All are related to various parts of rhythmic units in the titanomagnetite gabbro series. The types 2, 3, 8, and 9 are distinguished by significant amounts of orthopyroxene. They

dominate in the norite series (types 8, 9 of Group 3) and also occur in the lower part of the titanomagnetite gabbro series (types 2, 3 of Group 2). A special place is occupied by rocks of type 10, representing monzodiorite from the eastern upper and marginal parts of the pluton (Group 2).

Thus, all the types correspond to certain petrographic varieties of the rocks from the Chiney pluton (Groups 1–3), generally characterized by irregular distribution in the vertical sequence of the pluton (Figs. 5.16 and 5.18). Only some of them are strictly confined to certain stratigraphic levels: chineyite of type 4 is localized at the boundary between titanomagnetite gabbro and leucogabbro series; types 11 and 12 occur in the central part of the latter series; types 8 and 9 are characteristic of norite series; and type 1 (xenoliths of Group 1) occurs in the upper part of titanomagnetite gabbro series (Fig. 5.17).

The rock types, selected on the basis of clustering analysis, apparently represent certain stages of fractionation of an initial magma, irrespective of possible additional injections, and may be used for the reconstruction of phase and chemical compositions of a magma source and evolved residual melts. These problems were solved using numerical simulation of melt-crystal equilibria (Ariskin and Barmina 2000).

6.3   Temperature and Composition of Magmas

We used numerical simulation of equilibrium crystallization of the melts, corresponding to the selected petrochemical types, to estimate the temperature and composition of initial melts in the intrusive rocks of the Chiney pluton (Table 6.5). This method was described by Ariskin and Barmina (2004) as a version of geochemical thermometry of intrusive mafic rocks based on the analysis of temperature-composition trends of model liquids and compositions of major rock-forming minerals. It helps to estimate a chemical composition of intercumulus melts, which are in equilibrium with the phases close in composition to the mean compositions of minerals in various rock types (http://www.geokhi.ru/~dynamics/docs/papers/p01_057.pdf).

The geochemical thermometry is a method developed in the 1980s (Frenkel' et al. 1987). It helps to estimate the original temperatures and composition of entrapped (intercumulus) melts (Barmina et al. 1988). Sometimes, this approach gives phase composition of the intruded magma and the original composition of the intratelluric phases (Barmina et al. 1989; Chalokwu et al. 1993).

The computer simulation with the use of the COMAGMAT computer program was applied during the study of the inner structures of intrusions inside the Siberian flood basalt province on the basis of the convection-cumulation mechanism of in-chamber differentiation (Frenkel' et al. 1985, 1987,

Table 6.5 Petrochemical rock types distinguished in the Chiney pluton based on cluster analysis

Type	Group	Series	SiO₂	TiO₂	Al₂O₃	FeO	MnO	MgO	CaO	Na₂O	K₂O	P₂O₅
1	1		45.3	1.37	4.15	20.27	0.22	13.44	14.62	0.37	0.2	0.05
2	2	TMG	45.69	2.53	8.72	24.77	0.27	5.84	2.51	1.23	0.16	0.05
3	2	TMG	37.04	4.47	7.43	34.28	0.27	9.5	5.61	1.06	0.29	0.01
4	2	TMG	16.49	9.17	4.98	59.34	0.27	5.84	2.51	1.23	0.16	0.02
5	2	TMG	46.5	2.08	14.13	15.19	0.17	6.21	11.13	3.03	0.72	0.08
6	2	TMG	44.39	2.49	12.58	18.46	0.2	7.26	11.64	2.5	0.43	0.05
7	2	TMG	46.8	1.98	15.23	16.98	0.15	5.88	10.15	2.3	0.45	0.08
8	3	H	49.57	1.36	13.1	16.2	0.2	9.72	6.82	2.32	0.65	0.07
9	3	H	52.75	0.84	16.32	11.35	0.17	7.59	7.63	2.63	0.65	0.07
10	2	LG	52.31	1.75	14.83	13.88	0.2	5.52	6.1	2.66	2.4	0.35
11	2	LG	51.04	0.91	23.68	7.54	0.09	1.45	10.36	4.01	0.78	0.13
12	2	LG	45.89	2.34	18.77	17.34	0.14	2.33	9.47	3.17	0.46	0.08
13	2	TMG	36.37	4.59	16.25	27.9	0.14	2.88	7.76	3.25	0.51	0.05

Note: (1) Rock series: TMG, titanomagnetite gabbro; LG, leucogabbro; H, norite; (2) oxide concentrations are given in wt%

Table 6.6 Representative analyses of rocks of various petrochemical types (with measured concentrations of FeO and Fe₂O₃)

N	Type	№ sample	SiO₂	TiO₂	Al₂O₃	Fe₂O₃	FeO	MnO	MgO	CaO	Na₂O	K₂O	P₂O₅	LOI	Total
1	1	58012500	44.17	1.36	4.23	3.86	17.78	0.23	13.48	13.68	0.46	0.25	0.02	0.07	99.59
2	2	11622840	46.64	2.25	7.79	7.85	17.1	0.29	11.21	6.06	1.31	0.43	0.04	0.02	100.99
3	3	11622730	36.1	4.61	6.93	15.11	19.8	0.29	8.96	7.89	0.96	0.26	0.03	0.03	100.97
4	4	11063800	16.53	8.58	4.42	30.74	28.45	0.23	7.04	2.38	0.64	0.12	0.02	0.31	99.46
5	6	11020400	42.4	2.12	13.2	10.13	10.34	0.17	6.29	11.88	2.7	0.38	0.03	0.25	99.89
6	7	11013800	45.89	1.02	17.87	4.85	9.23	0.02	6.77	10.77	2.36	0.23	0.17	0.18	99.36
7	8	11089500	50.96	0.95	14.98	3.14	10.02	0.17	8.8	7.44	2.19	0.67	0.07	0.27	99.66
8	9	11092050	48.55	1.45	9.69	4.88	9.27	0.18	9.61	13.37	1.76	0.57	0.09	0.37	99.79
9	11	11029500	48.84	1.25	22.48	2.08	7.95	0.11	1.45	10.16	3.42	0.56	0.14	0.39	98.83
10	12	11053260	45.79	2.28	20.34	6.74	8.84	0.12	1.95	8.83	3.12	0.46	0.1	0.84	99.41
11	13	11045860	37.88	5.89	18.83	5.29	19.45	0.14	0.91	7.12	2.7	0.37	0.07	0.49	99.14

Note: (1) Sample numbers: the first two digits correspond to the hole number, and the third through eighth digits correspond to the depths (in cm); (2) types correspond to the petrochemical types of rocks (Table 6.5); (3) oxide concentrations are given in wt%; LOI means lost on ignition

1988) for the Kuz'movka and Vavukan intrusions (Ariskin and Barmina 2000) and then for the Talnakh intrusion (Krivolutskaya et al. 2001). In the latter case, we successfully determined the phase and chemical composition of the parental magma and crystal phases that produced taxitic and picritic gabbro-dolerites of the lower zone of the intrusion.

To carry out these thermometric calculations, the independent data on pressure, contents of volatile components, and redox conditions of magma crystallization are necessary. Insignificant alumina content in the Chiney orthopyroxenes (0.8–1.3 wt% Al_2O_3) indicates hypabyssal conditions of pluton crystallization at a pressure no higher than ~1.5 kbar (Konnikov 1986). Absence of early magmatic hydroxyl-bearing minerals in the main rock types means its undersaturation in H_2O of initial magma (at a low pressure, a water content in melt was hardly higher than ~0.5 wt%). That is why we performed calculations under dry conditions and a total pressure of 1 kbar. To estimate redox parameters of magma evolution under these conditions, the simulation of equilibrium crystallization of melts, corresponding in composition to 11 representative rocks with the known Fe^{3+}/Fe^{2+}

ratio, has been preliminarily carried out (Table 6.6). The calculations with COMAGMAT-3.5 computer model (Ariskin and Barmina 2004) demonstrated a probable temperature range of intercumulus melts within 1100–1150 °C. The crystallization redox conditions for all types of rocks correspond to the NNO ± 0.5 buffer equilibrium (Fig. 6.9). The solid lines characterize the evolution of modeling melts with an allowance for the given Fe_2O_3 and FeO contents (the system is closed in respect to oxygen). The dotted line shows the temperature range of calculated cotectics including Ol + Pl ± Mag ± Cpx and probable redox conditions of magma crystallization (NNO ± 0.5).

The calculations were performed as a series of crystallization steps of 1 mol%. The calculation terminated at a high degree of model system crystallization, when crystalline phase content reaches 65–90%. The sequences of melt crystallization for the main rock types are shown in Fig. 6.10. It is noteworthy that olivine occurs in these sequences as the first crystalline phase (types 1–3, 8, 9; in five cases) and in cotectic assemblages together with plagioclase, pyroxenes, and magnetite olivine (in five other cases). The sequences of

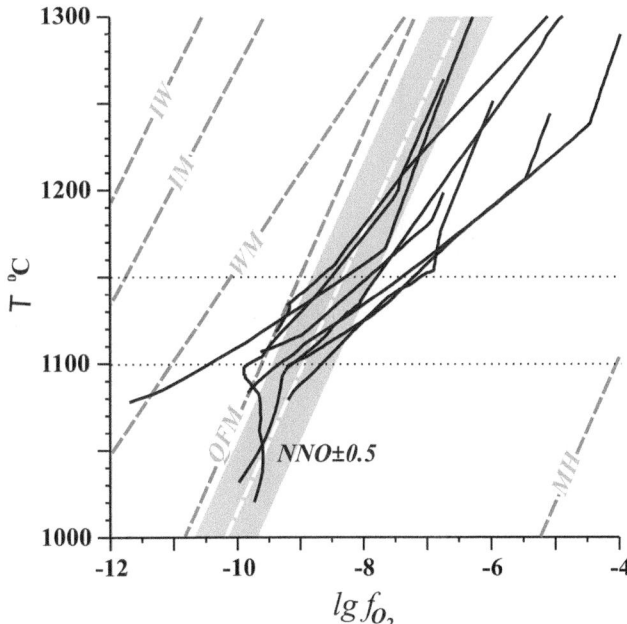

Fig. 6.9 Diagram T-log f_{O_2} with results of simulations of the equilibrium crystallization of melts corresponding to representative rocks (Table 6.6) of the Chiney pluton

Solid lines characterize the evolution of the modeled melts with regard to the specified Fe_2O_3 and FeO concentrations (the system is closed with respect to oxygen). Dotted lines show the temperature range of the calculated cotectics with olivine + plagioclase ± magnetite ± clinopyroxene. The overlap area of these characteristics (shaded) corresponds to the probable redox parameters, at which the Chiney magma is crystallized (NNO ± 0.5)

mineral crystallization for types 2, 3, 8, 9, 12, and 13 show that olivine occurs in reaction relationships with melt and completely dissolve as a result of peritectic reaction, replacing with pyroxenes and magnetite at $T = 1110–1150\ °C$. Olivine does not crystallize in monzodiorite (type 10). This implies a silica-saturated or slightly oversaturated character of the initial melts and elevated silica content in the intercumulus systems. The oversaturation with SiO_2 determines the evolution of the melt toward the Ol/Opx reaction point. The peritectic reaction in the cumulus-intercumulus-liquid system is responsible for the lack of modal olivine and occurrence of norites in the pluton.

The results of modeling allow to subdivide the process of crystallization into two modes: (1) conditionally high-aluminous mode, typical of the leucogabbro compositions, and (2) crystallization of the magnetite-plagioclase cotectics (petrochemical types 6, 7, 11–13) (Gongalsky et al. 2008).

The latter is distinguished by the early appearance of plagioclase (above 1250 °C) and magnetite (~1200 °C). The highest temperature of magnetite crystallization (1365 °C) is demonstrated by chineyite (type 4) and the adjacent rocks of type 3 (1252 °C). These rocks are extremely enriched in iron (59.3 and 34.3 wt% Fe). These attributes and similar crystallization temperature of augite as the third cotectic phase (1108–1112 °C) allow us to com-

bine compositions 4, 11, 12, and 13 into the integral group of plagioclase-magnetite cumulates (consistent with petrographic observations), which apparently represent similar stages of Chiney magma fractionation. A different temperature of plagioclase and magnetite appearance on liquidus can be explained as a result of variable proportions of these minerals in cumulus.

The second mode of sequences, which can be called magnesian, is characterized by early crystallization of olivine-magnetite and olivine-plagioclase cotectics. Types 1–3, 8, and 9 belong to this group. The most primitive petrochemical types 8 and 9 are represented by norite and distinguished by high-Mg composition of the liquidus olivine (Fo_{83} and Fo_{84}, respectively). The liquidus temperature of these melts is 1230–1250 °C. Augite or low-Ca pyroxene is the third phase, crystallizing at a temperature of 1150 °C. For compositions 1–3, plagioclase is the late phase, which crystallizes below 1110 °C. This differentiates the given compositions from primitive types 8 and 9 with the appearance of plagioclase at 1160–1180 °C. Type 5 that adjoins these compositions is characterized by joint crystallization of olivine and plagioclase within the same temperature range. Magnetite and augite join these minerals at the late stage of crystallization (1140–1125 °C). The calculation results have been used for estimating the temperature and composition of intercumulus melts for rocks of various geochemical types. The problem has been solved using the technique described in detail for the Skaergaard, Kiglapait, and Talnakh layered intrusions (Ariskin and Barmina 2004). This approach is based on the enumeration of model melt sequences for each starting composition and the search for those liquids in the composition-temperature coordinates, which are in equilibrium with mineral assemblage close to natural minerals. This implies a comparison of real and model compositions of mineral phases and makes it possible to estimate the temperature range of intercumulus melts with an accuracy of 10–20 °C. To solve these problems, we used representative microprobe compositions of plagioclase and pyroxene from the samples characterizing specific petrochemical types (Table 6.7). The temperature interval, which was estimated in this way, does not commonly exceed 20 °C. For the case of type 8, it was 1110–1126 °C, if plagioclase compositions vary within interval of $An_{50–60}$. In the further analysis of calculation results, we used a mean value of the intercumulus temperature range and composition of modeling residual melt, corresponding to this mean temperature (Table 6.8). In contrast to the preceding table, the latter has been rearranged in the order of a consecutive drop of the calculated temperature. Such a presentation implies the crystallization control of intercumulus liquid evolution in the course of solidification and differentiation of residual melt in the Chiney pluton.

The above data indicate a relatively narrow crystallization temperature interval of the Chiney magmas (1140–1080 °C) (Fig. 6.11). The depletion of model melts in MgO and CaO

Fig. 6.10 Sequences of crystallization, temperatures, and the compositions of major minerals for the petrochemical rock types of the Chiney pluton

Numbers 01 through 13 mean the compositions of the petrochemical types in Table 6.5. Numbers near the crystallization lines of minerals demonstrate the dependence on temperature

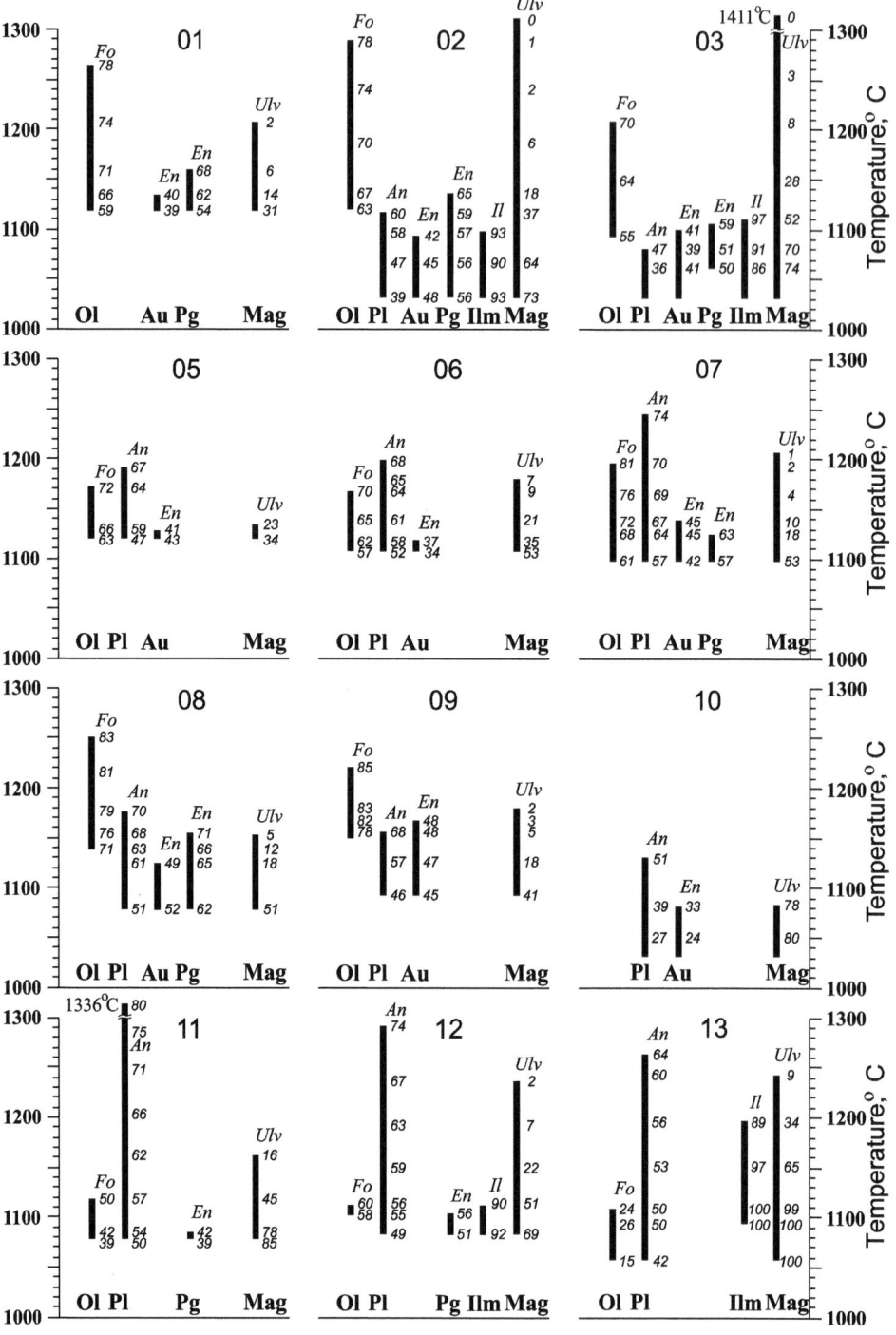

against the background of enrichment in silica and alkali metals characterizes a trend of fractionation from basalt to dacite. High FeO and TiO$_2$ (~3 wt%) contents emphasize the ferrobasaltic composition of parental magmas of the Chiney pluton. The absence of distinct and monotonous trends on variation diagrams is caused by difficulties emerging in the calculation of phase equilibria in the presence of a great amount of magnetite at a temperature of about 1100 °C (Ariskin and Barmina 2000) and by possible injections of the

magmas of similar temperature but different in the phase composition of initial magmas. This possibility probably explains a rather separate position of composition 10 (1080 °C), which represents products of equilibrium crystallization of monzodiorite (Fig. 6.11).

The phase composition of magma (Table 6.8, Fig. 6.11) suggests that the parental magmas of the Chiney pluton were the melts, containing 4–5 wt% MgO and heated to ~1130 °C at the moment of injection. Intercumulus melts of petro-

Table 6.7 Temperatures of crystallization for melts determined based on the composition of minerals from the rocks

| Type | Measured mineral compositions (mol%) | | | Composition of model crystals at specified temperature, mol% | | | | | | | | | |
| | An | OPx | CPx | ΔT, °C cryst | Tav | Cpx | | | | | Opx | | |
						Fo	An	En	Fs	Wo	En	Fs	Wo
1			44–48, Wo	1147–1124	1135	–	71	44	17	38	64	26	9
2		29–42, Fs		1119–1122	1120	–	–	–	–	–	58	38	4
3		29–32, Fs		1097–1086	1092	–	–	44	20	35	59	32	10
5			38–42,En	1128–1124	1126	65	57	41	16	43	–	–	–
6	54–58			1118–1122	1120	65	56	41	16	43	–	–	–
7	52–56			1103–1092	1098	–	54	45	21	34	58	33	9
8	50–60			1126–1110	1118	62	58	–	–	–	60	32	8
9	57–59			1100–1114	1107	–	58	47	21	32	60	32	8
10	44–47			1072–1087	1080	–	45	53	19	28	–	–	–
11	47–57			1142–1030	1136	–	57	–	–	–	–	–	–
12	54–57			1126–1114	1120	–	56	–	–	–	–	–	–
13	50–52			1135–1100	1128	–	53	–	–	–	–	–	–

Note: *An* anorthite, *Cpx* clinopyroxene, *Opx* orthopyroxene, *Fo* forsterite, *En* enstatite, *Fs* ferrosilite, *Wo* wollastonite, T is a range of crystallization temperatures of the mineral

Table 6.8 Model melt compositions for the major petrochemical rock types

Type	T, °C	Phase, %	Ol	Pl	Au	Pg	Ilm	Mag	SiO_2	TiO_2	Al_2O_3	FeO	MgO	CaO	Na_2O	K_2O	P_2O_5
11	1136	58		57				1	50.19	2.43	14	16.15	3.98	7.86	3.32	1.83	0.24
1	1135	35	9			26			50.85	2.43	12.71	15.68	5.28	10.15	1.62	0.92	0.23
5	1126	34	7	22	3			2	47.73	2.66	12.13	17.09	4.72	10.62	3.45	1.27	0.15
6	1120	36	9	23				4	47.02	3.15	11.16	17.89	4.98	11.59	3.1	0.83	0.11
2	1120	44	1			33		10	58.18	3.09	11.13	18.41	3.42	3.43	1.8	0.23	0.07
12	1120	58		48				10	49.81	3.35	11.93	17.79	4.04	9.06	2.84	0.82	0.16
8	1118	57		30	2	22		3	52.91	2.36	13.26	15.29	4.03	7.34	3.02	1.41	0.17
9	1107	73		15	52			7	56.98	2.02	14.06	12.78	3.16	5.67	2.94	1.89	0.24
7	1098	85	6	50	13	9		7	55.75	2.85	12.83	14.12	3.18	6.53	2.98	1.31	0.26
3	1092	46			15	2	5	25	59.66	3.34	12.52	12.45	2.23	5.01	3.35	1.06	0.38
10	1080	18		15	1		1	1	62.92	1.5	16.78	5.98	1.56	2.58	3.64	4.24	0.64
IM	1130								49.37	2.88	12.12	17.02	4.44	9.65	3.1	1.08	0.15

Note: (1) (1)–(3) and (5–12)—simulated compositions of the petrochemical types
(2) IM is the initial melt of the Chiney pluton (calculated as an average of simulated melt types 5, 6, 8, and 12)
(3) Oxide concentrations (wt%) in solid phases
(4) Sol. phase is the total amount of crystals in melt, including *Ol* olivine, *Pl* plagioclase, *Au* augite, *Pg* pigeonite, *Ilm* ilmenite, *Mag* magnetite

chemical types 5, 6, 8, and 12 fit these parameters. These four compositions (Table 6.8) were used to calculate a mean composition of the possible primary melt, corresponding to the liquid parental magma(s), which filled a chamber of the Chiney pluton. The fractionation path of this melt was calculated following the COMAGMAT model at the same physicochemical parameters as those used in thermometric calculations. The calculated lines of fractional crystallization coincide, in general, with compositional trends of model liquids of various petrochemical types. This similarity can be regarded as an evidence for cognation of various rock types and the reality of the estimated temperature and composition of the primary Chiney melt, despite the preliminary nature of the performed calculations. It is important to emphasize that the result obtained cannot be regarded as an argument in favor of a momentary filling of the chamber with homogeneous magma of certain composition. One can only suggest

that at the moment of intrusion into and filling of the magma reservoir, the melt was in equilibrium with a significant amount of crystalline phases represented by olivine, plagioclase, and magnetite, nonuniformly distributed throughout the intrusive chamber.

The absence of reliable data on the average composition of the main structural units of the Chiney pluton restrains an accurate estimation of the relative amount of solid phases in magmas.

Nevertheless, we suppose that the simulation results allow us to consider the parental magmas in main blocks of pluton as products of evolution of the same primary magma, which initially was a mush of olivine, plagioclase, and magnetite crystals at a temperature of ~1130 °C. Gravity separation of these phases, the mutual migration of magmatic suspensions differing in phase compositions, and the duration of magma emplacement could have led to the nonuniform distribution

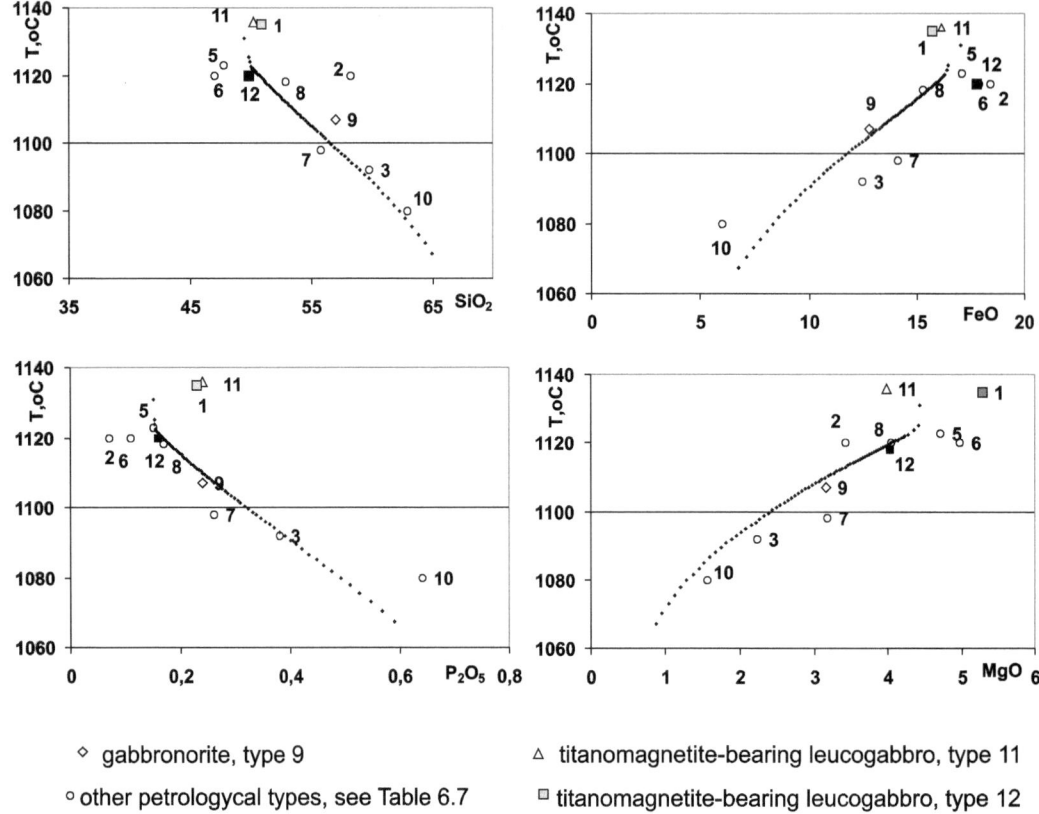

◇ gabbronorite, type 9   △ titanomagnetite-bearing leucogabbro, type 11

○ other petrologycal types, see Table 6.7   □ titanomagnetite-bearing leucogabbro, type 12

Fig. 6.11 Trends of fractionation crystallization for the modeled melt of the Chiney pluton and the compositions of the modeled melts of the major petrochemical types (numbers)

of a solid phase throughout the volume, which was expressed afterward in the heterogeneous phase composition of rocks in different blocks of pluton. Additional COMAGMAT calculations focusing on more primitive compositions of the Chiney rocks were performed by Ariskin and Nikolaev (Gongalsky et al. 2016).

6.4   Conclusions

1. The rocks of the Chiney intrusion, enriched in Fe and Ti, are represented mostly by gabbro, gabbronorite, and leucogabbro. Norite and pyroxenite take a subordinate position. They form four groups. Inside rhythmic units, SiO_2, Na_2O, and K_2O increase from the bottom to the top, while Fe_2O_3 decreases.

2. The rare element distribution in rocks and the results of calculations using the COMAGMAT-3.5 computer program show that rock groups of the layered mafic-ultramafic complex were formed as a result of emplacements of several portions of magma, consisting of suspension of olivine+plagioclase+magnetite crystals in ferrobasaltic melt at a temperature of 1130 °C.

3. The crystallization of olivine, plagioclase, and magnetite in the melt took place before its intrusion into the cham-

ber. These minerals were dispersed irregularly in different parts of the modern chamber during the subsequent emplacement of various portions of the initial magma. This distribution of minerals is reflected in the structure of different blocks of the Chiney pluton. The main rock groups, the second and third composing the main volume of the pluton, differ from each other by the dominance of plagioclase and magnetite crystals in the second group (gabbronorite and leucogabbro series in the Western and Southeastern blocks) and orthopyroxene in the third group (norite series, Central block).

References

Ariskin AA, Barmina GS (2000) Modeling of phase equilibria during the crystallization of basaltic magmas. Moscow, Nauka, 363 (in Russian)

Ariskin AA, Barmina GS (2004) COMAGMAT: development of magma crystallization model and its petrological applications. Geochem Int 42(Suppl. 1):1–157

Barmina GS, Ariskin AA, Koptev-Dvornikov EV, Frenkel' MY (1988) Estimation of the compositions of primary cumulative minerals in differentiated traps. Geokhimiya 8:1108–1119

Barmina GS, Ariskin AA, Frenkel' MY (1989) Petrochemical types and crystallization conditions of plagiodolerites in the Kronotsky

Peninsula. Eastern Kamchatka Geokhimiya 2:192–206 (in Russian)

Chalokwu CI, Grant NK, Ariskin AA, Barmina GS (1993) Simulation of primary phase relations and mineral compositions in the Partridge River Intrusion, Duluth Complex, Minnesota: Implications for the parent magma composition. Contrib Mineral Petrol 114:539–549

Frenkel' MYa, Yaroshevsky AA, Koptev-Dvornikov EV, Ariskin AA, Kireev BS, Barmina GS, Ariskin AA, Koptev-Dvornikov GS (1985) Crystallization mechanism of layering in layered intrusions. Zap Vses Mineral O–va part CXIV 3:257–274 (in Russian)

Frenkel' MYa, Ariskin AA, Barmina GS, Korina MI, Koptev-Dvornikov EV (1987) Geochemical thermometry of igneous rocks: principles and examples of application. Geokhimiya 11:1546–1562 (in Russian)

Frenkel' MYa, Yaroshevsky AA, Ariskin AA, Barmina GS, Koptev-Dvornikov EV, Kireev BS (1988) Dynamics of the chamber differentiation of basic magmas. Moscow, Nauka, 216 (in Russian)

Gongalsky BI (2015) Deposits of the unique metallogenic province of Northern Transbaikalia. Moscow, VIMS 248 (in Russian)

Gongalsky BI, Krivolutskaya NA (1993) Chiney layered pluton. Novosibirsk, Nauka, 184 (in Russian)

Gongalsky BI, Krivolutskaya NA (2009) The Udokan-Chiney ore-magmatic system, Russia. Northwest Geol 42:180–184

Gongalsky BI, Krivolutskaya NA, Ariskin AA, Nikolaev GS (2008) Internal structure, composition and genesis of the Chiney anorthosite–gabbronorite pluton, Northern Transbaikalia. Geochem Int 46:637–665

Gongalsky BI, Krivolutskaya NA, Ariskin AA, Nikolaev GS (2016) The Chiney gabbronorite-anorthosite layered massif (NorthernTransbaikalia, Russia): its structure, Fe-Ti-V and Cu-PGE deposits, and parental magma composition. Mineral Deposita 51(8):1013–1034

Hofmann AW (1988) Chemical differentiation of the Earth: relationship between mantle, continental crust, and oceanic crust. Earth Planet Sci Lett 90:297–314

Hofmann AW (2003) Sampling mantle heterogeneity through oceanic basalts: isotopes and trace elements. Treatise Geochem 2:61–101

Joron JL, Treuil M (1988) Hydro magma file element distributions in oceanic basalts as finger of partial melting and mantle heterogeneities: a specific approach and proposal of an identification and modeling method. In: Magmatism in Ocean Basins. Geol Soc Spec Publ 42:277–299

Konnikov EG (1986) Differentiated ultrabasic–basic Precambrian complexes in Transbaikalia. Novosibirsk, Nauka, 224 (in Russian)

Krivolutskaya NA, Ariskin AA, Sluzhenikin SF, Turovtsev DM (2001) Geochemical thermometry of rocks of the Talnakh Intrusion: assessment of the melt composition and the crystallinity of the parental magma. Petrology 9(5):389–414

Ward JH (1963) Hierarchical grouping to optimize an objective function. J Am Stat Assoc 58(301):236–244

Yaroshevsky AA (2004) Geochemical structure of magmatic complexes: an example of the Kivakka layered olivinite–norite–gabbronorite intrusion, Northern Karelia. Geochem Int 42:1107–1125

Yaroshevsky AA, Bolikhovskaya SV, Koptev-Dvornikov EV (2006) Geochemical structure of the Yoko-Dovyren layered dunite–troctolite–gabbro–norite massif, Northern Baikal Area. Geochem Int 44(10):953–964

Titanomagnetite Ore in the Chiney Pluton

Abstract

There are two Fe-Ti-V deposits in the Chiney pluton—Magnitnyi and Etyrko. They represent the largest deposits of commercially viable vanadium ore in Russia, with average content of 0.5 wt% V_2O_5 and 6.3 wt% TiO_2 in ores. The Etyrko deposit hosts stratabound, mostly disseminated titanomagnetite, whereas the Magnitnyi deposit consists predominantly of cross-cutting ore veins and irregular bodies in gabbronorite. The Magnitnyi deposit contains about 1.5 billion tons of ore in several orebodies. The major orebodies account for 89.1% of the total ore reserves and 86.1 to 92% of mineable metal reserves. The individual orebodies are between 5.7 m and 26.5 m thick, with the total iron content gradually decreasing westward from 39.4 wt% to 28.9 wt%. Titanomagnetite and ilmenite (the amount of the latter rarely exceeds 10% of the total) are the major ore minerals. According to the Mössbauer spectroscopy, magnetite in the titanomagnetite ore is close to stoichiometric, and titanium is mainly contained in ilmenite (11–13%).

7.1 Titanomagnetite Ore

Titanomagnetite as a rock-forming mineral occurs in all varieties of gabbro in the Chiney pluton (Petrusevich 1946). But in some cases, its aggregates have economic value. There are two Fe-Ti-V deposits related to the layered Chiney pluton. The economic titanomagnetite concentrations are inherent to high-Ti gabbro rocks in the western part of the pluton (Etyrko deposit) and in its eastern part (Magnitnyi deposit) (Fig. 7.1). The Magnitnyi and Etyrko Fe-Ti-V deposits are the two most significant vanadium accumulations in the Chiney pluton (30 Gt of ore resources). They represent the largest deposits of potentially economic vanadium ore in Russia, with average grades of 0.49 wt.% V_2O_5 in ores (maximum values of 1–1.5 wt.% V_2O_5) and 6.3 wt.% TiO_2. The spatial separation of titanomagnetite ore partly coincides with the subdivision of them into different morphological and genetic groups. Disseminated ore, syngenetic to the host gabbro, dominates in the west (Etyrko deposit). The vein orebodies superimposed on host gabbro rocks are localized in the east (Magnitnyi deposit). In other words, the Fe-Ti-V mineralization is represented by the early and late magmatic types of deposits (Fig. 7.2). Contact metasomatic mineralization was identified, but it does not have an economic value (Lebedev 1962; Kulikov et al. 1981; Konnikov 1986).

7.2 Morphology

The deposits are hosted by titanomagnetite-rich gabbro. The Etyrko deposit hosts stratabound (Fig. 7.2a), mostly disseminated titanomagnetite, whereas the Magnitnyi deposit consists predominantly of cross-cutting veins and irregular bodies in gabbronorite that also contains xenoliths of the country rocks (Fig. 7.3a). Rounded and striated titanomagnetite aggregates are frequently encountered at the contacts between the veins and leucogabbro and other gabbro (Magnitnyi deposit; Fig. 7.3a, b).

The stratiform orebodies of two types are characteristic of the Etyrko deposit: (1) disseminated orebodies with titanomagnetite directly participating in the fine layering and microrythmicity in the titanomagnetite gabbro series (this ore is called finely layered); (2) thin (~0.5 m) interlayers of massive ore in association with anorthosite of the leucogabbro series (forming chineyite, Gongalsky and Krivolutskaya 1993).

The first type of ore is the most important and comprises the first two types of titanomagnetite mineralization recognized by Lebedev (1962): (1) accessory type with subhedral titanomagnetite and ilmenite grains and (2) early magmatic segregation type of massive and disseminated ore. The orebodies were traced for several kilometers along the strike.

© Springer Nature Switzerland AG 2019
B. Gongalsky, N. Krivolutskaya, *World-Class Mineral Deposits of Northeastern Transbaikalia, Siberia, Russia*,
Modern Approaches in Solid Earth Sciences 17, https://doi.org/10.1007/978-3-030-03559-4_7

Fig. 7.1   General view of the Etyrko (**a**) and Magnitnyi (**b**) deposits

Fig. 7.2   Disseminated (**a** and **b**) and massive (**b** and **c**) titanomagnetite ore
(**a** and **b**) Etyrko deposit; (**c** and **d**) Magnitnyi deposit

Fig. 7.3 Rounded
segregations of
titanomagnetite in gabbro at
the Magnitnyi deposit. (**a**)
Sample, (**b**) thin section

Fig. 7.4 Xenolith of carbonate rocks (skarn) in titanomagnetite ore (**a**) and stream-like titanomagnetite aggregates (**b**)

Their strike and dip coincide with those of gabbro rocks, with the dip angle varying from 2° to 20°. The orebodies pinch out in the left wall of the Right China river valley.

The boundaries of orebodies are, as a rule, diffuse. They were identified based on sampling. The disseminated ore is sharply dominant. The massive ore occurs only as thin (few centimeters) layers at the base of microrhythms, when titanomagnetite content exceeds 80 vol%. The layers of massive ore in association with chineyite, leucogabbro, or anorthosite are characteristic of the upper parts of units, composed of high-Ti gabbro rocks. The ore layers are commonly a few decimeters thick, but their lateral extent reaches several kilometers. The boundaries with under- and overlying rocks are distinct, often emphasized by fracture systems parallel to the general layering of the pluton.

The veins, lenses, and irregular in shape orebodies dominate at the Magnitnyi deposit in the eastern part of the Chiney pluton (Gongalsky et al. 1995). They can be morphologically separated into (1) extended lenticular lodes; (2) lenses; (3) large isometric and irregular bodies; and (4) small round, oval, or irregular in shape titanomagnetite segregations (Fig. 7.4b).

The extended lenticular lodes, localized between the finely layered ore and lenses, are related to pinch-out of titanomagnetite gabbro rocks. Their strike is conformable to general layering in the given area (dip azimuth is 345°NW, and dip angle is <10°). These orebodies have distinct bound-

aries with mineralized host gabbro rocks. The transition from high-grade ore, containing up to 80% of titanomagnetite, to gabbro occurs over a distance of 10–15 m. The orebodies are often surrounded by a halo of disseminated titanomagnetite. The ore itself is also disseminated and less frequently massive. In the southern part of pluton, the orebodies are hosted in the low-Ti gabbro rocks as large blocks (xenoliths), dip in southward in contrast to the general northward dip.

The ore lenses are hosted in slightly layered low-Ti gabbro. Their dimensions are (1–3) × (0.7–1.5) m. The contacts with host rocks are sharp, emphasized by fractures oriented parallel to the lens surface and clearly expressed in topography. The host rocks are commonly barren and contain only ~5% titanomagnetite.

Small segregations of massive titanomagnetite, which are round, oval, or irregular in shape, mainly occur in the southeastern and southern parts of the pluton (Fig. 7.3). They have sharp boundaries with host gabbro rocks. Stream-like titanomagnetite segregations in barren gabbronorite (Fig. 7.4b) of the leucogabbro series are noteworthy (Gongalsky 2010, 2015). Their mean size is a few centimeters; the maximum size is few decimeters. This type of ore is not economically viable on its own.

The isometric bodies of titanomagnetite ore occur in the contact zone between high-Ti and low-Ti gabbro rocks at the source of the River Right China. This area is characterized by multiple xenoliths of sedimentary country rocks, mainly carbonate, which are transformed into skarn and skarnoid rocks. Layering in gabbro and gabbronorite is obliterated by numerous orebodies and xenoliths, occasionally reaching 70–80% of the total rock volume. The mean size of titanomagnetite bodies is a few meters, occasionally reaching 15–20 m in size. Konnikov (1986) referred these orebodies to the reaction type. Following Kulikov et al. (1980), the contact metasomatic rocks related to carbonate-clastic xenoliths have been described. The thickness of titanomagnetite rims around xenoliths and their morphology depend on size and shape of the xenoliths. Konnikov (1986) paid special attention to this type of titanomagnetite mineralization, elucidating the role of assimilation of carbonate rocks in the formation of titanomagnetite ore. Based on the μCa versus μFe plot for real parageneses of the contact rocks of the Chiney pluton, he has shown that Fe activity steadily increased during the formation of magnesian skarn, reaching a maximum in the contact rim of gabbro rocks and then μFe dropped. In addition to the increase of chemical Fe activity that ensures release of iron from melt, suitable redox conditions are needed for titanomagnetite crystallization. Konnikov (1986) supposed that such conditions had been created by transvaporization of CO_2 and H_2O derived from country sedimentary rocks heated by high-temperature magmatic melt with abrupt increase in PCO_2.

Without refuting the theoretical possibility of the proposed mechanism, it hardly could have led to the formation of economic mineralization. A number of geological observations do not support this model:

1. In many cases, titanomagnetite rims do not surround xenoliths immediately, and their morphology does not correspond to the shape of xenoliths.
2. Frequently, the thickness of titanomagnetite ore is several times more than the size of xenoliths.
3. Most carbonate xenoliths are not surrounded by titanomagnetite segregations at all.
4. Xenoliths occur in massive titanomagnetite ore (Fig. 7.4a).

The third comment is proved by the geological structure of the eastern part of pluton in the basin of the River Right China. The large xenoliths of carbonate rocks in gabbro occur on both sides of the river. They dominate in number and size among massive low-Ti gabbro rocks on the left bank, which is completely free from titanomagnetite mineralization, whereas the right bank is distinguished by abundant iron orebodies of variable morphology, but carbonate xenoliths are partly devoid of titanomagnetite rims. The ore veins in this area are controlled by faults. This was noted by Chechetkin (1966) and other authors.

Similar idea on the origin of titanomagnetite lenses, which are widespread in the southern part of the pluton, was also developed by Shabalin and Sharapov (1981), although only sporadic xenoliths of carbonate rocks occur at the outcrop, chosen by these authors, and these xenoliths are unrelated to the titanomagnetite lenses. In their opinion, assimilation of carbonate rocks by magmatic melt exerts effect on ore deposition indirectly rather than in the form of titanomagnetite rims. The excess of carbon dioxide as a product of limestone dissolution facilitates the breakdown of melt into two immiscible liquids with formation of titanomagnetite ore.

Thus, the authors assume a combination of two separation mechanisms—the directed crystallization (overall) and the local liquid immiscibility (in domains enriched in calcium carbonate); however, it is not easy to interpret the published factual material in terms of this model. The xenolith-like angular ore fragments, with banding oriented at an angle to the layering in host rocks, and further destruction of them into angular clasts are described. These are titanomagnetite-pyroxene rocks; titanomagnetite is idiomorphic relative to silicates in contrast to the ore veins, where titanomagnetite grains are xenomorphic. It should specify that the above relationships were observed in low-Ti gabbro rocks, which are, in our opinion, products of late magmatic injections. This sequence of events is confirmed by xenoliths of high-Ti rocks in the barren low-Ti gabbroids, which are observed in the southern part of the Chiney pluton.

Diversity of ore veins and lenses in the contact zone between high-Ti and low-Ti gabbro rocks in the eastern part of pluton, as well as occurrence of lenticular lodes at the base of low-Ti gabbro members, can be readily explained by the interaction of high-Ti rocks with subsequent low-Ti melts. Agreeing with arguments proposed by Shabalin and Sharapov (1981), we nevertheless assume that the hybrid melt, formed as a product of assimilation of not only carbonate rocks, but also xenoliths of high-Ti gabbro, has been broken into two immiscible liquids. The enrichment of the melt in fluids owing to the assimilation of carbonate material facilitates origination of liquid immiscibility.

The pegmatoid type of mineralization has been revealed and described by Lebedev (1962) in various parts of pluton, mainly in the north. The content of ore minerals in the giant-grained pegmatites reaches 26 vol%, and this value fits the composition of the magnetite-silicate eutectic system (Ostrovsky and Ol'shansky 1956). Titanomagnetite segregations reach 10 cm across.

7.3　The Magnitnyi Deposit

The Magnitnyi deposit (Fig. 7.5) contains almost 1.5 billion tons of titanomagnetite ore, hosted in several orebodies (Gongalsky et al. 2016). The major orebodies contain 89.1% of the total ore reserves and 86.1 to 92% of mineable metal reserves (Chechetkin and Kharitonov 2002). The orebodies form lower (11 major orebodies) and main (23 minor orebodies) clusters (Chumachenko et al. 2000). The apophyses of major orebodies represent minor orebodies.

Table 7.1 Chemical composition of titanomagnetite ore from the Magnitnyi deposit, wt.%

N	N orebody	Fe$_{tot}$	Fe$_{Mag}$	TiO$_2$	V$_2$O$_3$
1.	1	29.95	23.92	5.6	0.409
2.	3	38.09	32.07	6.87	0.469
3.	14	36.58	29.96	7.44	0.565
4.	6	28.88	20.93	5.29	0.424
5.	4	39.39	34.47	6.7	0.571
6.	3	32.64	25.92	6.21	0.587
7.	3	36.3	25.78	5.90	0.466

Note. Fe$_{tot}$—total Fe (FeO + Fe$_2$O$_3$) content in rock. Fe$_{Mag}$, Fe in oxide minerals; TiO$_2$, V$_2$O$_3$, content in rock (after Chumachenko et al. 2000)

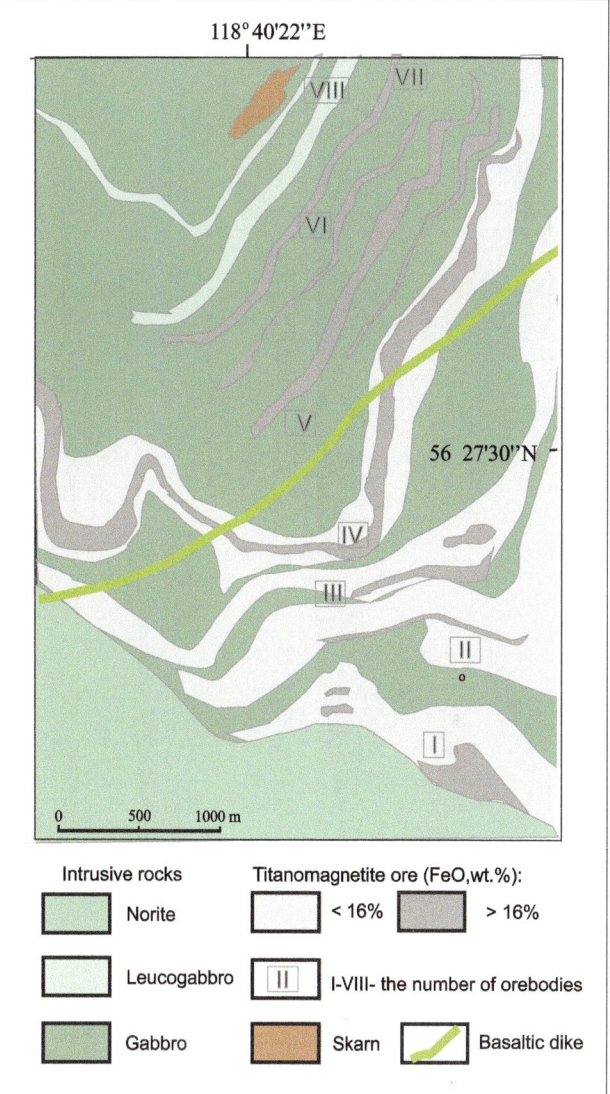

Fig. 7.5 Schematic map of the Magnitnyi deposit (After Chumachenko et al. 2000)

The ores form massive veins and lenses with sharp contacts with the host rocks. The size of lenses is about 1–3 × 0.7–1.5 m. Globular, oval, and irregularly shaped titanomagnetite bodies reach up to a dozen centimeters in width and length. The thicknesses of ore lenses vary between 5.7 and 26.5 m. The iron content changes from 39.4 to 28.9 wt% (Table 7.1). The carbonate xenoliths, several tens of meters in diameters, occur in ore, especially abundant in the southern block. The main ore cluster extends for 4.5 km; its thickness is around 400 m. Some of the orebodies are found over 2–3 km to the west of the main cluster (Fig. 7.5).

7.4   The Etyrko Deposit

The Western block of the Chiney pluton comprises the Etyrko deposit (Fig. 5.2), whose resources remain classified. There are two ore types: (1) disseminated titanomagnetite-bearing ore and (2) thin layers of massive titanomagnetite ore associated with anorthosite (Fig. 5.7a). Golev et al. (1987) recognized eight orebodies grouped into a 100–700-m-thick ore-bearing package. The mineralization was traced over a distance of 12 km, generally parallel to the magmatic layering. The orebodies dip at 20° toward the center of the layered mafic-ultramafic complex.

The thickness of individual orebodies varies from 2 to 50 m (20–30 m in average). Besides major orebodies, there are several long minor orebodies. Their composition is summarized in Table 7.2. The upper parts of layered units (micro-rhythms, rhythms, macrorhythms) contain plagioclase-titanomagnetite rocks (10–90 vol.% Pl) (Gongalsky et al. 2016).

7.5   Chemistry of Titanomagnetite Ore

The Chiney iron deposits were classified as a titanomagnetite type (Kuznetsov 1964). The ilmenite-titanomagnetite ore is obviously predominant, although magnetite-ilmenite ore

was also noted (Fedotova et al. 1977). In general, the ore from the Chiney pluton is Fe-rich, with significant vanadium concentrations.

The positive Fe-Ti and Fe-V correlations for 662 samples from three sites (average Fe-Ti and Fe-V correlation coefficients are 0.885 and 0.720, respectively; Kulikov et al. 1981) show that the overwhelming majority of these metals is incorporated into oxides and only an insignificant amount is contained in other minerals.

The more detailed characterization of chemical composition of oxide ore localized in the Chiney pluton is given in Table 7.2, for massive ore, and in Table 7.3, for disseminated ore. The main components of ore are Fe, Ti, and V, enriched in Ni, Cu, and Pt + Pd (Table 7.4). The ore is low-phosphorus as compared with other titanomagnetite deposits, containing

Table 7.2 Representative chemical composition of massive titanomagnetite ore from Chiney pluton, wt%

Ore type	n	SiO_2	TiO_2	Al_2O_3	FeO	MnO	MgO	V_2O_5	Total
Titanomagnetite drops	3	0.17	11.68	3.43	80.31	0.25	1.20	2.02	99.35
Massive titanomagnetite	8	0.22	11.74	4.03	78.92	0.22	1.99	1.99	99.37
Disseminated ore in pyroxenite (kosvite)	27	0.26	11.77	2.43	81.86	0.16	0.31	1.43	98.54
Disseminated ore in anorthosite (chineyite)	16	0.21	11.68	3.23	82.11	0.25	0.12	1.77	99.59
Disseminations in breccia	14	0.28	6.73	0.67	89.56	0.21	0.09	1.17	99.22

Note. Analyses were carried out at Chita Institute of Natural Resources, SB RAS, analyst NS Balyev

Table 7.3 Chemical composition ore at Etyrko deposit

N Orebody	Thickness, n	Fe_{tot}	TiO_2	V_2O_5 wt%	Cu	Ni	Co	Pt ppm	Pd
1.	15.50	17.36	2.93	0.35	0.190	0.023	0.012	0.030	0.060
2.	20.30	16.05	2.57	0.28	0.400	0.019	0.008	0.042	0.066
3.	22.00	16.99	2.92	0.28	0.040	0.020	0.009	0.100	0.286
4.	12.40	17.21	2.57	0.26	0.150	0.026	0.010	0.152	0.191
5.	13.20	17.43	2.97	0.32	0.050	0.022	0.009	0.076	0.172
6.	49.20	19.68	3.95	0.30	0.070	0.014	0.011	0.016	0.008
7.	24.00	16.25	2.91	0.26	0.040	0.011	0.007	0.030	0.046
8.	91.10	19.27	3.86	0.37	0.050	0.013	0.008	0.063	0.064
9.	93.40	18.82	3.44	0.34	0.080	0.014	0.009	0.080	0.010
9a.	14.40	22.24	4.36	0.40	0.120	0.013	0.009	0.079	–
9b.	16.40	17.16	2.91	0.31	0.080	0.011	0.007	0.065	0.008
10.	5.50	30.84	6.95	0.57	0.750	0.008	0.011	–	0.003
11.	6.90	16.45	2.62	0.22	0.130	0.006	0.010	0.024	0.017
12.	12.20	18.01	3.23	0.26	0.090	0.012	0.009	0.017	–
13a.	17.60	19.88	4.13	0.35	0.070	0.022	0.010	0.030	0.057
13.	64.00	23.87	4.69	0.48	0.100	0.016	0.012	0.010	0.010
14.	12.00	21.70	3.59	0.25	0.180	0.035	0.012	0.026	0.006

Note. Analyses were carried out at PGO Chitageologia

Table 7.4 Composition of disseminated titanomagnetite ore from the Chiney pluton

№	N sample	TiO_2	MnO	Cu	S	Ni	Co	Zn	Cr	V_2O_5
1.	11/619.91	5.74	0.27	0.12	0.06	376	395	246	679	4500
2.	11/619.95	6.59	0.25	0.15	0.09	400	410	271	712	5100
3.	11/619.99	8.15	0.26	0.17	0.84	410	489	286	884	6400
4.	11/622.68	8.20	0.27	0.08	0.16	255	406	253	775	2216
5.	11/622.73	4.40	0.27	0.06	0.12	158	227	142	309	1435
6.	11/989.6	2.34	0.12	0.09	0.19	141	119	115	119	2444
7.	11/992	2.34	0.12	0.06	0.15	162	95	109	116	2647
8.	83/1327.3	2.03	0.12	1.01	9.29	1100	327	221	357	2292
9.	Et-66-1	6.75	0.26	0.09	0.22	132	358	219	441	1180
10.	Et-05-1	0.51	0.05	1.30	2.97	668	136	54	26	198
11.	Et-08	0.67	0.05	1.20	2.74	530	129	45	32	262
12.	Et-08-1	0.83	0.05	1.29	2.90	561	131	49	45	336
13.	Et-08-11	0.83	0.06	1.45	2.92	624	130	55	43	358
14.	9/210	0.89	0.10	0.91	1.77	546	79	74	45	569
15.	9/219.7	1.03	0.12	0.71	1.43	398	98	104	46	530
16.	66/41.5	0.69	0.18	0.14	0.18	161	92	90	103	537
17.	66/92	0.85	0.29	0.04	0.11	183	161	162	163	1232
18.	66/130.6	2.44	0.12	0.17	0.46	165	100	81	118	2012
19.	Mgn-674	1.60	0.09	1.30	1.84	855	222	56	117	757
20.	119/155	8.90	0.15	0.15	0.48	199	365	292	537	8700

Note. Analyses: 1–13, Etyrko; 14–15, Kontaktovyi; 16–20, Magnitnyi. N sample—11/619.91—number of borehole/depth (m). Concentrations of TiO_2, MnO, Cu, and S are given in %; Ni, Co, Zn, Cr, and V_2O_5, in ppm. Analyses were carried out by XRF at IGEM RAS, analyst AI Yakushev

Table 7.5 Rock composition of the microrhythm 11/622, wt%

Dept, cm	SiO$_2$	TiO$_2$	Al$_2$O$_3$	Fe$_2$O$_3$	Feo	MnO	MgO	CaO	Na$_2$O	K$_2$O	P$_2$O$_5$	Total	LOI
62,268	15.83	7.85	3.62	24.22	26.22	0.26	5.98	3.08	1.59	0.18	0.01	88.84	-1
62,269	17.59	8.23	3.4	26.77	26.44	0.29	6.2	3.71	1.39	0.18	0.01	94.21	−0.85
62,271	41.65	3.38	7.25	10.18	16.99	0.26	10.02	9.4	0.99	0.37	0.08	100.57	0.21
62,273	36.1	4.75	6.93	15.37	20.18	0.29	8.96	7.89	0.96	0.26	0.03	101.72	−0.08
62,274	30.85	5.75	6.04	19.43	21.98	0.30	9.13	5.31	1.01	0.3	0.03	100.13	−0.62
62,275	20.6	7.82	4.17	25.47	26.72	0.31	6.78	3.15	0.95	0.23	0.03	96.23	−0.89
62,276	16.14	8.58	3.22	28.37	28.09	0.30	6.55	2.03	1.36	0.18	0.01	94.83	−1.18
62,278	32.93	4.95	5.27	16.39	22.05	0.31	10.12	4.06	0.9	0.28	0.05	97.31	−1.06
62,279	28.59	6.73	5.74	21.17	23.85	0.31	9.2	3.7	0.94	0.25	0.02	100.50	−0.89
6228	23.5	7.48	4.74	26.63	25.08	0.32	7.27	3.19	0.94	0.23	0.02	99.40	−1.3
62,281	19.23	8.18	4.26	26.45	26.87	0.29	7.15	2.75	0.83	0.2	0.01	96.22	−0.9
62,283	48.69	1.54	18.02	1.04	12.43	0.13	4.05	8.8	2.3	0.48	0.06	97.54	−0.24
62,284	46.64	2.59	8.79	8.09	17.6	0.29	11.21	6.06	1.31	0.43	0.04	103.05	−1.01
62,285	45.28	3.04	8.1	10.79	18.02	0.31	11.46	5.93	1.14	0.35	0.06	104.48	−1.52
62,286	36.79	4.41	5.33	13.68	21.77	0.32	11.54	4.11	0.98	0.24	0.05	99.21	−1.2
62,287	29.36	5.95	3.99	18.59	24.04	0.31	10.18	3.48	0.95	0.19	0.02	97.06	−1.44
62,289	48.82	1.97	14.39	8.43	9.7	0.19	6.6	8.53	2.15	0.49	0.08	101.35	0.42

Note. Depth: borehole 83/622 m, 68–89 cm. Analyses were carried out at Chita Institute of Natural Resources, analyst Mikhailova. Negative LOI is explained by oxidation of Fe during analysis

Fig. 7.6 TiO2, SiO$_2$, and Al2O$_3$ distribution inside of microrhythms. See data in Table 7.5. Red, Pt; blue, Pd

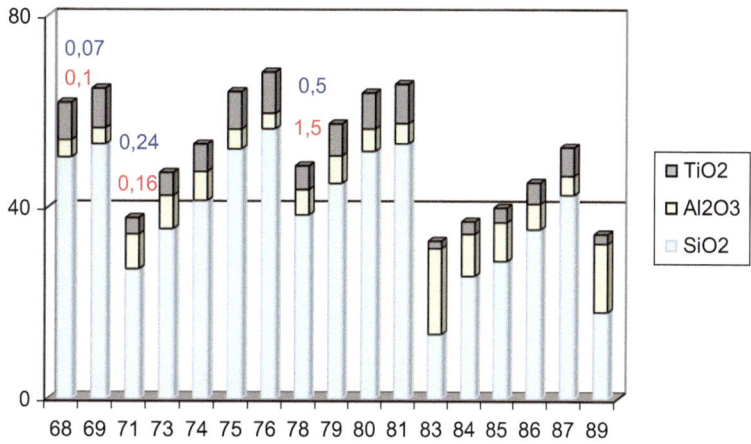

10–15% P$_2$O$_5$, e.g., in the Kruchina pluton in Transbaikalia region or Gremyakha-Vyrmes deposit in the Kola Peninsula, both in Russia. The Chiney deposits can be regarded as platinum-bearing titanomagnetite deposits (Dodin et al. 2003), similar to Pudozhgora layered mafic-ultramafic complex (Karelia, Russia) and Guseva Gora deposit (Urals, Russia).

The best-studied part of the Etyrko deposit has been penetrated by boreholes to a depth of 1.5 km (Figs. 5.16, 5.17 and 5.18). Variations of ore components inside the microrhythm 11/622 are demonstrated in Table 7.5.

Within the layered stratum, this mineral is distributed uniformly throughout the rocks and accumulates in the lower parts of layers, forming the microrhythms. The distribution of oxides is demonstrated in Fig. 7.6, where three rhythms are diagnosed based on iron contents.

7.6 Mineral Composition of Ore

7.6.1 Titanomagnetite and Ilmenite

Titanomagnetite and ilmenite (the amount of the latter rarely exceeds 10%) are major ore minerals. The grade, size, and morphology of titanomagnetite grains are widely variable. The morphology of grains is close to polygonal in aggregates (Fig. 7.7) and oval among silicates. In case of low concentrations, titanomagnetite makes up the segregations, 5–6 mm in length and 2–3 mm wide, oriented parallel to the layer bottom. As the amount of mineral increases, the segregations merge into larger aggregations (15 × 5–6 mm) (Fig. 7.8).

Rock-forming minerals occur in ore segregations as round grains. Rough cracks propagate from them, intersecting titanomagnetite grains. In some cases, a reaction rim, consist-

Fig. 7.7 A sample of massive ore from the Magnitnyi deposit

ing of fine-grained mica aggregate, is localized between oxide and silicate. The thickness of these rims increases at the contact with plagioclase, where biotite, amphibole, and garnet are identified, and decreases at the contact with pyroxenes. The composition of titanomagnetite from these rocks is given in Table 7.6.

The mean size of titanomagnetite grains is 1.5–2.0 mm, ranging from 0.5 to 4.0 mm. Both separate grains and laminar screens along the cleavage of pyroxene are noted, with a length of the latter reaching 0.01 mm. The morphology and size of titanomagnetite and ilmenite grains are shown in Fig. 7.9. The titanomagnetite composition is summarized in Tables 7.6 and 7.7, and ilmenite composition is given in Table 7.8. The grade of elements varies essentially, especially Fe and Ti. Besides major components, titanomagnetite contains V and Mn, Mg, Al, and Si (Table 7.9). The variations of titanomagnetite composition in the microrhythm 11/622 are shown in Table 7.10. In comparison with titanomagnetite, ilmenite is enriched in Mn and depleted in V, Al, and Mg.

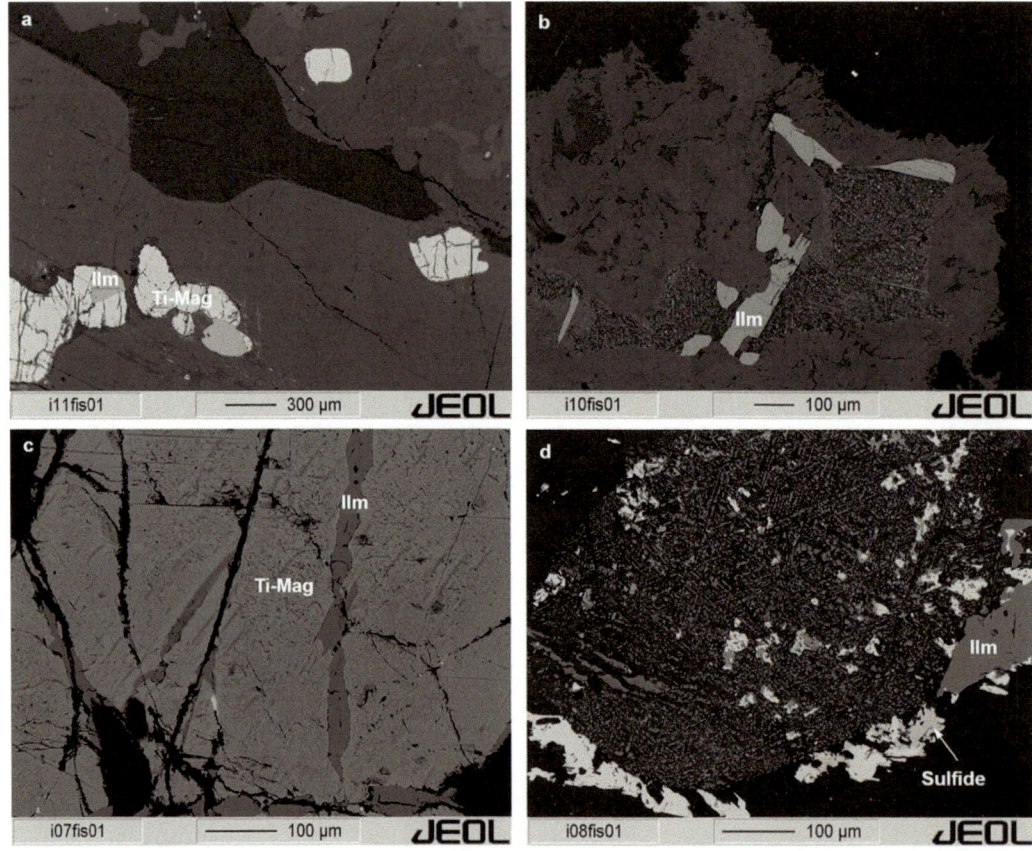

Fig. 7.8 BSE images of titanomagnetite ore from the Magnitnyi and the Etyrko deposits. (**a**) sample 11/97, small grains of titanomagnetite and ilmenite; (**b**) sample 9/130, ilmenite in silicate matrix, (**c**) sample 9/142, titanomagnetite with lamellae of ilmenite, (**d**) resorbed titanomagnetite with sulfides. *Ti-Mag* titanomagnetite, *Ilm* ilmenite. BSE images were taken at Max-Planck Institute of Geochemistry, Mainz, Germany (Analyst DV Kuzmin)

Table 7.6 Representative chemical composition of titanomagnetite from the Chiney pluton, wt%

Ore type	n	SiO$_2$	TiO$_2$	Al$_2$O$_3$	FeO	MnO	MgO	V$_2$O$_5$	Total
Titanomagnetite drops	3	0.17	11.68	3.43	80.31	0.25	1.20	2.02	99.35
Massive titanomagnetite	8	0.22	11.74	4.03	78.92	0.22	1.99	1.99	99.37
Disseminated ore in pyroxenite (kosvite)	27	0.26	11.77	2.43	81.86	0.16	0.31	1.43	98.54
Disseminated ore in anorthosite (chineite)	16	0.21	11.68	3.23	82.11	0.25	0.12	1.77	99.59
Disseminations in breccia	14	0.28	6.73	0.67	89.56	0.21	0.09	1.17	99.22

Note. Analyses were carried out at IGEM RAS, analyst EV Kovalchuk

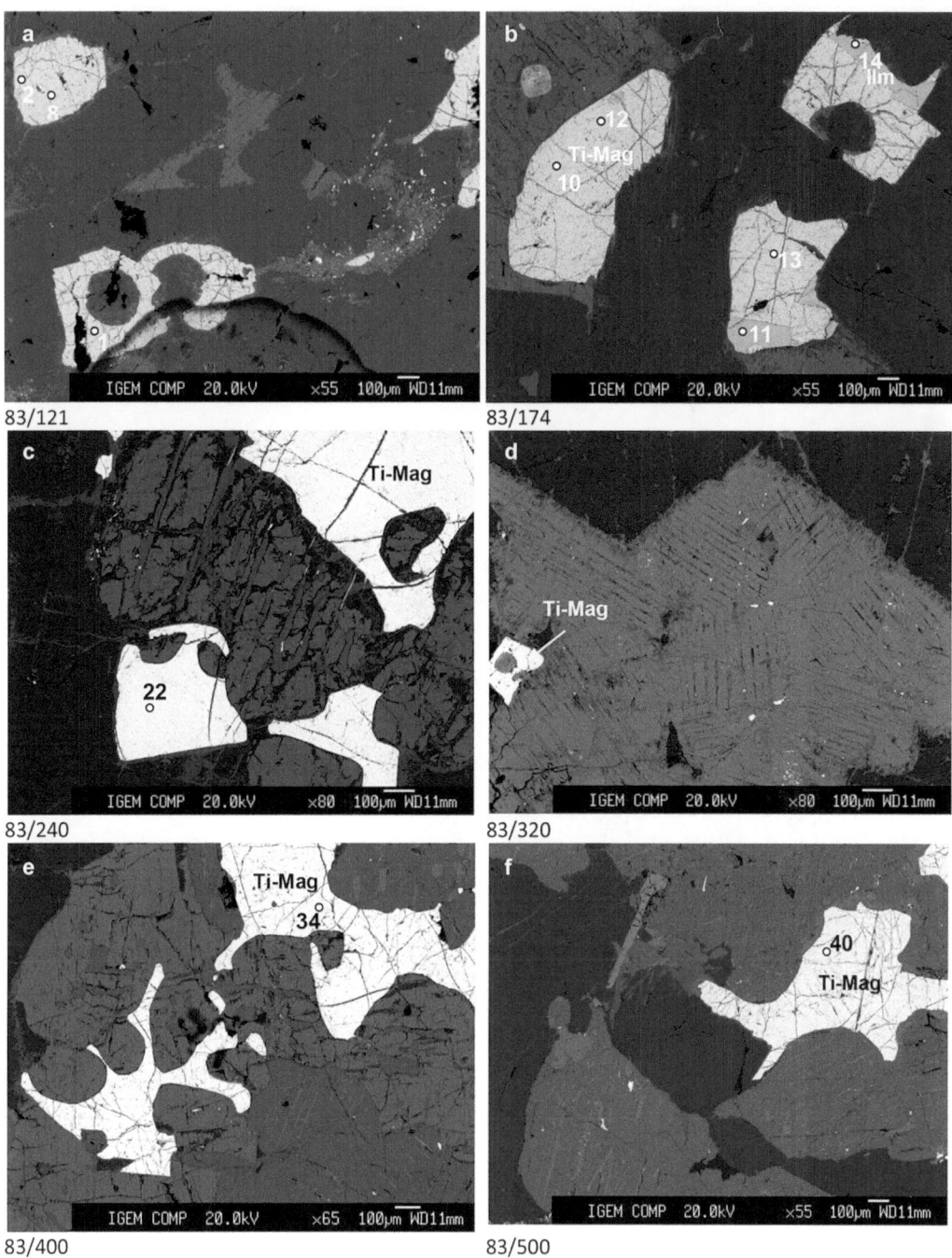

Fig. 7.9 BSE images of disseminated ore at the Etyrko deposit. Samples from drillcore of borehole 83, number of sample = borehole 83/depth (m): (**a**) 83/121, (**b**) 83/174, (**c**) 83/240, (**d**) 83/320, (**e**) 83/400, (**f**) 83/500, (**g**) 83/600, (**h**) 83/760, (**j**) 83/848, (**k**) 83/1101, (**l**) 83/1101, (**m**) 83/1285. Ti-Mag, titanomagnetite; Ilm, ilmenite. Number point in Figure = number point in Table 7.7 Analyses and images were taken at IGEM, analyst EV Kovalchuk

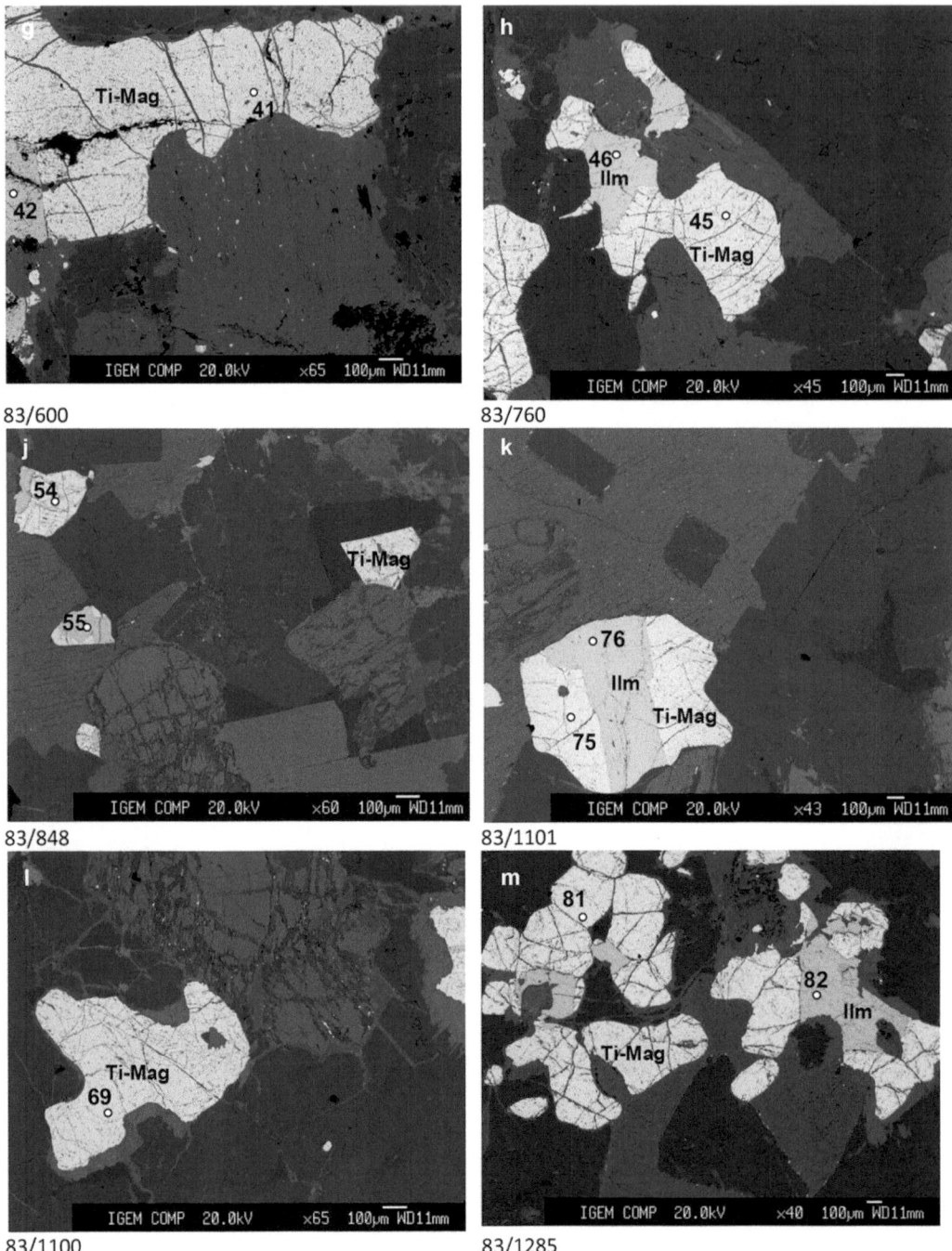

Fig. 7.9 (continued)

7.6.2 Internal Structure of Titanomagnetite Grains

As is known, titanomagnetite is a series of magnetite $FeFe_2O_4$-ulvospinel $FeTi_2O_4$ solid solutions in the FeO-TiO_2-Fe_2O_3 system. The phase relations in this system are described by means of two phases: (1) between the homogeneous magnetite-ulvospinel solid solution and pleonaste and (2) between magnetite and ulvospinel (Patnis and McConnel 1983). At a relatively high temperature, pleonaste is the first to be released from the solid solution, leaving the homogenous magnetite-ulvospinel matrix at a lower temperature. Thus, the exsolution texture of the latter will be much thinner.

The relationships between various phases are a source of information in respect to the sequence of the processes, occurring within a wide temperature interval. Precisely, these microstructures are inherent to titanomagnetites in

Table 7.7 Titanomagnetite compositions from disseminated ore at the Etyrko deposit, wt%

N	N sample	N point	MgO	Al₂O₃	SiO₂	TiO₂	Cr₂O₃	MnO	FeO	V₂O₃	Total
1.	83/121	1	0.02	0.13	0.01	5.87	0.06	0.10	87.65	1.50	95.40
2.	83/121	2	0.18	3.23	0.13	12.98	0.04	0.32	79.25	1.22	97.80
3.	83/121	3	0.03	12.43	0.12	7.12	0.24	0.19	77.78	1.77	99.74
4.	83/121	7	0.13	0.14	0.09	23.29	0.18	0.54	69.82	1.17	95.55
5.	83/121	9	0.03	0.18	0.18	11.93	0.27	0.23	81.01	1.19	95.09
6.	83/174	10	0.03	0.69	0.04	11.85	0.13	0.23	82.85	0.90	96.79
7.	83/174	12	0.05	1.83	0.04	12.70	0.08	0.39	80.20	0.91	96.28
8.	83/174	13	0.05	0.99	0.08	4.76	0.04	0.15	89.76	0.81	96.72
9.	83/200	19	0.00	0.20	0.08	8.95	0.03	0.50	84.20	0.96	94.99
10.	83/240	22	0.06	3.97	0.14	11.18	0.01	0.48	79.28	1.08	96.28
11.	83/400	34	0.01	0.88	0.04	2.49	0.05	0.09	90.14	1.38	95.17
12.	83/500	40	0.05	1.28	0.01	3.99	0.04	0.11	90.45	1.47	97.49
13.	83/600	41	0.02	0.76	0.06	9.13	0.01	0.24	84.65	1.42	96.45
14.	83/760	45	0.03	0.49	0.05	4.40	0.10	0.13	88.85	1.53	95.64
15.	83/848	54	0.00	0.39	0.04	0.24	0.19	0.02	92.17	1.23	94.42
16.	83/848	56	0.02	0.26	0.03	0.23	0.29	0.03	92.67	1.10	94.75
17.	83/1100	69	0.01	0.22	0.09	0.54	0.13	0.03	93.01	1.05	95.12
18.	83/1101	75	0.01	0.34	0.04	1.42	0.17	0.06	92.01	1.22	95.34
19.	83/1285	81	0.04	0.25	0.03	0.52	0.27	0.04	92.17	1.38	94.80

Note. Here and in Table 7.8, analyses carried out at IGEM RAS, analyst EV Kovalchuk. N sample = N borehole/depth (m)

Table 7.8 Ilmenite composition from disseminated ore of borehole 83, wt%

N	N sample	N point	MgO	Al₂O₃	SiO₂	TiO₂	Cr₂O₃	MnO	FeO	V₂O₃	Total
1.	83/121	8	0.26	0.13	0.17	48.14	3.13	0.86	45.38	0.80	99.06
2.	83/174	11	0.11	0.00	0.08	51.27	0.07	1.10	49.37	0.06	102.10
3.	83/174	14	0.09	0.00	0.01	51.48	0.00	1.10	48.40	0.05	101.18
4.	83/600	42	0.04	0.01	0.01	47.96	0.02	1.25	50.41	0.24	100.01
5.	83/760	46	0.08	0.02	0.04	50.67	0.02	1.47	48.73	0.21	101.24
6.	83/848	55	0.13	0.04	0.05	46.87	0.05	1.80	52.84	0.24	102.05
7.	83/1101	76	0.22	0.07	0.02	47.77	0.00	1.39	51.96	0.19	101.72
8.	83/1285	82	0.09	0.06	0.00	43.06	0.04	1.58	56.27	0.13	101.23

Note. Sample, N = borehole number/depth, m

the Chiney pluton, especially those making up the finely layered ore.

7.6.2.1 Texture and Structure of Titanomagnetite

Titanomagnetite of this ore type is characterized by exsolution textures of several orders (Fig. 7.10). Under a microscope, the mineral has a net fabric caused by the combination of isometric blocks that extinct simultaneously due to parallel arrangement of ilmenite tablets. Boundaries between blocks, 0.03–0.1 mm in size, are diffuse. They grow through one another and differ in the internal structure, orientation of ilmenite lamellae, and character of exsolution textures of the second and third orders. The most typical is a net fabric that represents the primary structure of grains. According to the indicated dimensions, the solid solution structures are subdivided into coarse (Fig. 7.10c, e, f), medium (Fig. 7.10a, d), and fine (Fig. 7.10b). They reflect complex and multistep cooling and the transformation of protomagnetite.

A low solubility of ilmenite in the spinel phase has been established in experiments. Therefore, most authors (Patnis and McConnel 1983; Khisina 1987; Kudryavtseva et al. 1982) assume that this type of structure is not a product of a single-phase solid solution, cooling to the temperature of stable solvus, but arises due to the oxidation of ulvospinel in solid state (so-called oxidative exsolution when oxygen fugacity is higher than the equilibrium value for specific magnetite-ulvospinel composition).

The areas, irregular in shape with large (up to 0.1–1.0 mm) parallel ilmenite lamellae, are observed against the background of the lattice texture of titanomagnetite. The ilmenite lamellae are localized along fractures or at grain margins. Spinel grains enlarge with the formation of a spindle-shaped segregation. These facts show that the aforementioned structures were formed along fractures at the post-magmatic stage as a result of matter redistribution in weakened zones with enlargement of grains, thickening of ilmenite lamellae up to

Table 7.9 Compositions of titanomagnetite and ilmenite from the Etyrko deposit, wt%

N	N sample	MgO	Al$_2$O$_3$	SiO$_2$	TiO$_2$	MnO	FeO	V$_2$O$_3$	Total
1.	11/282-1-1			0.217	2.81		86.40	1.850	91.23
2.	11/282-1-2		0.40		4.64		86.59	1.66	93.23
3.	11/282-1S	0.40	2.14	0.234	11.74		80.16	1.60	96.28
4.	11/282-2S	0.41	3.15	0.245	11.79	0.430	78.42	1.43	95.88
5.	11/320-1S	0.39	3.84		10.14		80.12	1.23	95.72
6.	11/320-2S	0.28	2.35		10.31		79.96	0.85	93.76
7.	11/320-4S	0.29	3.43	0.736	9.19		74.07	1.11	88.83
8.	11/620		0.82	0.234	4.77		84.72	1.86	92.41
9.	11/620-1S		0.91	0.208	10.23		81.89	1.75	94.98
10.	11/620-2S	0.50	3.07		11.29		79.71	1.33	95.89
11.	11/0-2-1S		0.47		5.66		86.64	1.15	93.92
12.	11/0-2-2S		0.25	0.239	4.25		86.55	1.24	92.53
13.	11/0-2S		0.32	0.273	5.22		86.60	1.15	93.56
14.	11/0-1S		0.68		7.50	0.35	83.37	0.98	92.87
15.	11/0-3		0.85		11.21	0.27	79.52	0.99	92.84
16.	11/0-3S		0.42	0.371	8.87	0.25	82.28	1.08	93.23
17.	11/0-4		1.02	0.17	8.27	0.27	81.73	0.98	92.45
18.	11/0-5		0.66	0.858	3.97		78.28	0.82	84.59
19.	11/0-6		0.50		7.84		83.33	0.83	92.50
20.	11/0-3S		0.26		5.69		86.46	1.12	93.58
21.	11/282-1-3				50.34	1.85	47.43	0.56	100.17
22.	11/0R-1				51.19	1.79	47.01		99.99
23.	11/0-2S			0.258	48.20	1.67	49.90	0.62	100.63

Note. Analyses were carried out at IEM RAS, analyst AN Nekrasov

N sample = borehole 11/depth (m). N 11/0—the samples were taken in 80 m to SW form borehole 11

Sample 11/0—magmatic breccia (Group 4). 11/0-3S—analyses were carried out over 100 × 100 mkm

Table 7.10 Titanomagnetite composition from microrhythm 11/622, wt%

N	SiO$_2$	TiO$_2$	Al$_2$O$_3$	FeO	MnO	MgO	CaO	Na$_2$O	K$_2$O	Cr$_2$O$_3$	V$_2$O$_5$	Total
1.	0.21	10.35	2.51	84.01	0.17	0.11	0.01	0	0	0.6	2.05	99.42
2.	0.44	12.61	3.78	77.86	0.18	0.41	0	0.61	0	2.88	1.24	97.13
3.	0.4	13	3.41	79.03	0.17	0.24	0.05	0	0.05	2.54	1.11	97.46
4.	0.12	10.95	1.8	83.17	0.21	0.16	0.08	0.5	0.08	0.97	1.96	99.03
5.	0.36	11.78	2.24	78.08	0.2	0.23	0.07	0.27	0.01	5.8	0.97	94.21
6.	0.17	11.25	0.65	85.86	0	0	0.2	0	0	2.38	0.58	98.53
7.	0.18	10.64	0.92	81.94	0.16	0.18	0.04	0.05	0.03	2.18	0.69	97.83
8.	0.19	10.08	1.62	84.25	0.17	0.1	0.98	0.31	0	1.06	1.22	98.92
9.	0.02	11.07	3.87	81.73	0.12	0.36	0.02	0.01	0.02	1.38	1.4	98.62
10.	0.16	10.2	3.23	83.51	0.27	0.3	0.08	0	0.06	1.37	0.82	98.63
11.	0.18	8.97	0.68	85.14	0.21	0.03	0.11	0	0.01	1.61	1.06	98.39
12.	0.18	12.56	3.71	81.52	0.23	0.39	0	0.21	0.01	1.11	0.1	98.91
13.	0.19	10.94	2.76	82.34	0.17	0.44	0.02	0.51	0	1.016	1.48	98.85
14.	0.18	12.97	3.9	78.91	0.32	0.55	0	0.78	0.07	0.76	1.57	99.25
15.	0.9	13.28	1.02	81.05	0.08	0.58	0	0.11	0.06	2.26	0.67	97.75
16.	0.15	9.85	1.41	85.71	0.05	0.41	0.3	0.12	0.05	1.41	0.82	98.6
17.	0.11	12.72	3.55	80.27	0.05	0.43	0	0.35	0	1.66	0.85	98.33
18.	0.15	13.27	0.77	82.42	0.31	0.02	0.06	0	0.03	1.79	1.2	98.23
19.	0.08	13.02	3.8	78.28	0.32	0.45	0.01	0.45	0	1.28	2.31	98.72

Note. Analyses were carried out at IEM RAS, analyst AN Nekrasov

Fig. 7.10 BSE images of textures of titanomagnetite from the Chiney pluton (ilmenite and spinel in magnetite matrix). (**a**, **c**, and **d**) medium; (**b**) fine; (**e** and **f**) coarse. Samples: (**a**) 63/73, (**b**) 63/85, (**c**) 57/38, (**d**) 57/11, (**e**) 11/620, (**f**) 11/620 (N sample = N borehole/depth, m) Here and in Fig. 7.11, analyses were carried out at IEM, RAS (Analyst AN Nekrasov)

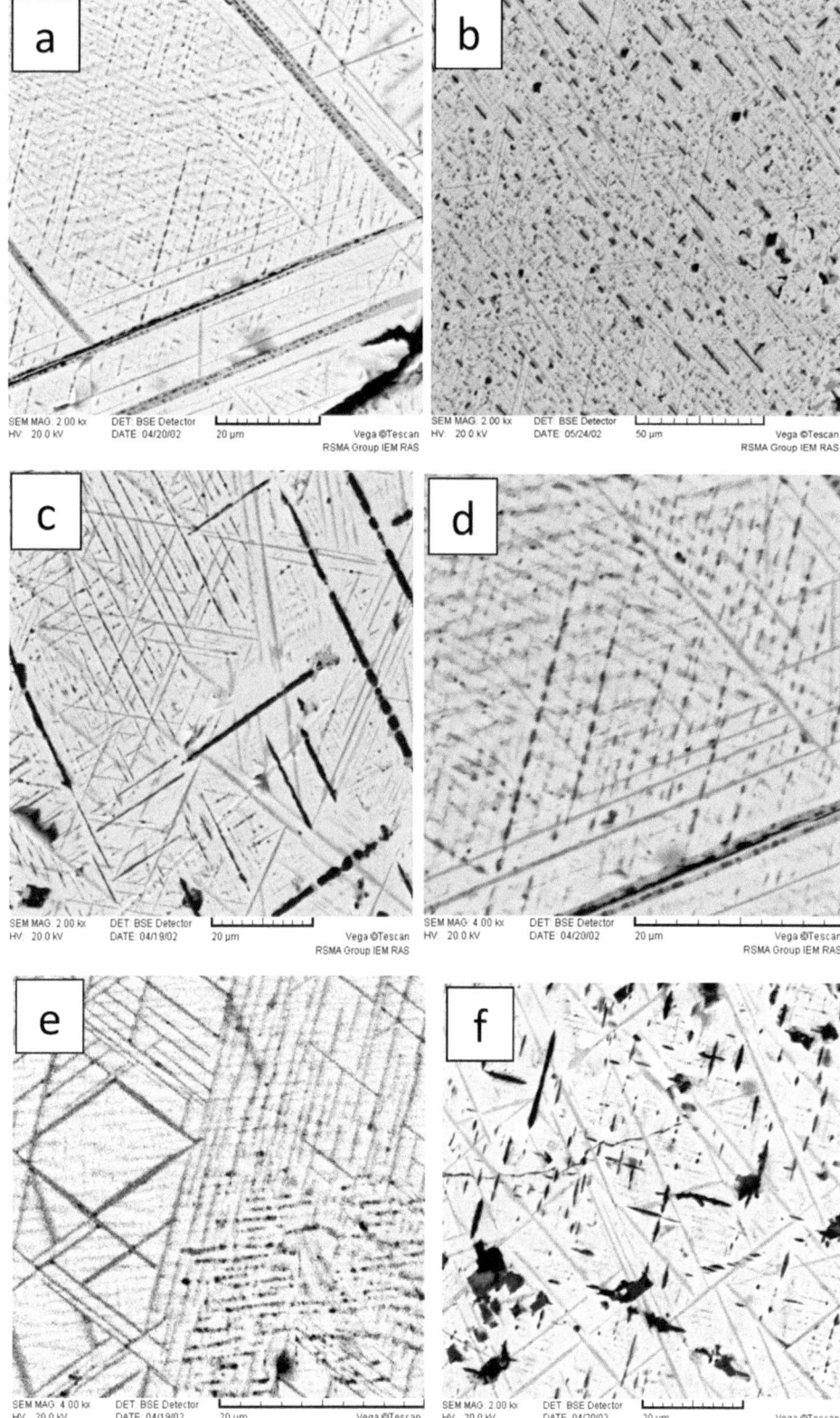

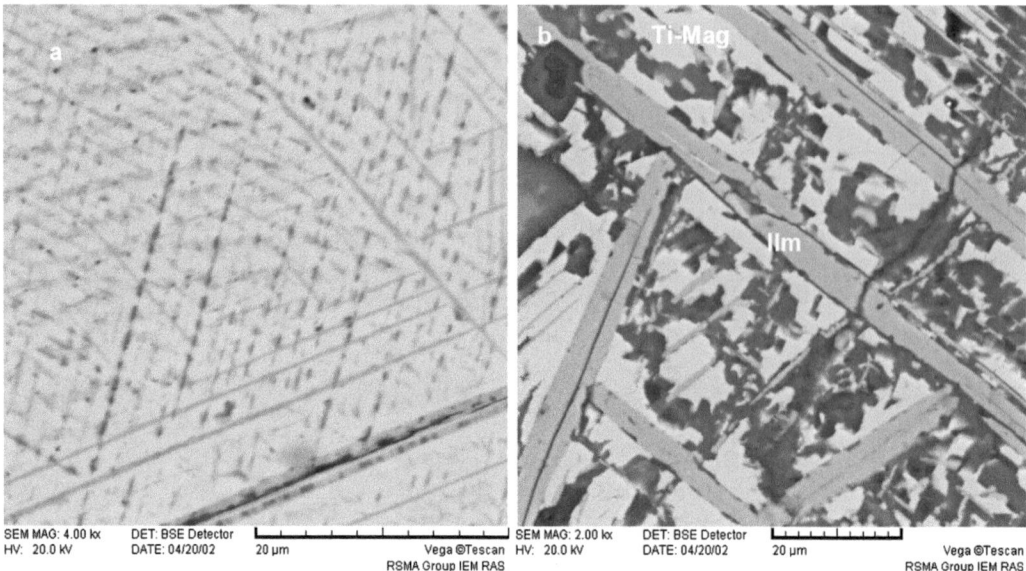

Fig. 7.11 BSE images of titanomagnetite textures
(**a**) Titanomagnetite matrix, (**b**) titanomagnetite relicts with ilmenite

the formation of the grains irregular in shape, and the streaming of spinel down to cracks with the appearance of empty zones. The superposition of these processes on magmatic processes makes the general pattern of titanomagnetite structure is difficult to interpret.

The coarse exsolution textures were formed at a high temperature, thin structures at the lower temperature, and the tiny structures in ulvospinel and magnetite which are detected only using electron microscope, in the process of exsolution by spinodal decomposition (Khisina 1987). Their total thickness does not exceed several thousand angstroms. Afterward, in the course of subsolidus cooling and the increase in oxygen fugacity, ulvospinel was partly oxidized with the formation of ilmenite. The subsequent cooling gave rise to the formation of fractures and weakened zones and to matter redistribution along these zones (refinement of magnetite matrix and enlargement of ilmenite lamellae). This probably happened before the complete crystallization of rock, because fractures are not traced in the adjacent silicate minerals. The sulfide minerals were mostly localized within these fractures or along boundaries of the largest phases in titanomagnetite, i.e., at the very early stages of exsolution texture formation and, thus, at the highest temperature.

Ilmenite in ore occurs as products of the decomposition of titanomagnetite solid solution or as independent grains. In the exsolution textures, ilmenite is represented by (1) large lamellae, intersecting titanomagnetite grain as a whole (up to 2 mm long); (2) medium-size lamellae, comparable in dimensions with blocks about 0.1 mm in size; and (3) short lamellae within blocks (0.01 mm), approaching lenses and irregular segregations in morphology. The average amount of free ilmenite in this type of ore is 10%, and habit of its segregations varies from large euhedral

grains to xenomorphic aggregates between titanomagnetite grains.

The euhedral tabular and polygonal ilmenite grains have sharp rectilinear boundaries and exceed in size titanomagnetite grains, reaching 4–5 mm in length. They often reveal twinning; thickness of tablets varies from 0.1 to 1.0 mm. The lattice texture arises occasionally as a result of simultaneous twinning along all (1011) planes. Most grains, however, have a simple structure. At a high magnification (>2000) of a microscope and in immersion, exsolution textures are always seen in ilmenite. Fine hematite lenses 0.001–0.019 mm long are localized as parallel lines. They often are concentrated at fracture's ends, so that their configuration is transformed thereby into isometric. The hematite content can reach 50% of the host mineral volume. It should be noted that the titanomagnetite series of solid solutions remains stable in a more reductive setting than in the ilmenite-hematite series.

Morphology of ilmenite grains of the second variety (generation) is determined by the shape of surrounding titanomagnetite grains. Ilmenite is traced in the form of cord-like grains along the contact of titanomagnetite crystals and rims the sulfide clusters. Ilmenite is more resistant to alteration. That is why it is kept in rocks better than titanomagnetite (Fig. 7.11b).

7.6.2.2 Magnetic Characteristics of Titanomagnetite

The Fe-Ti oxides in massive ore are close in properties to the minerals described above, although a relative amount of free ilmenite is somewhat lower. Comparison of the chemical composition of titanomagnetite and ilmenite from disseminated finely layered and massive ore of vein type (data of Zhevagin; Gongalsky and Krivolutskaya 1993) reveals some distinctions. Titanomagnetites from disseminated ore are characterized by elevated Ti (17–19 wt% TiO_2) as compared with massive ore (<13 wt% TiO_2) and a respective decrease in the Fe content. A more striking difference is established for trace elements. The contrast in Mg concentrations is the most distinct. Similar results were obtained by Konnikov (1986). The difference in Mn and Cr contents is not so strong. The chemical composition of titanomagnetites from finely layered ore (microrhythm from core 11, depth 622) is given in Table 7.10.

Magnetite sporadically occurs in ore in addition to titanomagnetite and ilmenite. These are dispersed grains, most frequently in association with newly formed ilmenite or pyrite (<0.25 mm in size). Clusters of magnetite crystals are occasionally observed.

The factual data on finely layered disseminated and massive ore of vein type do not allow us to oppose them to each other. The composition and structure of minerals are quite similar. Similar results have been obtained by Zhevagin (unpublished data) on the basis of thermomagnetic study. The Curie and Vervey points for disseminated finely layered and massive ore overlap (551 and 579 °C; −142 and −152 °C, respectively, Gongalsky and Krivolutskaya 1993). The absence of a maghemite-hematite transition point in thermograms shows that titanomagnetites of these ore types are either nonoxidized or only slightly oxidized.

Insignificant decrease in Curie point values of the measured titanomagnetites as compared with pure stoichiometric mineral is caused by development of exsolution textures in natural grains. Heating due to diffusion by homogenization of magnetite-spinel phases results in enrichment of ferromagnetic component in Mg and Al, and this drops the Curie temperature. The shift of Vervey points toward low temperature is caused by a wide range of trace elements in titanomagnetite, primarily Co, Ni, Cu, and Zn.

7.6.2.3 Mössbauer Spectroscopy

The amount of ilmenite in titanomagnetite was estimated using the Mössbauer spectroscopy at the Institute of Geology, Russian Academy of Sciences in Syktyvkar (Lyutoev et al. 2017).

The quantitative characteristics of the distribution of iron were obtained by the Mössbauer method spectroscopy. Usually, as a measure for relative iron content in various structural positions of the mineral phases, the relative area under the corresponding spectral component is used. For quantitative estimates of the distribution of iron ions over minerals and the positions in them, it is also necessary to take into account the difference in the probabilities of the Mössbauer effect of the iron ions in different structural states. This information is given for a large number of iron-containing mineral phases, including presence in titanomagnetite ores (Eeckhout and De Grave 2003; De Grave and Van

Alboom 1991). Mössbauer spectra of the Chiney pluton sample sand an example of their decomposition into spectral components are shown in Fig. 7.12. The spectra contain a pair of intense sextets (SA and SB) with hyperfine splitting, quadrupole splitting, and isomeric shift characterized for stoichiometric magnetite, doublet D0 of Fe^{2+} for ilmenite ions (IS = 1.06, QS = 0.65 mm/s), and doublets D2–D5 from Fe^{2+} and Fe^{3+} in other mineral phases. A sextet with a large hyperfine magnetic field (SA) occurs from Fe^{3+} ions in tetrahedral positions. The second sextet (SB) is due to bound electronic exchange of Fe^{2+} and Fe^{3+} in the octahedral position. For stoichiometric magnetite $[Fe^{3+}]$ $[Fe^{2+}$ $Fe^{3+}]$ O_4, the ratio of the areas of SA and SB sextets (A/B in Table 7.11) in the Mössbauer spectrum is ~0.53. The same ratio, taking into account the error, is determined in the samples 11/619.78, 11/638.00, 2101-2, and 6041-3, so the magnetite in these samples is close to stoichiometric with a small content of impurities. In the sample 11/603.3 and, especially, in M-2, the ratio exceeds the stoichiometric value, which indicates the defectiveness of the magnetite lattice with isomorphous replacement of iron by titanium (Zalutskiy et al. 2015).

Since titanium enters only in the octahedral positions of the magnetite structure (Gunnlaugsson et al. 2008), then the values of the hyperfine fields and the ratios A (SA)/A(S2) determined by us correspond to only 0.05–0.08 titanium units in the mineral. Thus, magnetite in the titanomagnetite ore of the Chiney layered mafic-ultramafic complex is close to stoichiometric, and titanium is mainly contained in ilmenite. In the spectrum of the sample of massive ore M-2, 89% of the area of the spectral contour correspond to magnetite, while 11% refer to ilmenite. In other samples, where the total fraction of magnetite and ilmenite is reduced to 50–80%, the spectral fraction of ilmenite remains at the same level (11–15%). In the doublet part of the Mössbauer spectrum, in addition to the characteristic doublet of Fe^{2+} ions, there are in general three more doublets D2, D3, and D4 of Fe^{2+} ions with a large quadrupole splitting QS = 2.6–2.8, 2.0–2.1, 1.7–1.8 mm/s and doublet D5 Fe^{3+} ions with QS = 0.5–0.8 mm/s. The isomer shift of Fe^{2+} doublets is approximately the same (IS = 1.10–1.15 mm/s) and corresponds to the octahedral oxygen coordination of Fe^{2+}, characteristic of silicates (Menil 1985). In the silicate part of sample 2102–2, according to X-ray and X-ray diffraction data, amphibole (actinolite-tremolite) dominates. The doublet D2 with QS = 2.81 mm/s of the Mössbauer spectrum of this sample corresponds to the M1 position, i.e., corresponds to orthopyroxene-enstatite-ferrosilicate series (Dyar et al. 2006; Goldman 1979). The second, more intense and broadened, doublet D3 (QS = 2.01 mm/s) can be interpreted as the averaged component from M2 and M3 positions in amphibole with possible contribution of iron to spectrum in the M2-M1 positions of clinopyroxene (augite). In sample 11/697.8, the silicate part of which consists of orthopyrox-

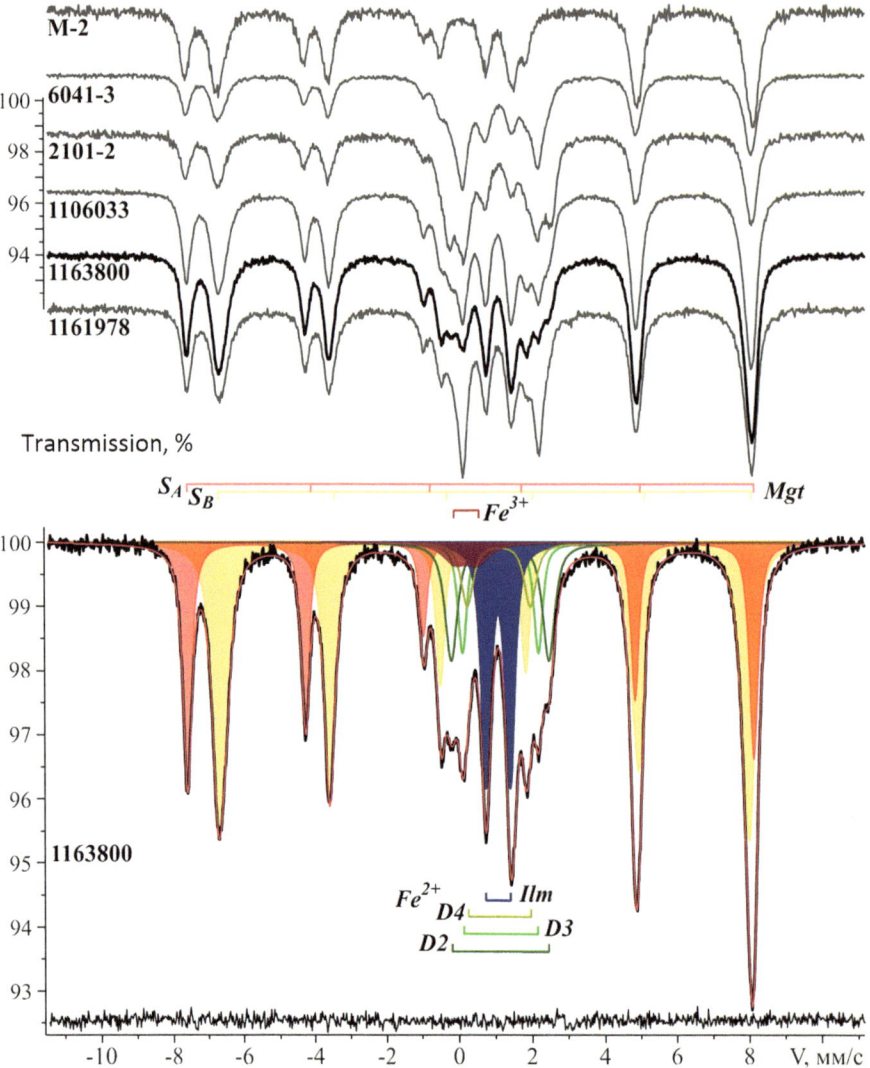

Fig. 7.12 Mössbauer spectra of ores from the Chiney pluton and interpretation of spectrum of sample 11/6800 (1168000 in figure)

ene, the doublet D2 has a much smaller value of QS = 2.58 mm/s and is close to the corresponding value of the position M1, i.e., corresponds to orthopyroxene-enstatite-ferrosilicate series (Dyar et al. 2006; Wiedenmann et al. 1986). Doublets D2 and D3 probably contain components of both pyroxene and amphibole in the spectra of other samples. The low-intensity D4 Fe^{2+} doublet, with the smallest QS and the greatest degree of distortion, most likely refers to the M4 position of the amphibole (Dyar et al. 2006; Goldman 1979). The doublet D5 Fe^{3+} has a large line width which can be the result of superposition of resonances from

different positions of iron in these minerals. The Fe^{3+} fraction in silicates (Fe^{3+}/SFe) is defined as the ratio of the area under doublet D5 to the total area of iron doublets in silicate. The total area of D2 + D3 + D4 is increased by 15% in accordance with the average difference in the resonance probability for Fe^{3+} and Fe^{2+} (De Grave and Van Alboom 1991; Eeckhout and De Grave 2003) and an average is 0.18, varying from 0.11 (11/638.00) to 0.25 (sample 6041–3). Traces of the doublet Fe^{3+} in the sample of massive ore M2 with a high content of hydrotalcite are possibly associated with iron impurities in this mineral.

Table 7.11 Components of Mossbauer spectra of ore samples from the Chiney pluton

Component type	Parameter	M-2	6041–3	1,106,033	1,161,978	1,163,800	2101–2	Interpretation
S_A	IS, mm/c	0.273_2	0.274_5	0.282_1	0.274_2	0.280_2	0.275_4	Magnetite
	QS, mm/c	-0.008_4	-0.016_5	-0.007_2	-0.014_3	-0.011_2	-0.006_6	
	H, кЭ	490.1_2	487.8_4	488.0_1	487.0_2	488.0_1	488.8_4	
	G, mm/c	0.33_1	0.32_1	0.312_3	0.316_5	0.308_4	0.33_1	
S_B	IS, mm/c	0.657_3	0.66_2	0.660_2	0.666_7	0.664_5	0.663_8	$[Fe^{3+}]_{tet}$
	QS, mm/c	-0.012_4	0.00_4	-0.00_2	0.00_2	0.01_3	-0.00_2	$[Fe^{2+} - Fe^{3+}]_{okt}$
	H, кЭ	457.5_3	456_3	454_3	453_4	454_4	457_2	
	G, mm/c	0.51_1	0.43_5	0.48_7	0.50_5	0.43_3	0.42_8	
	A/B	0.62_2	0.55_5	0.57_3	0.52_3	0.53_2	0.53_2	
	ΣA, %	86	52	64	64	68	47	
DO	IS, mm/c	1.06_2	1.063_5	1.069_2	1.065_3	1.070_2	1.054_7	Ilmenite
	QS, mm/c	0.65_4	0.70_1	0.660_3	0.658_5	0.657_3	0.67_1	Fe^{2+}
	G, mm/c	0.34_2	0.35_1	0.306_3	0.27_1	0.30_1	0.28_2	
	A, %	13	9	11	8	12	6	
D2	IS, mm/c	–	1.133_5	1.134_2	1.15_1	1.130_4	1.141_2	Amphibole,
	QS, mm/c		2.61_2	2.682_7	2.58_3	2.67_1	2.811_6	Pyroxene Fe^{2+}
	G, mm/c		0.37_2	0.35_1	0.37_3	0.40_2	0.32_1	
	A, %		6	7	5	8	13	
D3	IS, mm/c		1.14_2	1.144_2	1.142_2	1.14_1	1.151_3	
	QS, mm/c		2.062_5	2.092_4	2.071_4	2.08_1	2.01_1	
	G, mm/c		0.32_1	0.32_1	0.30_1	0.30_2	0.45_2	
	A, %		18	10	17	6	24	
D4	IS, мм/с		1.14_2	1.106_7		1.10_1		
	QS, мм/с		1.74_5	1.78_3		1.72_4		
	G, мм/с		0.44_2	0.35_3		0.40_4		
	A, %		4	3		4		
D5	IS, mm/c	0.4_1	0.38_2	0.25_4	0.40_4	0.2_1	0.38_1	Silicate
	QS, mm/c	0.6_2	0.81_2	0.46_5	0.68_5	0.6_2	0.75_4	Fe^{3+}
	G, mm/c	0.4_2	0.69_9	0.8_1	0.8_1	0.8_2	0.68_8	
	A, %	1	11	5	6	2	10	

Note. S, sextet; D, doublet; IS, QS, H, G, isomeric shift, quadrupole splitting, hyperfine magnetic field on the nuclei, and halfwidth of the Mossbauer lines, respectively. A—relative area under spectral components. The total value of relative areas of sextets S_A and S_B and their ratio A/B are given for magnetite. The index in the record of quantities is the error of determination of the last sign. M2, 6041-3…—sample numbers

7.6.3　The Chiney Fe-Ti-V Deposits and Other Deposits in Russia

Titanomagnetite ore is the main source of titanium and vanadium. The problem of their exploration is discussed in the publications of Bykhovsky et al. (2001, 2007), Pakhomov et al. (2010), Tigunov et al. (2005), Korolenko (2001), Reznichenko and Shabalin (1986), and Reznichenko et al. (1997, 2010). In the USSR, these metals were extracted from the Kusinskoe deposit in the Urals and from deposits in Ukraine (Laverov et al. 1997). In Russia, all important deposits are located in the Karelia-Kola area, the Urals (Smirnov et al. 2004) and Siberia (Polyakov 1967; Polyakov and Krivenko 1981), including Transbaikalia (Shabalin 2010). Nowadays, the main sources of titanomagnetite are located in the Kachkanar deposit in the Urals (Popov and Nikiforova 2004) and Pudozhgorskoe deposit in Karelia (Trofimov and Golubev 2008). Sometimes, titanomagnetite ore can be a source of noble metals as well (Anikina et al.

2002). Titanomagnetite deposits of Transbaikalia are not being mined, including the Chiney deposits. Their characteristics are given in publications of Konnikov (1978, 1986). A book of Deryabin et al. (1999) was devoted to the prospects of processing the Chiney ore.

Shabalin (2010) developed a classification of titanomagnetite deposits based on the requirements of mining industry. According to the ratio of total iron to titanium dioxide, ores were subdivided into low-Ti ($Fe/TiO_2 > 8$), medium-Ti ($Fe/TiO_2 = 5$–8), high-Ti ($Fe/TiO_2 = 2.3$–5), and extremely high-Ti ($Fe/TiO_2 < 2.3$) ore.

There is a classification of titanomagnetite deposits that takes into account a composition of titanomagnetite, i.e., its TiO_2 contents. The low-Ti deposits are not economic. This was technologically studied for the Kachkanar and Pervouralsk ore with ($Fe/TiO_2 = 10$) in the Urals. Their ores were melted in blast furnaces without extraction of titanium. For the medium-titanium ores, with values of $Fe/TiO_2 = 5$–8, the parameter 5 was taken as starting from this value, TiO_2

increases in linear proportion in ores due to occurrence of free ilmenite and, consequently, the growth of titanium in the collective concentrate. High-Ti ores are characterized by Fe/TiO_2 = 2.3–5. This ore shows a markedly sharp increase in titanium and vanadium into collective concentrate, where TiO_2 quantity varies from 15 to 31 wt%. Extremely high-Ti ores are characterized by high weight of titanium, because ilmenite dominates in the ores (as perovskite or rutile). This requires different beneficiation technology. According to this classification, the Chiney ore can be attributed to the low-Ti class (Table 7.1).

Three groups were distinguished among titanomagnetite deposits in the Kola Peninsula: (1) TiO_2-rich (11.9–19.04%) deposits in the central-type alkali gabbro complexes (Kovdor), (2) medium TiO_2 deposits (10.41–12.44%) with hysteromagmatic type of mineralization in gabbro-anorthosite complexes, and (3) low TiO_2 deposits (7.31–10.4%) in alkaline ultramafic intrusions (Zak et al. 1994). Titanomagnetite from the Chiney pluton is close to those from the Tsagin pluton (11–12% TiO_2) in the Kola Peninsula. Because both deposit types are genetically related to gabbro rocks, the composition of titanomagnetite is determined by geochemical specialization of host intrusive rocks (low-Ti deposits are linked to ultramafic rocks).

Another classification of titanomagnetite deposits was proposed by Karpova (1979) based on the titanomagnetite composition. Titanomagnetite from the Urals deposits is classified into high-Ti type (Kopan deposit), with 10–15% TiO_2; medium-Ti type (Volkovskoe deposit), with 4.5–8.0% TiO_2; and low-Ti type (Kachkanar deposit), with 2–6% TiO_2 (Karpova 1979). In this parameter, the Chiney deposit falls into the first type, with 11.45 to 17.64% TiO_2. In general, high-Ti titanomagnetite is typical of the Chiney pluton. The highest TiO_2 contents are typical of titanomagnetites that crystallized at shallow depth and high temperature. The highest TiO_2 in titanomagnetite (21.3–21.9%) is known in titanomagnetite from massive ore related to gabbro-labradorite intrusive complexes (Karpova 1979).

The Tsagin titanomagnetites are different from the Chiney counterparts by concentrations of trace elements, in particular, by enrichment in the elements inherent to ultramafic rocks (0.99–3.28% MgO (as compared to 0.4% in the Chiney analog), up to 0.29% Cr_2O_3 and 0.06% NiO), whereas the concentrations of these elements in the Chiney titanomagnetites are below detection limit of the microprobe. At the same time, the Tsagin titanomagnetites contain 0.2% Mn, as compared to 0.5–0.6% Mn in Chiney titanomagnetites. They are devoid of detectable Co in contrast to 0.06% Co in the Chiney titanomagnetites.

Comparison of the latter with those from other plutons in the northern Transbaikalia region shows that they are close in composition to the titanomagnetites from the Slyudyanka pluton, primarily, in high TiO_2 content (Balykin et al. 1983).

The TiO_2 content in titanomagnetite from the Yakut pluton is lower, and MgO contents, in general, are not high in this mineral. For example, titanomagnetites from the rocks of a gabbro-pyroxenite complex in Eastern Sayan contain 3–6% MgO (Balykin et al. 1983). The concentrations of both elements in titanomagnetites increase with rising temperature.

Vanadium, which is recoverable from ore, plays a special role among minor elements. Significant concentrations of this element have been found in titanomagnetite of the Chiney pluton. Nevertheless, these concentrations are much lower than in the deposits in the northern Baikal region, where the highest vanadium content reaches 3%. In titanomagnetites with exsolution textures, V and Cr are mainly incorporated into magnetite matrix, whereas Mn is contained in ilmenite lamellae and partly in ulvospinel.

The temperature of mineral formation, volatile components, and the depth of intrusive rock formation are also among crucial factors (Fominykh 1976; Gongalsky and Krivolutskaya 1993; Gongalsky et al. 2008, 2016). For the estimation of redox conditions during magma crystallization, we used the Ballhaus oxybarometer. A new version of Ol-Opx-Spoxy barometer for melts, saturated with olivine and spinel, was suggested (Nikolaev et al. 2016), and redox conditions were estimated for Yoko-Dovyren layered mafic-ultramafic complex in Northern Transbaikalia (Ariskin et al. 2017). This approach should be applied to the basic-ultrabasic intrusions during the future research.

7.7 Conclusions

1. The huge resources of vanadium ore (30 Gt) are concentrated in titanomagnetite ore of the Chiney pluton. Titanomagnetite aggregates form two deposits, the Magnitnyi and Etyrko, located in the eastern and western parts of the pluton.

2. The ore at the Etyrko deposit is mostly a disseminated variety, whereas at the Magnitnyi deposit it is massive. Veins, lenses, and irregular bodies dominate in the eastern part of pluton. Rounded and striated titanomagnetite aggregates are frequently encountered at the contacts between the veins and leucogabbro with leucogabbro and other gabbro.

3. Titanomagnetite is characterized by exsolution textures of several orders. The mineral has a network fabric, caused by the combination of isometric blocks that extinct simultaneously due to parallel arrangement of ilmenite tablets. The ilmenite lamellae are localized along fractures or at grain margins. Spinel grains enlarge with the formation of a spindle-shaped segregation.

4. The Curie and Vervey points for disseminated finely layered and massive ore overlap (551 and 579 °C; −142 and −152 °C). According to data of Mössbauer spectroscopy,

magnetite in the titanomagnetite ore of the Chiney layered mafic-ultramafic complex is close to stoichiometric, and titanium is mainly contained in ilmenite. In the spectra of the titanomagnetite, spectral contours mostly correspond to magnetite, while 13% refer to ilmenite.

5. The Chiney titanomagnetite is a high-Ti mineral, similar to the Tsagin pluton in the Kola Peninsula.

References

Anikina EV, Rusin IA, Filippov VN, Pushkarev EV, SYa B (2002) Noble metal mineralization in ultramafic rocks of Volkovsky intrusion, Middle Ural: minerals and mineral parageneses. In: Ezhegodnik-2002. Institute Geology and Geochemistry, Yekaterinburg, pp 250–260 (in Russian)

Ariskin AA, Fomin IS, Zharkova EV, Kadik AA (2017) Redox conditions during crystallization of ultramafic and gabbroic rocks of the Yoko–Dovyren massif (Based on the results of measurements of intrinsic oxygen fugacity of olivine). Geochemistry International 55(7):595–607

Balykin PA, Rudnev SN, Izokh AE (1983) Petrology and ore potential of the Yakut gabbroic pluton, northwestern Transbaikalia Region. OIGGM. Siberian Division USSR Acad Sci, Novosibirsk, pp 57–96 (in Russian)

Bykhovsky LZ, Tigunov LP, Zubkov LB (2001) The development of the material base of titanium as an actual task of the mining industry. Mineral Resources of Russia 4:25–36 (in Russian)

Bykhovsky LZ, Pakhomov VF, Turlova MA (2007) Complex ores of titanomagnetite deposits in Russia as a large mineral-raw base of ferrous metallurgy. Razvedka I Okhrana Nedr 6:20–23 (in Russian)

Chechetkin VS (1966) Some features of Cu–Ni mineralization in the Chiney stratified gabbronorite pluton. In: Geology and mineral resources of Transbaikalia. ZabNII, Chita, pp 54–65 (in Russian)

Chechetkin VS, Kharitonov YF (2002) Geology and mineral deposits of the Chita Segment of BAM. Administration of Chita, Chita, p 63 (in Russian)

Chumachenko NM, Nikitina N, Bykov VY, Fedorov VP (2000) Structure of a deposit of V-bearing titanomagnetite ore. Razvedka I Okhrana Nedr 1:33–36 (in Russian)

De Grave E, Van Alboom A (1991) Evaluation of ferrous and ferric Mossbauer fractions. Phys Chem Miner 18:337–342

Deryabin YA, Smirnov LA, Deryabin AA (1999) Prospects for processing of the Chiney ore. Sredneuralskoe Izdatelstvo, Yekaterinburg, p 367 (in Russian)

Dodin DA, Landa EA, Lazarenkov VG (2003) Platinum-bearing chromite and titanomagnetite deposits. Geoinformmark, Moscow, p 408 (in Russian)

Dyar MD, Agresti D, Schaefer MW, Grant CA, Sklute EC (2006) Mössbauerspectroscopy of earth and planetary materials. Annu Rev Earth Planet Sci 34:83–125

Eeckhout SG, De Grave E (2003) Evaluation of ferrous and ferric Mossbauer fractions. Part II. Phys Chem Miner 30:142–146

Fedotova VM, Chechetkin VS, Savchenko AA, Kuzmina LS (1977) Fe-Ti mineralization in the Chiney gabbronorite pluton, Transbaikalia. Soviet Geol 4:136–141 (in Russian)

Fominykh VG (1976) Temperature of titanomagnetite formation based on magnetite-ilmenite thermometer for the Urals deposits. In: Shteinberg DS (ed) Problems of biomineral thermometry. IGG, Sverdlovsk, pp 58–69

Golev VK, Gongalsky BI, Davy MN, Zinoviev YuI, Krivenko VA, Narkekyun LF, Pereyaslovsky IV, Rutstein IG, Sunkinzyan VS,

Trubachev AI, Chechetkin VS (1987) Excursion metallogeny of Siberia. In: Krendelev FP (ed) Guidebook of the 11th All-Union Metallogenic conference: Novosibirsk, 81 (in Russian)

Gongalsky BI (2003) A place of chineites (plagioclase–titanomagnetite rocks) in formation of the Chiney layered pluton, Northern Transbaikalia. Bull MOIP Sect Geol 68(2):83–88 (in Russian)

Gongalsky BI (2010) Basic magmatism of the Udokan–Chiney ore district, Northern Transbaikalia. Litosfera 3:87–94 (in Russian)

Gongalsky BI (2015) Deposits of the unique metallogenic province of Northern Transbaikalia. VIMS, Moscow, p 248 (in Russian)

Gongalsky BI, Krivolutskaya NA (1993) The Chiney layered pluton. Nauka, Novosibirsk, p 184 (in Russian)

Gongalsky BI, Krivolutskaya NA, Goleva NG (1995) Ore deposits of the Chiney pluton. In: Laverov NP (ed) Mineral deposits of the Transbaikalia, vol 1. Geoinformmark, Moscow, pp 20–28 (in Russian)

Gongalsky BI, Krivolutskaya NA, Ariskin AA, Nikolaev GS (2008) Inner structure, composition, and genesis of the Chiney anorthosite-gabbronorite massif, Northern Transbaikalia. Geochem Int 46(7):637–665

Gongalsky BI, Krivolutskaya NA, Ariskin AA, Nikolaev GS (2016) The Chiney gabbronorite-anorthosite layered massif (NorthernTransbaikalia, Russia): its structure, Fe-Ti-V and Cu-PGE deposits, and parental magma composition. Mineral Deposita 51(8):113–1034

Karpova OB (1979) Typomorphic features of titanomagnetite as an indicator of ore formation conditions. In: Smirnov VI (ed) Formation conditions of magmatic ore deposits. Nauka, Moscow, pp 171–210 (in Russian)

Khisina NP (1987) Subsolidus transformations of solid solutions of rock-forming minerals. Nauka, Moscow, p 207 (in Russian)

Konnikov EG (1978) Precambrian titanium-bearing gabbroids of Northern Transbaikalia. Nauka, Novosibirsk, p 118 (in Russian)

Konnikov EG (1986) Precambrian differentiated mafic–ultramafic complexes in Transbaikalia. Nauka, Novosibirsk, p 224 (in Russian)

Korolenko NV (2001) Raw materials for ferrous metallurgy in Russia. Titanium Razvedka I Okhrana Nedr 11–12:24–28 (in Russian)

Kudryavtseva GP, Garanin VK, Zhilyaeva VA, Trukhin VI (1982) Magnetism and mineralogy of natural ferrimagnetics. MSU, Moscow, p 294 (in Russian)

Kulikov AI, Kryukov VK, Morozova NN, Grechishnikov DN (1980) Ore types of the Chiney titanomagnetite deposits and their compositions. Geol Ore Deposits 22(5):85–88 (in Russian)

Kulikov AI, Golev VK, Grigor'ev VM, Kryukov VK (1981) Geology and titanomagnetite ore of the Chiney gabbroic pluton. In: Vakhromeev SA (ed) Geology, Prospecting and Exploration of Ore Deposits. Irkutsk Polytechnical Institute, Irkutsk, pp 26–35 (in Russian)

Laverov NP, Patyk-Kara NG, Benevolsky BI, Bykhovsky LZ (eds) (1997) Placer deposits of Russia and other CIS countries. Nauchny Mir, Moscow, p 479 (in Russian)

Lebedev AP (1962) The Chiney gabbro–anorthosite pluton, Eastern Siberia. USSR Akad Sci, Moscow, p 100 (in Russian)

Lyutoev VP, Gongalsky BI, Makeev AB, Lysyuk AY, Magazina LO, Taskaev VI (2017) Titanomagnetite ores: mineral composition and Mössbauer spectroscopy. Fortschr Mineral 2:43–65 (in Russian)

Menil F (1985) Systematic trends of the ^{57}Fe Mössbauer isomer shifts in (FeOn) and (FeFn) polyhedra. Evidence of a new correlation between the isomer shift and the inductive effect of the competing bonds T–X ($\rightarrow$Fe) (where X is O or F and T any elements with a formal positive charge). J Phys Chem Solid 46(7):763–789

Nikolaev GS, Ariskin AA, Barmina GS, Nazarov MA (2016) Test of the Ballhaus–Berry–Green Ol–Opx–Sp oxybarometer and calibration of a new equation for estimating the redox state of melts saturated with olivine and spinel. Geochem Int 54(4):301–320

Ostrovsky IA, Ol'shansky YO (1956) System fayalite – magnetite. Doklady Earth Sci 107(6):881–883 (in Russian)

Pakhomov FP, Tigunov LP, Bykhovsky LZ (2010) Titanomagnetite deposits of Russia. VIMS, Moscow, p 138 (in Russian)

Patnis A, McConnel JDS (1983) Basic features of the behavior of minerals. Mir, Moscow, p 304 (in Russian)

Petrusevich MN (1946) The Chiney titanomagnetite deposit. Soviet Geol 10:91–94 (in Russian)

Polyakov GV (1967) On the regularities of location and formation of magnetite deposits in connection with magmatism (on the example of the central regions of the Altay-Sayan mountain region). In: Dymkin AM (ed) Geology and genesis of magnetite deposits in Siberia. Nauka, Moscow, pp 16–47 (in Russian)

Polyakov GV, Krivenko AP (1981) Features of composition and conditions of formation of gabbroid massifs with titanomagnetite ore specialization. In: Dymkin AM, Baklaev YP (eds) Geology and genesis of ore deposits. IGG, Sverdlovsk, pp 35–40 (in Russian)

Popov VS, Nikiforova NF (2004) Ultramafic, mafic rocks and titanomagnetite ore of the Kachkanar deposit (Middle Ural: integrated petrological model). Geochem Int 42(1):11–25

Reznichenko VA, Shabalin LI (1986) Titanomagnetite. Deposits, metallurgy and chemical technology. Nauka, Moscow, p 295 (in Russian)

Reznichenko VA, Sadykhov BG, Karyazin IA (1997) Titanomagnetite as a material for new model of technology. Metals 6:3–8 (in Russian)

Reznichenko VA, Averin VV, Olyunina TV (2010) Titanates: scientific foundations, technology and production. Izd RAS, Moscow, p 267 (in Russian)

Shabalin LI (2010) Titanomagnetite deposit: geology, genesis and perspectives of exploration. SNIIGIMS, Novosibirsk, p 174 (in Russian)

Shabalin LI, Sharapov VN (1981) Elements of differentiation dynamics of the Chiney gabbroic pluton. In: Velinsky VV (ed) Issues of genetic petrology. Nauka, Novosibirsk, pp 163–180 (in Russian)

Smirnov LA, Tigunov LP, Maslovsky PA, Bykhovsky LZ (2004) Kuranakh ilmenite-titanomagnetite deposit. Sredneuralskoe Izd-vo, Yekaterinburg, p 310 (in Russian)

Tigunov LP, Bykhovsky LZ, Zubkov LB (2005) Titanium ore of Russia: state and perspectives of development. Miner Mater Geol Econ Ser VIMS 17:104 (in Russian)

Trofimov NN, Golubev AI (2008) Pudozhgorskoe noble metal titanomagnetite deposit. Karel Sci Center, Petrozavodsk, p 123 (in Russian)

Wiedenmann A, Regnard JR, Fillion G, Hafner SS (1986) Magnetic properties and magnetic ordering of the orthopyroxenes $Fe_xMg_{1-x}SiO_3$. J Phys C Solid State Phys 19:3683–3695

Zak Z, Unfried P, Giester G (1994) The rare earth elements: fundamentals and applications. J Alloys Compd 205:235–242

Zalutskiy AA, Zalutskaya AA, Sedmov NA, Kuzmin RN (2015) The Mössbauer analysis of iron oxyhydroxides in soils of Earth and Mars. Lithol Miner Dep 50(4):370–298 (in Russian)

Sulfide Mineralization Related to the Chiney Pluton

Abstract

There are five main sulfide deposits related to the Chiney pluton: Kontaktovyi, Skvoznoe, Pravoingamakitskoe, Verkhne-Chineyskoe, and Rudnoe. The high Cu/Ni ratio (Cu:Ni:Co = 76:7:1) is a distinguishing feature of the sulfide mineralization. The highest PGE tenor (up to 355 ppm PGE, Pd/Pt ~3–4) typically occurs in the massive ore. In contrast, the average grades of Pd and Pt in the disseminated sulfide are ~2 and ~0.5 ppm, respectively. The Rudnoe and Kontaktovyi deposits have the most important potential economic value. The Rudnoe is localized in gabbro rocks of the endocontact zone in the eastern offset of the Chiney pluton and in the exocontact zone with sedimentary rocks of the Udokan Supergroup. The disseminated sulfide in the endocontact is frequently related to magmatic breccia. Chalcopyrite dominates among sulfides, whereas pentlandite is a subordinate and even a rare mineral. Pyrrhotite and pyrite are also major sulfides, especially in the Kontaktovyi deposit. The other sulfide minerals are sphalerite, siegenite, galena, violarite, millerite, cobaltite, gersdorffite, safflorite, loellingite, argentopentlandite, michenerite, merenskyite, sperrylite, and other platinum group minerals.

8.1 General Characteristics

8.1.1 Sulfides

The sulfide minerals occur in almost all rock types of the Chiney pluton. However, the highest concentrations of sulfide minerals were established near the basal contact of the intrusion. The ores were divided into endocontact and exocontact types. There are five main deposits (from west to east): Kontaktovyi, Skvoznoe, Pravoingamakitskoe, Verkhne-Chineyskoe, and Rudnoe (Fig. 5.2). The sulfide mineralization was not continuously traced along the contact

zone. A segment between the Verkhne-Chineyskoe and Skvoznoe deposits is noteworthy in this respect. The chalcopyrite veins hosted in the country rocks are clustered here within a narrow zone oriented at an angle to the intrusive contact, extending for 4 km away from the contact. The most distant veins and stringer-disseminated mineralization returned the highest grade and make up the Pravoingamakitskoe deposit (see Fig. 4.1 in Chap. 4).

The sulfide stringers and disseminations also occur at the Magnitnyi and Etyrko titanomagnetite deposits located in the central part of the Chiney pluton. The metals could be recovered as by-products during mining of the Fe-Ti-V oxide ore. However, sulfides are mostly hosted in the endo- and, especially, in the exocontact zones.

The diversity of sulfide minerals in mode of occurrence includes the following situations: (1) low-grade (up to 3%) disseminations in cleavage fractures of silicate minerals in diorite and monzodiorite; (2) isometric and round segregations (5–8 mm) in monzodiorite with a low grade of sulfides (10–20%) within orebodies; (3) large pockets and veinlets (3–5 cm) in monzodiorite, quartz diorite, and magmatic breccia (30–40% sulfides) in high-grade parts of orebodies; (4) sulfide segregations (60–80%) in sideronite ore shoots; (5) sulfide lenses and veinlets in the marginal zones of magmatic breccia (5–20% sulfides); (6) fine-grained dissemination in sandstone xenoliths; (7) sulfide pockets (2–3 cm) in skarn; and (8) thin (1–3 mm) stringers in diorite, frequently with chlorite.

8.1.2 Chemical Composition of Ore

The predominance of copper in ore, the subordinate role of Ni, Co (Cu:Ni:Co = 76:7:1) and noble metals is a distinguishing feature of the sulfide mineralization in the Chiney pluton. The mean Cu/Ni ratio is 10, occasionally reaching 100. This feature drastically distinguishes the Chiney deposits from the global population of all magmatogenic deposits

© Springer Nature Switzerland AG 2019
B. Gongalsky, N. Krivolutskaya, *World-Class Mineral Deposits of Northeastern Transbaikalia, Siberia, Russia*,
Modern Approaches in Solid Earth Sciences 17, https://doi.org/10.1007/978-3-030-03559-4_8

Table 8.1 Metal concentrations at the deposits of the Chiney pluton

Metal	Rudnoe	Verkhne-Chineyskoe	Skvoznoe	Kontaktovyi
Cu, %	0.73	0.28–0.63	0.20–1.0	0.39–2.62
Ni, %	0.05	0.03–0.13	0.04–1.7	0.03–0.2
Co, %	0.008	0.003–0.018	0.008–0.018	0.005–0.014
Pt, ppm	0.26	0.06–0.37	0.02–0.9	0.02–0.21
Pd, ppm	1.24	0.1–0.83	0.14–1.0	0.07–0.79
Au, ppm	0.2	0.01–0.12	0.02–0.27	0.01–0.07
Ag, ppm	3.41	2.3–3.7	1.6–5.3	1.5–4.9

Note. Data after (Chechetkin and Kharitonov (2002))

Table 8.2 Chemical composition of ore from Rudnoe deposit (trench 400)

Interval, m	Cu	Ni	Co	Pt	Pd	Au
	wt %			ppm		
0–1	0.75	0.027	0.005	0.10	1.00	0.15
1–2	1.16	0.080	0.005	0.20	2.00	0.15
2–3	0.40	0.037	0.005	0.30	3.00	0.20
3–4	0.65	0.035	0.050	0.10	2.00	0.05
4–5	1.50	0.047	0.005	8.40	29.60	3.60
5–6	11.00	0.125	0.007	30.40	142.00	9.60
6–7	12.50	0.260	0.005	32.00	147.00	3.00
7–8	16.75	0.235	0.005	72.00	255.00	5.10
8–9	6.00	0.150	0.010	23.20	78.50	1.80

Note. Channel samples collected by Udokan Expedition; analyses carried out at Central Institute of Geological Exploration for Base and Precious Metals (TsNIGRI), Moscow

Table 8.3 Results of core sampling from Hole 9 (interval 174.4–226.9 m) at the Kontaktovyi deposit

Number of sample	Sample length, m	Cu, wt%	Pt, ppm	Pd, ppm	Au, ppm
1902	1.065	0.27	0.4	1.13	0.13
1903	2.35	0.34	0.3	1.0	0.07
1904	3.1	0.07	0.1	0.2	0.015
1905	2.0	1.00	0.1	0.5	0.02
1906	3.1	0.53	0.2	1.0	0.11
1907	2.45	0.56	0.2	1.0	0.05
1908	1.65	.030	0.2	1.0	0.11
1909	2.55	0.52	0.15	1.0	0.03
1910	2.75	0.51	0.4	0.75	0.08
1911	1.85	0.90	0.2	1.0	0.03
1912	2.1	0.80	0.4	1.1	0.16
1913	2.1	0.80	0.4	1.13	0.18
1914	1.05	0.80	0.15	0.86	0.07
1915	1.05	1.05	0.35	0.47	0.15
1916	2.35	1.13	0.4	1.05	0.15
1917	1.85	0.82	0.5	1.28	0.32
1918	2.4	1.02	0.35	1.07	0.15
1919	2.3	1.35	0.4	1.31	0.2
1920	2.65	1.17	0.4	1.18	0.14
1921	1.8	0.48	0.25	2.0	0.05
1922	2.05	0.75	0.1	0.5	0.02
1923	2.75	0.80	0.4	1.13	0.2
1924	2.40	0.17	0.2	1.00	0.2

Note. Analyses were carried out at the Central Chemical Laboratory of Chitageologiya Expedition

Table 8.4 Sulfide minerals from the ore at the Chiney pluton

Main	Major	Rare
Chalcopyrite CuFeS$_2$	Cubanite Fe$_4$Ni$_5$S$_8$	Arsenohauchecornite Ni$_9$BiAsS$_8$
Pyrrhotite hex. FeS	Cobaltite CoAsS	Hauchecornite Ni$_9$BiSbS$_8$
Pyrrhotite mon. FeS	Gersdorffite -NiAsS	
Throilite FeS	Galenite PbS	Tsumoite BiTe
Pentlandite (Fe,Ni,Co)$_9$S$_8$	Linnaeite Co$_3$S$_4$	Maucherite Ni$_{11}$As$_8$
Pyrite FeS$_2$	Millerite NiS	Nikelite NiAs
Bornite Cu$_5$FeS$_4$	Sphalerite ZnS	Safflorite (Co,Fe)As$_2$
	Siegenite (Co,Ni)$_3$S$_4$	
	Violarite FeNi$_2$S$_4$	

related to mafic-ultramafic rock complexes that bear Ni-Cu mineralization (Chechetkin 1966; Mel'nikova et al. 1983; Gongalsky and Krivolutskaya 1993; Gongalsky 2015). A certain similarity is noted only for some intrusions, including Tsagin in Kola Peninsula and Volkovsky in the Urals, both in Russia (Murzin et al. 1988; Volchenko et al. 1998; Zaccarinni et al. 2004), Kevitsa in Finland (Yang et al. 2013), and Okiep in South Africa (Maier et al. 2013). In the Chiney pluton, only the Skvoznoe deposit is similar to typical Ni-Cu mineralization in nickel enrichment. The sulfide ores also contain high amounts of Au, Ag, and PGE (Table 8.1). The highest PGE grade (up to 355 ppm PGE, Pd/Pt ~3–4) typically occurs in the massive ore (Table 8.2). In contrast, average concentrations of Pd and Pt in the disseminated sulfide are ~2 and ~0.5 ppm, respectively (Table 8.3).

8.1.3 Mineral Composition of Ore

Our studies (Krivolutskaya 1986) of the abovementioned proportions of metals reflect predominance of chalcopyrite among sulfides, whereas pentlandite is a subordinate and even a rare metal. Pyrrhotite and pyrite are also major sul-

fides (Tables 8.4, 8.5). In addition to the aforementioned minerals, 86 other mineral species have been identified to date in the copper ore, including minerals of platinum group elements (see below).

The compositional, textural, and structural features of ore are variable. Seven mineralogical subtypes were recognized: (a) pyrrhotite ore (>80% Po), (b) chalcopyrite ore (>80% Ccp), (c) chalcopyrite-pyrrhotite ore, (d) pyrrhotite-chalcopyrite ore (50–80% of pyrrhotite or chalcopyrite), (e) chalcopyrite-

Table 8.5 Chemical composition of sulfides from the deposits related to the Chiney pluton, wt %

No	As	S	Fe	Ni	Co	Se	Cu	Total	Deposit
Pyrrhotite									
1	0.21	38.76	58.01	1.75	0.12	0.14	0.00	98.99	Kontaktovyi
2	0.17	38.76	57.79	1.62	0.15	0.10	0.00	98.59	"
3	0.18	38.27	58.02	1.29	0.09	0.13	0.03	98.01	"
4	0.16	37.93	59.15	0.74	0.13	0.21	0.00	98.32	"
5	0.27	38.25	58.83	0.72	0.26	0.18	0.00	98.52	Rudnoe
6	0.19	38.09	58.94	0.72	0.26	0.14	0.01	98.36	"
7	0.26	37.82	59.24	0.64	0.14	0.14	0.01	98.24	"
8	0.13	37.67	59.85	0.62	0.24	0.11	0.01	98.63	"
9	0.21	37.79	59.91	0.56	0.15	0.19	0.00	97.81	"
10	0.09	37.18	60.06	0.54	0.11	0.18	0.00	99.16	"
11	0.20	37.36	60.15	0.44	0.10	0.15	0.00	98.41	"
12	0.14	37.26	60.12	0.36	0.10	0.10	0.00	98.07	"
13	0.20	36.77	60.65	0.34	0.10	0.12	0.02	98.20	Verkhne-Chineyskoe
14	0.22	35.57	61.87	0.28	0.11	0.16	0.01	98.22	"
15	0.19	37.36	60.82	0.25	0.10	0.20	0.00	98.92	"
16	0.04	39.74	59.65	0.25	0.43	0.09	0.07	100.07	"
17	0.08	37.10	60.52	0.22	0.10	0.15	0.00	98.16	"
18	0.17	37.09	60.90	0.20	0.10	0.16	0.02	98.63	"
19	0.19	37.91	60.04	0.15	0.10	0.15	0.28	98.82	"
Pyrite									
20	0.20	52.51	46.80	0.02	0.56	0.14	0.00	100.22	Kontaktovyi
21	0.17	52.79	46.83	0.01	0.36	0.13	0.00	100.29	"
Cubanite									
22	0.20	35.73	41.39	0.16	0.09	0.20	21.27	99.04	Verkhne-Chineyskoe
23	0.00	33.53	40.98	0.02	0.07	0.07	22.89	97.56	"
24	0.22	34.14	40.79	0.00	0.07	0.11	23.32	98.65	"
25	0.15	33.76	40.62	0.03	0.07	0.19	23.42	98.23	Rudnoe
26	0.19	31.67	38.53	0.06	0.07	0.21	24.06	70.73	"
Chalcopyrite									
27	0.20	34.54	29.52	0.01	0.05	0.17	33.67	98.17	Rudnoe
28	0.21	34.13	29.75	0.45	0.05	0.19	33.78	98.56	"
29	0.26	34.49	30.23	0.01	0.05	0.20	33.79	99.03	"
30	0.20	34.16	30.54	0.00	0.06	0.21	33.93	99.10	"
31	0.18	34.39	29.70	0.03	0.06	0.17	33.95	98.48	"
32	0.18	33.69	30.37	0.01	0.05	0.21	34.02	98.53	"
33	0.18	33.87	30.52	0.02	0.06	0.18	34.18	99.00	"
34	0.17	34.35	30.09	0.01	0.06	0.26	34.20	99.04	"
35	0.16	33.52	30.52	0.01	0.05	0.23	34.21	98.71	"
36	0.22	33.45	30.93	0.04	0.07	0.27	34.22	99.20	"
37	0.22	33.45	30.93	0.04	0.07	0.27	34.22	99.20	"
38	0.15	34.68	30.18	0.00	0.06	0.15	34.27	99.48	Skvoznoe
39	0.24	34.44	30.34	0.00	0.05	0.25	34.29	99.62	"
40	0.14	34.63	30.50	0.01	0.05	0.26	34.29	99.88	"
41	0.22	34.50	30.29	0.00	0.05	0.23	34.31	99.60	"
42	0.23	34.03	30.24	0.01	0.05	0.16	34.33	99.05	"
43	0.23	34.11	30.13	0.00	0.05	0.23	34.35	99.09	Kontaktovyi
44	0.10	34.70	30.30	0.01	0.05	0.17	34.37	99.69	"
45	0.16	34.15	30.44	0.00	0.05	0.21	34.37	99.38	"
46	0.20	34.11	30.45	0.02	0.06	0.24	34.38	99.45	"
47	0.17	34.59	30.15	0.00	0.06	0.18	34.38	99.53	"
48	0.19	34.25	30.29	0.00	0.05	0.23	34.39	99.40	"
49	0.15	34.49	30.25	0.00	0.05	0.22	34.39	99.54	"
50	0.23	34.10	30.28	0.00	0.05	0.18	34.40	99.24	"

(continued)

Table 8.5 (continued)

No	As	S	Fe	Ni	Co	Se	Cu	Total	Deposit
51	0.19	34.17	30.33	0.00	0.06	0.15	34.41	99.30	"
52	0.13	34.47	30.28	0.01	0.06	0.20	34.45	99.59	"
53	0.15	33.49	30.36	0.00	0.05	0.19	34.45	98.69	Verkhne-Chineyskoe
54	0.19	34.59	30.42	0.01	0.05	0.21	34.46	99.92	"
55	0.24	34.33	30.28	0.01	0.06	0.24	34.46	99.61	"
56	0.23	33.31	30.40	0.00	0.05	0.13	34.46	98.58	"
57	0.22	33.76	30.42	0.01	0.04	0.21	34.47	99.13	"
58	0.09	34.16	30.42	0.00	0.05	0.20	34.47	99.39	"
59	0.15	34.10	30.42	0.00	0.05	0.29	34.51	99.53	"
60	0.21	34.21	30.41	0.00	0.05	0.14	34.52	99.54	"
61	0.22	33.56	30.28	0.00	0.05	0.22	34.54	98.87	"
62	0.17	34.43	30.36	0.01	0.05	0.24	34.57	99.83	"
63	0.16	33.80	30.50	0.00	0.05	0.18	34.74	99.44	"
64	0.12	34.08	30.40	0.00	0.05	0.18	34.75	99.58	"
Bornite									
65		25.73	11.8	0.95	0.00		62.57	101.05	Rudnoe, d-60
66		25.76	12.86	0.72	0.00		59.72	99.06	"
67		25.88	9.61	2.43	0.00		60.53	98.45	"
68		25.88	11.33	0.26	0.00		62.82	100.29	"
Millerite									
69	0.34	34.87	1.00	62.36	1.44	0.12	0.00	100.14	"
70		35.34	3.56	61.65	0.26		1.14	101.95	"
71		35.16	2.56	61.52	0.17		0.23	99.64	"
72		34.85	0.87	61.16	0.78			100.18	Rudnoe, d.h.20/133
73		35.37	0.76	62.44	0.6			99.17	Rudnoe, d.h.20/138
74		35.35	0.75	63.14	0.52		0.18	99.95	"
75		34.78	0.72	63.17	0.5			99.17	"
76		33.79	0.83	63.26	0.63		0.17	98.68	"
77		35.24	1.05	63.69	0.72			100.69	"
78		35.98	1.00	60.84	1.34			99.16	Rudnoe, L-2
79		34.87	1.00	62.36	1.44	0.12	0.00	100.14	Rudnoe, L-2

Note. d.h. = drill hole 20; L-2 is a lens 2; d − 60 is a trench 60; here in Tables 8.6, 8.7, 8.8, and 8.9 analyses carried out by EPMA (CAMECA SX 100) at Vernadsky Institute of Geochemistry and Analytical Chemistry, RAS, analyst NN Kononkova

pyrite ore, (f) millerite-chalcopyrite ore, and (g) chalcopyrite-bornite ore. a–e types were found in the endocontact zone in the western part of the layered mafic-ultramafic complex, while f–g types are widespread in the exocontact zone. From the endo- to exocontacts of the layered mafic-ultramafic complex, the composition of ore changes, and pyrrhotite and chalcopyrite-pyrrhotite associations are replaced by pyrrhotite-chalcopyrite, chalcopyrite, and bornite-chalcopyrite.

Based on spatial occurrence, we distinguish four major types of sulfide mineralization (Bognibov et al. 1995; Krivolutskaya 1986; Krivolutskaya et al. 1997): (1) endocontact-disseminated sulfides in gabbronorite, gabbro-diorite, and monzodiorite; (2) exocontact-disseminated sulfides in sandstone; (3) veins and lenses of massive sulfides in sandstone; and (4) disseminated sulfides (up to 20%) in layered titanomagnetite ore. In general, disseminated and veinlet-disseminated mineralization is more common than massive ore, the latter being always located in sandstone of the Udokan Supergroup, 20–60 m below the lower contact of the Chiney pluton.

At the Rudnoe deposit, the disseminated chalcopyrite-pyrrhotite is hosted in the endocontact zone, whereas substantially chalcopyrite disseminations are combined with veins and lenses of massive chalcopyrite and chalcopyrite-bornite in the exocontact zone. Almost only pyrite-chalcopyrite occurs at the Kontaktovyi deposit, whereas the Verkhne-Chineyskoe and Skvoznoe deposits are close to the Rudnoe deposit in sulfide mineralogy.

Sulfide mineralization is also known within the main body of the Chiney pluton (Gongalsky et al. 1995), where it is hosted in both leucogabbro and gabbro rocks enriched in titanomagnetite in the form of numerous veinlets, conformable and crosscutting relative to the layering of host intrusive rocks.

The Rudnoe and Kontaktovyi deposits (Figs. 5.2, and 8.1) have the most important potential economic value (Table 8.2,

Table 8.6 Pentlandite and minerals of linnaeite group compositions from deposits of the Chiney pluton, wt. %

N	N sample	S	Fe	Ni	Co	Cu	Total
Pentlandite[a]							
1	11/541	31.64	22.33	27.62	18.38		99.97
2	11/542	32.83	56.01	30.11	13.10		101.4
3	11/542	32.32	25.40	30.63	12.49		100.85
4	11/542	32.03	25.36	30.9.	13.20		101.31
5	11/542	32.01	25.40	30.07	13.84		101.31
6	11/542	32.17	24.74	27.87	16.01		100.39
7	11/542	33.94	24.77	23.98	15.92		98.62
8	11/543	32.69	24.83	31.03	10.65		99.20
9	11/543	32.24	25.97	31.83	11.97		101.84
10	11/541	33.11	23.18	27.31	13.45		97.06
11	11/541	31.63	24.88	28.61	13.19		98.32
12	11/541	31.91	24.61	28.06	13.63		98.21
13	11/542	32.84	25.98	32.78	9.26		100.87
14	11/542	32.28	26.34	32.98	9.57		101.17
15	11/542	31.86	24.62	32.21	11.91		100.60
16	11/542	31.88	23.63	32.88	11.91		99.31
17	11/542	32.00	25.11	31.74	12.30		102.17
18	11/542	32.50	26.40	32.78	8.20		99.88
19	MP-2	33.16	17.94	29.48	17.01		97.59
20	MP-2	33.77	23.71	29.23	10.94		97.65
21	MP-2	32.73	23.55	26.56	16.02		98.85
22	MP-2	33.86	24.70	24.30	15.42		98.27
23	MP-2	33.18	24.25	23.87	16.47	0.09	97.86
24	MP-2	33.82	24.20	23.45	16.15		97.61
25	MP-2	33.34	24.32	25.75	15.15	0.13	98.68
26	39/105	31.59	28.73	38.14	2.08		100.54
27	39/106	32.15	28.91	36.05	3.38		100.49
28	39/130	31.48	28.95	35.68	3.39		99.50
29	39/106	32.64	29.33	36.21	2.37		100.57
30	39/91	32.12	26.64	34.13	6.14		99.04
31	39/91	32.17	25.62	33.80	6.72		98.32
33	39/91	31.62	27.78	35.74	3.45		98.58
34	39/93	32.40	27.07	35.15	4.53		99.14
35	39/147	33.71	25.64	33.96	5.02		98.33
36	39/92	32.99	25.90	34.41	5.30		98.60
37	39/93	33.46	25.46	34.10	5.56		98.58
38	d-59	33.46	25.00	38.81	2.46	0.26	100.03
39	d-59	33.25	28.50	34.73	3.02	0.08	99.58
40	d-59	32.47	27.88	35.69	3.60		99.74
41	d-59	32.64	30.37	30.68	2.67	1.08	99.44
42	d-59	32.46	30.47	33.40	2.37	0.50	99.18
43	d-59	32.64	30.37	30.68	2.67	1.08	97.44
44	9/450	32.82	27.79	36.93	1.68		99.58
45	9/450	32.86	28.21	36.66	1.99		100.10
46	66/358	33.42	32.09	33.59	0.15		99.76
47	66/358	32.27	28.78	31.78	6.43		99.73
48	66/370	32.34	28.64	31.61	6.54		99.54
49	66/372	32.34	28.66	31.60	6.54		99.56
50	66/373	32.45	28.10	31.48	7.45		99.92
Minerals of linnaeite group[b]							
1	MP-2	40.88	9.32	23.10	26.20		99.50
2	MP-2	40.80	8.63	23.06	26.69		99.19
3	MP-2	40.38	8.91	23.91	24.55		97.75
4	39/91	40.33	9.19	24.70	23.70		97.30

(continued)

Table 8.6 (continued)

N	N sample	S	Fe	Ni	Co	Cu	Total
5	39/91	40.87	11.45	24.46	22.76		99.53
6	39/93	41.33	8.16	14.13	24.28		97.89
7	39/93	41.83	8.04	24.27	27.77		98.91
8	39/ 94	40.79	11.13	23.65	26.78	0.08	102.42
9	39/94	41.61	4.78	21.89	31.47		99.75
10	39/94	38.81	42.00	13.19	4.82	0.19	99.01
11	39/94	38.82	41.65	13.01	6.92	0.19	100.57
12	39/94	38.82	41.49	13.93	6.05	0.19	99.47
13	20/133	41.81	4.85	34.66	18.12		99.44
14	20/138	41.64	4.76	34.83	15.76	0.24	97.23
15	20/138	41.97	4.76	34.87	16.19	0.45	98.22
16	20/138	42.53	4.90	33.49	16.69	0.33	97.94
17	20/152	41.54	23.24	30.73	2.07	0.92	98.67
18	20/151	39.72	11.57	36.00	2.60	6.33	97.12
19	20/151	40.65	11.78	36.41	2.47	7.75	99.05
20	20/151	41.06	11.56	36.26	2.33	6.73	97.94
21	36/150	41.74	16.53	27.14	4.00	10.09	99.49
22	27/169	41.73	6.12	21.47	31.31		100.64
23	27/169	41.66	5.35	21.00	31.79		99.80
24	39/130	38.12	18.74	39.31	1.58		97.75
25	d-59	40.33	17.80	32.88	2.87	5.64	99.52
26	d-59	40.62	22.73	32.03	3.27	0.07	98.72
27	d-59	42.02	10.61	24.68	22.11		99.29
28	d-59	40.05	10.62	31.94	2.65	12.70	97.96

Note[a]. 1–25, titanomagmetite ores (Etyrko deposit); 26–29, exocontact sulfide-disseminated ores (Rudnoe deposit); 30–37, endo-contact sulfide-disseminated ores (Rudnoe deposit); 38–43, massive ores (Rudnoe deposit); 44–45, endocontact sulfide-disseminated ores (Kontaktovyi deposit); 46–50, endocontact sulfide-disseminated ore (Verkhne-Chineyskoe deposit). Here and in Tables 8–11, all analyses are after Gongalsky and Krivolutskaya (1993)

Note[b]. Minerals of linnaeite group from 1–3, disseminated titanomagnetite; 4–16, endocontact sulfide disseminated ore, Rudnoe deposit; 16–24, outer-contact-disseminated sulfide, Rudnoe deposit; 25–28, massive ores, Rudnoe deposit; N sample = borehole N/depth (m)

Ni-Co-Fe-Cu diagram for pentlandites and linnaeite group minerals

Table 8.7 Composition of sphalerite from endocontact ores of the Rudnoe deposit (borehole 39), wt. %

N	N sample	S	Fe	Zn	Cd	Total
1	39/91	32.08	10.17	58.62	0.40	100.63
2	39/134	32.27	6.78	59.89	0.26	99.20
3	39/91	34.41	5.89	59.53	0.02	99.83
4	39/93	32.86	5.98	59.49	0.49	98.82
5	39/93	33.83	6.94	59.17	0	99.94
6	20/147	33.31	7.19	58.21	0.36	99.07
7	39/92	33.09	5.76	58.34	0.36	97.56
8	39/93	32.98	9.83	56.27	0.46	99.54
9	39/134	33.35	6.67	58.32	0.32	98.65

fide minerals reach 2–3 mm in size, whereas in coarse-grained rocks, the size of sulfide grains exceeds 5–8 mm.

The 10–15-m-thick zone of the nearest endocontact is highly mineralized. The sulfide content reaches 60–70% of rock volume and is 20–30%, on average. The endocontact sulfide impregnation is frequently related to magmatic breccia. The breccia, where the number of fragments reaches 40–50 vol % and cement occupies 50–60 vol %, is the most favorable for mineralization. In this case, sulfide minerals may occupy up to 40% of the total volume of rock. This is characteristic of the internal zones of orebodies.

and 8.3). The different types of mineralization are described in more detail for these sulfide deposits.

Textural and structural features are determined by the host rocks. The disseminated, stringer-disseminated, and patched ore are hosted in medium-grained monzodiorite, quartz diorite, pegmatoid varieties of these rocks, and in magmatic breccia. Roughly and finely layered structures are sporadic. The size of sulfide grains directly depends on the granular structure of rocks: in fine-grained gabbro rocks, sul-

8.2 Deposits

8.2.1 The Rudnoe Deposit

The Rudnoe deposit is best studied in comparison with other sulfide deposits of the Chiney pluton (Fig. 5.2, 8.2) due to its economic value. This deposit is localized in gabbro rocks of the endocontact zone in the eastern offset of the Chiney pluton and in the exocontact zone composed of sedimentary

Table 8.8 Composition of arsenides and sulfoarsenides from the Rudnoe deposit, wt. %

N	N sample	As	S	Fe	Ni	Co	Cu	Total
1	L-1	48.46	0.14	1.54	47.70	1.12	0.95	100.33
2	L-2	46.10	20.82	0.17	32.94	0.36	0.06	100.76
3	39/105	47.25	0.20	0.54	49.58	0.32	0.29	98.58
4	39/105	46.09	19.29	4.46	13.14	15.64	0.46	99.04
5	39/108	69.30	0.43	17.87	4.59	7.74		99.93
6	39/105	70.20	0.41	5.63	0.90	24.69		101.81
7	39/105	45.85	18.58	0.92	1.04	34.43		100.82
8	39/130	46.11	18.01	9.07	20.69	3.96	0.82	98.66
9	39/130	43.82	18.30	6.64	15.12	14.79	0.87	99.55
10	39/130	48.52	18.41	8.28	21.56	3.85		100.62
11	39/130	47.30	18.01	8.29	22.07	3.92		99.60
12	39/130	46.47	18.33	6.74	20.34	8.46		99.34
13	39/130	46.67	17.38	6.88	14.78	13.12		99.83
14	39/130	46.56	17.56	6.76	14.37	13.42		98.67
15	39/106	47.01	19.90	1.22	1.42	27.55		97.10

N sample: L-2; lens 2, 39/105; borehole 39/depth, m

Note. N, ores: 1–2 massive, 3–15 disseminated. Minerals: 7,15 cobaltite; 8, 10–12 gersdorffite; 4, 9, 13,14 cobaltite-gersdorffite; 2 arsenopyrite; 1, 3 loellingite; 5 rammelsbergite; 6 safflorite

Table 8.9 Chemical compositions of rock-forming minerals from the microrhythm 11/638, wt %

No.	SiO$_2$	TiO$_2$	Al$_2$O$_3$	FeO	MnO	MgO	CaO	Na$_2$O	K$_2$O	Cr$_2$O$_3$	Total	Mineral
1	54.54	0.021	0.688	11.554	0.149	17.808	12.511	0.112	0.01	–	97.40	Cpx
2	54.36	0.042	0.585	13.707	0.269	18.121	9.677	0.105	0.016	–	96.88	Cpx
3	52.83	0.236	1.378	11.738	0.179	16.652	12.018	0.246	0.024	–	95.30	Cpx
4	52.23	0.332	1.211	21.788	0.453	22.869	1.634	0.015	–	0.01	100.54	Opx
5	51.80	0.336	1.157	21.782	0.46	22.623	1.923	0.052	–	0.007	100.14	Opx
6	54.54	0.06	27.951	0.326	0.003	0.016	10.427	5.538	0.215	–	99.08	Pl
7	53.25	0.346	1.231	20.892	0.441	23.451	1.763	0.014	0.003	0.014	101.40	Opx
8	54.22	0.054	27.607	0.334	–	–	10.362	5.563	0.252	0.003	98.40	Pl
10	52.58	0.318	1.153	21.091	0.45	23.482	1.368	0.018	0.015	0.003	100.48	Opx
11	52.18	0.328	1.282	21.175	0.456	22.701	1.707	–	0.001	0.004	99.84	Opx
12	0.18	52.134	0.061	48.83	0.814	0.336	–	0.016	–	0.017	102.39	Ilm
13	52.58	0.326	1.171	20.968	0.447	23.294	1.551	0.031	0.014	0.003	100.38	Opx
14	52.26	0.315	1.269	21.031	0.456	23.152	2.081	0.043	–	0.001	100.61	Opx
15	54.21	0.043	0.638	21.818	0.545	18.983	0.805	0.074	0.016	0.005	97.14	Opx
16	52.44	0.301	1.03	21.66	0.486	23.01	1.431	–	–	–	100.36	Opx
17	52.15	0.297	1.257	23.576	0.484	20.985	1.871	0.069	0.002	–	100.69	Opx
18	51.23	0.327	1.75	23.16	0.457	19.698	2.945	0.044	0.449	0.019	100.08	Opx
19	53.66	0.06	28.301	0.304	–	–	11.352	5.128	0.194	0.001	99.00	Pl
20	54.10	0.059	28.019	0.477	–	–	10.564	5.771	0.096	–	99.09	Pl
21	54.77	0.065	27.469	0.322	0.001	–	10.365	5.354	0.449	–	98.80	Pl
22	54.56	0.045	27.609	0.242	0.007	–	10.333	5.487	0.377	0.005	98.67	Pl
23	53.57	0.066	28.326	0.308	0.003	–	11.202	4.927	0.373	–	98.78	Pl
24	54.08	0.065	27.929	0.317	0.003	0.019	10.67	5.314	0.447	–	98.85	Pl
25	48.63	0.704	3.541	25.088	0.331	10.208	9.33	0.346	0.072	0.007	98.26	Cpx
26	54.14	0.073	28.129	0.314	0.004	–	10.995	5.235	0.378	0.005	99.27	Pl
27	54.44	0.067	27.974	0.346	–	–	10.738	5.354	0.337	–	99.26	Pl
28	54.33	0.067	27.91	0.287	0.002	–	10.639	5.247	0.432	–	98.91	Pl
29	51.19	0.556	2.194	11.309	0.274	13.173	20.326	0.405	0.001	0.01	99.44	Cpx
30	53.39	0.069	28.754	0.358	–	–	11.669	4.908	0.264	–	99.41	Pl
31	53.89	0.059	28.078	0.297	0.012	0.009	10.978	5.078	0.367	–	98.77	Pl
32	51.73	0.339	1.418	12.373	0.251	13.165	19.253	0.302	0.01	0.002	98.84	Cpx
33	54.20	0.06	27.413	0.329	–	–	10.334	5.327	0.552	0.005	98.22	Pl

(continued)

Table 8.9 (continued)

No.	SiO$_2$	TiO$_2$	Al$_2$O$_3$	FeO	MnO	MgO	CaO	Na$_2$O	K$_2$O	Cr$_2$O$_3$	Total	Mineral
34	53.03	0.062	28.302	0.337	0.003	–	11.301	4.956	0.423	–	98.41	Pl
35	51.49	0.218	2.147	27.074	0.43	12.348	4.229	0.361	0.016	0.012	98.32	Cpx
36	51.61	0.567	2.47	10.836	0.266	13.159	21.299	0.385	0.01	–	100.60	Cpx
37	53.18	0.075	28.337	0.269	0.008	–	11.304	4.919	0.336	0.007	98.43	Pl
38	53.34	0.063	28.226	0.82	–	0.048	11.074	5.194	0.301	0.003	99.06	Pl
39	51.58	0.438	1.588	11.641	0.292	13.855	19.952	0.278	–	0.018	99.64	Cpx
40	53.25	0.059	28.243	0.272	0.013	–	11.356	4.788	0.5	0.003	98.48	Pl
41	49.11	0.268	1.182	22.147	0.445	18.698	2.598	0.045	0.004	0.005	94.51	Opx
42	54.97	0.076	27.004	0.319	0.003	–	9.84	5.568	0.713	0.003	98.49	Pl
43	53.66	0.073	28.084	0.312	–	–	11.182	4.836	0.424	–	98.57	Pl
44	51.91	0.324	1.113	23.121	0.465	21.82	1.661	0.015	–	0.002	100.43	Opx
45	51.87	0.341	1.187	23.262	0.482	21.749	1.292	–	0.003	0.01	100.19	Opx
46	51.69	0.332	1.131	22.937	0.451	21.666	1.843	0.028	–	0.004	100.08	Opx
47	52.35	0.326	1.068	23.195	0.503	21.255	1.934	0.036	0.013	0.004	100.68	Opx
48	54.28	0.049	28.384	0.294	0.003	0.003	11.095	4.873	0.526	–	99.51	Pl
49	54.12	0.06	28.373	0.226	0.003	–	11.053	5.195	0.484	–	99.51	Pl
54	51.73	0.395	1.44	11.297	0.277	13.731	20.407	0.3	0.002	0.008	99.59	Cpx
55	44.18	1.872	6.233	7.063	0.121	17.405	0.52	0.146	3.687	–	81.23	Hb
56	31.14	10.557	12.132	30.471	0.303	3.094	8.801	2.279	0.49	0.095	99.36	Hb
57	52.04	0.196	1.601	28.225	0.525	12.606	3.849	0.208	0.007	0.004	99.26	Px
58	36.13	1.857	10.873	28.157	0.17	2.104	10.66	2.144	1.035	0.387	93.52	Hb
59	34.34	4.663	15.458	28.494	0.24	2.072	10.208	2.577	0.62	0.156	98.83	Hb
60	53.78	0.067	28.086	0.495	0.009	–	11.047	5.093	0.381	–	98.96	Pl
61	51.20	0.49	2.217	11.183	0.276	13.039	20.874	0.35	0.005	–	99.63	Cpx
62	54.86	0.077	27.645	0.311	0.014	–	10.173	5.418	0.467	–	98.97	Pl
63	27.45	13.765	9.094	34.167	0.352	3.135	7.718	1.963	0.43	0.25	98.32	Hb
64	53.06	0.048	0.984	24.804	0.417	13.481	5.606	0.197	0.021	0.006	98.62	Px
65	54.71	0.055	1.091	15.708	0.301	18.012	7.794	0.155	0.029	0.003	97.86	Px
66	51.78	0.325	1.366	23.297	0.467	21.752	1.299	0.048	0.006	0.017	100.36	Opx
67	36.54	0.038	19.396	21.813	0.189	4.379	11.268	2.59	0.394	–	96.61	Hb
68	38.41	0.053	16.294	24.033	0.201	5.125	10.972	2.679	0.355	0.013	98.13	Hb
69	54.75	0.044	27.557	0.318	–	–	10.263	5.483	0.435	–	98.85	Pl
70	51.66	0.526	2.126	10.776	0.267	12.975	21.451	0.377	–	0.008	100.16	Cpx
71	54.38	0.071	28.053	0.289	–	–	10.672	4.816	0.723	–	99.00	Pl
72	52.27	0.316	1.134	23.851	0.484	21.288	1.425	–	–	0.003	100.77	Opx
73	52.78	0.312	1.14	22.158	0.472	22.247	1.7	0.006	–	–	100.82	Opx
74	51.23	0.46	1.636	16.489	0.356	15.867	13.876	0.173	–	0.002	100.09	Cpx
75	53.90	0.075	28.096	0.433	–	–	11.057	5.229	0.365	–	99.15	Pl
76	51.78	0.541	2.131	9.388	0.254	14.234	21.172	0.326	–	0.022	99.85	Cpx
77	51.96	0.32	1.248	21.462	0.472	22.694	1.684	0.035	–	–	99.88	Opx
78	52.25	0.307	1.267	21.795	0.486	22.974	1.257	0.039	–	–	100.38	Opx

Notes. *Cpx* clinopyroxene, *Opx* orthopyroxene, *Px* pyroxene, unspecified, *Hb* hornblende, *Pl* plagioclase. Dash denotes concentrations below detection limit. Analyses were carried out at IGEM RAS, analyst E.V. Kovalchuk

rocks of the Udokan Supergroup. In general, the mineralized zone is stratiform. It can be traced along the entire exposed contact of gabbro intrusion (~9 km), which dips to the northwest at 6–12°. The ore zone, 3–65 m thick, has been penetrated by boreholes down a depth of a few to 240 meters. Copper is the main valuable component; Pt, Pd, Au, Ag, Ni, and Co are of secondary importance. The content of sulfide minerals reaches 50–60% in gabbro rocks and monzodiorite with a sideronite structure and to 20–30% in sandstone and 100% in veins. The Rudnoe deposit comprises three types of

sulfide mineralization: (1) endocontact-disseminated sulfides, (2) exocontact-disseminated sulfides, and (3) veins of massive sulfides in the exocontact rocks.

8.2.1.1 Endocontact-Disseminated Ore

Endocontact mineralization is extremely nonuniform. Ore shoots, sharply enriched in ore minerals, stand out against the background of low-grade mineralization. In general, mineralization intensity increases upward (toward the contact surface) and abruptly pinches out in country rocks at a

Fig. 8.1 General view of the Rudnoe (**a**) and Kontaktovyi (**b**) sulfide deposits of the Chiney pluton

distance of 5–6 m. At the Rudnoe deposit, where the bottom of pluton is nearly horizontal, the amount of sulfides increases sharply. The orebodies were delineated from the sampling data. They are represented by nearly horizontal stratiform and lenticular lodes; the average thickness is 15–10 m, and a maximum thickness is 50–60 m. The lodes extend for a few hundred meters.

Most of the ores consist predominantly of chalcopyrite and pyrrhotite in variable proportions (Fig. 8.3). Additional minerals are pentlandite, sphalerite, pyrite, titanomagnetite,

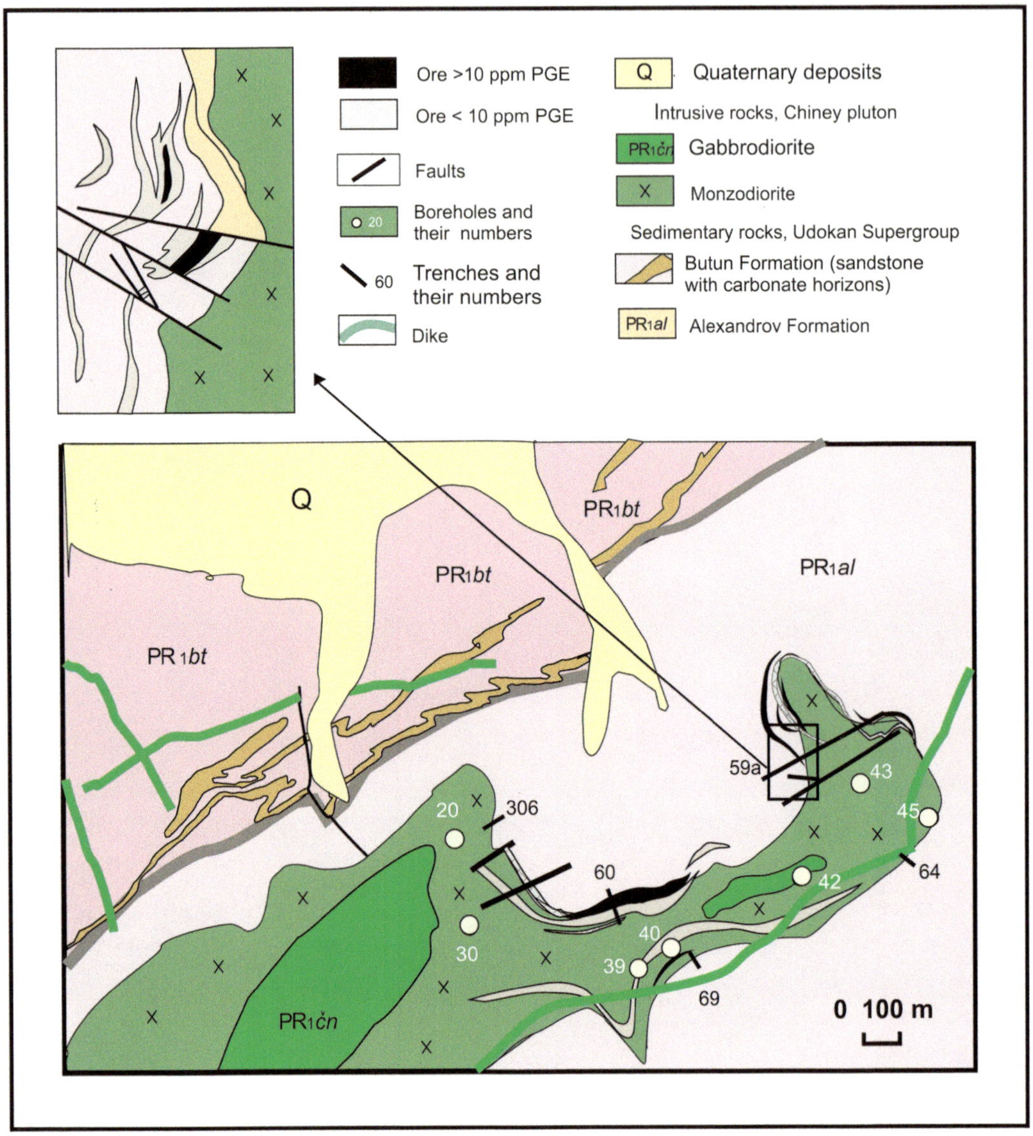

Fig. 8.2 Schematic geological map of the Rudnoe deposit. (Modified after Chitageolgia data)

Fig. 8.3 Disseminated ore in endocontact of the Chiney pluton, Rudnoe deposit

and ilmenite, with rare siegenite, galena, violarite, millerite, cobaltite, gersdorffite, safflorite, loellingite, argentopentlandite, michenerite, merenskyite, sperrylite, and other platinum group minerals. Two paragenetic mineral assemblages are distinguished: (1) chalcopyrite-pyrrhotite and (2) chalcopyrite with Co and Ni arsenides and sulfoarsenides.

The *chalcopyrite-pyrrhotite assemblage* consists of chalcopyrite and pyrrhotite in different proportions. As it was mentioned above, there are following mineral associations: (1) chalcopyrite (80% chalcopyrite), (2) pyrrhotite (>80% pyrrhotite), (3) pyrrhotite-chalcopyrite (50–80% chalcopyrite), and (4) chalcopyrite-pyrrhotite (50–80% pyrrhotite).

The boundaries between the above varieties are commonly gradual. The mixed mineral assemblages (pyrrhotite-chalcopyrite, chalcopyrite-pyrrhotite) are the most abundant (Fig. 8.4); the pure chalcopyrite and pure pyrrhotite varieties occur only locally (northeastern part

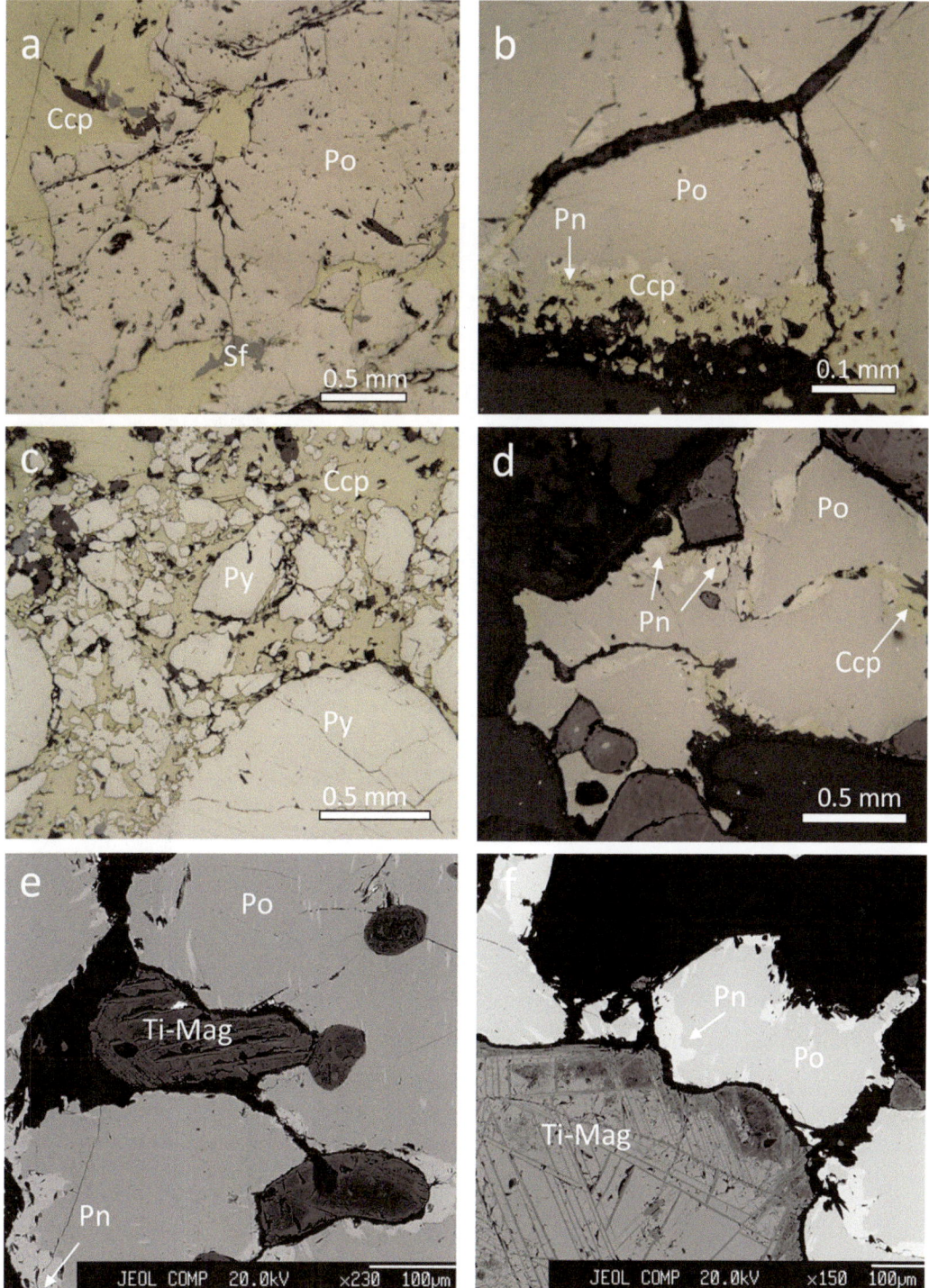

Fig. 8.4 Photomicrographs of sulfide ore
(**a**) pyrrhotite-chalcopyrite endocontact ore, Rudnoe deposit; (**b**) pentlandite-chalcopyrite-pyrrhotite ore, Rudnoe deposit; (**c**) chalcopyrite-pyrite ore, Kontaktovyi deposit; (**d–f**) chalcopyrite-pentlandite-pyrrhotitemineralization in titanomagnetite ore, Etyrko deposit
Po pyrrhotite, *Pn* pentlandite, *Ccp* chalcopyrite, *Py* pyrite, *Mag* titanomagnetite, *Sf* sphalerite

of the Rudnoe deposit). In single cases, boundaries between mineral assemblages are sharp. For example, in Borehole 45, the high-grade pyrrhotite ore (70–75% pyrrhotite) changes into chalcopyrite ore (80% chalcopyrite) within an interval of 5 mm, with the retention of textural and structural features: the coarse-grained sideronite structure and massive texture are characteristic of both varieties.

Fig. 8.5 Crystals of siegenite (*1*) and pentlandite (*2*) in association with pyrrhotite (*3*) and chalcopyrite (*4*) (**a, b**) image in backscattered electrons (*a* – composition, *b* – Relief), (**c–f**) Images in characteristic X-ray: (**c**), $Fe_{k\alpha}$; (**d**), $Co_{k\alpha}$; (**e**), $Ni_{k\alpha}$; (**f**), $S_{k\alpha}$. Here and in Figs. 8.6, 8.8, and 8.9, analyses were carried out at the Moscow State University (analyst NE Sergeeva)

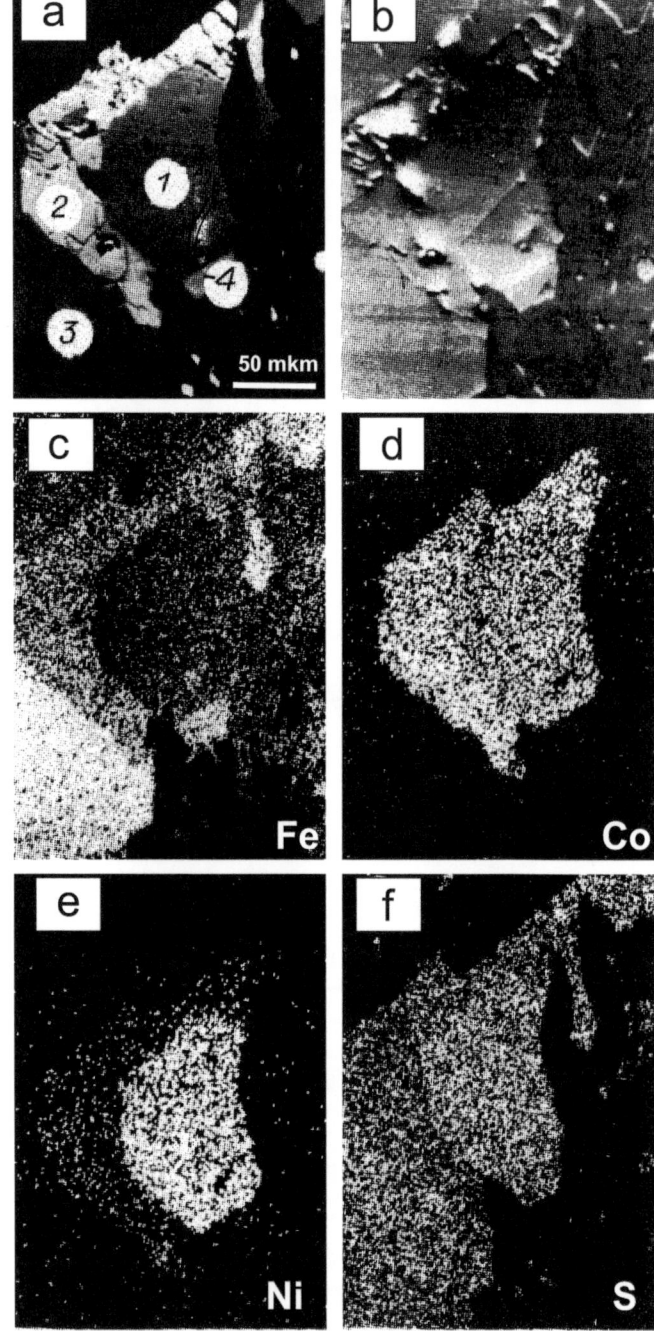

The typomorphic minerals for this association are cobalt-bearing pentlandite with an elevated Co content (up to 6–7 wt %, Table 8.6), Co-bearing pyrite, and the linnaeite group (Figs. 8.5, and 8.6; Table 8.6) of minerals and sphalerite (Krivolutskaya 1986). Many PGE minerals were described in this assemblage (Tolstykh et al. 2004; Tolstykh 2008; see below).

Commonly, ore minerals form pockets and veinlets in gabbro-diorites (Fig. 8.3), monzodiorite, and quartz diorite.

Chalcopyrite is comparable in abundance to pyrrhotite and occurs as grains, 0.001 to 3–4 mm in size. The largest grains are recorded in the chalcopyrite variety of ore. Segregations of chalcopyrite grains sometimes reach 6–8 cm across. Morphology of grains is controlled by boundaries of titanomagnetite and silicate minerals. Chalcopyrite grains are commonly irregular in shape and corrode pyrrhotite. The twinned structure of chalcopyrite can be distinctly seen in crossed polars; polysynthetic twins and the structure of "oleander leaves" are observed. Only tiny chalcopyrite disseminations appear homogenous. Chalcopyrite commonly contains sphalerite starlets (0.02 mm) or skeletal crystals (0.05 mm).

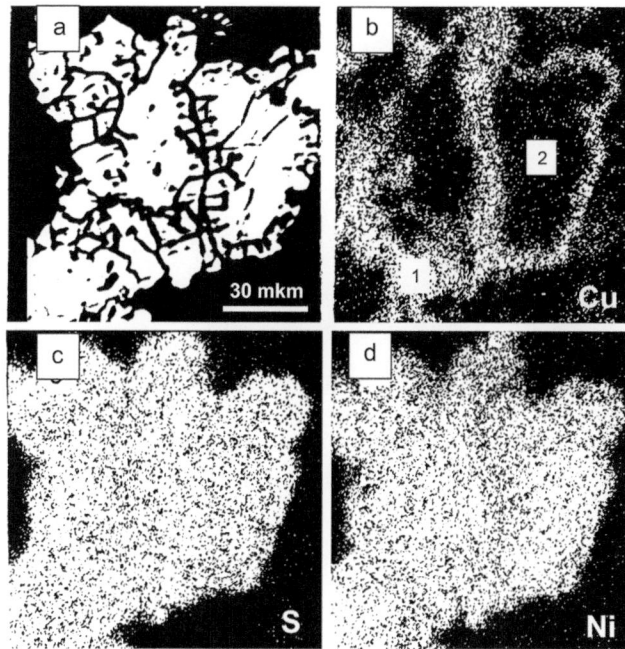

Fig. 8.6 Veinlets of Cu-violarite (*1*) in pentlandite (*2*)
(**a**) image in backscattered electrons (composition), (**b–d**) images in
characteristic X-ray: b, $Cu_{k\alpha}$; c, $S_{k\alpha}$; d, $Ni_{k\alpha}$.

Pyrrhotite aggregates (Fig. 8.4a, c) consist of prevailing
monoclinic and rare hexagonal intergrowths. The internal
structure of pyrrhotite is heterogeneous, with tabular and lenticular inclusions differing in optic properties. However, they
are identical in chemical composition, based on microprobe
measurements. Large pyrrhotite grains contain numerous
round or geometrically perfect pentlandite inclusions that
occupy up to 30% of the total grain volume. The lenticular
and flame-like pentlandite inclusions, inherent to the exsolution structure, are not characteristic of this type of mineralization. Oval, round, or irregular inclusions of sphalerite are
chaotically distributed through pyrrhotite grains. In this type
of mineralization, pyrrhotite is almost completely represented by monoclinic variety. The following trace elements
have been detected in pyrrhotite as isomorphic admixtures:
0.25–0.80 wt% Ni, 0.14–0.48 wt% Co, and 0.15 wt% Cu.

Pentlandite grains vary in size and morphology. In the
order of their relative abundance, they are observed as: (i)
large (up to 0.3 mm) euhedral grains near the margins of pyrrhotite crystals (Fig. 8.4b), (ii) flame-like and lenticular segregations within pyrrhotite, and (iii) oval segregations in
pyrrhotite. It is often associated with siegenite (Fig. 8.5). The
large pentlandite grains are replaced by violarite (Fig. 8.6),
pyrite, and magnetite along the cleavage fractures. Pentlandite
of this assemblage is characterized by elevated Co concentrations (Table 8.6; Figs. 8.6, 8.7a, b), and its composition is
close to $(Ni_{4.68} Fe_{3.53} Co_{0.65})_{8.86}S_{8.05}$. In some cases, the pentlandite grains have different composition: low-Co, high-Co,
and high-Ag (Fig. 8.7e).

Sphalerite is regarded as typomorphic mineral of endocontact sulfide disseminations (Fig. 8.7f). This mineral
occurs as emulsion in pyrrhotite and chalcopyrite, occupying
10–15% of host mineral volume, and also forms stellar and
skeletal crystals in chalcopyrite. The chemical composition
of sphalerite is characterized by variable Fe content (5.67–
12.7 wt%). Almost all sphalerite grains contain up to
0.49 wt% Cd (Table 8.7). The sphalerite starlets (0.02 mm)
or skeletal crystals of this mineral (0.05 mm) are incorporated into chalcopyrite grains. Vorobiev (1975) referred them
to the products of decomposition of solid solutions, enriched
in zinc. He has also established that the effect of ZnS on
exsolution temperature is insignificant because of its low
solubility in chalcopyrite (the ZnS content, reaching
40 mol%, shifts this temperature only by 35%). Thus, despite
skeletal sphalerite crystals contained in chalcopyrite from
endo-contact impregnation, the temperature of its formation
in alkaline rocks only little deviates from 550 °C, i.e., it is
wittingly above the temperature of transformation twinning
(Vorobiev 1972; Vorobiev and Borisovsky 1980).

Galena is a common mineral in endocontact ore, but it
forms very small grains (Fig. 8.7d, e).

Pyrite is subdivided by morphology and mode of occurrence into small euhedral crystals among silicates, rims
around pyrrhotite grains, and veinlets in rock-forming minerals. The small (<0.03 mm) pyrite crystals, quadrangular in
section view, are localized in marginal parts of large pyrrhotite grains or among silicate minerals nearby. The crystals
have a homogenous internal structure. The high Co (2.28–
2.46 wt%) and low Ni (below 0.01 wt%) contents are characteristic of pyrite (Table 8.5). Depletion in Ni is a common
feature, inherent to high-Co pyrites (Borishanskaya et al.
1981). The lacy rims, as well as oval or round "bird-eye"
structures, composed of pyrite and marcasite, are noted
around pyrrhotite grains as products of their desulfidation.
Pyrite of this type is devoid of Co and Ni admixtures in concentrations above hundredth fractions of percent and occasionally contains Cu (0.52 wt%). Pyrite veinlets (< 0.01 mm)
intersect silicate minerals and clusters of coarse crystalline
pyrrhotite. No admixtures in concentrations above hundredth
fractions of percent have been detected in pyrite from
veinlets.

Minerals of the *linnaeite group* (Table 8.6; Figs. 8.5, 8.6)
are frequently identified in this type of mineralization by
octahedral habit of euhedral crystals, their isotropy, relief,
pinkish hue, and reflectance value. The accurate identification of these minerals is possible only from their chemical
composition. The most abundant siegenite forms hexagonal
(in section) grains, 0.2–0.3 mm in size, hosted mainly among
silicate minerals and less frequently in chalcopyrite.
Siegenite is associated with high-Co pentlandite. In chemical
composition, the mineral does not match siegenite proper
and occupies a transitional position between $CoNi_2S_4$ and

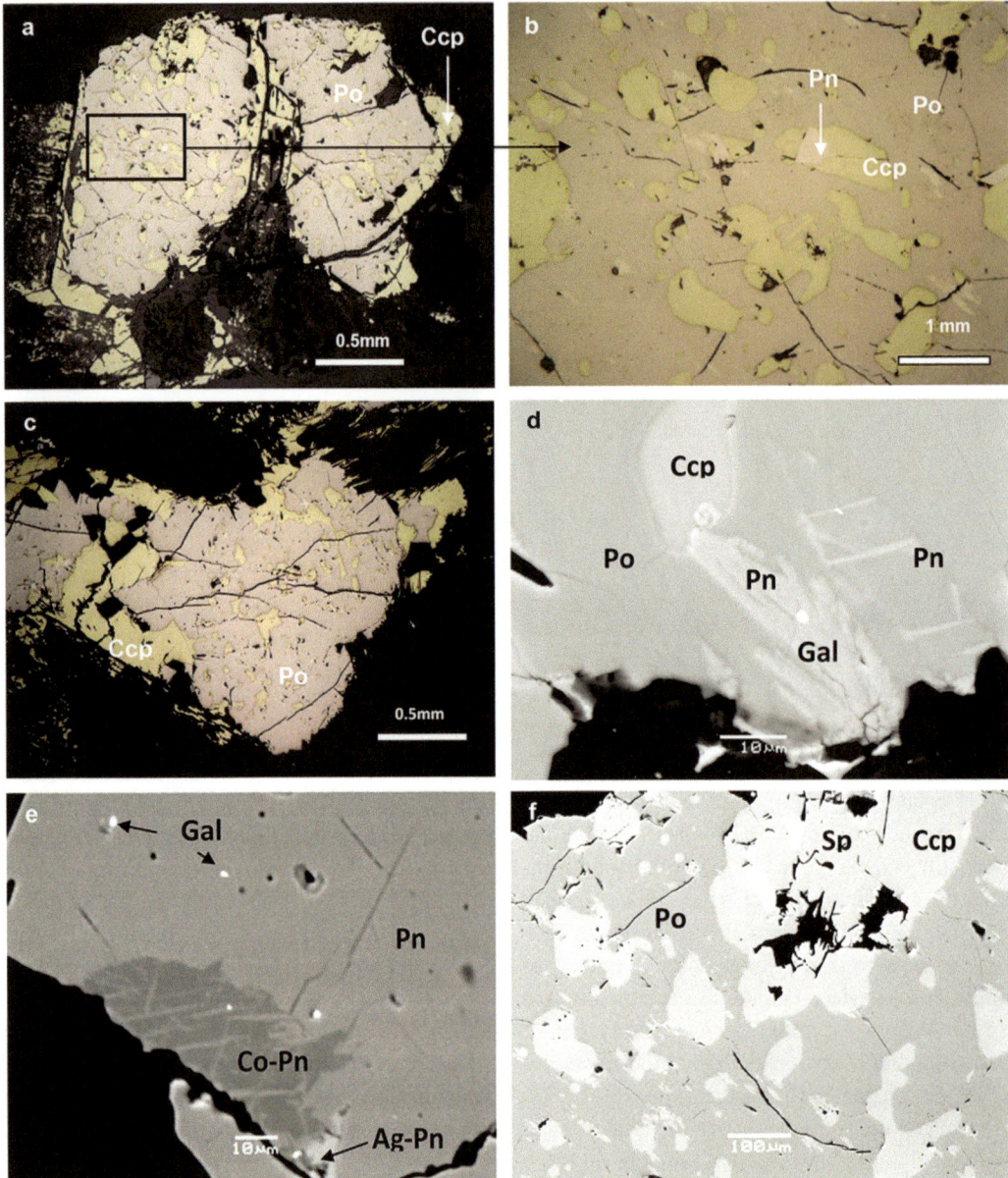

Fig. 8.7 Photomicrographs of the sulfide ore at the Chiney pluton
((**a-c**) - reflected light, (**d-f**) -BSM images) *Gal* galena, *Sp* sphalerite, *Ccp* chalcopyrite, *Pn* pentlandite, *Po* pyrrhotite

Ni-linnaeite NiCo$_2$S$_4$, with significant Fe admixtures (8.04–11.45 wt%). Linnaeite occurs sporadically and reveals a diversity of morphological forms, varying from lenticular to isometric grains with offsets up to 0.2 mm in size, which are always hosted in pyrrhotite. Violarite is found in the endo-contact zone much less frequently than in the exocontact zone. Its composition substantially differs from theoretical FeNi$_2$S$_4$ due to high Co concentrations, proportions between Ni and Fe, and, in general, shifts toward greigite.

Many minerals of platinum group were discovered in this assemblage, especially by (Tolstykh et al. 2004; Tolstykh 2008), see below.

The chalcopyrite with Co and Ni arsenides and sulfoarsenides assemblage is limited in abundance and consists almost of "pure" chalcopyrite with arsenides and sulfoarsenides of Co and Ni. It comprises the cobaltite-gersdorffite isomorphic series (Fig. 8.8), loellingite, and safflorite. Palladium minerals (michenerite, merenskyite, etc.) are associated with sulfoarsenides and arsenides. In particular, the Pd-bearing phases have been detected in niccolite (Fig. 8.9; Krivolutskaya 1986, 1989; Gongalsky and Krivolutskaya 1999, 2004; Tolstykh et al. 2004; Tolstykh 2008).

Euhedral crystals are typical of minerals, pertaining to the *cobaltite-gersdorffite series*. They have square sections or

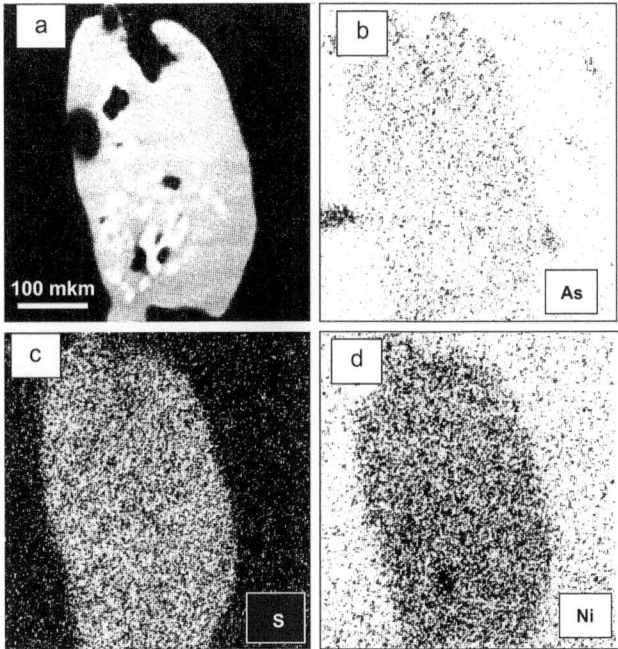

Fig. 8.8 Gersdorffite with inclusions of nickeline in millerite (**a**) image in backscattered electrons (composition), (**b–d**) images in images in characteristic X-ray: (**b**) – $As_{k\alpha}$, (**c**) – $S_{k\alpha}$, (**d**) – $Ni_{k\alpha}$

close to them. Cobaltite prevails over gersdorffite (Fig. 8.8) in abundance (Table 8.8). Sporadic monomineral veinlets, up to 5–7 mm thick, are composed of coarse-grained (up to 0.7–0.8 mm) cobaltite. Zoned cobaltite grains are unknown. The chemical composition of this mineral is close to theoretical CoAsS: $(Co_{0.97}Ni_{0.02}Fe_{0.03})_{1.02}As_{0.02}S_{0.96}$. Single light gray gersdorffite crystals contain Co, Fe, and Cu in addition to the major components. The chemical compositions that fall into the field of normal cobaltite and gersdorffite are combined with transitional compositions, corresponding to Ni-cobaltite or Co-gersdorffite.

Rare bright pink small (0.01 mm) *niccolite* grains have a chemical composition close to theoretical (49.58 wt% Ni, 47.25 wt% As), containing 0.54 wt% Fe, 0.32 wt% Co, 0.29 wt% Cu, and 0.20 wt%S.

Safflorite and *loellingite*, which were found for the first time among sulfides at the Chiney pluton, are also related to this mineral assemblage. Loellingite crystallizes as long prismatic grains among silicate minerals in the high-grade-disseminated chalcopyrite-cobaltite mineralization. This is a white, strongly anisotropic mineral pertaining to the Fe-Co diarsenides, enriched in Ni: $(Fe_{0.55}Co_{0.27}Ni_{0.16})_{1.08}(As_{0.89}S_{0.03})_{0.92}$. The second mineral of this assemblage is related to the same Ni-free isomorphic series of Fe-Co diarsenides and approaches safflorite.

Arsenopyrite, *glaucodot*, and some other minerals have been described in endocontact ore by our predecessors.

Formation temperature of the studied mineral assemblages can be estimated from a T– log_{s2} relationship. In the absence of pyrite and at an FeS mole fraction of pyrrhotite

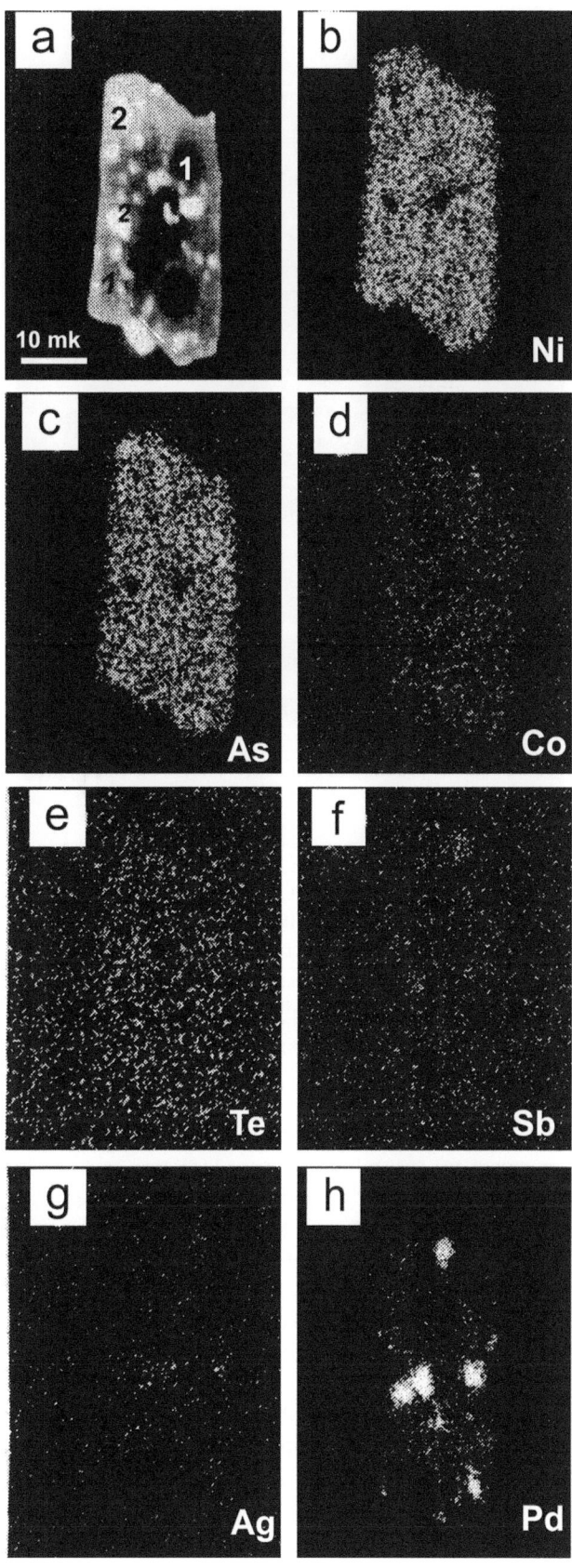

Fig. 8.9 Nickeline (*1*) with inclusions of Pd mineral (*2*) in chalcopyrite (**a**) image in backscattered electrons (composition), (**b–h**)– images in characteristic X-ray: (**b**) – $Ni_{k\alpha}$, (**c**) – $As_{k\alpha}$, (**d**) – $Co_{k\alpha}$, (**e**) – $Te_{k\alpha}$, (**f**) – $Sb_{k\alpha}$, (**g**) – $Ag_{k\alpha}$, (**h**) – $Pd_{k\alpha}$

Fig. 8.10 Sulfide in sandstone in the exocontact of the Chiney pluton, Rudnoe deposit

equal to 0.95, the temperature range of sulfide formation is estimated at 650–570 °C ($log_{a_{s2}} = -16$ to -5.1).

The general tendency consists of abrupt increase in chalcopyrite content relative to pyrrhotite by transition from the endocontact zone to the nearest exocontact. In certain cases, the chalcopyrite-pyrrhotite or pyrrhotite varieties give way to a pure chalcopyrite variety, e.g., in the eastern part of the Rudnoe deposit.

8.2.1.2 The Exocontact-Disseminated Ore

Exocontact ore is localized in the sedimentary rocks of the Udokan Complex and is represented by sandstone-hosted dissemination (Figs. 8.10, and 8.11). The orebodies are characterized by varying strike and dip and by wide variations in thickness (10–30 cm on average). The vertical quartz-chalcopyrite veinlets are noted as far as 250 m from the contact.

In contrast to endocontact ore, the exocontact ore is dominated by chalcopyrite and locally by chalcopyrite with bornite. Cubanite is a typical mineral as well (Fig. 8.12). According to Tolstykh (2008), a contribution of Cu to the Cu + Ni sum is greater than 90%, and the Cu/Ni ratio varies from 50 to 900.

Mineralization of this type is distinguished by prevalence of hexagonal *pyrrhotite* over a monoclinic variety and by a 2–3 wt% Co admixture in pentlandite. Cobaltite-gersdorffite

Fig. 8.11 Disseminated pyrrhotite in sandstone, Rudnoe deposit Samples Ch-92 (**a**) and Ch-92/1 (**b**)

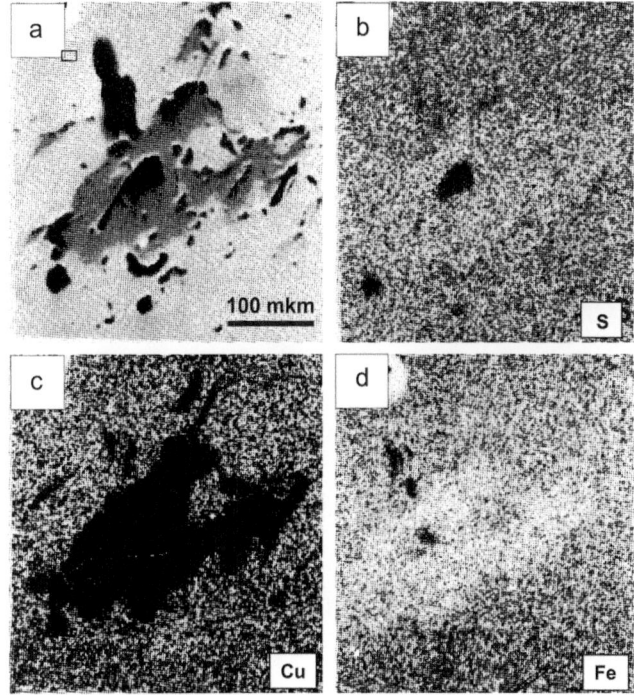

Fig. 8.12 Cubanite in pyrrhotite
(**a**) image in backscattered electrons (composition), (**b–d**) –images in characteristic X-ray: (**b**) – $S_{k\alpha}$ (**c**) – $Cu_{k\alpha}$, (**d**) – $Fe_{k\alpha}$.

sulfoarsenides often replace sulfide aggregates. Arsenohauchercornite, maucherite, nickeline, hessite, cubanite, mackinawite, millerite, Ni-Bi compounds, native lead, sperrylite, and numerous compounds of Pd with Bi, Te, Sb, Sn, and As (klockmannite, melonite, sudburyite, froodite) have been identified among rare minerals (Krivolutskaya 1986; Krivolutskaya et al. 1997; Tolstykh et al. 2004).

As in the endocontact disseminations, minerals from the linnaeite group are subdivided in composition into three varieties close to linnaeite, violarite, and siegenite (Table 8.6). Linnaeite occurs as flame-like segregations in pyrrhotite; siegenite ($Ni_{1.27}Co_{1.25}Fe_{0.58}S_{4.0}$) forms separate euhedral grains, up to 0.2–0.3 mm in gangue mineral aggregates. Violarite most frequently develops as veinlets, 0.2–0.3 mm thick, in pentlandite and sporadically as individual grains in the gangue matter.

Bornite is commonly limited in abundance. However, it locally prevails over chalcopyrite, making up a special chalcopyrite-bornite type of ore. The ore pockets have irregular or round shapes and reach 5–7 cm across. Bornite is most frequently associated with chalcopyrite and millerite. The chemical composition of bornite varies insignificantly and close to $Cu_{4.96}Fe_{1.01}Ni_{0.02}S_{4.04}$, with nickel only as admixture. The elevated content of this metal (2.43 wt%) is characteristic of bornite closely associated with millerite.

Millerite yields in abundance only to chalcopyrite and bornite in this type of ore. In the impregnated ore, it occurs

in the outer zone as segregations, with core composed of chalcopyrite (Table 3.8). In contrast to endocontact-disseminated and massive ore, millerite from sandstone is distinguished by high Co and Cu concentrations (up to 3.56 and 1.14 wt%, respectively).

Gersdorffite forms euhedral grains, clustered in millerite segregations or nearby, and reaches 0.5 mm in size. The composition of gersdorffite is close to theoretical: $Ni_{0.98}As_{1.00}S_{0.02}$.

8.2.1.3 Veins and Lenses of Massive Ore

Based on our observations and data, veins and lenses of massive ore (Fig. 8.13) are located in surrounding rocks (sandstone) at a distance of 20 and 30 m (sometimes 70 m) from the lower contact of the layered mafic-ultramafic complex (Gongalsky and Krivolutskaya 1993). Their thickness varies from 10 cm to 1.5 m. The veins extend laterally for 10–20 m (up to 50 m). Massive sulfides fill a system of horizontal fractures in the sandstones. These fractures are parallel to the contact of layered mafic-ultramafic complex with surrounding rocks, and their position is controlled by small folds developed in the exocontact zone of the layered mafic-ultramafic complex. The contacts of lenses of massive ore with country rocks are sharp. There is abundant breccia ore (Fig. 8.14) in this area, where sulfides (mostly chalcopyrite) cement fragments of gabbroic rocks (2–10 cm wide). In the western part of the pluton, massive ore occurs in the endocontact zone.

The ore in vein-shaped and lenticular bodies is composed of chalcopyrite (no less than 95%). The subsidiary minerals are millerite, pyrrhotite, pentlandite, minerals of linnaeite group, magnetite, mackinawite, gersdorffite, cubanite, and sphalerite; the rare minerals comprise nickeline, maucherite, arsenohauchercornite, and Pd minerals. The zoning, which is frequently observed in lenticular and vein-shaped bodies, is expressed in occurrence of selvages enriched in Ni minerals or composed of massive millerite. The thickness of millerite rims at pinch-out of orebodies (50–60 cm) reaches 7–8 cm.

Pentlandite occurs as small (0.03–0.05 mm) round and oval lentil-like grains. As a rule, they are distributed chaotically among large chalcopyrite grains. However, their clusters are elongated at an angle of 60° to planes of chalcopyrite twinning. In the breccia-type ore, pentlandite occurs as individual grains and inclusions in pyrrhotite. Larger grains (up to 0.3 mm) were detected in sandstone fragments. The euhedral tabular pentlandite crystals are contained in quartz; these tablets are close to perfect hexagons in shape. The larger segregations are intergrowths of several pentlandite grains, xenomorphic relative to gangue minerals. The rough rectilinear cleavage fractures with short curved offsets, located at a distance of 0.005 mm from one another, are widespread and filled with violarite. In small grains, violarite

Fig. 8.13 Vein (**a**) and lens of massive ore (**b**) in the exocontact of the Chiney pluton

develops as full pseudomorphs after pentlandite. Pentlandite in this type of ore is characterized by low Co contents (2–3 wt%), the elevated iron mole fraction, and Ag admixture up to 0.1 wt% (Table 8.6).

Millerite is a typomorphic mineral in vein-shaped and lenticular bodies, up to 8–10 cm long, where it is clustered along the selvages. The silvery white columnar crystals are intergrown. These aggregates are almost monomineralic and

Fig. 8.14 Breccias ores (**a, b, e**) and fluid inclusions (**c, d**) in quartz from breccia

consist of two types of grains, differing in size. Heazlewoodite grains are detected among small (up to 0.3 cm) millerite crystals.

Gersdorffite is commonly related to millerite clusters, where it is represented by hexagonal crystals or grains of similar morphology. These grains have straight even boundaries, located arbitrarily with respect to orientation of millerite and occasionally occur at boundaries of millerite grains along with chalcopyrite inclusions. Gersdorffite grains frequently contain small (0.0005 mm) nickeline inclusions. The chemical composition of gersdorffite is (wt %) 33.94 Ni, 0.36 Co, 0.17 Fe, 0.08 Cu, 20.82 S, and 42.40 As, 97.76 in total (Table 8.8).

Mackinawite is identified in chalcopyrite as small tabular and less frequent, irregular in shape, inclusions oriented at 45°, or parallel to planes of twinning. In addition to Fe and S, this mineral contains 5.56 wt% Ni and 0.25 wt% Co. Cubanite occurs as rather large (up to 0.3 × 0.1 mm) grains

in massive chalcopyrite. Cubanite segregations contain numerous tabular, oblong, or equant pyrrhotite grains, arranged parallel to one another and to boundaries of heterogeneous cubanite blocks.

Minerals of the *linnaeite group*, which occur in massive ore and as disseminations in sandstone fragments in breccia, are represented by siegenite, linnaeite, and violarite. Linnaeite in breccia ore is related to large pyrrhotite grains. Like pentlandite, it is identified as two morphological varieties. The first has a lenticular rhombic habit, and the second is characterized by large (up to 0.2 mm) euhedral crystals among the pentlandite grains replaced with violarite along fractures. Siegenite is localized in sandstone fragments in breccia as small euhedral crystals, up to 0.1 mm in size, associated with pentlandite and pyrrhotite hosted in gangue mass. Violarite develops along fractures in pentlandite and as disseminations in chalcopyrite. Maucherite is observed as large grains (0.2–0.5 mm), incorporated into massive chalcopyrite.

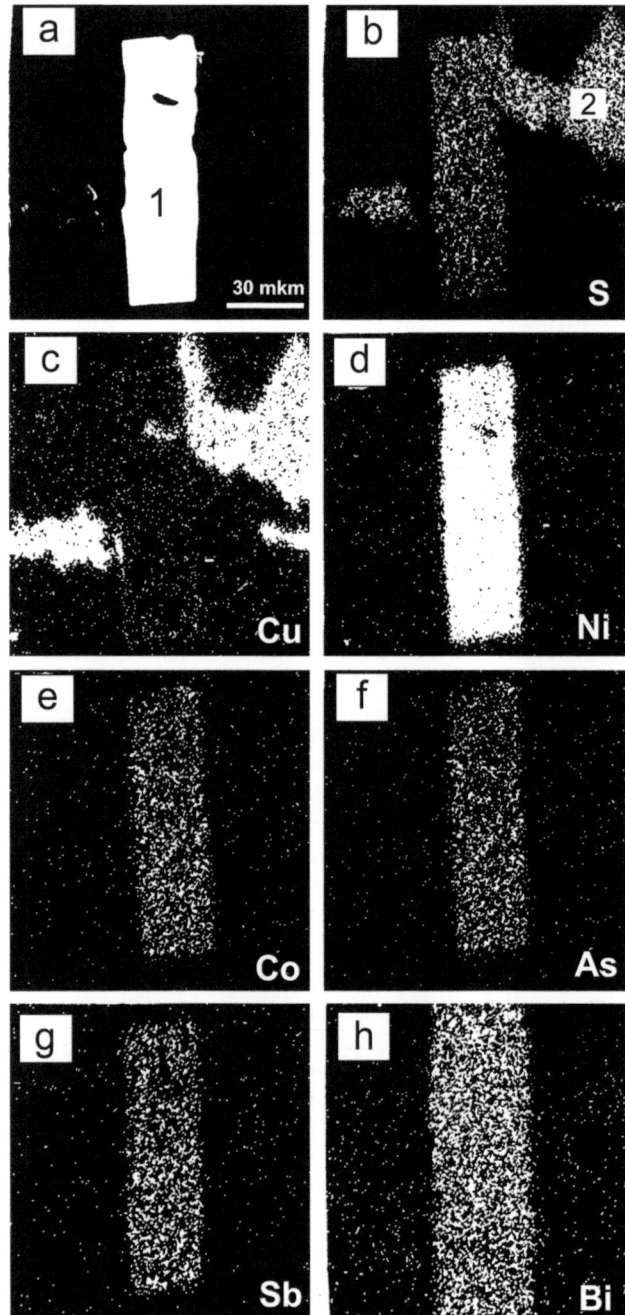

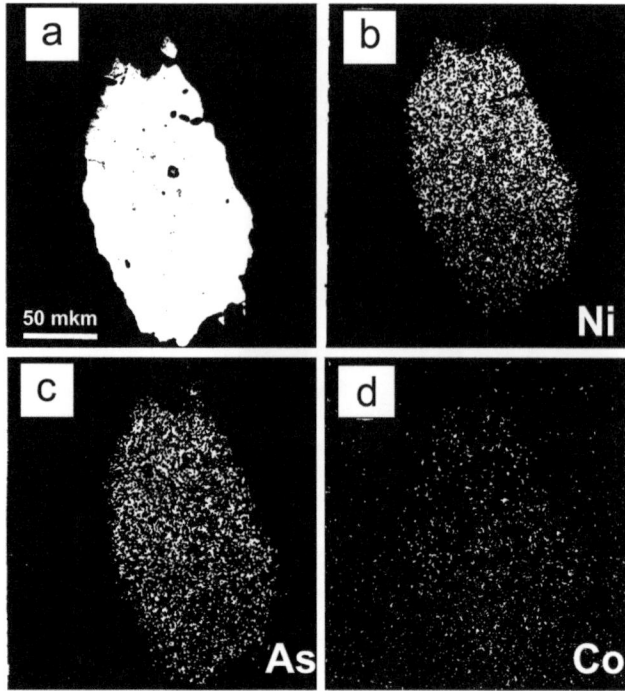

Fig. 8.16 Maucherite in chalcopyrite
(**a**) image in backscattered electrons (composition), (**b–d**) images in characteristic X-ray: (**b**) – $Ni_{k\alpha}$, (**c**) – $As_{k\alpha}$, (**d**) – $Co_{k\alpha}$

Fig. 8.15 Arsenohauchercornite (*1*) with chalcopyrite (*2*) in silicate minerals
(**a**) image in backscattered electrons (composition), (**b–h**) images in characteristic X-ray: (**b**) -$S_{k\alpha}$, (**c**) – $Ni_{k\alpha}$, (**d**) – $Co_{k\alpha}$, (**e**) – $As_{k\alpha}$, (**f**) – $Te_{k\alpha}$, (**g**) – $Sb_{k\alpha}$, (**h**) – $Bi_{L\,\alpha}$

Nickeline occurs as individual grains and as inclusions in gersdorffite. This mineral frequently contains numerous PGM inclusions that occupy 40% of its volume. Pd and Sb are the major components of these inclusions.

Arsenohauchercornite (Ni sulfobismuthite) is an extremely rare mineral. Its individual grains, up to 0.06 mm in length (Fig. 8.15), have been identified in sandstone fragments of breccia ore, where they are localized at the quartz-carbonate boundaries. Sporadic chalcopyrite inclusions are noted in arsenohauchercornite. The chemical composition of this mineral is (wt %) 42.28 Ni, 3.89 Co, 0.48 Cu, 0.77 Fe, 0.09 Ag, 21.31 Bi, 4.95 As, 3.49 Sb, and 22.93 S, in a total of 100.40. The high As content allows us to identify this mineral as arsenohauchercornite (Just 1980); a rather high Sb content marks its transition to hauchercornite Ni_9SbBiS. These intermediate members of isomorphic series with high As and Sb contents were not pointed out in the literature. The Co content is unusually high.

Maucherite was found in massive sulfide ore, inside the endocontact zone of chalcopyrite lens. Besides major components, it contains Co (Fig. 8.16).

Crystallization temperature of sulfides was estimated on the basis of Co content in pentlandite (Vaasjoki et al. 1974), exsolution structures in chalcopyrite (Petrovskaya 1973), and iron mole fraction in sphalerite (22 mol % Fe) in equilibrium with pyrrhotite. According to these data, these minerals crystallized at T = 620–550 °C and further drop down to 400 °C.

8.2.2 The Kontaktovyi Deposit

The Kontaktovyi deposit is the second in reserves after the Rudnoe deposit and is situated close to the western contact of the pluton (Fig. 5.2), where a maximum thickness of disseminated mineralization reaches 70–80 m. Large orebodies are represented by stratiform lodes, extending for hundreds of meters. The small bodies are lenticular in shape. They are conformable to the layering of gabbro rocks and parallel to their bottom, dipping toward 10–55° NE at up to 40°. The orebody boundaries are defined on the results of sampling. Two lodes have been encountered in this way: the first occurs at the bottom of the pluton, and the second is located 150 m up section.

In general, the mineralization of the Kontaktovyi deposit is lower in grade than the ore from the Rudnoe deposit, but it is more persistent in composition (Table 8.3). The Cu contents vary from 0.27 to 2.85 wt% (0.8 wt%, on average). The Pd content is 0.47–2.25 ppm (1.1 ppm, on average). The Pt content is 0.1–1.0 ppm (0.25 ppm, on average), and the Au content is 0.1–0.2 ppm.

The ore is mostly disseminated. Leucogabbro, spotty anorthosite, and typical leopard gabbro (spot size is 1.5–2.0 cm, on average) are host rocks. The composition of these rocks, their structure, and texture are persistent throughout the mineralized zone. Intense albitization, amphibolization, and epidotization are typical. Intensity of mineralization is correlated with the intensity of these secondary alterations. The spots are characterized by a micrographic intergrowth of plagioclase and pyroxene. In ore intervals, sulfides replace dark-colored minerals with retention of the primary rock texture. Thus, 70–80% of sulfide minerals concentrate in the spots that occupy 40–50% of rock volume. Sulfides are uniformly distributed through the host rocks. Except for pseudomorphs after pyroxene, they make up the segregations, 2–4 cm in size.

The ore mainly consists of chalcopyrite and pyrite in proportion of 1:1 (Figs. 8.4a, 8.17a). Pyrrhotite is a rare mineral. Titanomagnetite and ilmenite are second in abundance (Fig. 8.18). Magnetite, millerite, violarite, and sphalerite (Table 8.7) are rare. Composition of sulfide minerals and oxides from endocontact ore at the Kontaktovyi deposit is provided in Appendix.

Chalcopyrite occurs as grains, irregular in shape, while pyrite is represented by euhedral cubic crystals, somewhat larger than chalcopyrite grains (0.6–0.7 mm, on average). Pyrite crystals contain numerous chalcopyrite inclusions of irregular shape.

Pyrite also contains Co and Ni as admixtures, and the contents of these elements are twice as high than in chalcopyrite. Small (0.05–0.10 mm) pyrrhotite grains among gangue minerals are extremely rare.

Titanomagnetite occurs as very large (5–6 to 8 mm) crystals grouped into clusters composed of 3–4 grains. Internal titanomagnetite structure is commonly roughly latticed. The grains are strongly decomposed. Magnetite is replaced with a gangue mass, and only an ilmenite lattice is left among

rock-forming minerals and sulfides. The relative amount of ilmenite does not exceed 10 vol %.

8.3 Sulfide Mineralization in Titanomagnetite Ore and Fractures

In general, rocks of the titanomagnetite gabbro series are characterized by low contents of sulfide minerals (1–3%). However, some units with the highest titanomagnetite content (finely layered titanomagnetite ore) can be enriched in sulfides up to 20–30%. These units are a few meters in thickness and hundreds of meters in extent; their boundaries are rather distinct. The geometry of sulfide-rich units coincides with layering of host rocks; the strike varies from near longitudinal to near latitudinal; the dip angle is 10–15°, on average. The sulfide-rich units are known at the Magnitnyi and Etyrko deposits, where sulfides occur as disseminations and stringers.

A zone of disseminated sulfide has been studied in detail in Borehole 11 at a depth of 620–690 m (lower macrorhythm of 1 T member), where gabbro rocks are characterized by microrhythmic structure; the thickness of rhythms is 3–5 cm, on average. The amount of rock-forming minerals remains almost unchanged against the background of substantially variable oxide contents. In general, the host rock corresponds to gabbronorite (Fig. 8.19). In rhythm 11/638, for example, composition of rock-forming minerals is shown in Table 8.9. The boundaries between them are distinct and marked by fluctuations of titanomagnetite contents from one layer to another (20–25 to 80%); its composition is summarized in Table 8.10. Sulfides are distributed more or less uniformly. Local peaks of sulfide grade reach 20–30% of the total rock volume, against the background 10–15 vol %. Pyrrhotite and chalcopyrite dominate in this rhythm (Table 8.11).

Sulfides are clustered either in the lower melanocratic parts of microrhythms or in the upper leucocratic often pegmatoid rocks. The abundances of ore minerals decrease in the following order (Fig. 8.7b, c): titanomagnetite, pyrrhotite, chalcopyrite, ilmenite, pyrite, high-Co pentlandite, violarite, siegenite, linnaeite (Table 8.6, 8.9), hematite, spinel, and ulvospinel.

As compared with Fe-Ti oxides and silicate minerals, sulfide grains are smaller and xenomorphic. Their clusters are 1–2 mm in diameter; individual sulfide grains are even smaller. Sulfides frequently replace the magnetite component of titanomagnetite with the formation of pseudomorphs (Fig. 8.20).

In a mode of occurrence, sulfides are subdivided into three groups: (1) large (up to 4 mm) oval, lenticular, or irregular segregations; (2) fine disseminations and variously oriented stringers in silicate minerals, titanomagnetite, and ilmenite; and (3) thin veinlets (Fig. 8.21).

Fig. 8.17 Disseminated ore from the Kontaktovyi deposit (**a–f**) reflected light, (**g–h**) images in backscattered electrons (IGEM, analyst EV Kovalchuk)
Po pyrrhotite, *Py* pyrite, *Sp* sphalerite, *Ccp* chalcopyrite, *Pn* pentlandite, *Mag* magnetite

Large sulfide clusters are related to boundaries of titanomagnetite grains, especially to embayments of their surface. Sulfides replace silicates beginning from cleavage fractures and finishing by the formation of complete pseudomorphs, largely after pyroxenes; only occasional sulfide clusters have a geometrically perfect shape. Among large grains, pyrrhotite dominates (70 vol% of the total sulfides; Fig. 8.20a, b, c, and d), followed by chalcopyrite (30–50 vol%; Fig. 8.20e, f), and pentlandite is the least abundant (Fig. 8.20g, h). Other minerals are very rare, and their contents do not exceed

1–2 vol%. Fine sulfide disseminations are nonuniformly dispersed throughout the rock, concentrating within titanomagnetite grains and at their surfaces, with formation late magnetite (Fig. 8.21). At a distance from titanomagnetite, sulfides are less abundant. They are mostly localized in pyroxene and avoid plagioclase. The sulfide minerals are tenths and hundredths of a millimeter in size. They are irregular, and commonly elongated (tabular) grains are localized either along the cleavage in silicates or along the cracks or boundaries between ilmenite-magnetite phases in exsolution

Fig. 8.18 BSE images of disseminated ore at the Kontaktovyi deposit Analyses and images were taken at IGEM, Moscow (analyst EV Kovalchuk)
Po pyrrhotite, *Py* pyrite, *Ccp* chalcopyrite, *Pn* pentlandite, *Ti-Mag* titanomagnetite

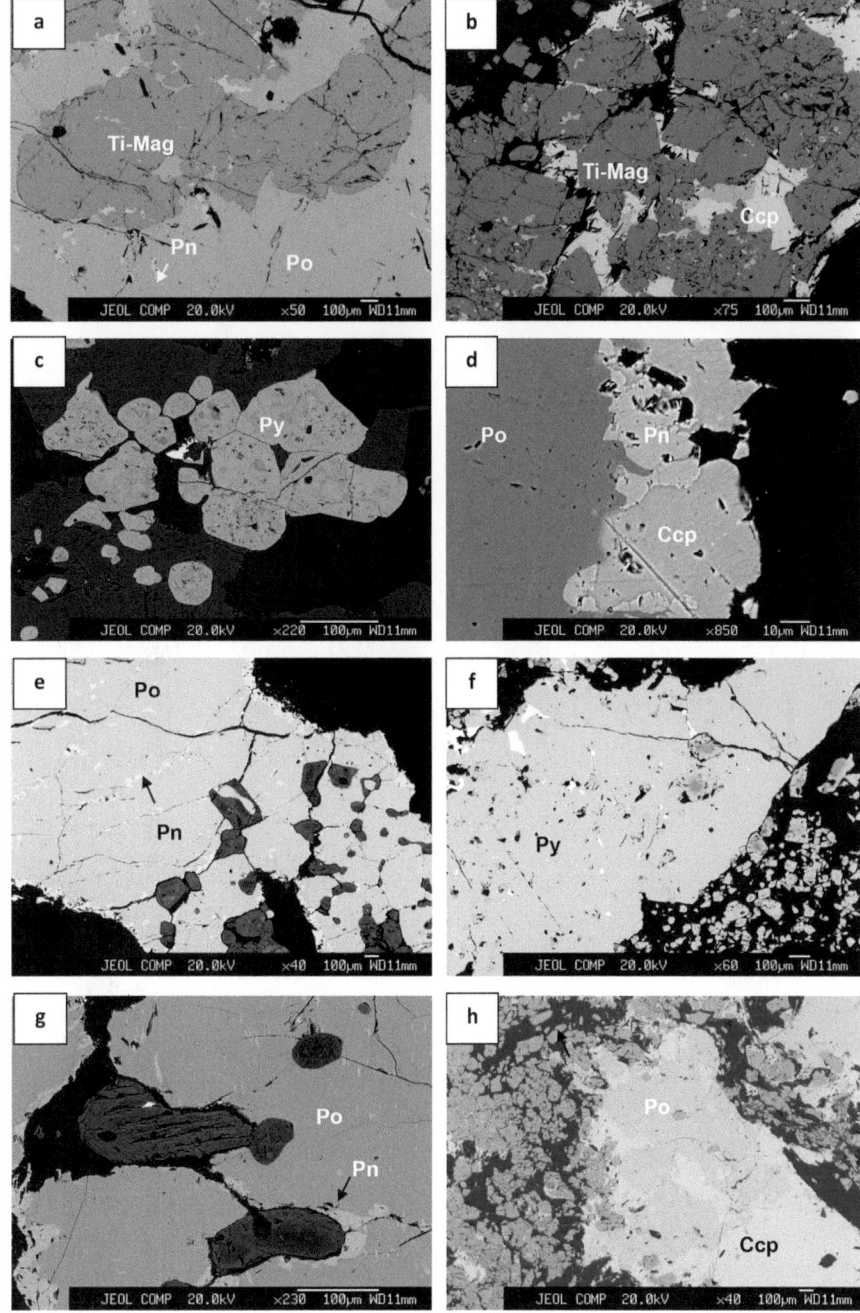

textures of titanomagnetite. The fine round sulfide disseminations are rare and resemble emulsion in silicate minerals.

Sulfide minerals not only penetrate into titanomagnetite grains along fractures but also replace them (Figs. 8.20c, f, 8.21). In the case of selective replacement, when ilmenite lamellae remain intact, the chalcopyrite-pyrrhotite aggregate acquires geometrically regular outlines. In rare cases, small titanomagnetite grains in silicate matrix are almost completely replaced with sulfides, micas, and secondary magnetite. Thus, the morphology and mode of occurrence of the sulfide minerals clearly show that they crystallized after silicate minerals and Fe-Ti oxides. The elevated Co content in

pentlandite of this ore shows that this mineral is a high-temperature variety (>650 °C).

In fine sulfide disseminations, impregnating titanomagnetite and rock-forming minerals, chalcopyrite is sharply dominant (80%), followed by pyrrhotite (20–30%) and pyrite (up to 5% of the total sulfide amount). The large *chalcopyrite* grains are diverse in morphology. This mineral is distinctly anisotropic; tabular and lenticular twins make up an oleander-leaf structure or transformation twins after Ramdohr (1980). As a rule, this mineral contains inclusions of pyrrhotite, pentlandite, linnaeite, and other rare minerals. Chalcopyrite is represented by a tetragonal variety. Its composition is

Fig. 8.19 A fragment of the microrhythm 11/638 with a profile of microprobe points and a fragment with titanomagnetite
Composition of rocks, oxides, and sulfides from this microrhythm is summarized in Tables 8.9, 8.10, and 8.11, respectively

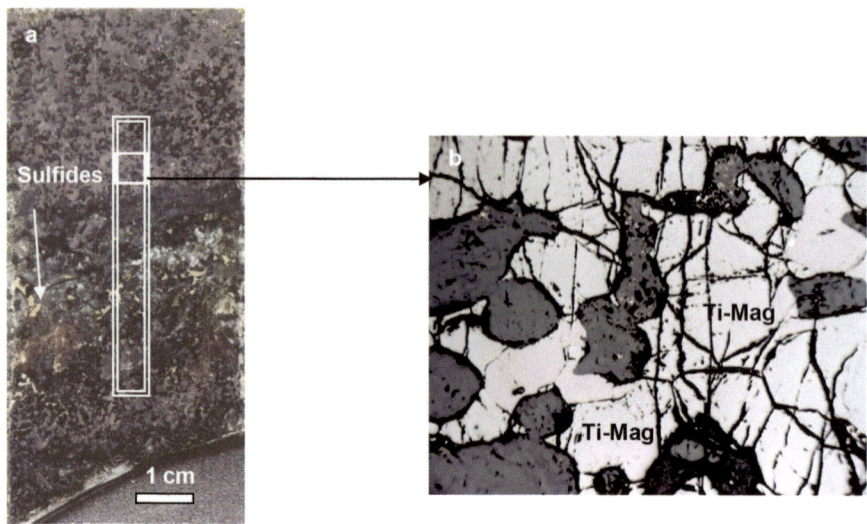

Table 8.10 Chemical compositions of oxide minerals from microrhythm 11/638, wt %

No.	SiO$_2$	TiO$_2$	Al$_2$O$_3$	FeO	MnO	MgO	CaO	NiO	ZnO	Cr$_2$O$_3$	V$_2$O$_3$	Total
Ilmenite												
4	0.164	51.007	0.238	47.083	0.925	0.13	–	0.003	0.008	0.027	0.597	100.18
7	0.17	50.874	0.196	47.435	0.709	0.211	0.002	0.012	–	0.009	0.537	100.16
8	0.166	50.812	0.21	47.338	0.66	0.177	–	0.004	0.012	0.035	0.649	100.06
9	0.178	51.155	0.207	47.392	0.662	0.206	–	–	–	0.023	0.571	100.39
10	0.188	49.591	0.227	48.44	0.813	0.034	0.008	–	0.012	0.025	0.623	99.961
11	0.18	50.568	0.162	47.986	0.818	0.041	0.001	0.001	–	0.02	0.589	100.37
18	0.15	50.462	0.259	47.825	0.786	0.272	–	–	0.01	0.011	0.484	100.26
20	0.135	51.1	0.192	47.11	0.808	0.316	–	0.005	0.004	0.003	0.498	100.17
31	0.177	50.003	0.246	48.273	0.685	0.131	–	0.009	0.005	0.016	0.462	100.01
Titanomagnetite												
5	0.207	9.455	2.339	80.149	0.175	0.192	0.025	0.024	0.16	0.115	1.147	93.988
6	0.177	11.989	2.786	77.429	0.197	0.133	0.007	0.021	0.187	0.128	1.138	94.192
14	0.206	11.416	1.685	79.085	0.197	0.131	–	0.009	0.085	0.128	1.193	94.135
15	0.209	14.586	1.759	76.171	0.258	0.154	0.006	0.016	0.076	0.131	1.051	94.417
16	0.24	13.977	3.242	75.989	0.265	0.312	–	0.022	0.122	0.158	1.04	95.367
17	0.235	77.685	3.267	81.344	0.13	0.204	0.005	0.018	0.118	0.152	1.114	94.272
21	0.221	11.66	2.241	79.2	0.215	0.359	–	0.018	0.067	0.137	1.085	95.203
22	0.199	13.761	5.407	73.703	0.26	0.536	–	0.009	0.169	0.14	1.008	95.192
23	0.185	13.388	3.733	75.348	0.237	0.413	0.008	0.023	0.134	0.139	1.045	94.653
24	0.187	13.467	5.104	74.272	0.232	0.537	0.006	0.023	0.183	0.142	0.985	95.138
25	0.205	16.905	2.424	74.024	0.308	0.277	–	0.02	0.066	0.127	1.011	95.367
26	0.192	9.981	0.791	81.154	0.182	0.152	0.003	0.021	0.012	0.141	1.162	93.791
27	0.202	6.728	3.375	81.178	0.129	0.268	0.012	0.02	0.103	0.15	1.116	93.281
28	0.214	12.261	3.829	76.793	0.233	0.42	0.005	0.01	0.119	0.147	1.075	95.106
29	0.274	14.956	2.492	75.262	0.256	0.383	–	0.014	0.075	0.142	1.037	94.891
30	0.178	12.207	4.49	75.619	0.22	0.391	–	0.015	0.154	0.143	1.04	94.457
32	0.21	12.058	3.143	77.85	0.21	0.374	0.004	0.019	0.104	0.132	1.056	95.16

Note. Dash denotes concentrations below detection limit
Analyses were carried out at MPI, analyst DV Kuzmin

Table 8.11 Chemical compositions of sulfide minerals from the microrhythm 11/638 (wt%)

№	As	S	Fe	Ni	Co	Se	Sb	Cu	Total
Chalcopyrite									
1	0.06	34.83	30.45	–	–	0.08	0.002	34.09	99.51
2	0.09	34.21	30.04	0.008	–	0.09	0.017	33.92	98.37
3	0.01	34.72	30.51	–	–	0.09	0.006	34.34	99.67
4	0.04	34.63	30.41	0.014	–	0.08	–	34.20	99.38
5	0.02	34.87	30.32	–	–	0.10	–	34.33	99.65
6	0.04	34.53	30.04	0.006	–	0.10	0.013	33.96	98.68
7	0.03	34.56	30.18	0.003	–	0.07	0.026	34.21	99.07
8	0.10	34.67	30.31	0.001	–	0.11	0.003	34.17	99.36
9	0.04	34.63	30.22	0.007	–	0.13	0.001	34.26	99.29
10	0.08	34.65	30.36	–	–	0.09	–	34.44	99.61
11	0.07	34.93	30.46	–	–	0.08	0.003	34.28	99.82
12	0.07	34.81	30.24	–	–	0.07	0.014	34.17	99.37
13	0.02	34.57	30.55	–	–	0.11	–	34.12	99.37
14	0.04	34.76	30.53	0.004	–	0.09	0.01	34.36	99.80
15	0.08	34.52	30.28	0.001	–	0.06	0.014	34.32	99.26
16	0.05	34.56	30.38	0.006	–	0.11	–	34.09	99.19
17	0.02	34.60	30.45	–	–	0.11	0.001	34.09	99.28
18	0.04	34.79	30.34	0.001	–	0.08	–	34.24	99.50
19	0.06	34.66	30.47	–	–	0.08	–	34.63	99.91
20	0.05	34.54	30.35	0.007	–	0.11	0.002	34.32	99.38
21	0.05	34.64	31.46	0.002	–	0.08	–	33.71	99.94
Pyrrhotite									
22	0.05	39.21	58.74	0.247	0.109	0.05	0.019	0.094	98.52
23	0.07	38.46	59.26	0.396	0.034	0.10	0.005	0.008	98.33
24	0.03	38.86	59.84	0.39	0.029	0.06	0.003	–	99.23
25	0.06	38.17	58.74	0.562	–	0.09	–	0.012	97.62
26	0.05	38.43	59.91	0.246	0.002	0.09	0.001	0.008	98.73
27	0.01	38.60	59.71	0.385	–	0.12	–	–	98.82
28	0.03	38.33	59.57	0.393	–	0.08	–	–	98.39
29	0.07	38.56	59.52	0.438	–	0.10	–	0.03	98.73
30	0.06	38.62	59.83	0.361	0.038	0.07	0.007	–	98.98
31	0.06	39.40	59.51	0.055	0.096	0.09	0.018	0.077	99.30
32	0.04	38.87	59.51	0.348	0.009	0.08	0.004	0.011	98.87
33	0.00	38.61	60.47	0.247	0.004	0.09	–	0.01	99.46
34	0.04	39.80	59.26	0.237	0.024	0.09	–	0.081	99.52
35	0.07	39.13	59.63	0.342	0.032	0.07	0.012	0.003	99.29
36	0.03	39.23	59.24	0.087	0.013	0.08	–	0.573	99.26
37	0.06	39.40	59.97	0.111	0.000	0.10	–	0.017	99.66

Note. Dash denotes concentrations below detection limit. Analyses were carried out at MSU, analyst Nat E Sergeeva

close to the theoretical CuFeS$_2$. The variations of major element contents are as follows (wt %): 32.3–33.76 Cu, 29.83–32.58 Fe, and 33.67–35.47 S. Co and Ni are the admixtures; their concentrations do not exceed hundredth fractions of a percentile except for boundaries with pentlandite grains, where the contents of these admixtures increase up to 0.30 and 0.43%, respectively.

Pentlandite grains vary in size and morphology. In order of abundance, these are large (up to 0.3 mm) euhedral grains in marginal parts of pyrrhotite crystals (Fig. 8.4d, e), flame-like and lenticular segregations within pyrrhotite as products of decomposition of a solid solution, and oval segregations in pyrrhotite (Fig. 8.4f). Pentlandite is always associated with pyrrhotite. Its findings within chalcopyrite are, as a rule, relics, remaining after the replacement of pyrrhotite-pentlandite aggregates with chalcopyrite. The large pentlandite grains are replaced with violarite, pyrite, and magnetite along cleavage fractures.

Violarite develops along fractures. *Pyrite* occurs as outer rims around pyrrhotite grains or as individual thin veinlets, 0.2 mm thick, in silicate minerals. *Linnaeite* occurs sporadically and is diverse in morphology, varying from lenticular to isometric grains with offsets up to 0.2 mm in size, which are always hosted in pyrrhotite. Siegenite is associated with high-Co pentlandite. The chemical composition of siegenite is constant based on microprobe analysis (wt %): 24.20–

Fig. 8.20 Photomicrographs
of sulfides in the
titanomagnetite gabbro series,
Etyrko deposit
Py pyrite, *Pn* pentlandite, *Po* pyrrhotite, *Ccp* chalcopyrite, *Ilm*
ilmenite
Samples: (**a, b**) 11/434; (**c, d**)
11/ 638; (**e, f**) 11/653; (**g, h**)
11/1209 (N sample = borehole
11/depth, m)

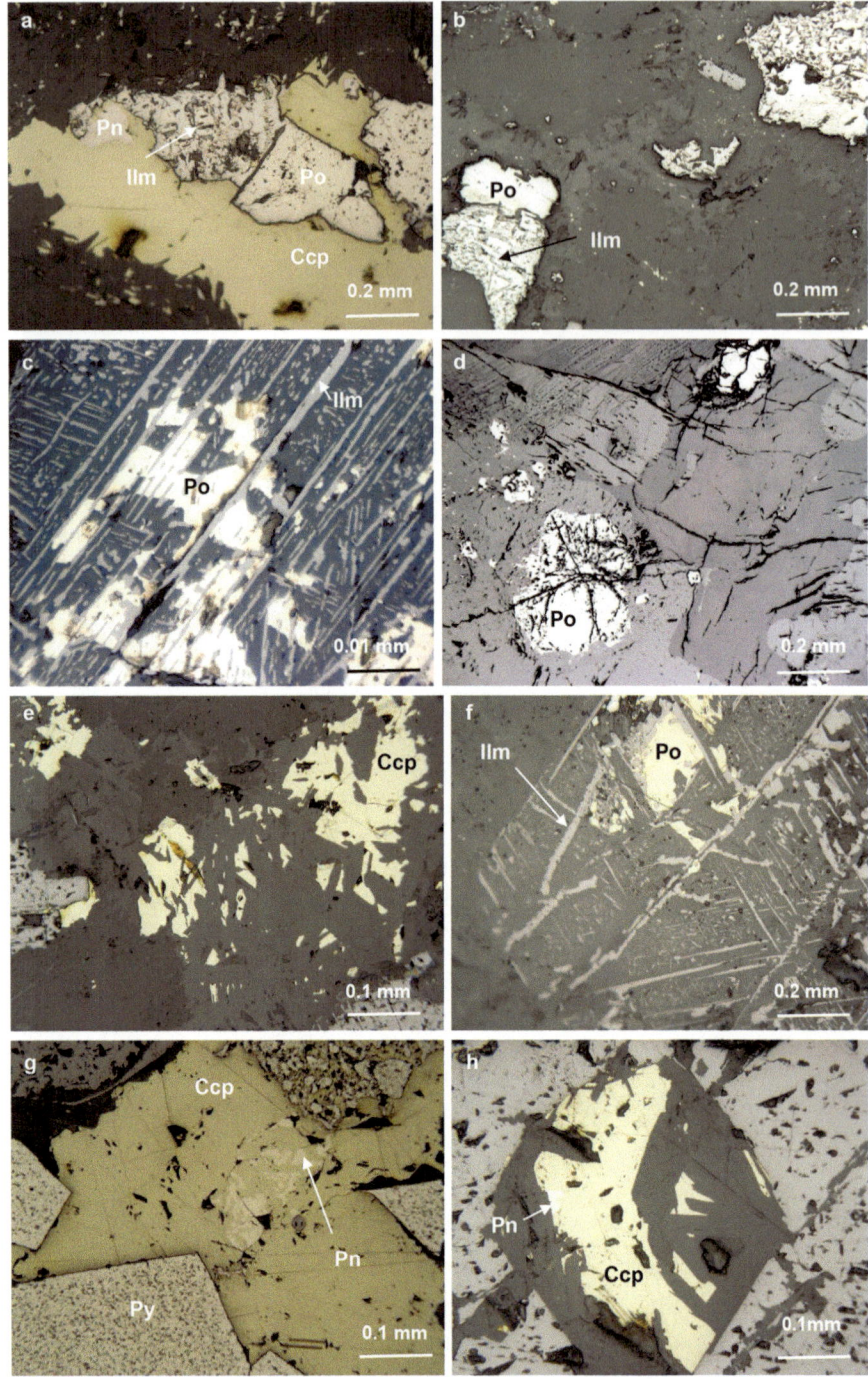

26.69 Co, 23.06–23.91 Ni, 8.63–9.32 Fe, and
38.80–40.88 S.

Sulfide minerals in titanomagnetite gabbro form small
veinlets (0.1–1 cm), consisting mostly of pyrrhotite
(Fig. 8.22).

The sulfide-rich zones mapped in the inner part of the
Chiney pluton are related to the superimposed stratiform and
crosscutting fracture and crush zones of variable orientation.
The stratiform tectonic zones are confined to boundaries
between layers in the central parts of the pluton. They are the
most frequent at the boundaries of leucogabbro and anortho-

site interlayers with underlying high-grade titanomagnetite
ore. Fracturing is advanced in iron ore as the most brittle
material. The elevated concentrations of sulfide minerals (up
to 5%) are typical of these interlayer zones, whose thickness
varies from a few decimeters to a few meters. Their boundaries are sufficiently distinct.

The disseminated, stringer-disseminated, and massive
types of sulfide mineralization are distinguished. The contents of sulfides reach 10–15%, with30–40% in particular
cases. In massive titanomagnetite, sulfide minerals fill interstices between Fe and Ti oxides making up sideronite struc-

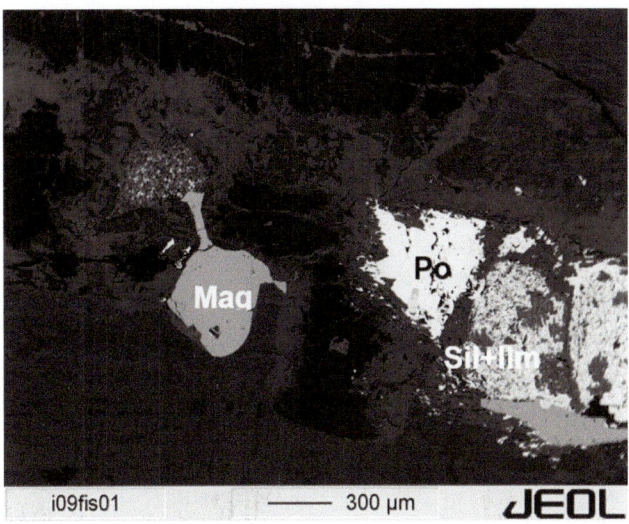

Fig. 8.21 Sulfides in relics of titanomagnetite grains from the Etyrko deposit
Sil silicate minerals, *Ilm* ilmenite, *Mag* magnetite. BSE images were taken at Max-Planck Institute of Geochemistry, Mainz, Germany (Analyst DV Kuzmin)

Fig. 8.22 Sulfide veinlets in titanomagnetite gabbro (polished sections)

ture. Chalcopyrite is a major ore mineral. Millerite and pentlandite are subordinate in abundance.

The crosscutting fractures were recognized in both the central and marginal parts of the pluton. In the first case (Peak Solnechnyi, right bank of the River Pravaya China), they are limited in thickness (a few meters) and extent (100 m). Their contacts are sharp. The zones of disturbances are almost vertical, crosscutting the layering. Crush and fracture zones are filled with sulfide minerals, the content of

which reaches 20–25%. Mineralization does not extend beyond the fracture zones.

A thick (40–50 m) zone of disseminated sulfide, oriented at an angle to layering, was traced in the western part of pluton. Its boundaries are diffuse. The amount of sulfide minerals gradually decreases from the central to marginal parts of the ore zone, which dips much steeper (85°) than layering (35–40°). Chalcopyrite is a major ore mineral.

8.4 Composition of Ore Minerals

The composition of the main ore mineral, chalcopyrite, is relatively constant for different types of the sulfide mineralization (Table 8.5, 8.11). Pyrrhotite is the second most widespread mineral in the deposits related to the Chiney layered pluton. Variations in its composition are equally insignificant (Table 8.5, 8.11); the Fe/S increases from monoclinic to hexagonal pyrrhotite and then to troilite (0.85, 0.92 and 1.00, consequently). In contrast to chalcopyrite, it contains some Ni (up to 1.75 wt%) and Co (up to 0.26 wt%).

Of greatest interest are those minerals that show variation in their composition. These include, first of all, pentlandite and the minerals of the linnaeite group.

8.4.1 Pentlandite

Two varieties can be distinguished among pentlandites: Ni-rich (35–38 wt%) and Ni-poor (30–33 wt%). The distinguishing feature of pentlandites from the Chiney pluton is the relatively high content of Co (Table 8.6). It is distinguished in this respect from this mineral from other Ni-Cu deposits, in particular, located in the Noril'sk ore district and the Kola Peninsula (with<3 wt% Co in pentlandite). The Co contents in pentlandite from the Chiney pluton vary from 1 to 18 wt %. Different types of sulfide mineralization differ appreciably from one another within narrow intervals of Co contents in pentlandite inherent to each type. The cobalt content in pentlandite is stepwise reduced from the sulfide mineralization located in the interior of the layered mafic-ultramafic complex to the endocontact and then to its exocontact. The most Co-rich pentlandites (8.20–18.38 wt% Co) were found in the titanomagnetite ores of the Etyrko deposit (Table 8.6). Thus, this mineral can be referred to as high-Co pentlandite (Shishkin et al. 1974), while pentlandites in the endocontact and exocontact sulfides have 5–7 and 2–3% Co, respectively.

In the triangular plot shown in Fig. 8.23a, the pentlandite compositions occupy a narrow zone, extending parallel to the Ni-Co side that reflects variations of Co content at almost constant Ni and Fe contents. Only pentlandites from vein ore reveal variations of Ni contents at 2–3 wt% Co, and their compositions extend parallel to the base of the triangle. They fall

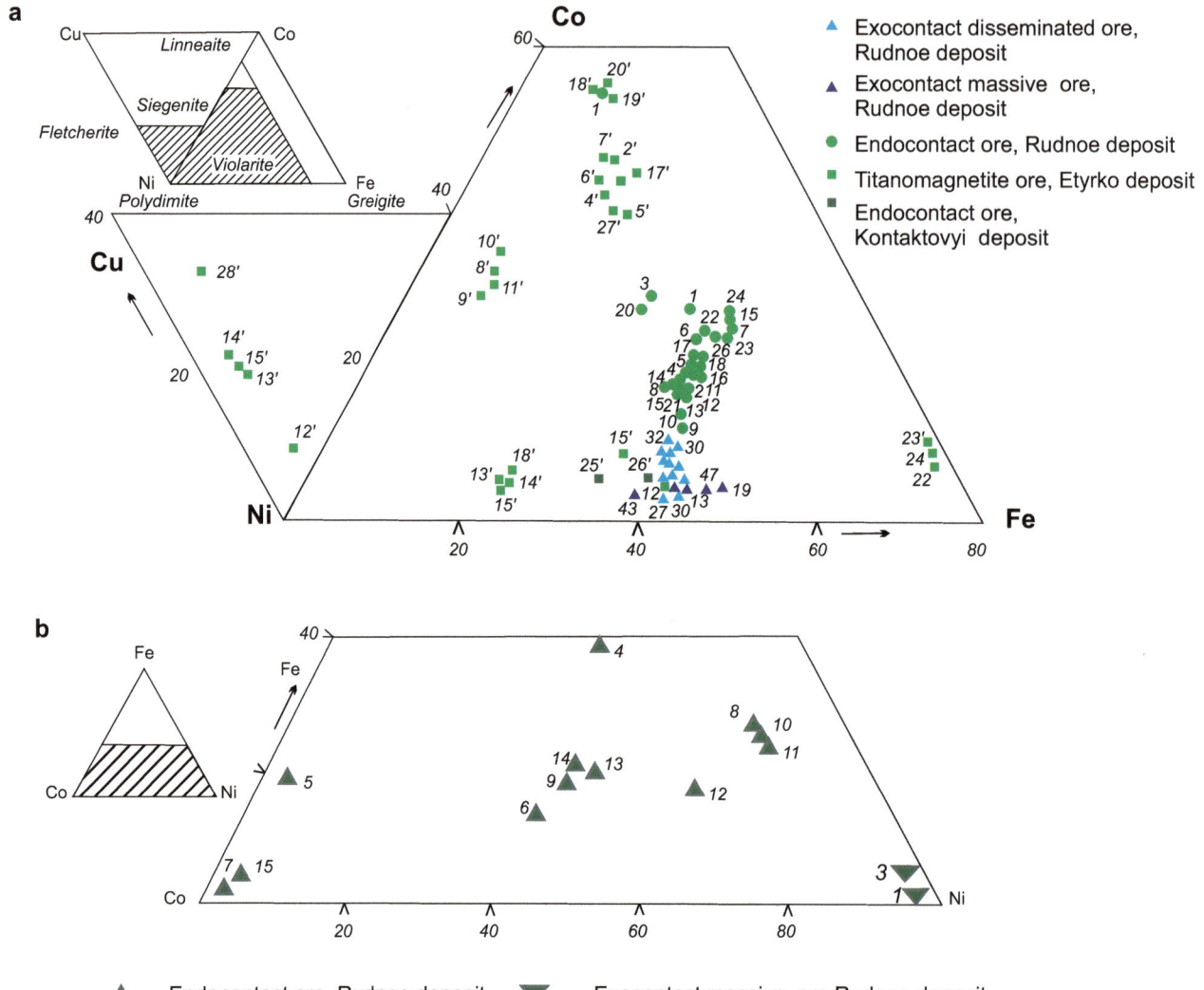

Fig. 8.23 Diagrams Fe-Ni-Co (Cu) for pentlandite and minerals of linnaeite group (**a**) and sulfoarsenides (**b**) from the Chiney pluton N point in diagram = N analyses in Tables 8.6 and 8.8

into the group of pentlandite (0.9 < Ni/Co <1.3) and Ni-pentlandite (Ni/Co >1.3). As has been established for many deposits, Ni-pentlandite is characteristic of the mineral assemblages with bornite, millerite, heazlewoodite, and monoclinic pyrrhotite. These minerals are found in the exocontact types of mineralization, where low-Co pentlandite occurs. The mineral assemblages typical of Fe-pentlandite (Ni/Fe > 0.9), including troilite, talnakhite, mooihoekite, and awaruite, are not known in the Chiney pluton, as well as Fe-pentlandite.

It is important to emphasize that the Co distribution in pentlandite from different types of sulfide mineralization shows a negative correlation with their economic value. For example, the high-grade ores from the outer-contact zone of layered mafic-ultramafic complex comprise pentlandites with the lowest Co contents. This feature can be used for estimation of economic value in other gabbroid layered mafic-ultramafic complexes in the Kodar-Udokan area.

8.4.2   Minerals of the Linnaeite Group

The chemical composition of minerals from the linnaeite group, occurring in various ore types of the Chiney pluton, is characterized by complex relationships between Ni, Co, Fe, and Cu. They are all far from theoretical composition and occupy a transitional position between conventional mineral species.

Minerals of the linnaeite group (linnaeite, siegenite, violarite) were described in all types of sulfide mineralization mentioned above (Table 8.6). Linnaeite group minerals, especially siegenite, typically form hexagonal 0.2–0.3 mm grains that are hosted mainly by silicate minerals and, less frequently, by chalcopyrite and pentlandite. Their composition varies somewhat from one ore type to another. The sulfides in the titanomagnetite (the Etyrko deposit) are dominated by a high cobalt species (Co = 24.20–26.69 wt%), intermediate between linnaeite and siegenite. These minerals in the endocontact disseminations, hosted in the rocks of elevated alkalinity, contain from 26.78 to 31.79 wt% Co; in the exocontact disseminations hosted in sandstone, from 16.69 to 18.12 wt%; and 22.11 wt% in the vein ore. The chemical composition of the studied siegenites deviates from the theoretical formula $CoNi_2S_4$ inherent to this mineral (Fig. 8.23a). They are clustered in three separate fields, depending on Co, and, to a much lesser degree, on Fe contents. The chemical compositions of siegenite and Ni-linnaeite are close to $NiCo_2S_4$; it is characterized by significant presence of Fe (8.04–11.45 wt%).

The composition of violarite, which is the second in abundance mineral of the linnaeite group in the Chiney ore, substantially differs from the stoichiometric $FeNi_2S_4$ due to high Co concentrations, variable proportions of Ni and Fe, and a general shift toward greigite $Fe^{2+}Fe_2^{3+}S_4$. These compositions are localized above the Ni-Fe side of the triangle (Fig. 8.23a), being clustered in two small separate fields differing in Ni-Fe relationships. A part of its composition falls into the field of Cu, with the highest Cu concentration reaching 10.09 wt%. Violarite, containing 12.7 wt % Cu, occupies a special position and approaches fletcherite discovered in the United States (Craig and Carpenter 1977). It differs from the latter in lower Co and higher Fe contents. This mineral, like others with elevated Cu contents, has been identified as Cu-violarite with formula $Cu_{0.63}Ni_{1.71}Co_{0.14}Fe_{0.60}S_{3.93}$, to which it is close in optic properties. Unfortunately, the small size of grains did not allow us to perform XRD study to specify the position of this mineral in the systematics of the linnaeite group. Although copper is not characteristic of violarite (Borishanskaya et al. 1981), this element is typical of violarite from the Chiney pluton.

The exocontact ores contain minerals with the lowest Co contents (2–3 wt %), belonging to the intermediate member of polydymite-greigite range.

8.4.3 Arsenides and Sulfoarsenides

Minerals of the cobaltite-gersdorffite group (27.55–34.43 wt% Co) are limited in abundance in the Chiney pluton. They occur as euhedral crystals and thin veinlets in all mineralization types, except for finely layered titanomagnetite ore. The frequency of abundance increases with transition from the endo- to exocontact ore. This group of minerals includes the cobaltite-gersdorffite isomorphic series. Minerals of this group identified among sulfides in the Chiney pluton are subdivided into three clusters (Table 8.8): cobaltite (15 wt% Co), gersdorffite (4 wt% Co), and intermediate members of this series (13–14 wt% Co).

As it was mentioned above, they often form their own association with chalcopyrite in the endocontact sulfide mineralization in the Rudnoe deposit. Cobaltite is more abundant than gersdorffite and has a composition close to theoretical CoAsS, also containing Cu, Ni, Pb, Sb, and Fe. Gersdorffite crystals (NiAsS) contain Co, Fe, and Cu in addition to the major components. The chemical compositions fall rarely into the field of normal cobaltite and gersdorffite (Fig. 8.23b); they are close to transitional compositions, which fit Ni-cobaltite (dzhulukulite, a preliminary name, not in www.mindat.org) or Co-gersdorffite (Borishanskaya et al. 1981). In the exocontact mineralization, gersdorffite grains dominate in the isomorphic series. It is characterized by 2–3 wt% Co, while gersdorffite from the veins is close to theoretical composition. Thus, one can see the same tendency in variations of cobaltite-gersdorffite isomorphic series as it was established for pentlandite, i.e., the Co content decreases from the center of the layered maficultramafic complex to its endo- and exocontacts.

Safflorite and loellingite also occur in the mineral assemblage in the endocontact ores. Loellingite, pertaining to the Fe-Co diarsenides, is enriched in Ni (Table 8.8). The second mineral of this assemblage is related to the same Ni-free isomorphic series of Fe-Co diarsenides and corresponds to safflorite in chemical composition ($CoAs_2$). Among arsenides and sulfoarsenides, nickeline and arsenopyrite were diagnosed. They have a chemical composition close to theoretical; nickeline sometimes contains Fe, Co, and Cu (Fig. 8.23b). In the exocontact ores, arsenohauchercornite was found. The chemical composition of this mineral (wt %) 42.28 Ni, 3.89 Co, 0.48 Cu, 0.77 Fe, 0.09 Ag, 21.31 Bi, 4.95 As, 3.49 Sb, and 22.93 S, 100.40 in total. The high As content allows us to identify this mineral as arsenohauchercornite (Just 1980). A rather high Sb content marks its transition to hauchercornite $Ni_9SbBi_8S_8$. These intermediate members of isomorphic series with high As and Sb contents are very rare in nature. The Co content is unusually high for this mineral species. The chemical compositions of bornite, millerite, and other minerals show little variation in comparison with their theoretical compositions.

8.5 Mode of Occurrence of Co and Ni in Copper Ore

The copper ore, hosted in the Chiney pluton, is complex. In addition to copper as a major component, the ore contains Co, Ni, and noble metals (Tables 8.1, 8.2, and 8.3), which are involved in the estimation of reserves and are recoverable from sulfides as by-products. The subordinate elements are distributed throughout the pluton extremely nonuniformly. This is especially typical of noble metals, the localization of which so far remains poorly studied.

Cobalt is contained in all types of sulfide mineralization, including subeconomic varieties, at the level of the hundredth and thousandth fractions of a percentile. The elevated concentrations of this element are characteristic of the exocontact ore at the Rudnoe deposit. Cobalt is concentrated in Co mineral phases and dispersed in all major, subordinate, and rare ore minerals. Pentlandite, cobaltite, and linnaeite group mineral phases are main cobalt-bearing minerals at this deposit.

Pentlandite is the most important Co mineral due to it wide presence. The Co content ranges from 1 to 18.38 wt% (Table 8.6, Fig. 8.23a). The second place in cobalt balance takes minerals of linnaeite group, because they occur often in different ore types and are characterized by high Co concentration in phases, changing from 8.04 to 31.79 wt%. Cobaltite contains high Co content (Fig. 8.23a, Table 8.8), but it is a rare mineral.

In addition to cobalt mineral phases, Co is incorporated into other minerals as an admixture. The following concentrations of this metal have been detected under a microprobe: 2.46 wt % in pyrite, 1.1–2.0 wt % in nickeline, 3.89 wt % in arsenohauchercornite, 3.56 wt % in millerite, and 0.81 wt % in maucherite.

The major ore-forming minerals are depleted in Co. Only single chalcopyrite and pyrrhotite grains associated with pentlandite are characterized by elevated Co (0.30 and 0.12 wt %, respectively). In general case, the Co contents in major minerals are a hundredth fraction of a percentile. It has been established that only pyrite is the richest in this respect (up to 0.05 wt%), but it occurs only in the layered titanomagnetite ore and at the Kontaktovyi deposit (where it contains 0.009 wt% Co). It was established the same tendency that was found for pentlandite: Co concentration decreases from the inner to outer zone of the pluton.

Nickel is contained in all types of sulfide mineralization known from the Chiney pluton, revealing correlation with copper. However, Ni concentrations are insignificant. Like cobalt, nickel occurs as Ni mineral phases, and it is dispersed in major ore-forming and rare sulfides. Mineralogy of nickel is much more diverse than that of cobalt. Pentlandite, gersdorffite, violarite, and millerite are major Ni minerals.

As it was noted (Anonymous 2007), pentlandite is a main Ni mineral in many deposits. Firstly, pentlandite is a source of nickel in the Noril'sk deposits (Genkin et al. 1981; Distler et al. 1988; Krivolutskaya et al. 2011). It was described in massive ore related to ultrabasic layered mafic-ultramafic complex as well (Melekestseva 2006). The variations of pentlandite composition in the pluton were considered above (Table 8.6). Here, we have to emphasize once again that variations in Ni contents are appreciable in pentlandites (from 23.45 to 38.81 wt %). The inverse Ni-Co correlation is noted for this mineral: the lowest-Ni pentlandite (23.75 wt % Ni) from the finely layered titanomagnetite ore is enriched in Co (up to 18 wt %). In the endo- and exocontact ore, the Ni content in pentlandite increases along with the drop of Co content. This relationship is somewhat distorted by the behavior of Fe, although the range of Fe contents is much narrower (from 22.33 to 30.49 wt %), except for one analysis, where the Fe content abruptly falls to 17.94 wt %.

Minerals of the cobaltite-gersdorffite series, enriched in Ni up to 22.07 wt %, occur in the vein and disseminated exocontact ore (Table 8.8). Among minerals of the linnaeite group, violarite and siegenite are Ni-rich (36.90 and 21.0–24.7 wt % Ni, respectively). The minerals close to greigite in composition contain only 12.93–13.19 wt % Ni.

Rare Ni minerals (arsenohauchercornite, maucherite, heazlewoodite, nickeline) occur in the exocontact zone of the Chiney pluton. Nickel is detected as an admixture (>1 wt %) less frequently than cobalt. We succeeded in detecting Ni only in bornite (up to 2.43 wt %) and mackinawite (7%).

The major ore minerals generally contain low nickel concentrations comparable with those of cobalt. In single pyrrhotite grains coming into contact with pentlandite, the Ni content reaches 1.42 wt %, while chalcopyrite contains 0.43 wt % Ni. In general case, the Ni concentrations correspond to a n*0.01–0.001% fraction. The major ore minerals are arranged based on Ni contents in the following order (from higher to lower): pyrite-pyrrhotite-chalcopyrite-titanomagnetite. The same minerals, pertaining to various types of ore, are characterized by approximately equal nickel content in contrast to cobalt, the concentrations of which decrease beginning from the first type.

The selvages of veins and lenses, consisting of massive ore in the exocontact zone, are frequently enriched in Ni or consist of millerite. According to the results of laser analysis, the Ni content in chalcopyrite increases from the center to the margin of the orebodies (from 0.008 to 0.08 wt %). In other words, there is a cryptic mineralogical and geochemical zoning in the lenses of massive ore.

8.6 Platinum Group Elements, Au and Ag

The first data on platinum group elements (PGE) were obtained in the 1980s (Mel'nikova 1981). In general, they are related to copper ore. Correlation of PGE with Cu and Ag is noted; Pd/Pt ratio is equal to ~2.5 (Truneva et al. 1979). Data on PGE distribution were obtained in the 1990s (Gongalsky and Krivolutskaya 1994, 1999; Tatarinov et al. 1998; Gongalsky 2011, 2015). In this subsection, we report the data for deposits of the Chiney pluton and layered mafic-ultramafic complexes of the Chiney Complex (Table 8.12). The Pd/Pt ratio in ore and mineralized zones is widely variable (more than three orders of magnitude). In most samples, where concentrations of Pd and Pt are low (Tables 8.1, 8.2, 8.3, and 8.12), this ratio is close to 1. As far as Pd concentration increases, the Pd/Pt ratio grows up to 30–110. At a depth of 985–995 m in Borehole 11, a sharp prevalence of Pt over Pd (Pd/Pt = 0.01–0.03) was established in titanomagnetite leucogabbro with sulfide disseminations.

8.6.1 Geochemistry of PGE in Rocks

The available data show that rocks of the Chiney pluton are enriched in platinum, the concentrations of which are generally above the global mean value for mafic rocks (Gongalsky and Krivolutskaya 1994, 1999; Gongalsky 2011). Of 723 samples analyzed with fire assay, Pt has been found in 457 samples (60%), whereas Pd and Au were detected only in 274 (38%)

Table 8.12 Pt, Pd, and Au contents in rocks and ores at the Chiney pluton, ppm

No	Pd	Pt	Au	No	Pd	Pt	Au
1	0.060	0.010	–	41	0.100	0.100	–
2	0.070	0.030	–	42	12.260	0.480	0.420
3	0.030	0.010	–	43	14.760	4.710	0.720
4	0.040	0.070	–	44	0.950	0.041	0.140
5	0.010	0.040	–	45	1.360	0.199	0.190
6	0.020	0.010	–	46	0.810	0.041	0.130
7	0.020	0.030	–	47	0.990	0.043	0.070
8	0.020	0.020	–	48	2.840	0.053	0.030
9	0.070	0.120	–	49	2.010	0.270	0.079
10	0.040	0.040	–	50	7.400	0.220	–
11	0.050	0.150	0.250	51	0.740	0.160	–
12	0.200	0.180	0.210	52	1.710	0.072	0.350
13	0.162	0.041	0.020	53	2.750	1.350	–
14	2.050	0.041	0.470	54	0.360	0.300	–
15	1.750	0.048	0.380	55	0.110	0.190	–
16	4.510	0.041	0.500	56	0.030	0.150	–
17	3.510	0.099	0.210	57	0.070	0.120	–
18	0.300	0.090	–	58	0.130	0.210	–
19	0.500	0.130	–	59	0.050	0.060	–
20	0.400	0.030	–	60	0.110	0.060	–
21	0.090	0.010	–	61	0.060	0.070	–
22	0.100	0.070	–	62	0.030	0.080	–
23	0.160	0.240	–	63	0.070	0.050	–
24	1.500	0.500	–	64	0.910	2.000	0.060
25	0.014	0.075	0.021	65	1.220	5.300	0.200
26	0.013	0.766	0.058	66	6.200	2.200	0.230
27	0.037	1.186	0.110	67	1.450	0.200	0.120
28	0.064	0.175	0.023	68	0.950	0.100	0.360
29	0.120	0.060	0.030	69	1.100	0.090	0.220
30	0.650	0.230	0.110	70	0.018	0.005	0.040
31	0.750	0.600	0.125	71	0.600	0.040	0.110
32	1.600	0.950	0.135	72	0.480	0.040	0.360
33	2.430	0.055	0.053	73	0.650	0.005	0.140
34	2.440	0.075	0.043	74	0.025	0.010	0.200
35	0.021	0.041	0.113	75	0.380	0.010	0.090
36	0.027	0.042	0.500	76	0.014	0.015	0.140
37	0.550	0.170	–	77	0.009	0.003	0.115
38	0.369	0.051	0.021	78	0.008	0.003	0.130
39	0.800	0.110	–	79	0.008	0.000	–
40	0.550	0.910	0.046	80	0.009	0.060	–

Notes. Analyses were carried out using the chemical-spectral method at IGEM RAS, analysts GE Belousov and VA Sychkova; Nos. 83–88 at Institute of Mineralogy, Geochemistry, and Ore Formation, Ukraine National Academy of Sciences with fire assay with spectral analytical ending, analyst AA Yushin. Deposits and prospects: (1–21) Kontaktovyi, (22–29) Etyrko, (30–32) Skvoznoe, (33–36) Magnitnyi, (37–53) Rudnoe, (54–63) Luktur, (64–65) Mylovskiy, (66–69) Bazaltovyi, (70–78) Pravoingamakitskoe, (79–80) Udokan. Dash denotes—not analyzed

and 207 (28%) analyses, respectively. The Pt/Pd ratio is above 1 for 124 samples and below 1 for 98 samples.

A rather distinct difference in PGE contents is noted between high-Ti and low-Ti gabbro rocks. The latter group is practically free of PGE that were revealed only in 49 of 152 samples, whereas Pt is detected more frequently and prevails over Pd. In the high-Ti gabbro rocks, Pt/Pd vary in interval 1.5–2.0, although in certain cases this ratio reaches 20. However, Pt and Pd are distributed nonuniformly even in this group.

The mean Pt/Pd ratio of titanomagnetite gabbro is equal to those of leucogabbro. However, abrupt fluctuations are

also noted: the 350–391 m interval is characterized by the 3–10 times prevalence of Pt over Pd, standing out against the background of Pt/Pd ratio of 1–2. This interval coincides with the central part of the titanomagnetite gabbro unit, occupied by low- and medium-grade titanomagnetite ore. At the base of the titanomagnetite gabbro unit, Pd dominates over Pt (Pt/Pd = 0.6), and this reflects the appearance of sulfide mineralization near the pluton's bottom. The macrorhythmicity in the high-Ti gabbro rocks is emphasized by variations in major oxides. Five members (1 T–5 T) were clearly distinguished in the titanomagnetite gabbro unit, penetrated by the deep (1352 m) Borehole 11 and in the reference series of Borehole 83. This tendency is not always seen in the field, being obliterated by fine rhythmicity.

The comparative analysis of PGE concentrations in members of the titanomagnetite gabbro unit shows that in members 1 T and 4 T (Fig. 5.16), Pt and Pd have been detected in 95% of samples (in member 2 T, in 80%). The Pt and Pd are distributed rather uniformly throughout the section. Members 3 T and 5 T contain Pt and Pd in smaller amounts. In the lower part of member 3 T, Pd has been detected in all samples, whereas the upper part is practically free of Pt. In member 5 T, Pt is contained in 100% of samples in its upper third and completely disappears below.

In order to provide insights into PGE distribution within the particular microrhythms, 43 samples have been collected over a distance of 1–2 cm from one another and analyzed using the semiquantitative scintillation technique at the Institute of Geochemistry, Siberian Branch of RAS (analyst SI Prokopchuk). The analysis brought to light two tendencies: (a) significant Pt and Pd concentrations (0.04–0.05 and 0.1–0.3 ppm, respectively) have been detected in the most samples, with their traces detected in leucocratic varieties with visible sulfide inclusions in the upper parts of the rhythms; and (b) a slight positive correlation between Pd and titanomagnetite contents was established in the lower parts of microrhythms.

At the same time, the detailed examination of core from Borehole 11 revealed a 7.5 m interval of rocks (bottom of member 2 M, Fig. 5.18), markedly enriched in Pt. The rocks of this interval are surprisingly constant in composition. They are represented by medium-grained chineyite (plagioclase-titanomagnetite rock with sporadic pyroxene). Sulfides occupy 3–5% of rock volume. The unit is hosted in pyroxenite and melanocratic norite. The abruptly elevated PGE contents (ppm), up to 1.19 Pt, 0.64 Pd, and 0.01 Rh and up to 0.1 Au in 14 samples analyzed with the chemical-spectral method (Institute of Mineralogy, Geochemistry and Crystal Chemistry of Rare Elements (IMGRE), analyst IS Razina), attracted our special attention to this interval. It is quite probable that low-sulfide PGE-bearing units can be discovered in the Chiney pluton by means of detailed sampling.

The bulk chemical composition and PGE concentrations of rocks indicate independent behavior of the latter; and no correlation with any rock-forming oxides has been established. The same pattern is observed with respect to the low grade of Cu, Co, and Ni. Only the increase in Cu content up to 0.2–0.4 wt % and higher is accompanied by the abrupt enrichment in Pd and to a lesser degree in Pt.

The sulfide-free altered gabbro rocks are almost completely devoid of PGE. These are amphibolized, chloritized, and cataclastic rocks with late hydrothermal veinlets, as well as silicified leucogabbro. The sulfide disseminations in the leopard gabbro and leucogabbro are concentrated in the western and partly central parts of the pluton. This type of

mineralization is characterized by weak vertical variability, which is reflected in the distribution of noble metals within ore zones. The Pt and Pd concentrations change insignificantly. Their mean contents are 0.02 and 0.90 ppm, respectively, and Pd/Pt = 45 remains almost unchanged. Michenerite and merenskyite have been identified here.

Another zone of this type in the central part of the pluton is closer in the mean PGE contents (0.35 ppm Pt and 0.95 ppm Pd). No PGE mineral species are so far known in this zone. To provide insights into the behavior of precious metals in the processing of ore, we have analyzed separate fractions of two samples of this ore type taken from boreholes 9 and 12. In the first sample, all PGE contents consecutively increase in the course of concentration: 0.14 ppm Rh and 0.01 ppm Ru have been detected in this bulk sample, with up to 0.10 ppm Ir in the heavy electromagnetic fraction. The second sample is enriched only in Rh (from 0.007 to 0.01 ppm), whereas Pt and Pd are dispersed. The variable behavior of PGE can be apparently explained by specific modes of their occurrence in the samples, taken from different parts of the lode. One can suggest that in the first case, the Pt and Pd minerals prevailed and then were concentrated in the heavy fractions, whereas in the second case, Pt and Pd occurred either as isomorphic admixtures in the major ore-forming sulfides or were dispersed in form of tiny inclusions, which were removed into tailings.

8.6.2 PGE in Ore

The Pd/Pt value in the sulfide phase is much higher than 1 and reaches 100, i.e., markedly exceeds this ratio in the endocontact ore. In addition, higher Ag concentrations, up to 160 ppm, are noted; Ag/Au ratio reaches 160. In general, precious metals in the exocontact ore are distributed extremely nonuniformly; the grade in the disseminated ore is lower than in massive ore. The correlations between metal contents differ from those in the endocontact zones, particularly for the ore with variable Cu concentrations. In contrast to the endocontact ore, PGE is negatively correlated with Cu (Tolstykh et al. 2004). The low Cu concentrations in the sulfide phase are positively correlated with Au + Ag, but an inverse relationship is noted for high concentrations. PGE contents are commonly correlated with Cu; therefore, Pd sharply prevails over Pt (Pd/Pt = 10–100). Other relationships are also noted between nonferrous and precious metals. The anomalous precious metal concentrations (15 ppm Pt, 124 ppm Pd, 14 ppm Au, and 345 ppm Ag) in the isometric orebodies do not correlate with copper (Tolstykh et al. 2004). Cubanite, millerite, and bornite occur in the exocontact ore in significant amounts in addition to the predominant chalcopyrite. Sperrylite crystals (Fig. 8.24, Table 8.13) have been recovered from a heavy fraction.

Fig. 8.24 Sperrylite crystals
(photo V Podgorbunskiy)

Table 8.13   Chemical composition of sperrylite at the Rudnoe deposit, wt %

No	As	S	Fe	Se	Rh	Co	Ni	Pd	Pt	Te	Sb	Total
1	42.115	0.057	0.038	0.525	0.398	0.057	0.196	0.016	55.657	0.007	–	99.09
2	42.109	0.068	0.031	0.515	0.6	0.076	0.484	0.02	55.048		0.065	99.02
3	41.857	0.053	0.025	0.544	0.293	0.069	0.157	0.139	56.283	0.014	–	99.44
4	41.592	0.054	0.023	0.49	0.743	0.076	0.31	–	55.135	–	0.049	98.47
5	42.016	0.034	0.015	0.535	0.422	0.05	0.227	–	55.756	0.001	–	99.11
6	41.942	0.074	0.045	0.478	0.37	0.041	0.088	0.05	55.393	0.016	0.024	98.52
7	42.533	0.049	0.035	0.527	0.521	0.065	0.236	–	55.983	–	–	99.96
8	42.012	0.027	0.016	0.485	0.4	0.042	0.117	–	56.054	–	–	99.17
9	41.644	0.071	0.029	0.547	0.483	0.09	0.16	–	56.038	0.005	0.067	99.20
10	41.892	0.073	0.007	0.515	0.465	0.039	0.129	0.076	55.772	–	0.025	98.99
11	42.322	0.065	0.038	0.534	0.482	0.034	0.296	0.137	55.785	–	0.022	99.72
12	42.42	0.054	0.056	0.483	0.332	0.061	0.172	0.273	55.637	0.003	–	99.56
13	41.834	0.072	0.027	0.546	0.513	0.061	0.252	–	55.966	0.03	–	99.36
14	41.953	0.057	0.003	0.522	0.597	0.042	0.308	0.022	55.556	0.004	–	99.07
15	42.091	0.032	0.012	0.386	0.581	0.058	0.256	0.008	56.209	–	0.023	99.68

Note. Dash denotes concentration below detection limit. Analyses were carried out at IGEM RAS, analyst EV Kovalchuk

8.6.2.1 PGE in the Endocontact Ore

The PGE distribution was studied at the *Rudnoe* deposit in detail. The mineral varieties of the endocontact ore are distinguished from one another by PGE contents; the maximum values are typical for the chalcopyrite variety. If the low-grade chalcopyrite disseminations hosted in diorite are characterized by low Pt content (0.041 ppm) and moderate Pd concentrations (0.997 ppm), i.e., Pd/Pt = 23, then impregnated chalcopyrite ore differs in high Pd content (up to 4.5 ppm), and Pd/Pt ratio reaches a maximum of 110 (55, on average). The Pt contents range from 0.041 to 0.099 ppm (0.07 ppm, on average), whereas the Pd contents vary from 2.43 to 4.51 ppm (3.22 ppm, on average). Such minerals as merenskyite, michenerite, potarite, polarite, and mayakite have been described for this type of mineralization. Small grains of these minerals are clustered mainly in chalcopyrite. Other mineralogical varieties, especially pyrrhotite ore, are distinguished by lower contents of both Pt and Pd.

In the endocontact disseminated ore at the *Etyrko* deposit, the contents of noble metals are generally not high: 0.08–0.20 ppm Pd, up to 0.08 ppm Pt, and up to 2 ppm Ag. All core samples within the interval of 174.4–226.9 m in Borehole 9 at the Kontaktovyi deposit yielded Pd contents above 1 ppm (Table 3.11) (data of Central Chemical Laboratory, Chitageologiya).

8.6.2.2 PGE in Exocontact Ore

As it was described above, the exo-contact ore of the *Rudnoe* deposit comprises two types of mineralization in sandstone: (i) disseminated ore and (ii) massive and breccia ore.

The *disseminated ore* in sandstone is also characterized by the nonuniform distribution of valuable components and high PGE concentrations, albeit to a somewhat lesser extent than in the preceding ore type. The mean Pd and Pt contents are 1.11 and 0.25 ppm, respectively. The chalcopyrite-bornite and substantially pyrrhotite ore varieties differ from each other in PGE geochemistry. The first variety is distinguished by higher PGE contents (2.01–2.84 ppm Pd and 0.053–0.27 ppm Pt) in comparison with the second one (0.10–0.55 ppm Pd and 0.0051–0.91 ppm Pt); the mean Pd/Pt = 30 and 4, respectively. The second ore variety therefore resembles ore disseminations hosted in monzodiorite of the endocontact zone. The elevated heavy REE contents can be expected here. Moreover, 0.08–0.14 ppm Rh and 0.05 ppm Ru have already been detected in two samples.

This type of mineralization contains Te, Se, Sb, and PGE minerals, in particular, sudburyite, froodite, sperrylite, and cupriferous platinum. The Ag content reaches 62.5 ppm and prevails over Au in all samples analyzed by Tolstykh et al. (2008, Tables 1, 2 therein). The lowest Ag/Au ratios are related to the layered gabbro that occurs far from the contact, whereas the highest Ag/Au ratios, reaching 258, are noted in massive sulfide ore from the exocontact zone.

The sulfide disseminations become denser toward the contact. However, concentration of precious metals, recalculated to 100% sulfide phase, remains comparable and insignificantly differs in endo- and exocontact ore (Tolstykh et al. 2008). Au and Rh dominate in the sulfide phase of endocontact ore in contrast to the dominance of Pd and Ag in the sulfide phase of exocontact ore. A peak of Pt content is reached in gabbro rocks, but the mean content in the sulfide phase of both ore types remains unchanged. The total Au + Ag positively correlates with the total Pt + Pd in the endocontact ore.

Veins and lenses of massive ore contain the highest PGE concentrations in comparison with all other ore types (up to 250–350 ppm). This is especially evident for palladium, the contents of which in anomalous samples reach hundreds of ppm. The mean Pd and Pt concentrations in samples from vein ore are 4.9 and 0.74 ppm, respectively.

This type of mineralization is characterized by the greatest variability and extremely nonuniform distribution of valuable components, in particular, PGE. The samples with anomalously high PGE concentrations are combined with almost barren samples. The PGE distribution in particular orebodies so far remains unknown. Comparison of chalcopyrite and millerite-chalcopyrite veins shows that the former veins are enriched in Pt and Pd. In the chalcopyrite veins, if the Pd concentrations range from 0.74 to 14.76 ppm (8.79 ppm, on average), and Pt concentrations vary from 0.16 to 1.71 ppm (1.14 ppm, on average), then in millerite-chalcopyrite veins, the Pd concentrations range from 0.25 to 1.36 ppm (0.87 ppm, on average), and Pt concentrations reach 1.199 ppm (0.045 ppm, on average). The latter variety undoubtedly needs additional in-depth study from the viewpoint of PGE geochemistry and the mode of PGE occurrence, because rare arsenides and sulfoarsenides, as well as gersdorffite occur here.

Table 8.14 Minerals of platinum group elements at the Chiney pluton

Major	Rare
Sperrylite PtAs2	Froodite $PdBi_2$
	Kotulskite PdTe
	Atokite $(Pd,Pt)_3Sn$
	Sudburyite PdSb
Michenerite PdBiTe	Hessite Ag_2Te
Sobolevskite PdBi	Hollingworthite RhAsS
	Majakite PdNiAs
	Menshikovite $Pd_3Ni_2As_3$
	Palladoarsenide Pd_2As
	Naldrettite Pd_2Sb
	Cooperite PtS
	Paolovite Pd_2Sn
	Stibiopalladinite Pd_5Sb_2
	Pt-Fe alloy
	Isomertieite $Pd_{11}Sb_2As_2$
	Mertieite II Pd_8Sb_3
	Pd-Sb and Pd-bi oxides
	Au-ag-hg alloy

8.6.3 Precious Metals Minerals in Ore

Two different types of noble-metal mineralization are combined in the Chiney pluton: (1) Pd-Pt-Ir (Au) mineralization in the Fe-Ti ore and (2) Pd-Pt-Rh-Au (Os, Ir, Ru, Ag) intermetallic-chalcogenic mineralization in the Ni-Co-Cu ore (Tolstykh et al. 2000). We provide information on the latter association below. They belong to the Pd-Bi-Te-Sb, Pd-As-Sb, Pd-Ni-As-Pt-Fe, and Au-Ag-Hg systems (alloys). The list of PGE minerals is given in Table 8.14.

Our data on PGE minerals (Table 8.15) are based on the study of polished sections under a JEOL electron microprobe at the Institute for Geology of Ore Deposits, Petrography, Mineralogy and Geochemistry, RAS (analyst EN Kovalchuk; voltage of 20 kV, current of 20–30 mA, and 30-second measurement on each analytical line).

Many PGE minerals have been identified by Tolstykh et al. (2004, 2008) in the polished sections, prepared for the grains from heavy fraction, using scanning electron microscope (analyst SV Letov). The chemical compositions of PGE minerals have been determined at the Institute of Geology and Mineralogy, Siberian Branch, Russian Academy of Sciences (Novosibirsk) on a Camebax-Micro microprobe, operating at accelerating voltage of 20 kV, current of 20–30 mA, and 10-second measurement on each analytical line (Table 8.16, analyst LN Pospelova).

Among the Chiney ore minerals, the Pd-Bi-Te-Sb group is the most abundant. These are froodite, sobolevskite, sudburyite, kotulskite, michenerite, stibiopalladinite, mertieite II, naldrettite, and isomertieite. The Pd(Bi,Te,Sb) solid solutions, with sobolevskite PdBi, kotulskite PdTe, and sudburyite PdSb as end members, occur in both exo- and endocontact

Table 8.15 Composition of PGE minerals from endocontact ore at the Rudnoe deposit, wt%

No	N point	Pt	Pd	Sb	Te	Bi	Fe	As	Total		
1	1		24.07	3.36	16.47	32.79	3.55		98.65	Michenerite ((PdBi(SbTe))	PdBi(SbTe)
2	2		37.35	4.99	15.81	40.83	0.86		99.84	Michenerite ((PdBi(SbTe))	
3	3		37.03	5.63	14.89	41.02	1.68		100.25	Michenerite ((PdBi(SbTe))	
4	3a		35.98	4.99	15.67	41.24	1.88		99.76	Michenerite ((PdBi(SbTe))	
5	4	52.08					2.19	43.76	99.19	Sperrylite (PtAs$_2$)	
6	5	48.72					1.58	42.19	99.51	Sperrylite (PtAs$_2$)	
7	6	51					3.63	43.51	99.63	Sperrylite (PtAs$_2$)	
8	7		25.6		48.98	25.42			100	Michenerite ((PdBi(SbTe))	
9	8		35.52	5.62	18.48	35.59			98.65	Michenerite ((PdBi(SbTe))	

Note. Analyses were carried out at IGEM, analyst E.V. Kovalchuk

Table 8.16 Chemical composition of Pd-bearing minerals at the Chiney pluton, wt%

N	Pd	Ni	Bi	Te	Sb	As	Sn	Total	Formula
1	20.09	0.00	78.07	0.00	0.00	0.00	0.00	98.16	$Pd_{1.00}Bi_{2.00}$
2	34.22	0.00	65.13	0.00	0.23	0.00	0.00	99.58	$Pd_{1.01}Bi_{0.98}Sb_{0.01}$
3	37.81	0.00	40.37	0.63	19.05	0.00	0.00	97.86	$Pd_{1.00}(Bi_{0.54}Sb_{0.44}Te_{0.01})_{1.00}$
4	37.34	0.00	36.30	13.13	11.19	0.00	0.00	97.96	$Pd_{0.98}(Bi_{0.48}Te_{0.29}Sb_{0.26})_{1.03}$
5	38.78	0.00	40.56	20.31	0.00	0.00	0.00	99.65	$Pd_{1.01}(Bi_{0.55}Te_{0.44})_{0.99}$
6	40.22	0.00	33.57	25.90	0.00	0.07	0.00	99.76	$Pd_{1.02}(Te_{0.55}Bi_{0.43})_{0.98}$
7	46.46	0.00	0.43	0.00	51.22	0.12	0.00	98.23	$Pd_{1.01}Sb_{0.99}$
8	39.86	0.00	31.42	0.00	27.13	0.00	0.00	98.41	$Pd_{1.00}(Sb_{0.60}Bi_{0.40})$
9	36.94	6.92	0.00	0.32	54.17	0.00	0.00	98.35	$(Pd_{0.76}Ni_{0.26})_{1.02}(Sb_{0.97}Te_{0.01})_{0.98}$
10	68.74	0.00	0.00	0.00	30.25	0.34	0.00	99.33	$Pd_{5.03}(Sb_{1.93}As_{0.03})_{1.96}$
11	70.79	0.00	0.00	0.00	26.10	2.79	0.00	99.68	$Pd_{7.97}(Sb_{2.57}As_{0.46})_{3.03}$
12	74.19	0.00	0.00	0.00	15.02	9.09	0.16	98.46	$Pd_{11.09}(Sb_{1.96}Sn_{0.02})_{1.98}As_{1.93}$
13	64.10	0.00	0.00	0.00	2.21	0.08	34.09	100.48	$Pd_{1.99}(Sn_{0.95}Sb_{0.06})_{1.01}$
14	68.20	0.00	0.00	0.00	0.98	7.36	25.04	101.58	$Pd_{2.00}(Sn_{0.66}As_{0.31}Sb_{0.03})_{1.00}$
15	65.85	0.00	0.00	0.00	14.79	0.12	19.03	99.79	$Pd_{2.06}(Sn_{0.53}Sb_{0.40}As_{0.01})_{0.94}$
16	23.00	0.00	46.62	28.66	0.00	0.00	0.00	98.28	$Pd_{0.98}Bi_{1.01}Te_{1.01}$
17	68.23	0.00	4.36	0.10	6.39	18.59	0.00	97.67	$Pd_{2.00}(As_{0.77}Sb_{0.16}Bi_{0.06})$
18	74.41	0.00	0.00	0.21	1.14	25.41	0.25	101.42	$Pd_{1.98}(As_{0.51}Sn_{0.41}Te_{0.09})_{1.01}$
19	25.32	1.29	12.60	58.44	1.03	0.04	0.00	98.72	$(Pd_{0.91}Ni_{0.08})_{0.99}(Te_{1.75}Bi_{0.23}Sb_{0.03})_{2.01}$
20	30.13	1.35	8.27	47.50	12.69	0.06	0.00	100.00	$(Pd_{1.03}Ni_{0.08})_{1.11}(Te_{1.36}Sb_{0.38}Bi_{0.14})_{1.88}$
21	46.14	24.13	0.00	0.00	0.48	30.71	0.00	101.46	$Pd_{1.03}Ni_{0.98}(As_{0.97}Sb_{0.01})_{0.98}$
22	48.70	17.80	0.00	0.00	0.00	33.37	0.00	99.87	$Pd_{3.04}Ni_{2.01}As_{2.94}$
23	62.85	0.00	7.41	4.38	25.65	0.00	0.41	100.7	$Pd_{2.03}(Sb_{0.73}Bi_{0.12}Te_{0.12})_{0.97}$
24	65.89	0.00	0.00	0.16	18.23	0.00	16.16	100.44	$Pd_{2.05}(Sb_{0.50}Sn_{0.45})_{0.95}$
25	76.81	0.00	1.28	0.19	0.22	16.40	2.84	98.15*	$(Pd_{2.96}Pt_{0.01})_{2.97}(As_{0.90}Sn_{0.10}Bi_{0.03}Sb_{0.01}Te_{0.01})_{1.05}$
26	63.58	0.00	27.46	9.47	0.12	0.04	0.00	100.67	$Pd_{2.97}(Bi_{0.65}Te_{0.37})_{1.02}$
27	76.60	0.00	0.78	0.35	0.00	18.37	2.73	98.83	$Pd_{5.06}(As_{1.72}Sn_{0.16}Bi_{0.03}Te_{0.02})_{1.93}$
28	70.89	0.00	0.00	0.00	25.38	3.06	0.00	99.33	$Pd_{5.09}(Sb_{1.59}As_{0.31})_{1.90}$
29	73.57	0.00	0.00	0.87	13.22	9.89	1.37	98.92	$Pd_{5.09}(As_{0.97}Sb_{0.80}Sn_{0.08}Te_{0.05})_{1.9}$

Note

1 froodite; 2–5 sobolevskite; 6 kotulskite; 7–9 sudburyite; 10 stibiopalladinite; 11 mertieite II; 12 isomertieite; 13–15 paolovite; 16 michenerite; 17, 18 arsenopalladinite; 19, 20 merenskyite; 21 majakite; 22 menshikovite; 23, 24 naldretteite; 25–29 unnamed phases; *in total included 0.42 Pt. After (Tolstykh et al. (2008))

ore of the Chiney pluton. They were discovered in the endocontact ore at the Rudnoe deposit (Fig. 8.25). The minerals form small grains (1–2 – 50μk, 20 μk on average).

Michenerite PdBiTe is the most abundant mineral in the endocontact ore, although it occurs in all rock types. Its grains were found as small grains in pyrrhotite (Fig. 8.25a, b), in association with chalcopyrite in silicate matrix

(Fig. 8.25c) and as very tiny inclusions in pentlandite (Fig. 8.25 h, j). It contains Sb and Fe in addition to major components (Table 8.15).

Sobolevskite PdBi, with insignificant Sb admixture, has been found in the exocontact ore as intergrowths with froodite PdBi$_2$, Ag-Au alloys, and mertieite II (Pd$_6$Sb$_3$) inclusions. Sb- sobolevskite (13–21 wt % Sb) is intergrown with

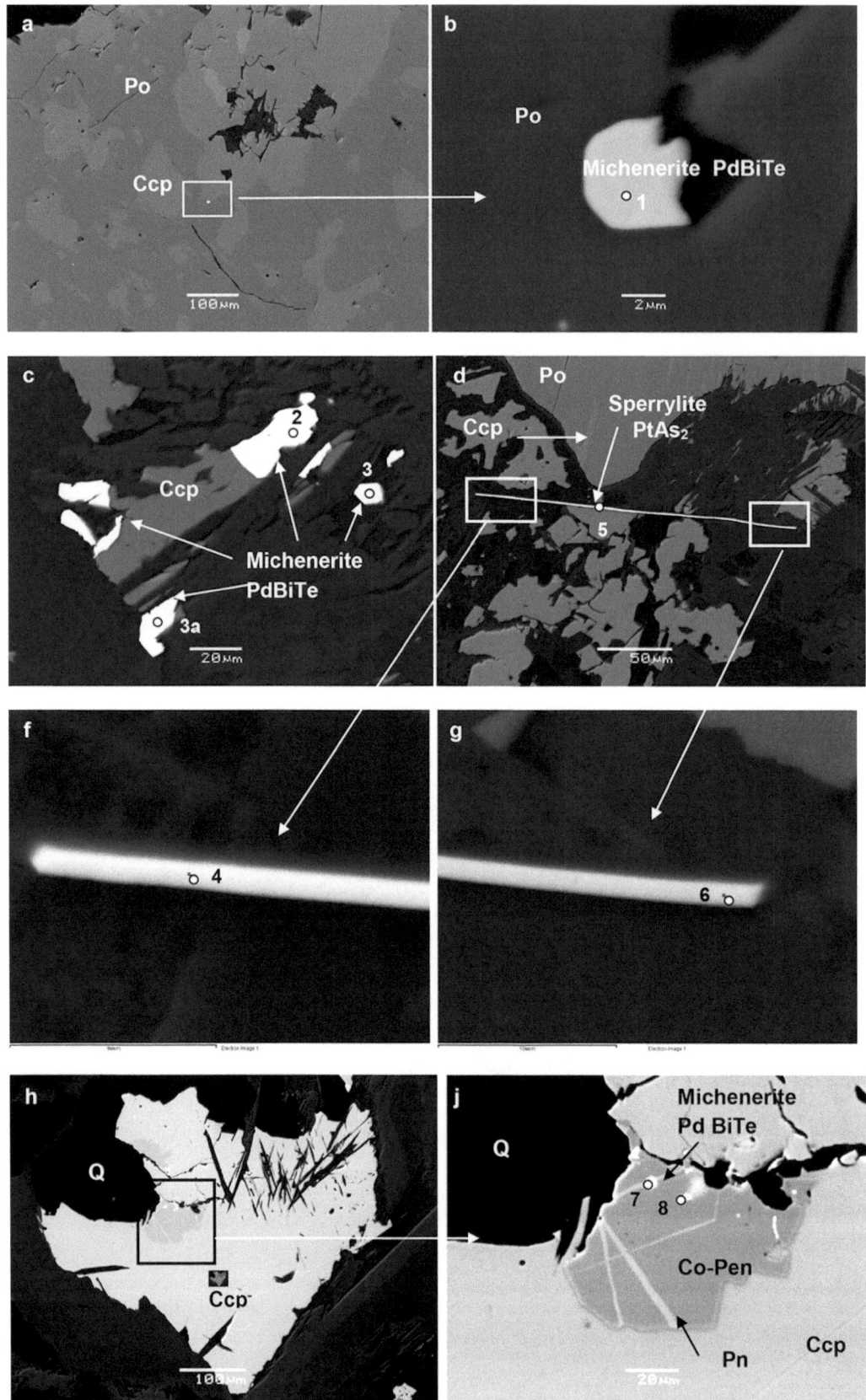

Fig. 8.25 BSE images (IGEM, analyst EV Kovalchuk) of platinum group element minerals in the endocontact ore at the Rudnoe deposit (**a**) – Michenerite in pyrrhotite; (**b**) – detail of Fig. 8.25a; (**c**) – michenerite in association with chalcopyrite; (**d**) – needle crystal of sperrylite, general view; (**f, g**) – details of Fig. 8.25d; (**h**) – co-pentlandite with michenerite, general view; (**j**) – detail of Fig. 8.25h

Fig. 8.26 BSE images (IGEM, analyst EV Kovalchuk) of platinum group element minerals (unnamed phase Pd (Te, Bi)$_2$) in silicate minerals of gabbro-diorite, Rudnoe deposit (**a**) – general view, (**b**) – Pd(Te,Bi)$_2$ in pyrrhotite.

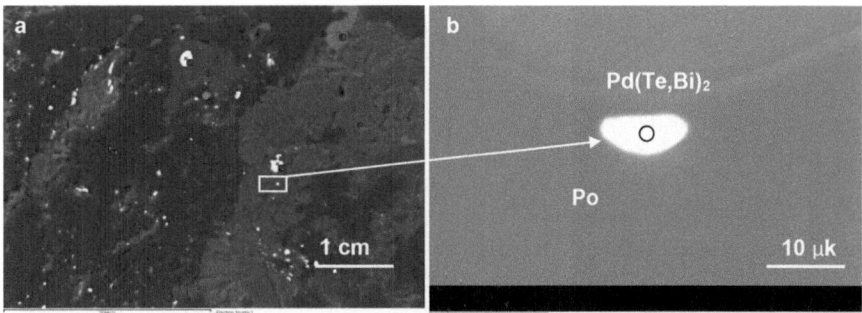

Sb-bearing paolovite (Tolstykh et al. 2000). The Te-bearing sobolevskite has been revealed only in the endocontact portion of the lode.

Kotulskite has been identified only in the endocontact zone. The Bi-bearing kotulskite is intergrown with palladoarsenide, paolovite, and michenerite (Table 8.16). Sb admixture does not exceed a tenth fraction of a percentile.

Froodite PdBi$_2$ occurs only in the exocontact massive chalcopyrite ore, where it is the leading palladium mineral. Froodite micro inclusions in sulfoarsenides of the cobaltite-gersdorffite series make up a stable assemblage.

Sudburyite PdSb occurs in massive and disseminated sulfide ore. In addition to end members of the isomorphic series, where Bi, Te, and As contents do not exceed a tenth fraction of a percentile, the Bi-sudburyite Pd (Sb,Bi) with up to 31.42 wt% Bi and Ni-sudburyite, with up to 10.1 wt% Ni, occur like in other deposits, where the Ni content in sudburyite reaches 10 wt% (Cabri and Laflamme 1974; Beaudoin 1990; Gervilla et al. 1998; Cabri 2002). A few grains of Te-bearing sudburyite are related to the sudburyite-kotulskite isomorphic series (Table 8.16).

Stibiopalladinite Pd$_2$Sb$_5$ is widespread in disseminated ore of the exocontact zone and also occurs in high-grade sulfide ore with arsenohauchercornite Ni$_9$BiAsS$_8$ and cobaltite.

Mertieite II Pd$_8$Sb$_3$ has been identified in sobolevskite in massive and disseminated ore of the exocontact zone.

Isomertieite Pd$_{11}$Sb$_2$As$_2$ has been identified in the microaggregate, consisting of sperrylite, hollingworthite, and menshikovite; intergranular space is filled with Rh-bearing cobaltite.

Naldrettite Pd$_2$Sb has been identified in a sample of oxidized ore as a euhedral inclusion in the matrix of partly corroded michenerite in association with stibiopalladinite.

Paolovite Pd$_2$Sn has been described in both endo- and exocontact ore. As-bearing paolovite corresponds to the formula Pd$_2$(Sn,As). These minerals are solid solutions of paolovite with antimony Pd$_2$(Sn,Sb) and arsenic Pd$_2$(Sn,As).

Palladoarsenide, including Sb- and As-bearing varieties, and Pd$_5$As$_2$-Pd$_5$Sb$_2$ are related to the Pd-As-Sb system.

Mayakite PdNiAs and *menshikovite* Pd$_3$Ni$_2$As, as two minerals of the Pd-Ni-As system, have been identified in the both endo- and exocontact ore.

The main Pt concentrations in gabbro rocks are sperrylite, Pt-Fe alloy (Pt$_{0.63}$Pd0.03Rh$_{0.01}$)$_{0.67}$(Fe$_{0.31}$Cu$_{0.02}$)$_{0.33}$, cooperite (Pt$_{0.87}$Pd$_{0.06}$)$_{0.93}$S$_{1.06}$, occasionally forming outer rims around platinum grains, and Pt-bearing atokite (Pd$_{2.23}$Pt$_{0.73}$)$_{2.96}$Sn$_{1.04}$.

Sperrylite PtAs$_2$ is a very often mineral. It was found as large crystals in Rudnoe Creek, the Rudnoe deposit (Fig. 8.24). It occurs as needle crystals (250 μk) in chalcopyrite-pyrrhotite aggregates with silicate minerals. It was formed the latest in comparison with sulfide minerals (Fig. 8.25d, h, g).

Pt-Fe alloys and *cooperate* PtS as rims around Pt-Fe alloys, as well as complete pseudomorphs or individual grains intergrown with chalcopyrite have been identified only among sulfide disseminations in gabbro and gabbronorite, far from the contact.

Au-Ag alloys are abundant in all ores of the Chiney pluton as individual grains and intergrowths with Pd minerals (Figs. 8.26, 8.27).

Hollingworthite RhAsS is a rare mineral that occurs as intergrowths with menshikovite, isomertieite, sperrylite, and cobaltite. Hollingworthite with Co admixture and Rh-bearing cobaltite make up the cobaltite-hollingworthite isomorphic series.

All Pd, Sb, and Bi oxides have been identified in the exocontact sulfide ore. Minerals of the Pd-Bi-Sb-O system have been described for the first time by Tolstykh et al. (2000, 2004).

8.7 Genesis of Ore

The origin of the Chiney sulfide mineralization is under discussion for several decades (Truneva et al. 1979; Truneva 1982; Konnikov 1986; Krivolutskaya 1986; Gongalsky and Krivolutskaya 1993; Tatarinov et al. 1998). The co-occurrence of two large copper deposits— Udokan and Chiney—arises a question of their relationships. Several models of ore origin have been proposed.

The first model regards the formation of copper ores of the Chiney pluton as a result of assimilation of Cu-sandstones by mafic magmas (Krendelev 1987; and partly Konnikov and Truneva 1982; Konnikov 1986). To the contrary, some geolo-

Fig. 8.27 Minerals of platinum group elements in ore of the Chiney pluton. (After Tolstykh et al. (2008)) (**a**) – Stibiopalladinite (Pd Sb), sudburyite-kotulskite-sobolevskite solid solution [Pd(Sb,Te,Bi)], and Te-bearing sobolevskite [Pd(Bi,Te)] included in cobaltite (CoAsS); (**b**) – intergrowth of sperrylite (PtAs2), michenerite (PdBiTe), sudburyite (PdSb), and hollingworthite (RhAsS) with cobaltite (CoAsS); (**c**) – stibiopalladinite (Pd Sb) is surrounded by a graded rim of CoAsS; (**d**) – tsumoite (BiTe) included in cobaltite (CoAsS) with maucherite (Ni As) grains; (**e**) – Bi-rich kotulskite and an Au-Ag alloy included in gersdorffite (NiAsS); (**f**) – inclusion of froodite (PdBi) in gersdorffite (NiAsS))

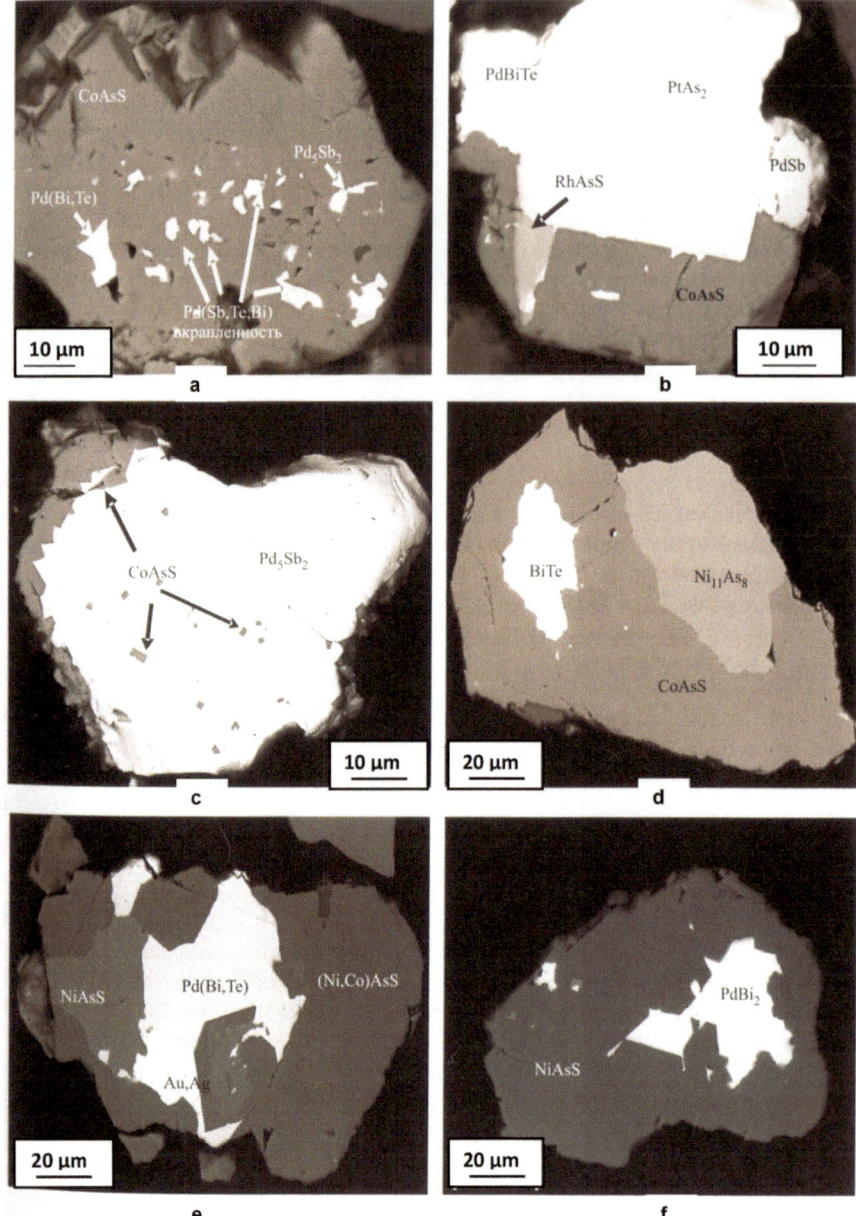

To answer this problem, sulfur isotopes were involved into discussion. Sulfur isotopic composition of minerals in sandstones shows a value, typical of almost all sandstone-hosted copper-ore occurrences ($\delta^{34}S$ = −16 to −27‰; Chukhrov 1969; Bogdanov and Golubchina 1971; Konnikov and Truneva 1982; Konnikov 1986). The sulfide ore of the Chiney layered mafic-ultramafic complex is characterized by $\delta^{34}S$ values ~0–4‰, which are similar to ores related to other mantle-derived Proterozoic intrusive complexes. It was concluded that these two types of Cu mineralization should have had different sulfur sources and different

gists (Volfson and Arkhangel'skaya 1987) proposed the sulfide appearance in sandstones due to hydrothermal Cu-rich fluids derived from the pluton.

origin. Nevertheless, isotope composition of sulfur can be changed dramatically in different geological processes. For example, if deep sulfur fluids from basic magmas penetrated through the crust and get into river basin with sulfate-reducing bacteria, the isotope composition of sulfide minerals will be different to magmatic rocks.

The next important question is coexistence of oxide and sulfide ores in one intrusive body. The Fe-Ti-V oxide precipitation seems to have been the main factor of the sulfide melt immiscibility in the Chiney magma chamber (during the middle and late stages of magma crystallization). Probably, the fO2 during the magma crystallization changed centrally and leads to reducing conditions favorable for the sulfide formation.

We demonstrated that there was a systematic change in high-temperature magmatic paragenetic sequences for sulfide minerals from the inner gabbroic part of the intrusion (where pyrrhotite dominated) toward the low-temperature paragenetic associations in the marginal monzodiorite zone and exocontact sandstones, where chalcopyrite was widespread (Gongalsky and Krivolutskaya 1993). The parameter of main types of mineralization was determined using mineral parageneses and composition, especially pentlandite and linnaeite group minerals. Tolstykh et al. (2008), Zhitova et al. (2004) came to the same conclusion based on distribution of PGE minerals. The distribution of PGE minerals depends on the thermal gradient, changing from the inner part of the intrusion to endo- and exocontact. In this direction, the size of PGE minerals decreases from 50–80 μm to 10 μm, and their composition changes as well.

Indeed, many authors suggest that a variation of mineralization from the central part of different plutons toward the endo- and exocontact zones is caused by a drop in temperature against the background of melt fractionation and enrichment in volatile components (Vortsepnev 1978; Gongalsky and Krivolutskaya 1993; Tolstykh et al. 2008).

8.8   Conclusions

1. There are five main deposits related to the Chiney pluton (from west to east): Kontaktovyi, Skvoznoe, Pravoingamakitskoe, Verkhne-Chineyskoe, and Rudnoe. The Kontaktovyi and Rudnoe are of the greatest interest.
2. The dominance of copper in ore, the subordinate role of Ni, Co (Cu:Ni:Co = 76:7:1), and precious metals is a distinguishing feature of sulfide mineralization in the Chiney pluton. The mean Cu/Ni ratio is 10, occasionally reaching 100Thatreflects prevalence of chalcopyrite among sulfides, whereas pentlandite is a subordinate and even a rare mineral. Pyrrhotite and pyrite are also major sulfides (Tables 8.4, 8.5). In addition to the aforementioned minerals, 86 other mineral species have been identified to date in the copper ore, including minerals of platinum group elements (see below), supplemented by sphalerite, pyrite, titanomagnetite, and ilmenite, with rare siegenite, galena, violarite, millerite, cobaltite, gersdorffite, safflorite, loellingite, argentopentlandite, michenerite, merenskyite, sperrylite, and other platinum group minerals.
3. Of greatest interest are those minerals that show variation in composition. These include, first of all, pentlandite ((Ni-rich (35–38 wt. %) and Ni-poor (30–33 wt. %)) and minerals of the linnaeite group. The distinguishing feature of pentlandites from the Chiney pluton is a relatively high content of Co. Different types of sulfide mineralization of pentlandite differ appreciably from one another in the narrow intervals of Co contents in pentlandite, inher-

ent to each type. The cobalt content in pentlandite is stepwise reduced from sulfide mineralization, located in the interior of the layered mafic-ultramafic complex to endocontact and then to its exocontact zones. The Co-richest pentlandites (8.20–18.38 wt % Co) were found in titanomagnetite ores of the Etyrko deposit (Table 8.6). Thus, this mineral can be referred to as high-Co pentlandite (Shishkin et al. 1974), while pentlandites in endo- and exocontact sulfide ore have 5–7 and 2–3% Co, respectively.

References

Anonymous (2007) Nickel and cobalt ores. Methodical recommendations for application of reserve classifications, vol 37. MinPrirod, Moscow (in Russian)

Beaudoin G (1990) Geological compilation map Northem Kokanee and Southern Goat Ranges, British Columbia, B.C. Minisfry of Energy, Mines and Petroleum Resources, Open File 1990-18

Bogdanov YV, Golubchina MN (1971) Isotopic composition of sulfur in stratified deposits of copper ores in Olekma-Vitim mountain area. Int Geol Rev 13:1405–1417

Bognibov VI, Krivenko AP, Izokh AE, Tolstykh ND, Glotov AI, Nesterenko GV, Balykin AP, Podlipskiy MJU, Glazunov OM, Mekhonoshin AS, Tsypukov MY, Konnikov EG, Kislov EV, Orsoyev DA, Gongalsky BI, Akhmetov RN, Buchko IV, Datsenko VM (1995) The mafic-ultramafic complexes of south Siberia enriched in platinum. Novosibirsk, Publisher of SIC UIGGM of SB RAS, 151 (in Russian)

Borishanskaya SS, Vinogradova RA, Krutov GA (1981) Minerals of cobalt and nickel, vol 221. MSU, Moscow (in Russian)

Cabri LJ (ed) (2002) The geology, geochemistry, mineralogy, mineral beneficiation of the platinum-group elements. Can Inst Min Metall Pet 54(2):13–129

Cabri LJ, Laflamme JHG (1974) Rhodium, platinum and gold alloys from the stillwater complex. Can Mineral 12:399–403

Chechetkin VS (1966) Some features of Cu–Ni mineralization in the Chiney stratified gabbronorite pluton. In: Lozovsky VI (ed) Geology and mineral resources of the Transbaikalia region. ZabNII, Chita, pp 54–65 (in Russian)

Chechetkin VS, Kharitonov (2002) Geological structure and commercial minerals of the Chita region, BAM Chita (in Russian)

Chukhrov FV (1969) Sulfur isotopic composition and genesis of ore at Dzhezkazgan and Udokan deposits. Geol Ore Deposits 11:18–25 (in Russian)

Craig JR, Carpenter AB (1977) Fletcherite Cu(Ni,Co)$_2$S$_4$, a new thiospinel from the Viburnum Trend (New Lead Belt), Missouri. Econ Geol 72(3):480–486

Distler VV, Genkin AD, Grokhovskaya TL, Evstigneeva TL, Sluzhenikin SF, Filimonova AA, Dyzhikov OA, Laputina IP (1988) Petrology of the magmatic ore-forming process. Moscow Nauka 232 (in Russian)

Genkin AD, Distler VV, Gladyshev GD, Filimonova AA, Evstigneeva TL, Kovalenker VA, Laputina IP, Smirnov AV, Grokhovskaya TL (1981) Sulfide copper–nickel ores of the Noril'sk deposits. Nauka, Moscow, p 281 (in Russian)

Gervilla F, Papunen H, Kojonen K, Johanson B (1998) Platinum-, palladium- and gold-rich arsenide ores from the Kylmäkoski Ni-Cu deposit (Vammala Nickel Belt, SW Finland). Mineral Petrol 64:163–185

Gongalsky BI (2011) PGE in rocks and ores at the deposits of the Udokan–Chiney district. Platinum of Russia 7:253–263 (in Russian)

Gongalsky BI (2015) Deposits of the unique metallogenic province of northern Transbaikalia. VIMS, Moscow, p 248 (in Russian)

Gongalsky BI, Krivolutskaya NA (1993) The Chiney layered pluton. Nauka, Novosibirsk, p 184 (in Russian)

Gongalsky BI, Krivolutskaya NA (1999) Mineralogy and geochemistry of platinum group metals in the Chiney pluton. Platinum of Russia 4:30–40 (in Russian)

Gongalsky BI, Krivolutskaya NA (2004) Noble metal sulfide mineralization in the Chiney pluton. Platinum of Russia 5:225–249 (in Russian)

Gongalsky BI, Krivolutskaya NA, Goleva NG (1995) Ore deposits of the Chiney pluton. In: Laverov NP (ed) Mineral deposits of Transbaikalia. Moscow, Geoinformmark 1(1): 20–28 (in Russian)

Just J (1980) Bismuthauchecornite – new name: hauchecornite redefined. Mineral Mag 43:873–876

Konnikov EG (1986) Precambrian differentiated mafic–ultramafic complexes in the Transbaikalia region, Nauka edn, Novosibirsk, p 224 (in Russian)

Konnikov EG, Truneva MF (1982) About the source of sulfide ores of the Chiney copper deposit (Northern Transbaikalia). Dokl Acad Nauk USSR 264:216–219 (in Russian)

Krendelev FP (1987) About genesis of sulfide ore of the Udokan cu sandstone deposit. Soviet Geol Geofiz 8:133–134 (in Russian)

Krivolutskaya NA (1986) Sulfide mineralization of the Chiney pluton. Geol Ore Deposits 28(5):94–100 (in Russian)

Krivolutskaya NA (1989) Mineralogical and geochemical features and the genesis of copper ores from the Chineiskoye deposit (Northern Transbaikalia). Author. on the competition Degrees Ph.D. Moscow 27.

Krivolutskaya NA, Gongalsky BI, Sergeeva NE (1997) Mineral composition of sulfide ore of the Chiney pluton. Gorny Zhurnal 7:187–201 (in Russian)

Krivolutskaya NA, Gongalsky BI, Svirskaya NM (2011) Mineralogical and geochemical features of rocks and ores of Maslovsky deposit (Norilsk ore district) Platinum of Russia Krasnoyarsk 7:342–350 (in Russian)

Maier WD, Andreoli MAG, Groves DI, Barnes S-J (2013) Petrogenesisof cu-Ni sulfide ores from O'Okiep and Kliprand, Namaqualand, South Africa: constraints from chalcophile metal contents. South Afr J Geol 115:499–514

Mel'nikova KM (1981) Conditions of ore localization and resource potential of the Chiney stratiform basic pluton. In: Polyakov GV (ed) Igneous rock associations in fold regions of Siberia: their origin. Ore resource potential and mapping. OIGGM, Novosibirsk, pp 203–205 (in Russian)

Mel'nikova KM, Kryukov VK, Belova NB, Kulikov AI (1983) Localization of ore in the Chineystratiform basic pluton, the Udokan Ore District. In: Andreev GV, Sobolev VS (eds) Endogenic processes and metallogeny in the BAM zone, vol 2. Nauka, Novosibirsk, pp 25–30 (in Russian)

Melekestseva IYu (2006) Ni-Co minerals at flanks at the Ivanovskoye and Dergamushskoye cobalt-bearing massive sulfide ore deposits associated with ultramafites (the south Urals).

Murzin VV, Moloshag VP, Volchenko VV (1988) Mineralogical paragenesis of PGE from Cu-Fe-V ores in Volkovsky type in the Urals. Dokl Earth Sci 300:1200–1202 (in Russian)

Petrovskaya NV (1973) Samorodnoe zoloto [Native gold]. Nauka, Moscow, p 347 (in Russian)

Ramdohr P (1980) The ore minerals and their intergrowth, 2nd edn. Pergamon Press, Oxford, p 1205

Shishkin NN, Karpenkov AM, Kulagov EA, Mitenkov GA (1974) On classification of pentladite group minerals. Dokl Akad Nauk USSR 217(1):194–197 (in Russian)

Tatarinov AV, Yalovik LI, Chechetkin VS (1998) Dynamometamorphic model for formation of layered basic massifs. Novosibirsk, 120 (in Russian)

Tolstykh ND (2008) PGE mineralization in marginal sulfide ores of the Chinei layered intrusion, Russia. Mineral Petrol 92:283–306

Tolstykh ND, Krivenko AP, Lavrent'ev YG, Tolstykh ON, Korolyuk VN (2000) Oxides of the Pd–Sb–Bi system from the Chiney massif, Aldan Shield, Russia. Eur J Mineral 12:431–440

Tolstykh ND, Krivenko AP, Krivolutskaya NA, Gongalskiy BI, Zhitova LM, Kotel'nikova MV (2004) Mineralization of noble metals of sulfide ores of the Chinei pluton. Platin Russ V:225–249 (in Russian)

Tolstykh ND, Orsoev DA, Krivenko AP, Izokh AE (2008) Noble-metal mineralization in mafic–ultramafic layered plutons in the southern Siberian platform. Novosibirsk, Papallel 194 (in Russian)

Truneva MF (1982) Evolution of the ore-forming process at the Chiney copper-sulphide deposit. Soviet Geo lGeofiz 7:59–66 (in Russian)

Truneva MF, Gurulev SA, Zhmodik SM, Konnikov EG (1979) Some features of genesis of sulfide ore at the Chiney deposit. In: Korzhinsky DS (ed) Contact processes and mineralization in gabbro-peridotite massifs. Nauka, Moscow, pp 97–107 (in Russian)

Vaasjoki O, Hakli TA, Tonitti M (1974) The effect of cobalt on the thermal stability of pentlandite. Econ Geol 69:549–551

Volchenko YA, Zoloev IN, Koroteev VA (1998) New prospective types of PGE mineralization in the Urals. Geology and Metallogeny of the Urals 1:238–255 (in Russian)

Volfson FI, Arkhangel'skaya VV (1987) Stratiform deposits of nonferrous metals. Nedra, Moscow, 255 (in Russian)

Vorobiev YK (1972) Temperature transformations in chalcopyrite. IGEM, Moscow, p 52 (in Russian)

Vorobiev YK (1975) Influence of ZnS on temperature of phase transformation in chalcopyrite. In: Chukhrov FV (ed) Isomorphism in minerals. Nauka, Moscow, pp 142–146 (in Russian)

Vorobiev YK, Borisovsky SE (1980) Phase transformation and composition of chalcopyrite. Izvestya AN USSR 8:86–101 (in Russian)

Vortsepnev VV (1978) Temperature, pressure, and geochemical condiitons of formation of the Talnakh copper–nickel deposit. Abstract Cand Dissertation Moscow, MSU 25 (in Russian)

Yang S-H, Maier WD, Hanski EJ, Lappalainen M, Santaguida F, Määttä S (2013) Origin of ultranickeliferous olivine in the Kevitsa Ni–Cu–PGE-mineralized intrusion, northern Finland. Contrib Mineral Petrol 166:81–95

Zaccarinni F, Anikina E, Pushkarev E, Rusin I, Garuti G (2004) Palladium and gold minerals from the Baronskoe-Kluevsky ore deposit (Volkovsky complex, Central Urals, Russia). Mineral Petrol 82:137–156

Zhitova LM, Tolstykh ND, Tsimbalist VG (2004) The feature of concentration of noble metals at the hillside deposits of Chiney pluton. Dokl Earth Sci 396:654–659 (in Russian)

Other Mafic-Ultramafic Intrusions of the Chiney Intrusive Complex

9

Abstract

The mafic-ultramafic rocks occur on the periphery of the Kodar-Udokan basin. They belong to the Chiney Intrusive Complex. The largest bodies are the Chiney and Luktur plutons, as well as the Upper Sakukan and Mylovskiy layered mafic-ultramafic intrusions. Gabbronorite dikes also belong to the Chiney Complex, with the Main Udokan Dike (up to 200 m thick) being the largest. The geochemical and geophysical data indicate that outcrops of mafic and ultramafic rocks at the margins of the Lurbun granite-granodiorite pluton, which cuts through the Chiney pluton, also belong to the Chiney Complex.

tically rearranged from the north-south trending Shamanka and Param ultramafic layered mafic-ultramafic complexes, the Olondo and other troughs with komatiites, into the near-west-east-trending units. The layered plutons of the Chiney Complex (Chiney, Luktur, Upper Sakukan) occur on the periphery of the Kodar-Udokan basin. Gabbronorite dikes of the Chiney Complex were identified within this trough. The Main Udokan Dike, up to 200 m thick, is the largest intrusion. The geochemical and geophysical data indicate that outcrops of mafic and ultramafic rocks at the margins of the Lurbun granite-granodiorite pluton, which cuts through the Chiney pluton, also belong to the Chiney Complex (Fig. 9.1; Gongalsky and Krivolutskaya 1993; Gongalsky 2015; Gongalsky et al. 2015a).

9.1 Introduction

The progress in the knowledge of geology and petrology of classic layered plutons facilitates comprehension of them as transitional magma chambers that arise in the course of emplacement of mantle-derived melts into the crust (Bogatikov et al. 2000; Kruger 2005; Tegner et al. 2006; Charlier et al. 2015). As a rule, they are regarded as roots of volcano-plutonic systems, where large magmatic reservoirs are connected by dike swarms, with volcanic plateaus localized above (Sharkov et al. 2002; Ernst and Buchan 2004; Ernst 2014). The vertical range of these systems reaches a few kilometers, and only subsequent vertical tectonic movements made it possible to study them (Wilson 1996; Naldrett 2004).

The Kodar-Udokan domain is an example of such a complexly built Proterozoic magmatic system, which developed between the Siberian Platform and younger fold belts, i.e., in the tectonically active zone permeable for mafic and ultramafic melts (Bognibov et al. 1995; Gongalsky 2015).

By the beginning of the Proterozoic, the main tectonic units at the southern margin of the Siberian craton were dras-

9.2 The Luktur Layered Mafic-Ultramafic Pluton

9.2.1 Brief Geology

This layered mafic-ultramafic pluton has a very restricted outcrop area (30 × 80 m), where gabbronorites with titanomagnetite were found during geological mapping in the 1980s (Fig. 9.2). Several hydrogeological boreholes provided primary information on geology of the Luktur layered mafic-ultramafic massif. The results obtained from the core of one of them (2T) are demonstrated in Fig. 9.3. The borehole penetrated 980 m of intrusive rocks, but did not intercept the contact of the layered mafic-ultramafic intrusion. The size of this intrusion (about 100 km^2) was estimated based on the geophysical data (Fig. 9.2b). Several years later, borehole 34 penetrated into 1980 m of coarse-grained norite and gabbronorite, but it did not intercept the contact of the pluton with the surrounding rocks as well. Layering in these rocks is not obvious, but it was inferred that layeres are angled at 40–45°, dipping to the northwest. In some cases ataxitic

© Springer Nature Switzerland AG 2019
B. Gongalsky, N. Krivolutskaya, *World-Class Mineral Deposits of Northeastern Transbaikalia, Siberia, Russia*,
Modern Approaches in Solid Earth Sciences 17, https://doi.org/10.1007/978-3-030-03559-4_9

Fig. 9.1 Geological map of the southern part of the Kodar-Udokan basin with positions of the massifs of the Chiney Complex. (After Tombasov et al. 2004)

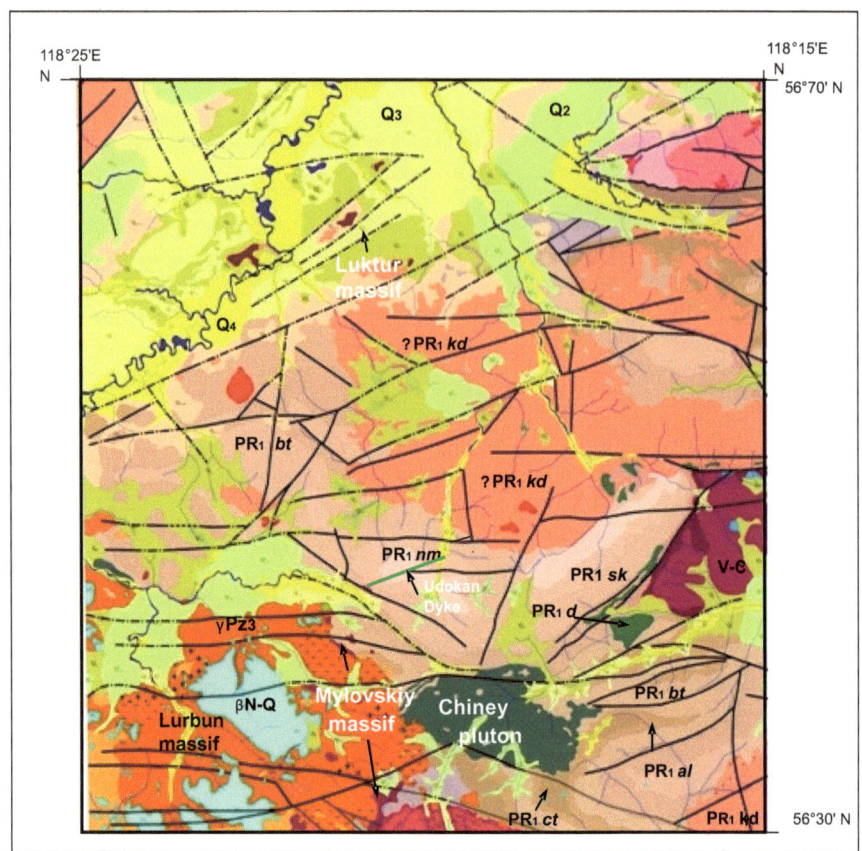

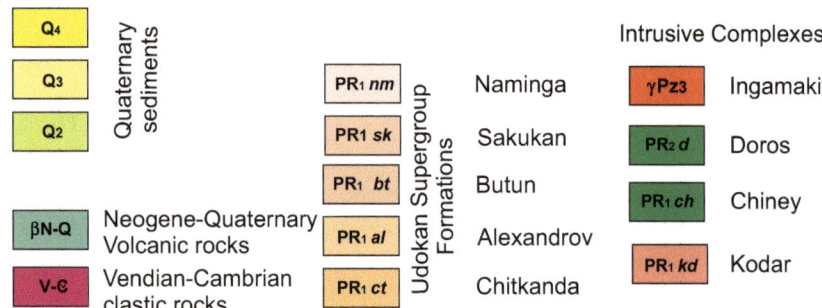

texture occurs in the rocks of the Luktur layered intrusion. Based on this fact, it was shown that rocks dip at 35–40°.

The Luktur intrusion consists mostly of norite, melanorite, and rarely gabbronorite, with layers of anorthosite and leucogabbro (Fig. 9.3), without sharp boundaries between them. They have coarse-grained structure, typical for gabbro rocks (Fig. 9.4). The composition of the leucocratic rocks is given in Table 9.1. The main rock-forming minerals are orthopyroxene, clinopyroxene, plagioclase, and titanomagnetite. The composition of clinopyroxene from the Luktur layered pluton is given in Table 9.2, together with the other rock-forming minerals from the rocks of the Chiney Complex. It is magnesium-rich in comparison with clinopyroxenes from other rocks.

9.2.2   Sulfide Mineralization

The mineralization is localized in coarse-grained gabbronorite and norite in the upper part of the Luktur pluton (Gongalsky et al. 2004). The ore is disseminated, with sulfides making up to 30 vol %. They are distinct from the deposits of the Chiney pluton by elevated Ni content and lower Cu/Ni ratio (0.5–0.8; Table 9.3). In this respect, it is close to the Noril'sk-type deposits (Krivolutskaya 2016). Chalcopyrite and pyrrhotite are major minerals; low-Co pentlandite (with <2 wt% Co) and pyrite are subordinate in abundance. As in the Chiney pluton, some units are strongly enriched in titanomagnetite and contain elevated concentrations of nonferrous and noble metals (up to 0.8 ppm PGE). In

Fig. 9.2 Schematic
geological map of the Chara
Trough, showing position of
the Luktur pluton (After
Pobedash et al. 1983,
unpublished)

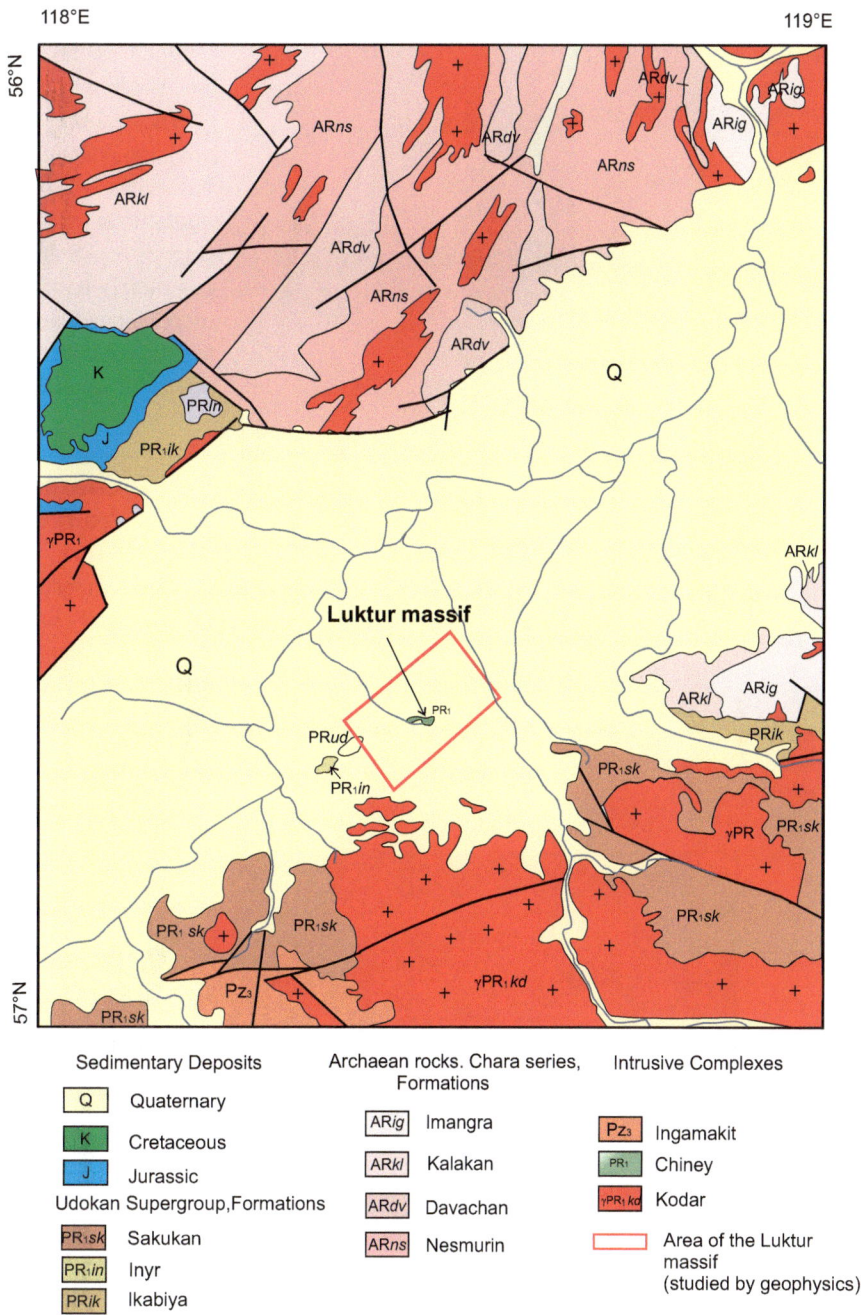

Fig. 9.2 Schematic geological map of the Chara Trough, showing position of the Luktur pluton (After Pobedash et al. 1983, unpublished)

the disseminated mineralization from the Luktur pluton, Pd and Pt grades are 0.03–0.3 and 0.05–0.36 ppm, respectively (Gongalsky and Krivolutskaya 1999, 2004; Tolstykh et al. 2004; Gongalsky 2011).

9.3   The Mylovskiy Layered Pluton

We attributed to the Chiney Complex gabbro rocks that occur in the peripheral part of the Lurbun granite pluton (Fig. 9.5), in the Skalisty prospect (outcrops of the Skalisty Brook, a tributary of the River Ingamakit), and along the Solokit

Brook, a tributary of the River Chukchudu (Gongalsky and Krivolutskaya 1993, 2009). The schematic map of this area (Fig. 9.6) was compiled by Dzyubenko, who kindly provided it to us.

In 2004, gabbro with Cu/Ni mineralization was found in the left tributary of the River Ingamakit to the west of the Udokan deposit. These rocks were previously recognized as the first phase of the Ingamakit Intrusive Complex (Fig. 9.7). During geological mapping, it was not unusual to consider gabbro as the earliest phases of granite complexes (Gongalsky et al. 2008). The recently applied geochemical methods helped to understand that local gabbro and granite belong to

Borehole 2T

Depth, m

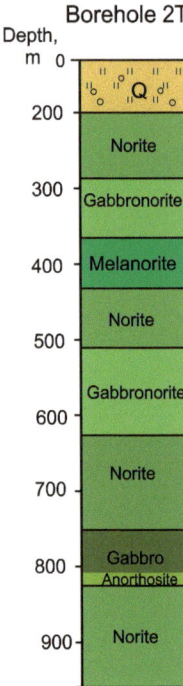

Fig. 9.3 Structure of the Luktur pluton based on drill core from the borehole 2T. (Based on data of the Udokan Expedition) Q Quaternary deposits

different intrusive phases, and on this basis, gabbro was outlined as the Mylovskiy massif of the Chiney Complex (see below).

The gabbro (Fig. 9.8a–c) is represented by fine- to medium-grained, rarely coarse-grained titanomagnetite gabbronorites. The xenoliths of olivine gabbro (n*cm–n*10 cm in diameter) were recognized in the Lurbun granite as well (Fig. 9.7c). Small xenoliths were altered and olivine was completely replaced by serpentine, while big xenoliths were recrystallized and altered along the contacts. Sometimes, taxitic gabbro with pyrrhotite-chalcopyrite mineralization occurs as xenoliths in granite (Fig. 9.8c).

9.4 Characteristics of Rocks from the Chiney Intrusive Complex

As was demonstrated above (Fig. 2.13), there is a large magnetic anomaly not only in the Chiney pluton area, but it covers much larger area west of the pluton and corresponds to the Mylovskiy pluton, which is overlain by thin plate of late Paleozoic granite. According to interpretation of the gravity field, the top of the Mylovskiy pluton is at 0.8–1.5 km depth

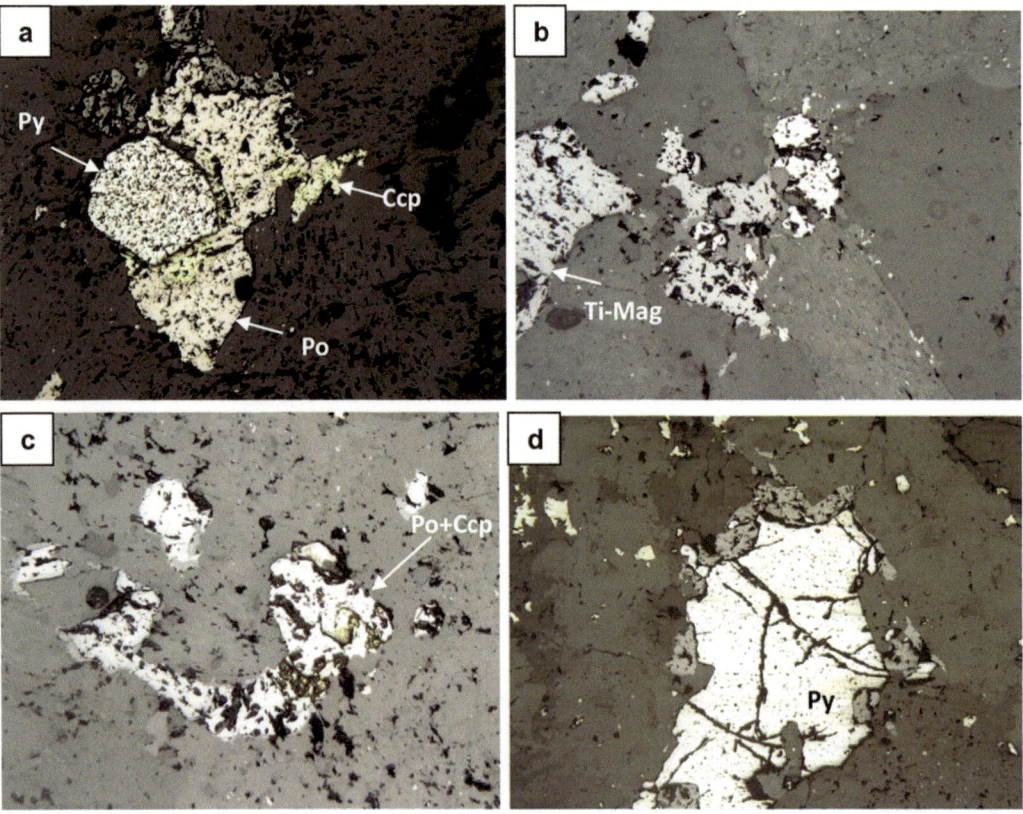

Fig. 9.4 Microphotographs of rocks from the Luktur pluton
Samples: (**a, b**) 2T/543; (**c**) 2T/608, (**d**) 2T/903. Sample number = borehole number/depth, m. *Ccp* chalcopyrite, *Po* pyrrhotite, *Ti-Mag* titanomagnetite, *Py* pyrite

Table 9.1 Composition of rocks from the Chiney complex

Oxide (wt%)					
Element (ppm)	11/988	308	5010	5008	9401
SiO_2	51.19	56.03	45.04	47.30	52.11
TiO_2	0.79	1.04	2.30	1.47	0.65
Al_2O_3	10.86	18.21	12.15	9.27	15.41
Fe_2O_3	15.63	7.54	20.28	17.16	11.99
MnO	0.23	0.14	0.18	0.20	0.17
MgO	11.74	4.21	5.93	8.28	7.42
CaO	6.55	8.14	10.95	13.65	8.39
Na_2O	1.80	2.86	2.38	1.69	2.74
K_2O	0.60	1.25	0.60	0.40	0.96
P_2O_5	0.05	0.25	0.08	0.08	0.11
Cs	1.06	1.44	0.30	0.50	0.99
Rb	21.45	37.9	8.46	22.8	35
Ba	19	539	113	220	395
Th	2.39	3.80	2.86	1.92	4.5
U	0.62	0.94	0.77	0.60	0.96
Nb	1.68	9.35	1.52	2.41	6.00
Ta	0.20	0.57	0.15	0.20	0.5
La	7.81	60.6	7.83	8.38	17
Ce	16.9	7.36	18.9	18.2	35
Pb	4.70		4.08	4.84	33
Pr	2.03	7.36	2.69	2.35	4.1
Sr	206	537	155	228	288
Nd	8.14	29.1	12.4	9.69	16
Sm	1.63	5.44	3.11	2.16	3.2
Zr	35.18	93.3	39.30	43.9	97
Hf	1.03	2.50	1.30	1.15	2.3
Ti	4623	5711	8478	12.3	4011
Eu	0.56	1.67	0.69	0.70	1.0
Gd	1.55	4.47	3.37	2.16	3.1
Tb	0.25	1.83	0.56	0.34	0.49
Dy	1.60	3.65	3.46	2.13	2.9
Y	9.19	22.4	19.44	13.2	20.0
Ho	0.34	0.73	0.72	0.44	0.62
Er	0.99	2.00	1.99	1.24	1.74
Tm	0.15	0.29	0.30	0.18	0.27
Yb	0.99	1.83	1.86	1.13	1.7
Lu	0.15	0.27	0.28	0.17	0.26

Note. N° 11/988 Chiney pluton; 308 Luktur pluton; 5010, 5008 Mylovskiy massif; 9401 Main Udokan Dike. Analyses were carried out at IGEM RAS (XRF, major components), analyst AI Yakushev, and at IEM RAS (ICP-MS, rare elements), analyst VK Karandashev

(VM Kravchenko pers. comm.; Gongalsky et al. 1995, 2015a). The three-dimensional model (Fig. 2.15) reflects the main gravitational structures: anorthosite-gabbronorite of the Chiney and Luktur layered mafic-ultramafic intrusions form positive (anticline) structures, between which there are synclinal structures (Unkur, Naminga).

Important evidence for these intrusive rocks, belonging to the common magmatic system, is their geochemical similarity. The distribution of trace elements, in particular, topology of their normalized patterns, which are independent of magma fractionation, is especially indicative. Various mafic and ultramafic rocks from the isolated outcrops within the geophysical anomalies were studied (Gongalsky et al. 2015a). In addition to the Chiney and Mylovskiy plutons, gabbro rocks of the Luktur layered pluton, Main Udokan Dike, as well as a number of smaller intrusive bodies were examined (Gongalsky et al. 2004). All of them are characterized by anomalous Ti concentrations, similar K_2O/Na_2O ratios regardless of the MgO concentration and similar configurations of the normalized multielement patterns (Fig. 2.14). This figure displays the distribution of trace elements normalized to the primitive mantle (Hofmann 1988) in gabbro rocks of the Luktur pluton and titanomagnetite gabbro of the Mylovskiy pluton, as compared with the chemical composition of the Chiney pluton. The latter is characterized by representative analyses of slightly fractionated Ti-magnetite-bearing gabbro and norite that differ from each other only in that the former shows a positive Ti anomaly. The monzodiorite of the Chiney pluton and biotite-bearing gabbronorite of the Luktur pluton contain the highest concentrations of all elements as a result of magma fractionation in the upper part of the intrusive bodies. The titanomagnetite gabbro of the Mylovskiy pluton is distinguished in depletion in all trace elements, especially LILE, and in positive Ti and Sr anomalies.

Despite the aforementioned differences, the trace element patterns for all rocks are quite similar in pronounced negative Ta and Nb and positive Pb anomalies and in La/Sm and Gd/Yb ratios, which indicate their similar origin. All rocks display evidence of crustal contamination, and this is confirmed by negative $\varepsilon_{Nd} = -4.4$ to -5.0 (Table 2.4; Gongalsky et al. 2008). In addition to petrographic and geochemical similarities between gabbro from the Chiney, Luktur, and Mylovskiy plutons, these rocks contain titanomagnetite and sulfide mineralization with anomalous PGE and Au concentrations.

Mineralization is represented by disseminations (up to 5%) and sulfide pockets, 2–3 cm in size, composed of pyrite and chalcopyrite, with some pyrrhotite. In sporadic sulfide segregations, the share of ore minerals in gabbro rocks increases to 15–20 vol %. The elevated concentrations of nonferrous and precious metals have been established in the sampled titanomagnetite-bearing gabbronorite: up to 2.7% Cu, 0.1% Ni, 0.9 ppm Pt, 2 ppm Pd, and 4.2 ppm Ag; one of the samples contains >0.28 wt% V. In chemical composition, proportions of nonferrous and precious metals, as well as in mineralogy, the mineralized zones are similar to the deposits of the Chiney pluton, especially to the Kontaktovyi deposit, which is characterized by disseminated pyrite-chalcopyrite ore with sporadic Pd reaching 2 ppm. It should be noted that pyrrhotite, and especially pyrite from the Mylovskiy pluton, is enriched in Ni, up to 2.5 wt%.

Correlation of major minerals in the ore from various plutons of the Chiney Intrusive Complex indicates certain variations for pyrrhotite and pentlandite. The maximum enrichment

Table 9.2 Composition of rock-forming minerals from the intrisions of the Chiney Complex, wt%

N°	1	2	3	4	5	6	7	8	9	10	11	12	13	14	15
SiO_2	33.35	33.99	33.94	53.11	52.00	53.92	53.74	53.76	51.75	52.00	52.11	52.86	51.70	51.49	51.72
TiO_2	0.01	0.00	0.00	0.00	0.57	0.24	0.32	0.29	0.22	0.17	0.35	0.29	0.45	0.30	0.26
Al_2O_3	0.00	0.00	0.02	0.08	2.12	1.36	1.21	1.11	1.03	0.58	1.08	1.29	1.95	0.50	0.45
Cr_2O_5	0.00	0.01	0.01	0.03	0.02	0.00	0.00	0.00	0.01	0.00	0.02	0.00	0.01	0.07	0.04
FeO	46.46	48.37	45.56	11.21	9.37	18.75	19.01	19.01	12.00	28.10	12.14	23.16	12.04	29.85	29.92
MgO	19.42	19.42	20.26	12.05	14.95	23.62	24.74	25.33	13.15	18.85	13.33	21.96	14.69	17.01	17.32
CaO	0.09	0.03	0.07	23.39	20.86	2.99	1.46	1.18	20.64	0.89	20.18	1.67	18.78	1.71	1.10
Na_2O				0.08	0.29	0.01	0.05	0.00	0.25	0.03	0.27	0.00	0.31	0.00	0.00
K_2O				0.01	0.00	0.01	0.00	0.00	0.01	0.02	0.00	0.00	0.01	0.01	0.00
Total	99.33	101.82	99.86	99.95	100.17	100.91	100.53	100.67	99.06	100.63	99.48	101.24	99.92	100.94	100.81
Fo, mol %	42.45	41.72	43.97												
Wo				47.57	42.06	5.90	2.87	2.29	42.35	1.80	41.34	3.31	38.09	3.48	2.26
En				34.11	41.95	64.83	67.50	68.51	37.54	53.10	38.01	60.47	41.47	48.33	49.33
Fs				18.03	14.93	29.23	29.47	29.20	19.17	44.98	19.66	36.22	19.31	48.19	48.41
Ac				0.29	1.05	0.04	0.16	0.00	0.94	0.12	0.99	0.00	1.13	0.00	0.00

Note. N° 1–3 olivine, Chiney pluton; 4,5, 9,11,13–14 clinopyroxene, Chiney pluton; 6–8,10, 12–15 orthopyroxene; 6–13, Chiney pluton; 14–15 Luktur massif.

Fo, mol% forsterite, *Wo* wollastonite, *En* enstatite, *Fs* fassaite, *Ac* acmite

of pyrrhotite in Ni is a characteristic for the Kontaktovyi deposit, whereas the minimum concentrations of this metal are typical of pyrrhotite hosted in the titanomagnetite ore at the Etyrko deposit (Gongalsky et al. 2004). Pyrrhotites from the Luktur ore are close to those from the Rudnoe deposit, but they differ distinctly from those in the Mylovskiy pluton (Fig. 9.9). The high-Co and therefore high-temperature pentlandites are hosted in the titanomagnetite ore of the Etyrko deposits, while the low-Co pentlandites are contained in the exocontact ore. A maximum Co content is established for pyrite from the Verkhne chineyskoe deposit, and minimum values correspond to pyrites from the Luktur pluton.

9.5   Geochemical Modeling of Magma Crystallization

Similar crystallization sequences of magmas in the Chiney and Mylovskiy layered mafic-ultramafic pluton and the Main Dike at the Udokan deposit are a fairly important factor in the identification of genetic links between their magmatic rocks (Fig. 9.10). The modeled crystallization sequences of minerals in these rocks are close to the aforementioned "magnesian" evolutionary trend of the Chiney magmas.

Crystallization starts from olivine, which then disappears owing to peritectic reaction with a melt at a lower temperature. The titanomagnetite-bearing gabbro of the Mylovskiy pluton and the most primitive gabbro of the Chiney pluton (type 9, see Chap. 6) reveal the closest similarity in this regard. Olivine Fo_{76} and Fo_{85} appear at liquidus. Both compositions are characterized by cotectic crystallization of augite, plagioclase, and magnetite within a narrow tempera-

ture interval (15–20 °C). The crystallization sequence of rocks from the Main Udokan Dike (sample 9401) is close to this sequence. However, plagioclase appears before augite, and low-Ca pyroxene crystallizes synchronously with magnetite. More significant differences are noted for phase relations of the second dike (sample 5068), where crystallization starts with plagioclase-clinopyroxene cotectics and is followed by magnetite. Additional investigations of more representative material are needed.

Rocks of the Chiney Complex can be compared with the upper zone of the Bushveld pluton (Gongalsky and Krivolutskaya 1993). The data points of rocks from the Luktur pluton lie in the compositional trend of the upper Bushveld zone (Fig. 9.11). The mean chemical composition of the Chiney pluton, based on a large dataset of analyses, is presented in Table 9.4, including average weighted compositions estimated for the above-described reference section across the pluton and for particular magmatic series.

Judging by mode of occurrence and composition, rocks of Group 1, primarily pyroxenite, were captured as xenoliths during the ascent of the main magma volume. In their geochemistry, they resemble Paleoproterozoic pyroxenite of the Mururin and New Katugin plutons. The independent younger lamprophyres of Group 4 are controlled by a system of near-horizontal faults at the boundary between basal gabbro rocks and sandstone of the Udokan Supergroup.

The problem concerning formation of the main body of the Chiney pluton is the most intractable. The formation of contrasting rock units can be explained by the main intrusive pulse (geologically very fast) of genetically associated and close in temperature, but somewhat fractionated, magmas, enriched in diverse intratelluric phenocrysts (to various

Table 9.3 Concentrations of ore elements in orebodies and mineralized zones

No.	Sample	TiO₂	MnO	V	Cu	S	Ni	Co	Zn	Cr
		wt %					ppm			
Chiney pluton, Etyrko deposit										
1	11/619,91	5.74	0.27	4500	0.12	0.06	376	395	246	679
2	11/619,95	6.59	0.25	5100	0.15	0.09	400	410	271	712
3	11/619,99	8.15	0.26	6400	0.17	0.84	410	489	286	884
4	11/622,68	8.20	0.27	2216	0.08	0.16	255	406	253	775
5	11/0622,73	4.40	0.27	1435	0.06	0.12	158	227	142	309
6	11/989,6	2.34	0.12	2444	0.09	0.19	141	119	115	119
7	11/992	2.34	0.12	2647	0.06	0.15	162	95	109	116
8	83/1327,3	2.03	0.12	2292	1.01	9.29	1100	327	221	357
9	Et-66-1	6.75	0.26	1180	0.09	0.22	132	358	219	441
10	Et-05-1	0.51	0.05	198	1.30	2.97	668	136	54	26
11	Et-08	0.67	0.05	262	1.20	2.74	530	129	45	32
12	Et-08-1	0.83	0.05	336	1.29	2.9	561	131	49	45
13	Et-08-11	0.83	0.06	358	1.45	2.92	624	130	55	43
Chiney pluton, Kontaktovyi deposit										
14	9/210	0.89	0.10	569	0.91	1.77	546	79	74	45
15	9/219,7	1.03	0.12	530	0.71	1.43	398	98	104	46
Chiney pluton, Magnitnyi deposit										
16	66/41,5	0.69	0.18	537	0.14	0.18	161	92	90	103
17	66/92	0.85	0.29	1232	0.04	0.11	183	161	162	163
18	66/130,6	2.44	0.12	2012	0.17	0.46	165	100	81	118
19	Mgn-674	1.60	0.09	757	1.30	1.84	855	222	56	117
20	119/155	8.90	0.15	8700	0.15	0.48	199	365	292	537
Mylovskiy pluton										
21	58/28	1.95	0.22	2838	0.07	0.22	413	279	269	473
22	Ml-08	1.47	0.20	419	0.13	1.16	151	101	82	132
23	Ml-03	0.88	0.05	189	2.70	4.68	854	172	118	29
24	Ml-25-21	0.52	0.12	170	1.45	4.38	2626	174	76	62
Luktur pluton										
25	34/441,8	0.27	0.07	123	0.05	3.05	4147	147	63	189
26	34/447,5	0.38	0.08	96	0.17	2.81	3939	173	86	161
27	34/510,6	0.38	0.19	196	0.05	1.09	1323	90	169	448
28	34/506,6	0.40	0.23	160	0.13	1.19	731	172	209	854
29	1 T./726	0.39	0.19	112	0.01	0.1	138	54	145	894
Pravoingamakitskoe deposit										
30	Pr-36	0.21	0.09	67	2.02	1.35	16	3	53	34
31	Pr-36-A	0.18	0.07	64	2.46	1.7	20	9	50	36
32	Pr-38-1	0.41	0.02	71	6.66	6.48	45	4	167	68
33	Pr-38-2	0.25	0.01	42	4.07	3.79	35	2	135	84
34	Pr-38-3	0.49	0.02	37	12.67	13.3	54	14	92	78
35	Pr-38-31	0.59	0.03	34	12.68	12.3	92	5	118	102
36	Pr-38-4	0.21	0.01	12	12.73	12.9	58	16	106	51
37	Pr-38-5	0.72	0.04	115	2.10	1.99	34	5	44	87
38	Pr-45-12	0.05	0.01	24	6.82	9.73	1420	171	121	38
39	Pr-45-2	0.23	0.02	83	7.04	9.71	7640	221	215	128

Note. Analyses were carried out at CHIPR, analysts TM Mikhailova, NS Baluev

degree), as well as by the migration of residual melts and injection of them into the already crystallized rocks. This mechanism is illustrated by self-dependent injections of residual melts in large plutons (so-called auto intrusions), which are readily identifiable in the field. On the other hand, the sharp contacts between rocks of Groups 2 and 3 provide evidence for formation intrusions separated in time and differing in magma compositions.

On this basis, it is suggested that Chiney pluton was formed as a result of the consecutive filling of its chamber

Fig. 9.5 Northern outcrops of gabbro (G) of the Mylovskiy pluton within the granite (Gr) of the Lurbun pluton

Fig. 9.6 Geological map of the Lurbun pluton. (Modified unpublished data of M Dzyubenko)

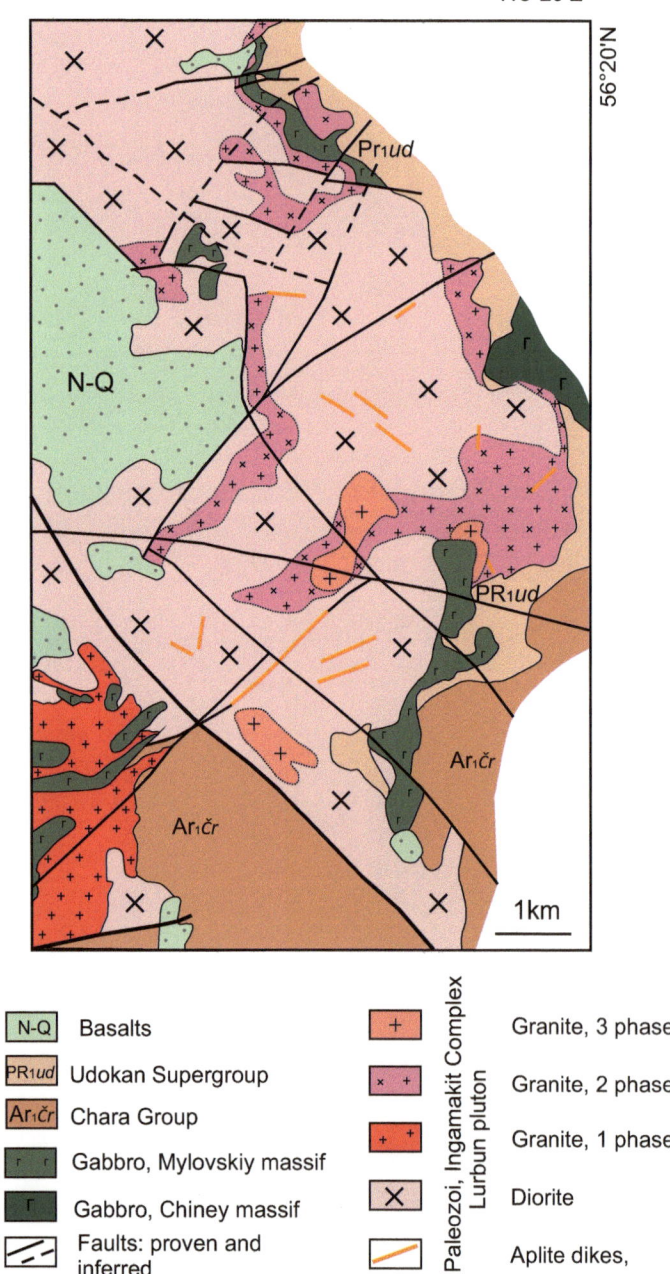

N-Q	Basalts	
PR₁ud	Udokan Supergroup	
Ar₁čr	Chara Group	
r r	Gabbro, Mylovskiy massif	
r	Gabbro, Chiney massif	
	Faults: proven and inferred	

Paleozoi, Ingamakit Complex Lurbun pluton

+	Granite, 3 phase
× +	Granite, 2 phase
+ +	Granite, 1 phase
☒	Diorite
	Aplite dikes,

Fig. 9.7 Outcrops of gabbro inside the granite of the Lurbun pluton (Skalistyi Creek)
(**a**) General view of the outcrop, (**b**–**d**) xenoliths of ultrabasic-basic rocks in granite. *G* gabbro, *Ol-G* olivine gabbro, *Gr* granite

Fig. 9.8 Samples of gabbro of the Mylovskiy pluton
(**a**) Contact gabbro (G) with granite (Gr) of the Lurbun pluton, (**b**) titanomagnetite leucogabbro of the Mylovskiy pluton, (**c**) taxitic gabbro with disseminated sulfides from the Mylovskiy pluton

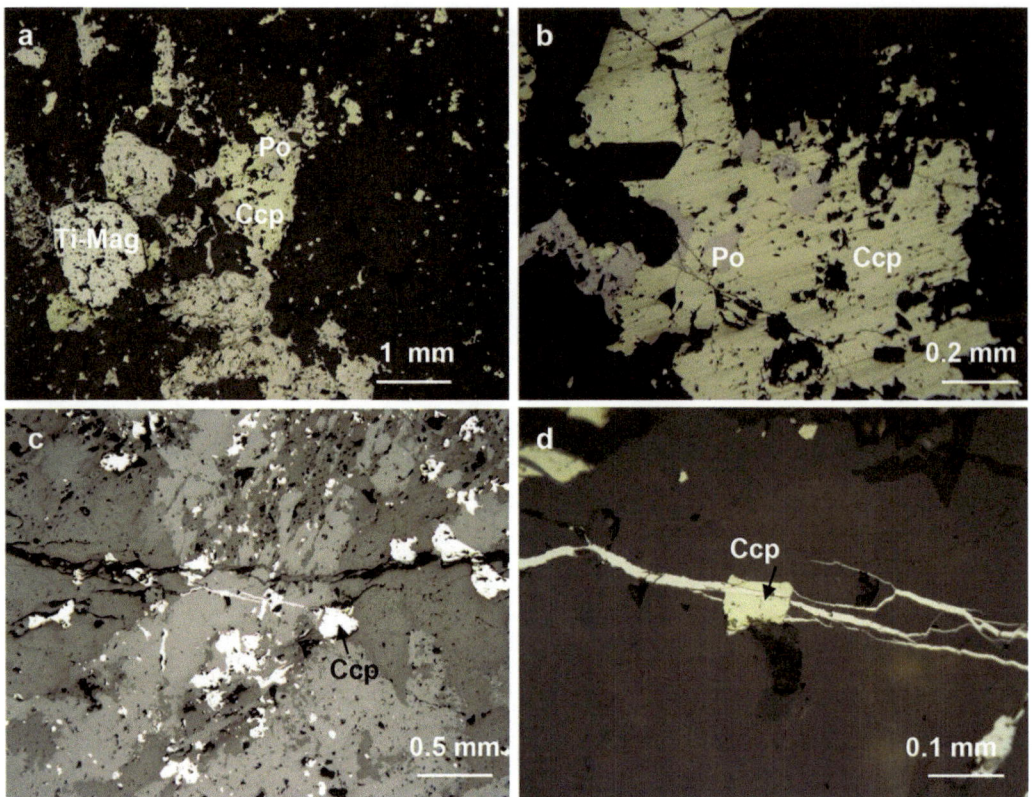

Fig. 9.9 Photomicrographs of disseminated ore (sample 5003) from the Mylovskiy pluton (reflected light). *Po* pyrrhotite, *Ccp* chalcopyrite, *Ti-Mag* titanomagnetite. (**a**) – Chalcopyrite-pyrrhotite association in titanomagnetite gabbro; (**b**) – chalcopyrite-pyrrhotite grains in intersticial spaces between silicate minerals; (**c**) – pyrite-chalcopyrite dissemination in gabbro; (**d**) – veinlets of pyrite and chalcopyrite grains in gabbro

Fig. 9.10 Temperatures and crystallization sequences of primitive rocks from the Chiney pluton (type 9), Mylovskiy pluton (samples 5008, 9401), and dikes at the Udokan deposit (samples 5068, 9401) (see Table 2.8 for mineral symbols)

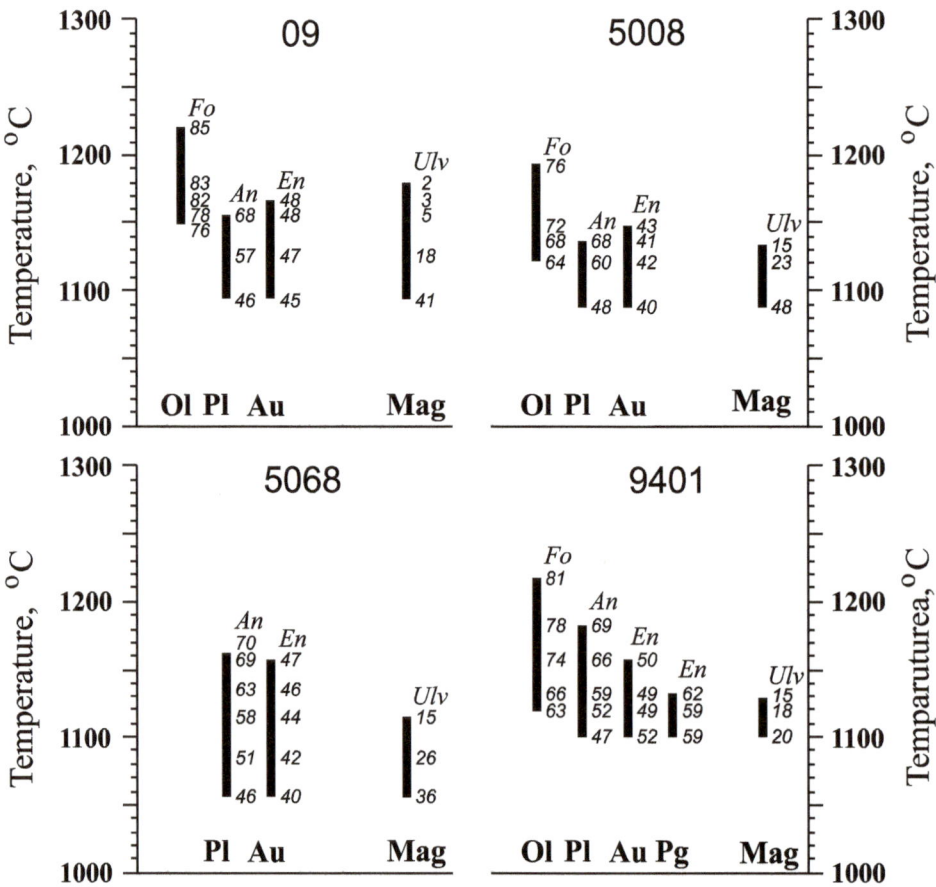

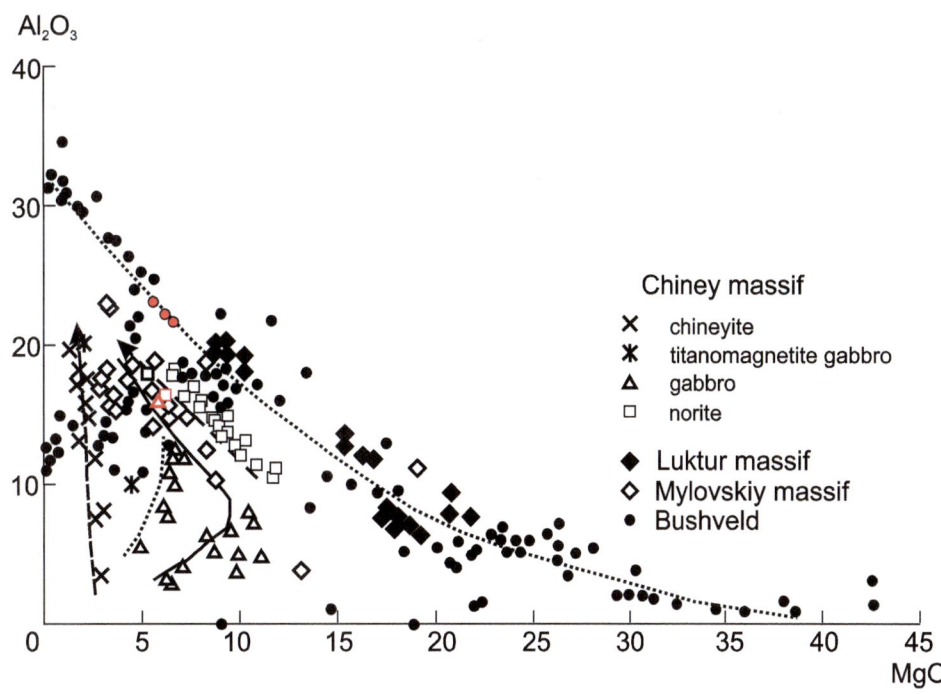

Fig. 9.11 Al₂O₃ vs MgO for rocks of the Chiney Complex and Bushveld pluton

Table 9.4 Average mean compositions of the rocks from the Chiney pluton, wt %

N	N analyses	S iO₂	TiO₂	Al₂O₃	Fe₂O₃	MnO	MgO	CaO	Na₂O	K₂O	P₂O₅	LOI	Сумма
1	987	44.81	2.38	14.26	20.41	0.17	5.95	8.53	2.32	0.66	0.07	0.25	99.81
2	16	49.78	1.12	13.43	14.71	0.20	9.31	8.00	2.52	0.83	0.07	0.67	100.62
3	31	43.20	2.61	14.11	21.32	0.17	5.72	9.42	2.72	0.73	0.05	0.44	100.49
4	20	44.65	2.29	14.22	17.95	0.17	6.09	11.10	3.06	0.54	0.08	0.25	100.40
5	85	44.20	2.38	14.37	19.71	0.17	6.18	9.34	2.84	0.68	0.07	0.45	100.38
6	2	44.41	2.14	14.58	19.42	0.2	6.2	10.12	2.4	0.31	0.08	0.17	100.03
7	2	47.57	1.15	15.4	12.72	0.14	6.54	7.68	1.44	2.57	0.23	3.70	99.14

Note. 1 average mean composition of rocks from the Chiney pluton, 2–8 average mean compositions, 2 gabbronorite series, 3 titanomagnetite gabbro series, 4 leucogabbro series, 5 rocks from drill core of borehole № 83, 6 cement of the magmatic breccia, 7 cement of fluid-magmatic breccia

with the suspension of crystals in a ferrobasaltic melt at a temperature of ~1130 °C (Chap. 6). Variable proportions of intratelluric crystals (mainly plagioclase and magnetite) predetermined distinct composition and structure of the main plutonic blocks, composed of different rocks: leuco-gabbro with titanomagnetite gabbronorite (rock Group 2) and gabbronorite (rock Group 3) with various kinds of lay-ering (rough, fine, cryptic) and various ranks of rhythmic-ity. The titanomagnetite gabbro (TIMAGG) and leucogabbro (LG) series are close in average weighted chemical compo-sition (Table 2.12). Leucogabbro and anorthosite layers occur in the upper part of TIMAGG series and lower part of the LG series. A possible mechanism of their formation was proposed by Dobretsov et al. (1984) in the assumption that the melt crystallized up to 80% acquired properties of the non-Newtonian liquid, which enables to sustain con-traction cracks, filled with residual anorthositic melts as a product of TIMAGG melt fractionation.

The contribution of intra-chamber crystal fractionation of the Chiney magma remains uncertain and requires further investigation.

9.6 Conclusions

1. The geological and geophysical data and petrologic simu-lation show that the Chiney pluton, as a complexly built magmatic body, is a small part of the large-scale Paleoproterozoic magma-feeding system, which com-prised a number of additional plutonic (Luktur, Mylovskiy) and subvolcanic bodies (Main Udokan dike, dikes near the Chiney pluton).

2. The pluton is composed of evolved igneous rocks, so that more primitive varieties should be expected. It cannot be ruled out that high-Mg gabbro rocks of the Luktur pluton belong to this category. All layered mafic-ultramafic bodies of the Chiney Intrusive Complex are composed of gabbro rocks with variable amounts of titanomagnetite, ranging from accessory to dominating rock-forming mineral up to massive Fe-Ti-V ore.

3. The sulfide ore and mineralization in the other plutons of the Chiney Intrusive Complex are generally comparable to those in the Chiney pluton. Their specific attributes depend on the composition of host gabbro rocks. For example, the ore from the Luktur pluton, which is hosted in high-Mg rocks, is enriched in Ni, whereas the mineralization from the Mylovskiy pluton is close to the spatially associated titanomagnetite mineralization in the Chiney pluton.

References

Bogatikov OA, Kovalenko VI, Sharkov EV, Yarmoluk VV (2000) Magmatism and geodynamics; terrestrial magmatism throughout the earth's history. Gordon and Breach Science Publishers, Amsterdam, 511 p

Bognibov VI, Krivenko AP, Izokh AE, Tolstykh ND, Glotov AI, Nesterenko GV, Balykin AP, Podlipskiy MJU, Glazunov OM, Mekhonoshin AS, Tsypukov MY, Konnikov EG, Kislov EV, Orsoyev DA, Gongalsky BI, Akhmetov RN, Buchko IV, Datsenko VM (1995) The mafic-ultramafic complexes of south Siberia enriched in platinum. Novosibirsk, Publisher of SIC UIGGM of SB RAS, 151 (in Russian)

Charlier B, Namur O, Latypov R (eds) (2015) Layered intrusions. Springer, Dordrecht, p 762

Dobretsov NL, Konnikov EG, Tsoi LA (1984) A new model for the formation of rhythmic stratification of basic plutons. Geol Geophys 2:3–11 ((in Russian)

Ernst RE (2014) Large igneous provinces. Cambridge University Press, Cambridge, p 666

Ernst RE, Buchan KL (2004) Large igneous provinces (LIPs) in Canada and adjacent regions: 3 Ga to present. Geosci Can 31–3:103–126

Gongalsky BI (2011) Platinum group metals in rocks and ores of deposits of the Udokan-Chiney region. Platinum Russ VII:253–263 (in Russian)

Gongalsky BI (2015) Deposits of the unique metallogenic province of northern Transbaikalia. Moscow, VIMS 248 (in Russian)

Gongalsky BI, Krivolutskaya NA (1993) The Chiney layered pluton. Nauka, Novosibirsk, p 184 (in Russian)

Gongalsky BI, Krivolutskaya NA (1999) Mineralogy and geochemistry of platinum group metals in the Chiney pluton. Platinum Russ IV:30–40 (in Russian)

Gongalsky BI, Krivolutskaya NA (2004) Noble metal sulfide mineralization in the Chiney pluton. Platinum of Russia V:225–249 (in Russian)

Gongalsky BI, Krivolutskaya NA (2009) The Udokan-Chiney ore-magmatic system. Russ Northwest Geol 42:180–184

Gongalsky BI, Golovatyi AS, Abushkevich SA (1995) Zonal ring structures of the Udokan range. Dokl Akad Nauk 343(1):80–82 (in Russian)

Gongalsky BI, Izokh AE, Krivenko AP, Krivolutskaya NA, Tolstykh ND (2004) Huge concentrations of copper in deposits of the Kodar-Udokan area (northern Transbaikalia). In: Rundkvist DV (ed) Large and superlarge deposits: regularities in formation and emplacement. IGEM, Moscow, pp 206–218 (in Russian)

Gongalsky BI, Krivolutskaya NA, Ariskin AA, Nikolaev GS (2008) Internal structure, composition and genesis of the Chiney anorthosite–gabbronorite pluton, northern Transbaikalia. Geochem Int 46:637–665

Gongalsky BI, Galyamov AL, Pavlovich GD, Petrov AV, Murashov KY (2015a) A 3D model of the head of a Paleoproterozoic plume in the southern Siberian craton. Large igneous provinces, mantle plumes and Metallogeny in the Earth's history. Publishing House of VB Sochava/Institute of Geography SB RAS, Irkutsk, pp 45–46 (in Russian)

Hofmann AW (1988) Chemical differentiation of the earth: relationship between mantle, continental crust and oceanic crust. Earth Planet Sci Lett 90:297–314

Krivolutskaya NA (2016) Siberian traps and Pt-Cu-Ni deposits of the Noril'sk area. Springer, Amsterdam, p 361

Kruger FJ (2005) Filling the bushveld complex magma chamber: lateral expansion, roof and floor interaction, magmatic unconformities, and the formation of giant chromitite, PGE and T-V-magnetite deposits. Mineral Deposita 40:451–472

Naldrett AJ (2004) Magmatic sulfide deposits: geology, geochemistry and exploration. Springer, Berlin/New York/Heidelberg, p 727

Sharkov EV, Smolkin VF, Chistyakov AV, Galkin AS, Kharrasov MK (2002) Geology and metallogeny of the Monchegorsk stratified ore-bearing complex. Russian Arctic: geological history, minerageny, geoecology. VNIIOkeangeologiya, pp 485–494 (in Russian)

Tegner C, Cawthorn RG, Kruger FJ (2006) Cyclicity in the Main and upper zones of the bushveld complex, South Africa: crystallization from a zoned magma sheet. J Petrol 47(11):2257–2279

Tolstykh ND, Krivenko AP, Krivolutskaya NA, Gongalsky BI, Zhitova LM, Kotel'nikova MV (2004) Mineralization of noble metals of sulfide ores of the Chiney pluton. Platinum Russ V:225–249 (in Russian)

Tombasov IA, Shemelina SF, Afonin GA, Drozdov SA (2004) Governmental geological map of the Russian Federation, scale 1–200000. 2nd edn. Udokan series. O-50-XXXVI (Katugin). Cards. Factory of A.P. Karpinsky Russian Geological Research Institute, St-Petersburg, Russia (in Russian)

Wilson AN (1996) The great dike of Zimbabwe. In: Cawthorn RG (ed) Layered intrusions. Elsevier, Amsterdam, pp 181–230

Abstract

The large Paleoproterozoic Nb-Ta-Zr-REE-Y Katugin deposit is situated in the southern part of the Kodar-Udokan district. Together with the Tomtor deposit in northern Siberia, Katugin deposit comprises the main Ta-Nb resources of the Russian Federation that can be developed in the future. The deposit occurs within a zone of fault-controlled alkaline alteration at the contact of rapakivi granite pluton. The ore-bearing quartz-albite-microcline alteration in a combination with alkaline granite form a steeply dipping lode exposed at the surface over an area of 5 km^2 and traced to a depth of 800 m. The ore contains 1.62 wt% ZrO_2, 0.026 wt% Ta_2O_5, 0.374 wt% Nb_2O_5, 0.16 wt% Y_2O_3, 0.22 wt% REE_2O_3, and 0.0078 wt% U_3O_8. Pyrochlore, zircon, gagarinite, REE-fluorite, monazite, and cryolite are the major minerals. Fergusonite, ilmenite, and ilmenorutile are widespread in the metasomatized country rocks. It was suggested that deposit was formed as a result of metasomatism, but nowadays there is an opinion on its magmatic origin.

10.1 Geological Setting

The Paleoproterozoic Nb-Ta-Zr-REE-Y Katugin deposit (Bykov and Arkhangel'skaya 1995; Larin et al. 2002; Arkhangel'skaya et al. 2004, 2012; Mashkovtsev et al. 2010) is situated in the southern part of the Kodar-Udokan district (Fig. 10.1). This type of mineralization is hosted in magnetite and copper sandstone of the Chitkanda and Alexandrov Formations of the Udokan Supergroup.

Together with the Tomtor deposit in northern Siberia, the Katugin deposit comprises main Ta-Nb resources of the Russian Federation that can be developed in the future. The Katugin deposit contains huge $\sum TR_2O_3$ reserves in around 28 Mt. of ore. Russia possesses the second largest REE min-eral endowment after China (Bykhovsky 2014; Lepeshkin 2014).

The deposit is related to a fault-controlled zone of hydrothermal alteration at the contact of rapakivi granite pluton (Fig. 10.2) that cuts through the Paleoproterozoic schist and gneiss (Arkhangel'skaya et al. 2004, 2012). The ore-bearing quartz-albite-microcline altered rocks in combination with alkaline granite occur as a steeply dipping lode, exposed at the surface over an area of 5 km^2 and traced to a depth of 800 m.

The orebody is controlled by the regional Kalar Fault over a distance of 10–12 km. It is erosionally separated into the western and eastern bodies. The orebodies correspond to alkaline granite and granodiorite. However, structure and texture of these rocks remind unreplaced or partly replaced remnants of crystalline schists, gneisses, hornfels, and dolerite, suggesting a metasomatic origin of granite and granodiorite (Fig. 10.3). All rocks contain nonuniformly distributed fine disseminations of tantaloniobates (mainly pyrochlore, columbite, and fergusonite), zircon, gagarinite, cryolite, and REE-Y-fluorite.

The western body is traced for 5–6 km along strike. The eastern body (Fig. 10.3) is faulted. In both bodies, the gneissic banding is oriented conformably to banding in the protolith. Both bodies are zonal. The quartz-feldspar pockets and veins, located in both the western and eastern lodes, are syngenetic to metasomatic rocks.

10.2 Chemical Composition of Rocks

Alkaline granite of the Katugin Complex consists of microcline, albite, quartz, biotite, aegirine, and amphibole (riebeckite-arfvedsonite) (Larin et al. 2015). The granite has extremely high iron (f = 0.93–0.99) and alkalis ($Na_2O + K_2O$ = 8.43–12.32 wt%) contents, with a high value of NK/A 0.96–1.48.It belongs to sodium alkaline type (K_2O/

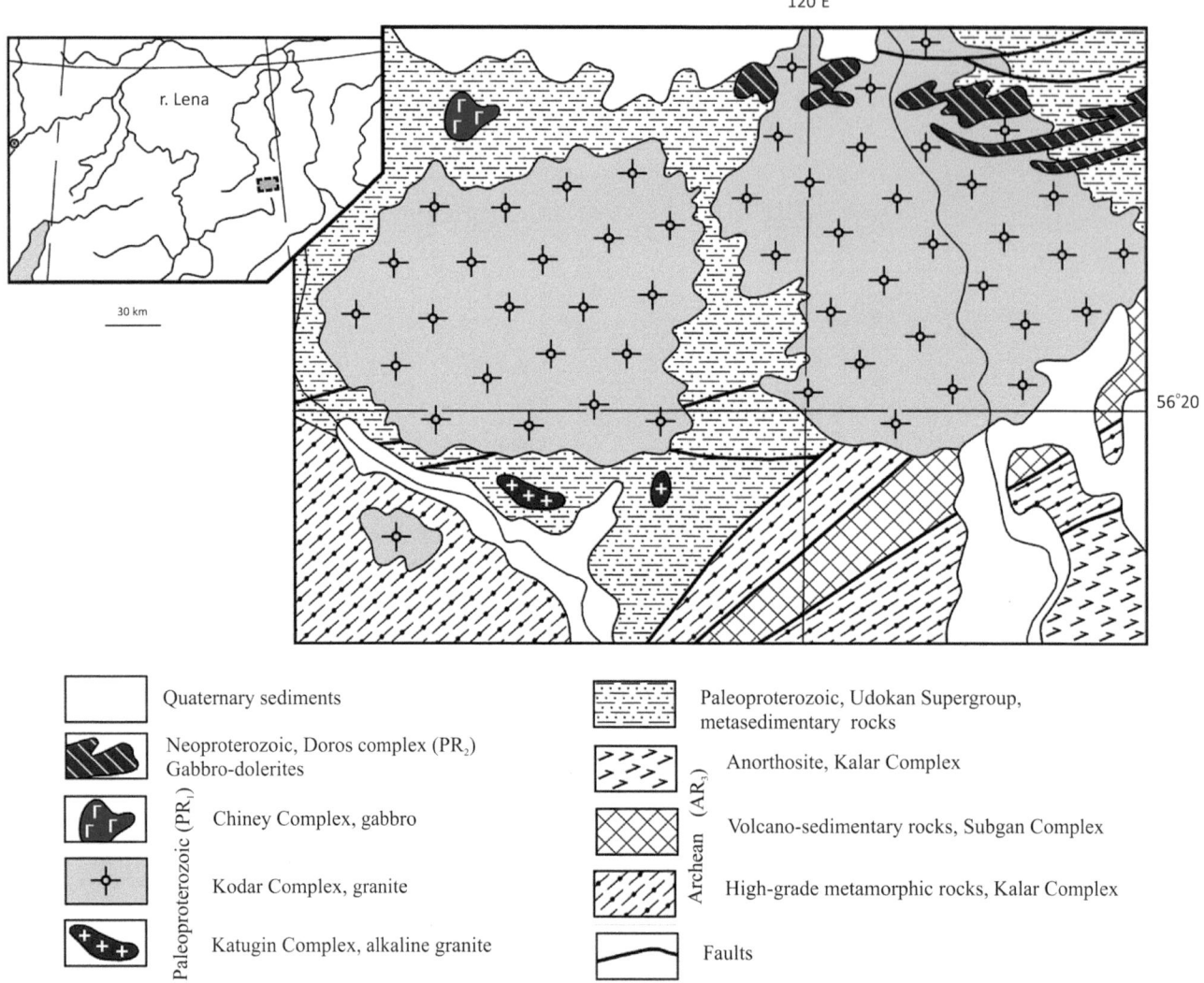

Fig. 10.1 Schematic map of the location of Katugin granite (After Larin et al. 2015)

Na_2O = 0.43–1.10). In addition, this granite contains extremely high Rb, Li, Y, Zr, Hf, Ta, Nb, Th, U, Zn, Ga, and REE (except Eu) and low Ca, Mg, Al, P, Ba, and Sr (Table 10.1).

The spectra of REE distribution in granite of the Katugin Complex are characterized by moderate LREE enrichment ((La/Yb) n = 3.40–5.30), sometimes with positive Ce anomalies, gentle slope of graphics in MREE-HREE ((Gd/Yb) n = 1.10–1.69) and a clear negative Eu anomalies (Eu/Eu * = 0.13–0.19).

The high-grade Ta-Nb ore is localized at deep levels of this deposit, where it is related to the aegirine-arfvedsonite, overlying arfvedsonite-bearing altered rocks, and underlying alkaline granite. The REE-Ta-Nb ore is hosted in arfvedsonite and annite-arfvedsonite, located above the aegirine-arfvedsonite variety.

The metasomatized rocks with Nb-Ta-Zr-REE-Y mineralization resemble gneiss and crystalline schist of their pro-

toliths in their structure but are distinguished by variable texture, qualitative and quantitative phase compositions, and distinct evidence for replacement and overgrowing of early generations of rock-forming minerals with late minerals. Metasomatized rocks are characterized by zoning (from the periphery to the center; Arkhangel'skaya et al. 1993): (1) biotite, (2) aegirine-arfvedsonite, (3) amphibole, and (4) arfvedsonite.

The Ca-Mg amphiboles of metasomatized country rocks are characterized by anomalous Sc, V, Ni, Co, and Cr, whereas the Na-Fe amphiboles of alkaline granite rocks are enriched in Li, Sn, and Zn. The magnesian micas of country rocks are distinguished by anomalous Sc, V, Ni, Cr, and Co contents, while Fe-rich micas from granite body and exocontact rocks bear Li, Rb, Ga, Zn, and Pb.

The aegirine-arfvedsonite rocks with the highest Ta, Nb, Zr, Ce, Zn, Pb, Sn, Li, and Rb contents are in the center of geochemical zoning. The aegirine-free alkaline granite

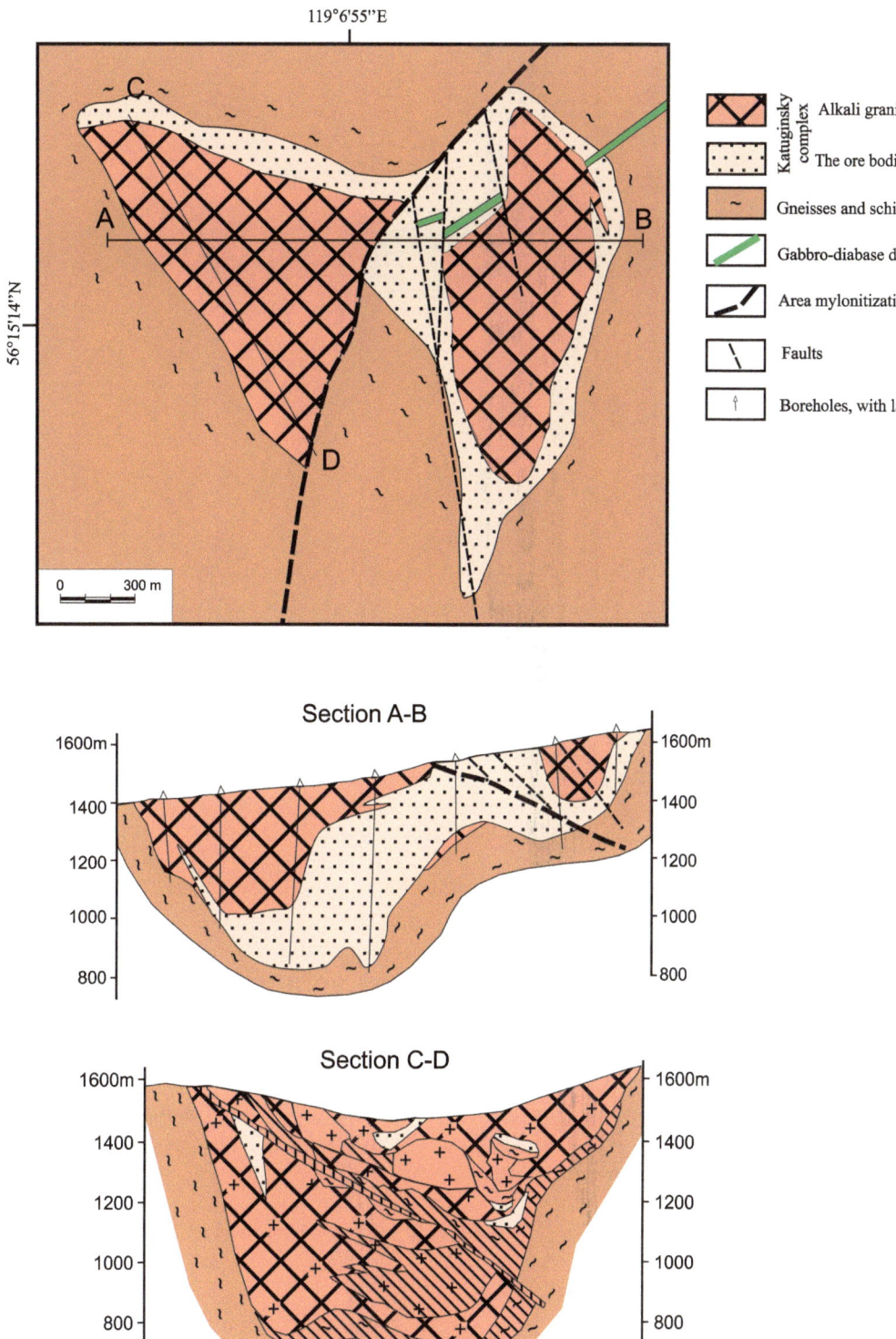

Fig. 10.2 Geological map of the Katugin deposit (After Arkhangel'skaya et al. 2004)

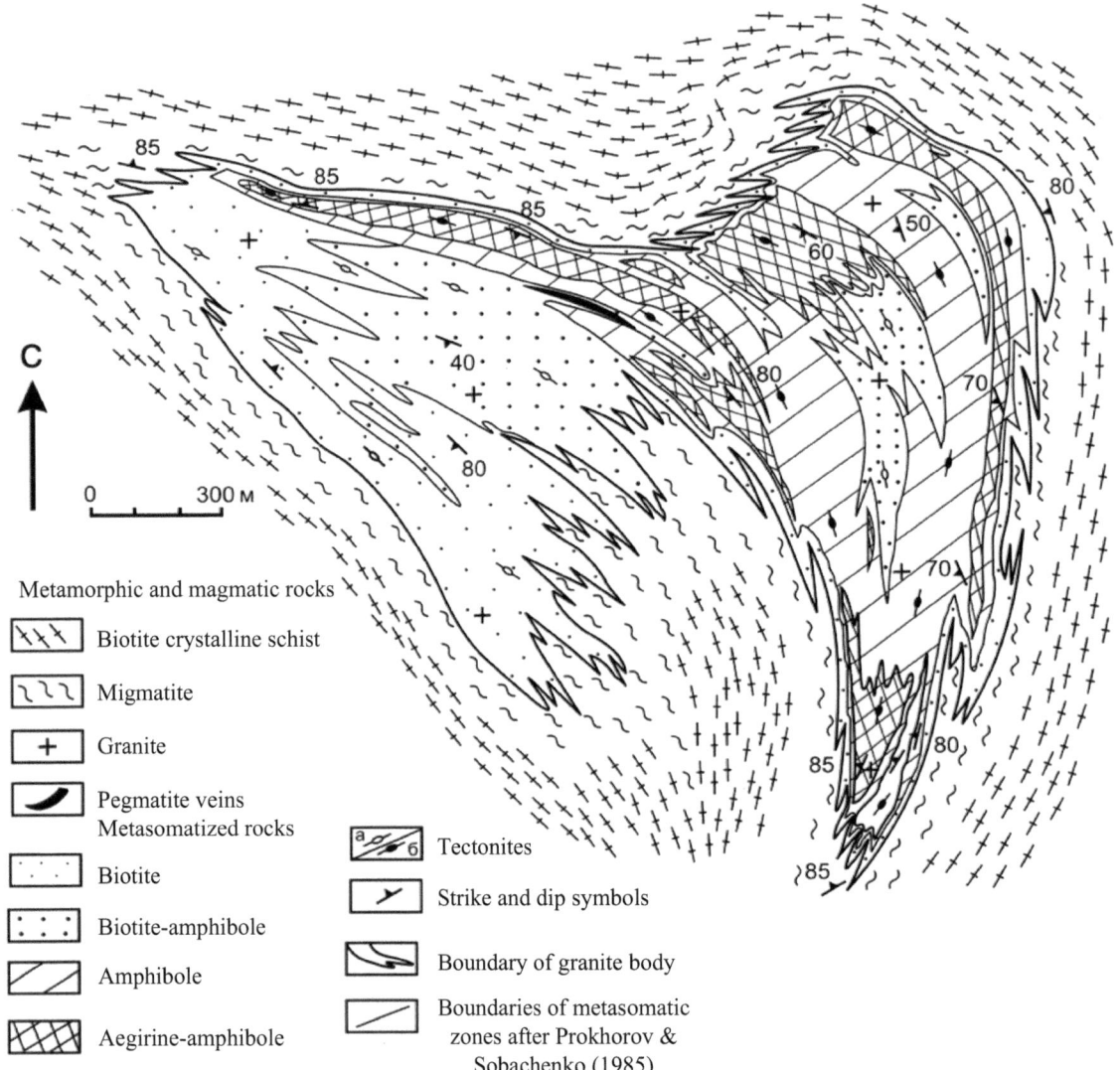

Fig. 10.3 Eastern orebody of the Katugin deposit (After Sklyarov et al. 2016)

rocks of the surrounding zone are characterized by the highest Y and Ga contents. The metasomatized country rocks are distinguished by the highest Mo and Be concentrations.

The vertical geochemical zoning of the Katugin deposit is broadly similar to the lateral zoning. However, the lowermost arfvedsonite rather than the overlying alkaline aegirine-arfvedsonite granite rocks are in its center.

The alkaline granite rocks in the western part of the Katugin deposit are similar to the upper levels of this deposit in mineralogy and geochemistry. This points to a high-grade rare-metal mineralization at depth. The trace elements are distinctly linked to certain minerals or their groups. Nb, Ta, Zr, Hf, Ln, Y, and Mo are almost completely concentrated in their typical

minerals, whereas Li, Rb, Ga, Sn, Sc, V, Ni, Co, and Cr are completely dispersed in the rock-forming minerals.

10.3 Chemical and Mineral Composition of Ore

10.3.1 Mode of Occurrences of Rare Elements in Ore

The mineralized system contains 1.62 wt% ZrO_2, 0.026 wt% Ta_2O_5, 0.374 wt% Nb_2O_5, 0.16 wt% Y_2O_3, 0.22 wt% REE_2O_3, and 0.0078 wt% U_3O_8. Pyrochlore, zircon, gagarinite, REE-fluorite, monazite, and cryolite are major minerals accumu-

Table 10.1 Chemical composition of representative samples from granite of the Katugin Complex

No	1	2	3	4	5	6	7
Sample	238	240	242	245	90_471	13,023	14,382
SiO_2	72.24	68.18	74.75	78.37	71.30	72.89	67.70
TiO_2	0.52	0.23	0.13	0.14	0.18	0.23	0.26
Al_2O_3	9.62	10.79	14.06	10.00	11.34	12.98	12.07
Fe_2O_3	5.80	7.85	1.20	3.70	6.19	3.5	5.59
MnO	0.09	0.07	0.02	0.05	0.02	0.06	0.08
MgO	0.21	0.19	0.10	0.18	0.16	0.16	0.13
CaO	0.12	0.11	0.60	0.18	0.08	0.43	0.05
Na_2O	4.30	7.10	4.95	3.17	5.12	4.46	8.13
K_2O	4.08	3.82	3.51	3.61	4.64	4.73	4.19
P_2O_5	0.01	0.01	0.02	0.02	0.01	<0.05	<0.05
LOW	1.17	1.07	0.45	0.11	0.55	0.80	1.80
Total	99.72	99.67	99.95	99.98	99.75	100.24	100.00
Be	15	6.6	21	4.4	3.7	9.3	16
Sc	0.56	1.3	0.71	0.43	0.60	1.3	0.65
Cu	20	7.8	10	35	7.0	7.3	18
Zn	1718	982	291	559	335	254	1153
Ga	51	47	30	45	60	39	56
Cs	0.76	1.6	0.80	0.38	2.5	2.1	2.5
Rb	1398	1144	301	947	914	376	1337
Sr	57	25	47	13.8	18	15	4.0
Ba	102	58	178	7.0	29	203	10
Y	1090	650	500	580	40	204	225
Zr	16,030	2290	1040	3760	1210	1013	3926
Hf	504	128	46	217	102	–	–
Nb	3758	1730	730	930	3270	399	1248
Ta	82	37	13	18	24	30	97
La	162	133	88	114	340	164	124
Ce	445	402	235	299	903	266	478
Pr	61	51	29	39	100	29	54
Nd	237	187	107	146	293	129	220
Sm	99	63	38	53	57	29	62
Eu	5.0	3.1	2.48	2.5	2.0	1.74	2.5
Gd	105	68	41	54	35	28	
Tb	30	19.0	12.0	14.9	5.5	5.3	10.9
Dy	204	130	83	95	26	33	65
Ho	41	29	18.0	20	4.0	7.4	13.3
Er	110	81	50	53	8.4	22	33
Tm	14.6	11.4	7.1	7.5	1.01	3.5	4.3
Yb	87	65	41	40	7.3	21	25
Lu	12.9	9.1	5.6	5.5	1.21	2.7	4.4
Pb	229	131	31	22	55	29	59
Th	406	213	72	61	52	45	54
U	83	49	29	22	105	6.5	74

Note. Oxide, in %; elements, in ppm. Sample No: 238, amphibole-aegirine granite; 240, aegirine granite; 242, albite leucogranite; 245, amphibole granite; 90_471, aegirine-astrophyllite granite; 13,023, amphibole granite; 14,382, amphibole aegirine (After Larin et al. 2015)

lating the metals. Fergusonite, ilmenite, and ilmenorutile are widespread in metasomatized country rocks. The list of minerals from the Katugin deposit is in Table 10.2.

The high Zr concentrations in rock-forming minerals are caused by numerous zircon inclusions. Gagarinite (Figs. 10.4 and 10.5), pyrochlore, monazite, yttrofluorite, and xenotime are the minerals concentrating REE and Y.

Nb, Ta, Zr, Hf, La, Y, and Mo are concentrated almost completely in their minerals. Molybdenite is the only mineral-carrier and concentrator of Mo. Li, Rb, Ga, Sn, Sc,

Table 10.2 Ore minerals from the Katugin deposit

Bastnaesite	$(Ce, La, Nd)(CO_3)F$
Pyrochlore	$A_2Nb_2(O,OH)_6Z$, A is Na, Ca, Sn^{2+}, Sr, Pb^{2+}, Sb^{3+}, Y, U^{4+}, H_2O, Z is OH, F, O, $(Ca,Na,REE,Y,Pb)_2Nb_2O_6(0H,F)$
Aegirine	$NaFe^{3+}Si_2O_6$
Arfvedsonite	$[Na][Na_2][Fe_4^{2+}Fe^{3+}]Si_8O_{22}(OH)_2$
Cryolite	$Na_3[AlF_6]$
Yttrofluorite	$(Ca,Y)F_2$
Thorite	$Th(SiO_4)$
Columbite	$FeNb_2O_6$ to $(Mn,Fe)(Nb,Ta)_2O_6$
Fluocerite-(Ce)	$(Ce,La, Nd)F_3$
Weberite	$Na_2Mg[AlF_6]F$
Chiolite	$Na_5Al_3F_{14}$,
Neighborite	$NaMgF_3$
Tveitite-(Y)	$(Y, Na)_6Ca_6Ca_6F_{42}$
Elpasolite	$NaK_2[AlF_6]$
Gearksutite	$Ca[Al(F,OH)_5(H_2O)]$
Prosopite	$CaAl_2(F,OH)_8$
Thomsenolite	$NaCa[AlF_6] \cdot H_2O$
Ralstonite	$Na_{0.5}(Al,Mg)_2(F,OH)_6 \bullet H_2O$
Jacobsite	$Mn^{2+}Fe_2^{3+}O_4$
Weberite	$Na_2Mg[AlF_6]F$
Pachnolite	$NaCa[AlF_6] \cdot H_2O$
Usovite	$Ba_2CaMgAl_2F_1$
Gagarinite	$NaCaYF_6$

V, Ni, Co, and Cr have no typical minerals and are completely dispersed in rock-forming minerals. Primarily, these are feldspar, dark-colored minerals, and pyrochlore. Sphalerite is a carrier of Zn, and galena bears Pb.

The REE ore forms fluorite stock works in the exocontact zone. Some rare minerals were found in the Katugin ore, such as ceriopyrochlore $A_2Nb_2(O,OH)_6Z$ and fluocerite-(Ce) $(Ce,La)F_3$ (Fig. 10.5).

10.3.2 Types of Mineralization

The study of ore minerals allowed to allocate and justify three main types of mineralization from granite (Sklyarov et al. 2016).

1. *Zirconium mineralization.* Zircon was found in all rocks, but its concentration is very variable, from rare accessory crystals to clusters with a total mineral amount of 15–20% of the rock volume. Zircon size varies widely: from a fraction of a millimeter to 1.5 cm. In most cases, zircon is free of inclusions of other minerals (with the exception of quartz and albite), but it often contains many small (2–15 micron) inclusions of different composition.

Fig. 10.4 Minerals from the Katugin deposit (photo AEvseev)
(**a**) Gagarinite-(Y) $NaCaYF_6$, red, sample from the Fersman Museum, RAS, Moscow (sample № 83195 of VI Stepanov); (**b**) thorite $Th(SiO_4)$, Fersman Museum, RAS, Moscow (№ 83194 of VI Stepanov); (**c**) sillimanite $Al_2(SiO_4)O$, RGGRU Museum, Moscow (№ 2018); (**d**) dravite $Na(Mg_3)Al_6(Si_6O_{18})(BO_3)_3(OH)_3(OH)$ (8 cm), RGGRU Museum (P-1273); (**e**) cryolite $Na_3[AlF_6]$, RGGRU Museum, Moscow (sample of AE Zadov); (**f**) cryolite $Na_3[AlF_6]$, RGGRU Museum, Moscow (№ 2013). http://geo.web.ru/druza/l-Katugin.htm

Fig. 10.5 Minerals of the Katugin deposit (photo P Kartashev)
(**a**) Ceriopyrochlore $A_2Nb_2(O,OH)_67$, $5 \times 7 \times 3,5$ cm, sample of P Kartashev; (**b**) fluocerite-(Ce) $(Ce,La)F_3$, $6 \times 4 \times 3$ cm, sample of TN Shuriga; (**c**) astrophyllite $K_2NaFe_7^{2+}Ti_2Si_8O_{28}(OH)_4F$, $10 \times 8 \times 4$ cm, sample of NI Frishman; (**d**) astrophyllite $K_2NaFe_7^{2+}Ti_2Si_8O_{28}(OH)_4F$, rosette is 2 cm, sample of P Kartashev (http://webmineral.ru/minerals/image.php=8961)

Bastnaesite, fluocerite, gagarinite, yttrofluorite, fluorite, thorite, and cryolite are recorded among the inclusions. In general, we observed a clear predominance of fluoride. Usually in the rocks, "clean" or rich inclusions of zircons can be found. The similar facts are observed in the Sn-Ta-Nb (REE, cryolite) Madeira deposit (Brazil), which is similar to the Katugin deposit in many ways.

2. *Ta-Nb-REE mineralization*. Minerals of this type of mineralization are concentrated mainly in the peripheral parts of the layered mafic-ultramafic complex. They occur (a) in the form of separate grains and their concentrations in granite splice ore, (b) as impurities in zircon, and (c) as interstitial grains among silicate minerals. Individual crystals or clusters are the most typical for pyrochlore,

which often contains inclusions of opaque minerals. The studies have shown that the majority of crystals and grains of pyrochlore have heterogeneous structure due to the variations in its composition and inclusions of other minerals. Bastnaesite, columbite, fluocerite, and hydrating products of pyrochlore are present near boundary of this mineral. The ore may be represented by various minerals; the most common of them are pyrochlore, columbite, fluocerite, ilmenite, and sometimes sulfides (sphalerite, galena, pyrrhotite, and pyrite). Interstitial minerals are represented by stacked fluorides and fluorcarbonates, often with a small amount of chlorite, and they have a very complex structure. In the bulk, tveitite, with numerous small oriented inclusions of fluocerite or early bastnaesite, prevails. Bastnaesite develops after fluorides, forming veins and individual clusters, sometimes symplectites with chlorite.

3. *Aluminofluoride mineralization.* Cryolite is usually found in the form of small segregations in granite, or it composes relatively thick veins or lenses (up to 30–50% cryolite). It is associated with the aluminum fluoride and fluoride (weberite Na_2MgAlF_7, chiolite $Na_5Al_3F_{14}$, neighbourite $NaMgF_3$, fluorine, tveitite-(Y), fluocerite-(Ce), gagarinite-(Y), yttrofluorite, elpasolite K_2NaAlF_6, simmonsite Na_2LiAlF_6) and products of their changes (gearksutite $Ca[Al(F,OH)_5(H_2O)]$, prosopite $CaAl_2(F,OH)_8$, thomsenolite $NaCaAlF_6 \cdot H_2O$, pachnolite $NaCaAlF_6 \cdot H_2O$, hydrokenoralstonite $Na_{0.5}(Al,Mg)_2(F,OH)_6 \cdot H_2O)$. In aegirine-arfvedsonite granite isolation, aluminum fluoride was found (mostly barium aluminum fluoride). The size of such segregates is up to 1 cm, they typically have a reddish color due to the presence of oxides and hydroxides of iron. In these phases of barium minerals were identified: $Ba_2CaMgAl_2F_{14}$, $BaAlF_4$ (OH), Ba-analogue jakobsite $CaAlF_4$ (OH)-$CaAlF_5$) and $BaCa_2AlF_9$. Note that barium minerals were not known previously in the Katugin ore, and the latter two phases are not known in nature, although their synthetic analogues are known. Barium phases are associated with cryolite, weberite, prosopite, pachnolite, thomsenolite, and fluorite and do not show features of their secondary origin for some minerals (Starikova and Sharygin 2015). Silicates in aluminofluoride show no separation.

Lithium-Na-Fe-amphibole, another new mineral, was recently discovered in the Katugin deposit (Sharygin et al. 2016). The species contain from 30 to 70 vol% of cryolite. The femic minerals are Fe-silicates (Li-Na-Fe-amphibole, Li-containing fluorannite, bafertisite), oxides (magnetite, ilmenite, pyrochlore, cassiterite), and sulfides (sphalerite, pyrite, chalcopyrite). Quartz, K-feldspar, polylithionite, RE-fluorides, and albite are present in minor amounts as accessory minerals. Variations of amphibole chemical composition are follows: 48.5–48.9 wt% SiO_2, 0.4–0.8 wt% TiO_2, 1.6–2.2 wt% Al_2O_3, 15.9–17.1 wt% Fe_2O_3, 17.6–18.4 wt% FeO, 0.8–0.9 wt% MnO, 0.3–1.1 wt% ZnO, 0.2–0.3 wt% MgO, <0.1wt% CaO, 8.4–8.7 wt% Na_2O, 1.4–1.5 wt% K_2O, 0.6–0.8 wt% Li_2O, 0.7–0.8 wt% H_2O, and 2.2–2.5 wt% F. The studied amphibole has a specific structure, which falls into a field of intermediate composition between F-Fe-member subgroup of sodic amphibole. Its

composition can be displayed as a ferro-ferri-fluoro-nyobite (40–45 mol%), ferro-ferri-fluoro-leakeite (40–45 mol%) ± fluoro-riebeckite-fluoro-arfvedsonite (10–20 mol%). The mineral is monoclinic. Its space group is C2/m, a = 9.7978 (2), b = 17.9993 (3), c = 5.33232 (13) Å, b = 103.748 (2) °, V = 913.43 (3) Å3, and Z = 2. The structural formula of Li-Na-Fe-amphibole: ($Na_{0.46}$ $K_{0.24}$ 0.30) $Na_{2.00}$ ($Fe^{2+}_{0.95}$ $Mg_{0.05}$)$_2$ ($Fe^{3+}_{0.95}Ti_{0.025}Mg_{0.025}$)$_2$ ($Li_{0.37}$ $Fe^{2+}_{0.48}Mn_{0.10}Zn_{0.05}$) [($Si_{0.91}Al_{0.09}$)$_4Si_4O_{22}$] ($F_{0.58}$ ($OH)_{0.42}$)$_2$. There are no data on Raman and Mössbauer spectroscopy for this amphibole.

In composition, usovite is close to the theoretical $Ba_2CaMgAl_2F_{14}$. Phase $BaAlF_4$ (OH), presented as prismatic or split beans, contains up to 5.5–6 wt% oxygen. Raman spectroscopy data indicate presence of OH groups. All spectra show peaks at the 3500–3600 cm^{-1}. For synthetic $BaAlF_5$, four modifications are known. Taking into account the ratio of minerals, phase transitions, and phase stability for $BaAlF_5$, it was assumed that $BaAlF_4(OH)$ phase from the Katugin granite corresponds to alpha- or beta-modifications. $BaCa_2AlF_9$ phase is present as a border around grains or aggregates with fluorite. According to a microprobe analysis, oxygen is absent in this mineral. Unfortunately, it was not confirmed or rejected the presence of OH groups or water in the structure of the mineral because of Raman laser strong luminescence in the region 2000–4000 cm^{-1}. Thermobarogeochemical results of alkaline granite from the Katugin Complex showed that many minerals, such as zircon, plagioclase, quartz, pyrochlore, and arfvedsonite, contain syngenetic melt inclusions. The homogenization temperature for melt inclusions from zircon is 760–780 °C.

Total fluid concentration in zircon and quartz from alkaline granite of the Katugin layered mafic-ultramafic complex is 400 and 30 mm^3 to 100 mg, respectively. Degassing of quartz occurs at $T = 600$–650 °C and of zircon at $T = 700$–850 °C.

10.3.3 Distribution of Rare Elements in Zircon

Distribution of trace and rare earth elements was studied by Levashova et al. (2014) in zircon from two principal types of ore-bearing rocks of the Katugin rare-metal deposit (arfvedsonite-aegirine and biotite-riebeckite rocks) and from zones of its recrystallization with nodular zircon clusters (Table 10.3). Rims and edge parts of zircon from the Katugin deposit are depleted (Fig. 10.6), compared with cores and

Table 10.3 Distribution of trace and rare earths' elements (ppm) in zircon from rocks of the Katugin deposit

	Aegirine-arfvedsonite rock						Biotite-riebeckite rock							
N sample	13/136				13/110					K5/9			17/166	
Element	1	2	3	5	6	7	8	9	10	11	12	13	15	14
	C	R	C	R	C	R	C	R	C	R	C	R	C*	C
La	1.83	0.07	0.06	0.09	0.72	0.05	10.6	0.53	0.49	0.21	1.18	0.17	0.20	0.44
Ce	54.0	0.92	0.90	0.45	8.28	0.73	53.8	4.02	16.4	5.12	9.47	1.55	8.99	2.91
Pr	8.48	0.01	0.07	0.03	0.31	0.01	3.52	0.18	0.24	0.07	0.72	0.06	0.10	0.15
Nd	49.6	0.04	1.04	0.09	1.24	0.05	18.8	0.90	1.26	0.35	4.19	0.31	0.70	0.62
Sm	41.6	0.11	3.67	0.11	1.88	0.08	11.0	0.81	1.55	0.60	3.98	0.33	2.28	0.68
Eu	2.38	0.02	0.40	b.d.l.	0.14	0.01	0.73	0.04	0.11	0.09	0.36	0.02	0.26	0.03
Gd	90.0	2.06	12.5	0.38	9.37	1.24	43.8	4.15	6.53	3.66	11.8	2.96	21.8	3.04
Dy	334	56.9	36.9	11.1	142	27.6	195	41.3	77.1	51.7	100	42.8	284	36.5
Er	384	227	51.1	46.7	355	116	375	136	347	194	387	152	851	165
Yb	441	378	85.2	76.7	378	261	554	389	1195	609	1168	431	1755	633
Lu	57.2	47.2	10.2	9.08	52.6	43.4	83.1	66.0	206	100	195	68.8	253	109
Li	146	15.9	49.1	5.70	141	8.06	22.8	10.7	80.3	29.2	119	6.79	31.7	10.0
p	117	40.2	32.9	bdl	bdl	4 62	81.7	34.7	128	113	282	81.7	281	89.2
Ca	34.8	37.4	6.03	10.7	24.1	9.87	21.7	10.7	13.7	8.64	15.9	7.15	20.5	10.9
Ti	37.7	0.57	5.11	1.99	90.7	0.73	15.3	1.72	12.6	1.21	7.53	1.44	4.49	0.56
Sr	0.91	28.4	0.53	0.43	1.57	0.94	1.43	0.97	1.08	1.16	1.55	1.12	2.23	1.46
Y	1154	409	146	78.5	823	206	2108	335	1347	482	1035	241	1576	187
Nb	533	134	52.7	26.0	520	22.0	170	38.0	290	49.2	119	21.7	82.1	36.4
Ba	8.29	1.36	2.14	1.11	2.76	2.18	2.99	1.46	6.64	7.97	12.7	1.91	2.72	1.32
Hf	8042	5248	13277	10854	20124	12755	6254	14833	13723	14019	10178	10511	15295	12820
Th	1053	53.8	76.2	8.14	135	13.4	179	36.8	337	125	269	22.4	264	51.8
U	390	39.5	217	6.68	143	8.88	316	75.4	721	166	889	27.4	298	52.7
Th/U	2.70	1.36	0.35	1.22	0.94	1.51	0.57	0.49	0.47	0.76	0.30	0.82	0.89	0.98
Eu/Eu*	0.12	0.10	0.18	n.d.	0.10	0.08	0.10	0.07	0.10	0.18	0.16	0.07	0.11	0.06
Ce/Ce*	3.32	9.12	3.34	2.15	4.22	9.43	2.13	3.17	11.6	10.2	2.49	3.74	15.4	2.76
ΣREE	1464	712	202	145	949	450	1349	643	1852	965	1881	700	3178	950
ΣLREE	114	1.04	2.07	0.66	10.6	0.84	86.6	5.63	18.4	5.75	15.6	2.09	9.99	4.12
ΣHREE	1306	711	196	144	937	449	1251	637	1832	958	1861	698	3166	946
Lu_N/La_N	302	6508	1516	1020	707	8172	76	1202	4017	4647	1596	3967	12212	2361
Lu_N/Gd_N	5.1	185	6.6	195	45	284	15	129	255	221	133	188	94	289
Sm_N/La_N	36	2.6	91	2.1	4.2	2.6	1.7	2.4	5.0	4.6	5.4	3.2	18	2.5
$T(Ti)$, °C	875	541	687	618	984	555	782	608	763	586	719	597	677	540

	Late zones of recrystallization								
	8/35					K2/80		4a	
Element	16	17	18	19	20	22	21	24	23
	C*	R	C	I	R	C*	R	C	R
La	0.09	0.16	3.37	0.99	0.83	0.05	0.04	14.0	1.09
Ce	3.55	1.85	49.2	7.10	3.51	18.6	7.50	69.5	7.48
Pr	0.02	0.04	8.23	0.88	0.19	0.08	0.03	8.51	0.65
Nd	0.13	0.12	42.4	3.62	0.66	1.91	0.47	31.4	4.19
Sm	0.55	0.21	41.5	3.45	0.37	10.6	2.27	10.8	3.53
Eu	0.07	0.01	2.20	0.26	0.03	1.19	0.25	0.51	0.23
Gd	7.97	1.76	60.9	4.72	2.59	78.2	21.6	25.3	13.8
Dy	168	33.4	574	58.7	42.6	811	239	118	107
Er	826	155	1741	302	219	1833	699	274	276
Yb	1548	487	3148	685	732	3320	1512	531	595
Lu	152	77.2	339	70.9	103	533	248	83.2	93.9
Li	52.9	10.6	27.9	25.7	16.9	29.9	10.6	101	64.8
P	302	64.3	754	102	108	496	265	101	125
Ca	7.88	8.19	44.2	13.7	28.0	6.81	5.21	321	30.4
Ti	0.38	0.77	38.5	6.11	2.13	2.93	2.64	34.7	2.01

(continued)

Table 10.3 (continued)

	Late zones of recrystallization								
	8/35					K2/80			4a
Element	16	17	18	19	20	22	21	24	23
	C*	R	C	I	R	C*	R	C	R
Sr	1.51	0.88	3.60	1.33	2.62	2.88	1.65	7.55	2.34
Y	1195	244	2884	384	314	3411	1173	638	585
Nb	95.8	35.6	458	103	60.3	209	154	329	60.8
Ba	1.14	1.26	5.01	2.52	2.89	0.78	1.32	17.9	4.64
Hf	14145	13226	6102	9259	17806	9073	10493	12215	12348
Th	232	17.8	1160	265	37.0	272	107	90.4	88.5
U	577	46.6	237	142	69.2	1131	243	257	222
Th/U	0.40	0.38	4.90	1.87	0.54	0.24	0.44	0.35	0.40
Eu/Eu*	0.10	0.03	0.13	0.19	0.11	0.13	0.11	0.09	0.10
Ce/Ce*	21.6	5.32	2.26	1.84	2.13	73.4	48.8	1.54	2.15
ΣREE	2706	757	6010	1137	1105	6608	2730	1166	1103
$\Sigma LREE$	3.78	2.17	103	12.6	5.19	20.6	8.04	123	13.4
$\Sigma HREE$	2702	754	5863	1120	1100	6576	2720	1031	1086
Lu_N/La_N	17003	4524	970	690	1204	107594	54814	57	831
Lu_N/Gd_N	155	354	45	121	323	55	93	27	55
Sm_N/La_N	10	2.1	20	5.6	0.7	354	84	1.2	5.2
$T(Ti),°C$	518	558	878	701	623	645	638	866	619

Note. C center of grain, (*C** recrystallized), *R* rim, *b.d.l.* below detection limit, *n.d.* not determined. (Levashova et al. 2014)

central parts of grains, in almost all trace elements, except hafnium. Compositional features indicate the magmatic origin of zircon cores, contrary to the point of view of the metasomatic origin of the deposit. Composition of zircon rims (similarity of the REE distribution spectra in rims and cores) assumes a possible post-magmatic transformation of the late-magmatic zircon or metamorphic origin of its rims.

10.4   Age of the Katugin Granite

The results of isotope-geochemical Sm-Nd studies of alkaline granite from the Katugin Complex are shown in Table 10.4 (Larin et al. 2015). They are characterized by negative and close to CHUR values $\varepsilon_{Nd}(t)$ 0.0 … −1.9 and, as a rule, very high $^{147}Sm/^{144}Nd$. Therefore, one can calculate the Nd age estimates only for two samples of alkaline granite ($t_{Nd(DM)}$ = 2.7 and 2.6 Ga). The values of the two-stage model Nd age ($t_{Nd (DM-2st)}$) in studied samples from alkaline granite suggest an age of 2.7–2.5 Ga.

The Katugin Complex does not exceed 10 km, while the formation of magmas of alkaline granites took place under much deeper-seated conditions ($P > 5$ kbar). Nevertheless, the age of mineralization is under discussion. The alkaline granite is dated at 2066 ± 6 Ma (U-Pb method on zircon)

(Larin et al. 2002). Kotov et al. (2015) analyzed a U-Pb system in zircon taken from ore in granite (Table 10.5). The ore in alkaline granite of the Katugin Complex consists mainly of zircon and pyrochlore. Zircon from the ore from aegirine-amphibole alkaline granite consists of transparent and translucent idiomorphic or subidiomorphic crystals. Habitus of zircon crystals varies from dipyramidal to prismatic. Microscopic study in transmitted light shows zonal structure of grains, with observed nucleus and rims (Fig. 10.7, V–X). The former are enriched in melt, fluid, and gas inclusions. Points of the isotopic compositions of the studied zircon form discordia, the upper intersection of which corresponds to the concordia with the age of 2055 ± 7 Ma and the bottom with 973 ± 110 Ma. This estimation of age obtained from zircon of the orebody in 2055 ± 7 Ma coincides within the error of the estimated age of Katugin alkaline granite.

10.5   Genesis of Mineralization

Despite a long period of study, the problem of the source of metals and the nature of the host rocks are still debatable (Arkhangel'skaya et al. 1993; Osokin et al. 2000; Sklyarov et al. 2016). It was suggested that deposit was formed as a result of metasomatic process (Prokhorov and Sobachenko

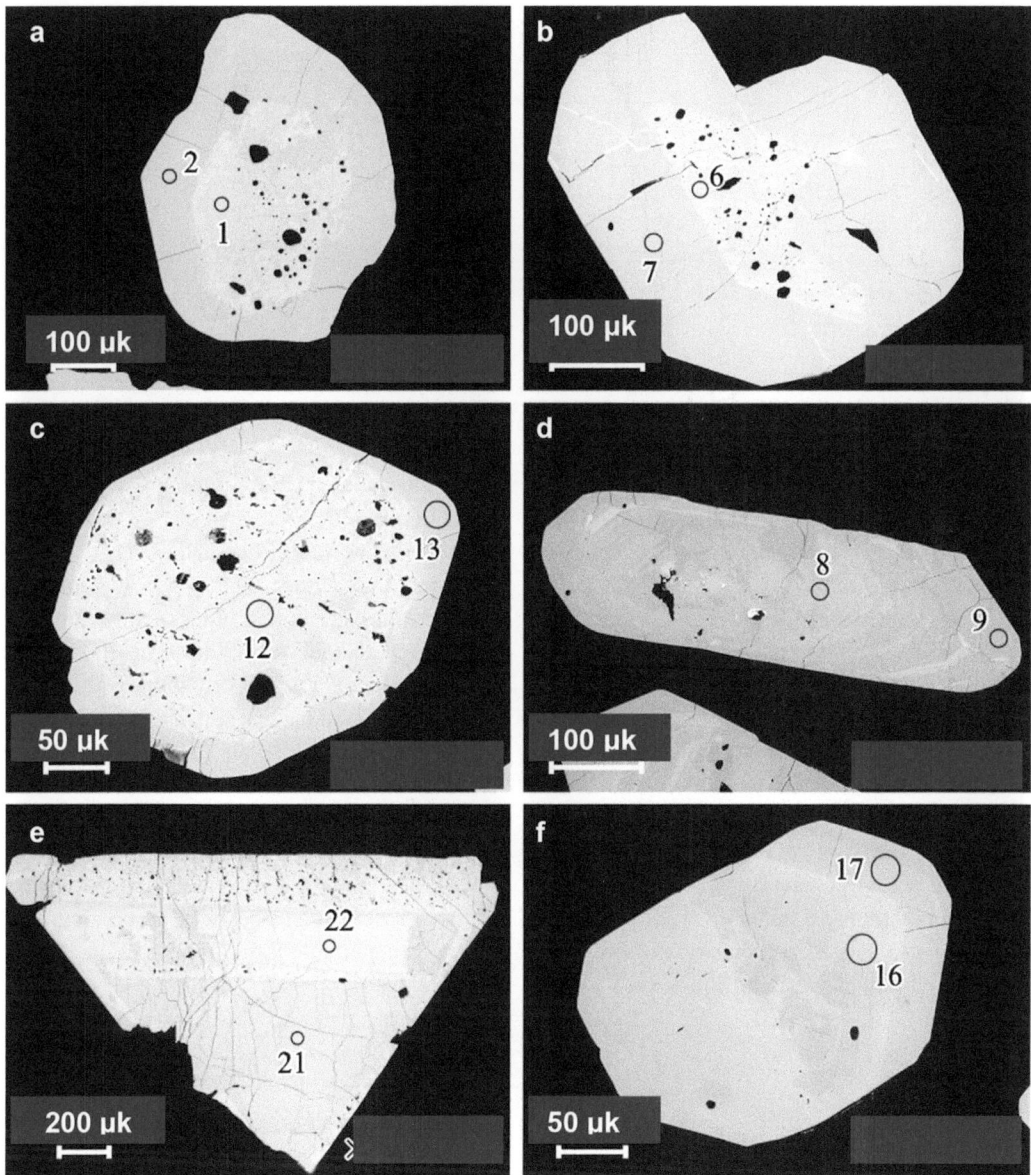

Fig. 10.6 Back scattered electron (BSE) images of zircon grains (after Levashova et al. 2014)
Samples: (**a**) 11/136, (**b**) 13/110, (**c** and **d**) K5/9, (**e**) K2/80, (**f**) 8/35. Number of point corresponds to number in Table 10.3

1985). Nowadays, there is an opinion on magmatic origin of this deposit. Sklyarov et al. (2016) believe that the mapped boundaries between metasomatic zones (Fig. 10.3) are boundaries between different granite types and that ore minerals crystallized from F-rich aluminum melt under liquation of it with silicate melt.

Gladkochub et al. (2015) and Larin et al. (2015) studied the mineralization of the Katugin deposit. Alkaline granites from the Katugin deposit mostly consist of quartz, albite, and K-feldspar. Femic minerals (arfvedsonite,

aegirine, astrophyllite) are present in significant quantities. The main minerals are pyrochlore, zircon, REE-fluorite, and cryolite. In the granite of the Katugin Complex biotite, amphibole-biotite (biotite-riebeckite), arfvedsonite, aegirine-arfvedsonite, and aegirine varieties were distinguished.

It was found that zircons from Katugin alkaline granite have distinct zonation. The cores of zircons are enriched in primary melt, gas, and fluid inclusions, while there are only a few fluid inclusions in the rim. The age of cores is 2065 Ma

Table 10.4 Sm-Nd isotopic data for the Katugin alkaline granite

Sample, No	Sm, mkg/g	Nd mkg/g	$^{147}Sm/^{144}Nd$	$^{143}Nd/^{144}Nd$ ($\pm2\sigma_{mes.}$)	ε_{Nd} (t)	$t_{Nd(DM)}$	$t_{Nd(DM-2st)}$
13023	30.1	143.9	0.1263	0.511656 ± 8	−0.5	2583	2548
14382	63.1	234	0.1630	0.512156 ± 5	−0.4		2547
C -54-8	38.8	155.7	0.1507	0.511919 ± 2	−1.8		2659
C-6-1	32.9	128.1	0.1551	0.512044 ± 2	−0.5		2554
C-15-1	78.5	167.4	0.2837	0.513723 ± 2	−1.9		2667
C38A-1	73.2	195.4	0.2266	0.513014 ± 2	−0.6		2558
C-54-36	55.5	216	0.1557	0.511994 ± 2	−1.7		2648
C-94-1	52.6	219	0.1452	0.511907 ± 2	−0.6		2559
C-54-27	153.7	671	0.1385	0.511780 ± 3	−1.3	2751	2616
C-121-1	184.9	642	0.1740	0.512331 ± 2	0		2507

Note. εNd (t) and two-stage model age (tNd(DM–2st)) were calculated for 2066 Ma (Larin et al. 2002). Data after Larin et al. (2015)

(U-Pb method, TIMS) and of rims is 1950 Ma (U-Pb method, TIMS; Kotov et al. 2015). In addition, alkaline granite from the Katugin layered mafic-ultramafic complex contains euhedral colorless crystals of zircon with an age of 1.85 Ga (U-Pb method, TIMS). Inclusions in the zircon crystals of this type were detected. The findings suggest that the rare metals from alkaline granites of the Katugin layered mafic-ultramafic complex are of magmatic origin.

The results of recent studies led to the following conclusions:

1. The vast majority of ore from the Katugin deposits associates with igneous stage of evolution of the granite complex, and the formation of ore minerals was due to two interrelated processes: (1) the crystallization of ore minerals directly from the melt and (2) the liquation of high-F granite melt into the aluminum silicate and fluoride or alumofluoride phases.
2. At the same time, the earliest stage of the liquation is recorded by numerous small inclusions of fluorides and fluorcarbonates in zircon. At the first stage, fluoride melts enriched in REE crystallized in the interstitial spaces, forming a collapse of structures with tveitite, fluocerite, gagarinite, and bastnaesite. At the second stage, globules and interstitial segregations of alumofluorides (including cryolite) were formed.
3. Gladkochub et al. (2015) concluded that formation of the Katugin rare-metal deposit was a result of the impact of mantle plume on the continental lithosphere at ~2.065 Ga. At ~1.95 Ga, a terrane, accommodating the deposit, was amalgamated into the Siberian craton.

Table 10.5 Results of geochronological U-Pb study of zircons from ore segregations of the Katugin granite

Size, mk	U/Pb*	Isotopic ratio						Age, Ma		
		206Pb/204Pb	207Pb/206Pba	208Pb/206Pba	207Pb/235 U	206Pb/238 U	Rho	207Pb/235 U	206Pb/238 U	207Pb/206Pb
100–150, 1	1.84	257	0.1227 ± 3	0.5928 ± 1	5.7273 ± 211	0.3384 ± 7	0.84	1936 ± 7	1879 ± 6	1997 ± 4
>200, 2	1.74	247	0.1254 ± 2	0.4684 ± 1	6.1623 ± 298	0.3565 ± 16	0.70	1999 ± 9	1965 ± 8	2034 ± 3
>200, 1	1.96	301	0.1252 ± 4	0.3296 ± 1	6.2436 ± 250	0.3618 ± 7	0.51	2011 ± 8	1991 ± 4	2031 ± 6
>200, 2	1.94	487	0.1257 ± 2	0.3764 ± 1	6.3099 ± 189	0.3640 ± 11	0.98	2020 ± 6	2001 ± 6	2039 ± 2
>200, 2	2.13	792	0.1256 ± 3	0.3553 ± 1	6.3254 ± 259	0.3654 ± 11	0.78	2022 ± 8	2008 ± 6	2037 ± 5

Note. A, isotopic ratios corrected on blank and usual Pb; *Rho*, coefficient of correlation for mistakes $^{207}Pb/^{235}U-^{206}Pb/^{238}U$; A = 20%, mount of zircon mass deleted during aeroabrasive clearning; >200, 2, size and amount of grains used for U-Pb age calculations; *, zircon mass was not detected; σ = last digits after point. Data after (Kotov et al. 2015)

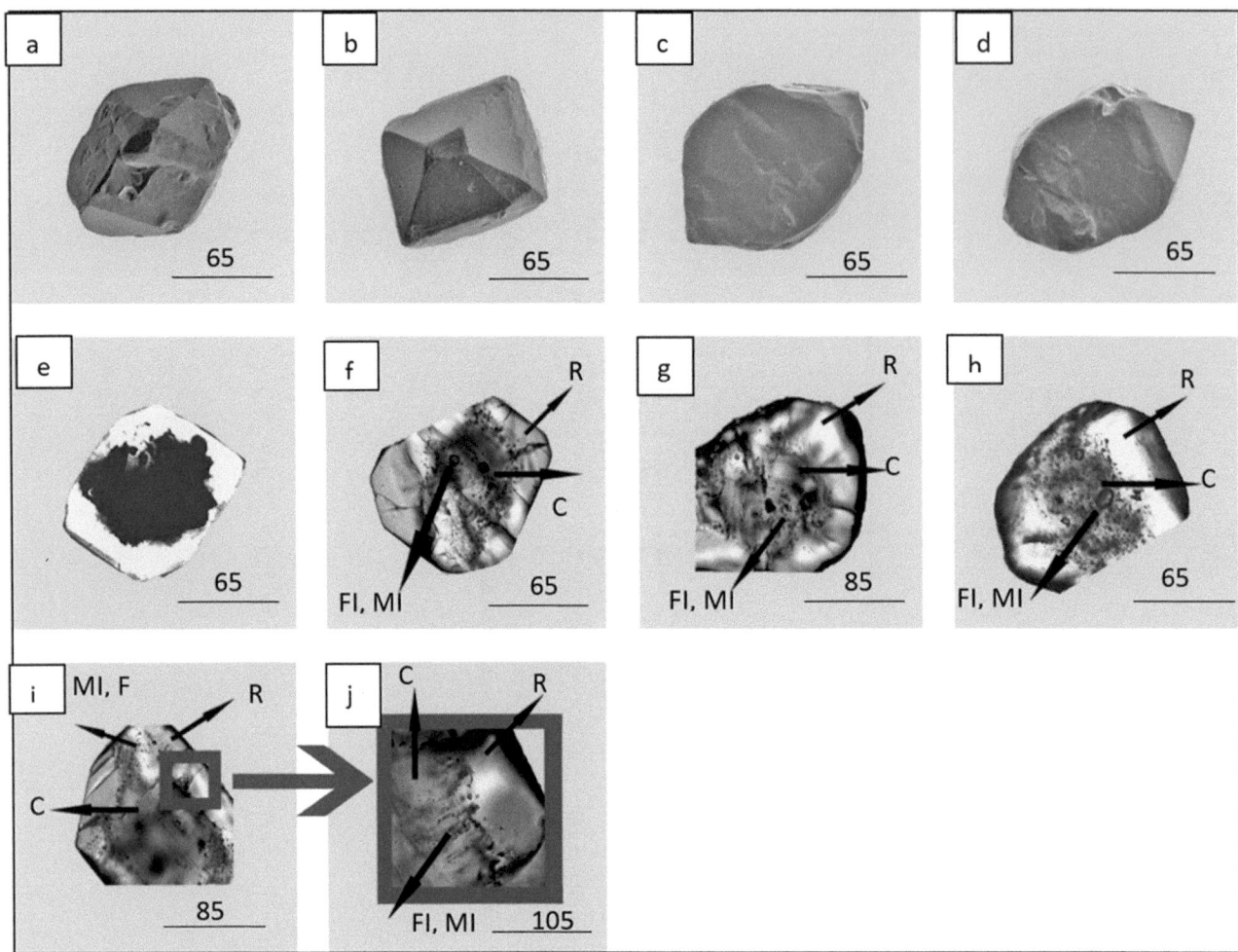

Fig. 10.7 Photomicrographs of zircon crystals (after Kotov et al. 2015). (**a–d**) In secondary electrons, electron microscope; (**e**) in cathodolumi-nescence; (**f–j**) in transmission light. C, core; R, rim; FI and MI, fluid and melt inclusions

10.6 Conclusions

1. The Paleoproterozoic Nb-Ta-Zr-REE-Y Katugin deposit is located in the south of the Kodar-Udokan district. The deposit occurs in a fault-controlled zone of alkaline metasomatism at the contact of rapakivi granite pluton that cuts through the Paleoproterozoic schist and gneiss.

2. The ore-bearing quartz-albite-microcline alteration in combination with alkaline granite form a steeply dipping lode, exposed at the surface over an area of 5 km^2 and traced to a depth of 800 m.

3. The ore contains 1.62 wt% ZrO_2, 0.026 wt%Ta_2O_5, 0.374 wt% Nb_2O_5, 0.16 wt% Y_2O_3, 0.22 wt% REE_2O_3, and 0.0078 wt% U_3O_8. Pyrochlore, zircon, gagarinite, REE-fluorite, and cryolite are major minerals.

4. The age of the Katugin granite is 2055–2066 ma.

5. The origin of the Katugin mineralization is debatable and is regarded as either metasomatic or magmatic.

References

Arkhangel'skaya VV, Kazansky VI, Prokhorov KV, Sobachenko VN (1993) Geology, zoning and condition of formation of the Katugin Ta-Nb-Zr deposit (Eastern Siberia). Geol Ore Dep 35(2):115–131

Arkhangel'skaya VV, Bykhov Y, Volodin RN, Narkelyun LF, Skursky VC, Trubachev AI, Chechetkin VS (2004) Udokan copper and Katugin rare metal deposits in the Chita region, Russia. Administration of Chita, Chita, p 519 (in Russian)

Arkhangel'skaya VV, Ryabtsev VV, Shuriga TN (2012) Geology and mineralogy of tantalum deposits in Russia. In: Mashkovtsev AG (ed) Mineralnoe Syr'e. VIMS, Moscow, p 318 (in Russian)

Bykhovsky LZ (2014) Real, potential and perspective sources of rare earths in Russia. Mineral resources of Russia. Econ Manag 4:2–8 (in Russian)

Bykov YV, Arkhangel'skaya VV (1995) Katugin rare metals deposit. In: Laverov NP (ed) Deposits of Transbaikalia, vol 1, pp 76–85 (in Russian)

Kotov AB, Vladykin NV, Larin AM, Gladkochub DP, Salnikova EB, Sklyarov EV, Tolmacheva EV, Donskaya TV, Velikoslavinsky SD, Yakovleva (2015) New data on the age of mineralization at the Katugin rare metals deposit (Aldan Shield). Dokl Earth Sci 463(2):187–190

Larin AM, Kotov AB, Sal'nikova EB, Kovalenko VI, Kovach VP, Yakovleva SZ, Berezhnaya NG, Ivanov VE (2002) Age of the Katugin Ta–Nb deposit, Aldan–Stanovoi Shield: evidence for the identification of the global rare-metal metallogenic epoch. Dokl Earth Sci 383:336–339

Larin AM, Kotov AB, Vladykin NV, Gladkochub DP, Kovach VP, Sklyarov EV, Donskaya TV, Veklikoslavinskii SD, Zagornaya NY, Sotnikova IA (2015) Rare metal granites of the Katugin complex (Aldan Shield): sources and geodynamic formation settings. Dokl Earth Sci 464:889–893

Lepeshkin SV (2014) Take back the Katugin deposit. The Rare Earth Magazine. http://rareearth.ru/ru/pub/20140411/01529

Levashova EV, Skublov SG, Marin YB, Lupashko TN, Ilchenko EA (2014) Trace elements in zircon from rocks of the Katugin rare metals deposit. Zap Vseros Min Ob-va 143(5):17–31 (in Russian)

Osokin ED, Altukhov EN, Kravchenko SM (2000) Criteria of typification, specifics of formation and localization of giant rare metals deposits. Geol Ore Deposits 42(4):389–396

Prokhorov KV, Sobachenko VN (1985) Structure-petrology and geochemical conditions of high-temperature Na-metasomatites. In: Tomson IN (ed) Internal structure of Precambrian faults. Nauka, Moscow, p 167 (in Russian)

Sharygin VV, Zubkova NV, Pekov IV, Rusakov VS, Ksenofontov DA, Nigmatulina EN, Vladykin NV, Pushcharovsky DY (2016) Li-bearing Na-Fe amphibole from cryolite rocks of the Katugin rare metals deposit. Russ Geol Geophys 57(8):1191–1203

Sklyarov EV, Gladkochub DP, Kotov AB, Starikova AE, Sharygin VV, Velikoslavinsky SD, Larin AM, Mazukabzov AM, Tolmacheva EV, Khromova EA (2016) Genesis of the Katugin rare metals deposit: magmatism contra metasomatism. Russ J Pac Geol 10:155–167

Starikova AE, Sharygin VV (2015) Alumofluorite minerals from the rocks of the Katugin rare metals deposit. In: Structure and geodynamics. IZK, Irkutsk, pp 180–181 (in Russian)

Abstract

Uranium mineralization in the Kodar-Udokan zone was discovered in 1950 in the rocks of the Udokan Supergroup (the Chitkanda prospect). But new occurrences of uranium mineralization were found in the rocks of the Chiney pluton and surrounding rocks only in the beginning of the twenty-first century. Mineralization is related to metasomatized rocks and is represented by brannerite, uraninite, and other minerals. Inside the Chiney pluton, many rare earth minerals were found as well. In this chapter, we report the first results (Gongalsky BI, Krivolutskaya NA, Makariev LB, Murashov KYu, Pavlovich GD, Timashkov AN: Concentration of U + REE at the final stage of operation of the Udokan-Chiney ore-magmatic system. Alkaline magmatism and associated deposits of strategic metals. Proceedings XXXIII international conference of Moscow, GEOKHI RAS, pp 28–29 (in Russian), 2016).

Uranium mineralization in the Kodar-Udokan zone was discovered in sedimentary rocks of the Udokan Supergroup in 1950 by Sosnovskaya Expedition. Geologists of the Udokan Expedition found U-bearing rocks inside the Chiney pluton in the end of 1980. Finally, complex uranium and PGE mineralization was studied in the surrounding rocks of the Chiney pluton in the beginning of this century. We describe these three occurrences of uranium mineralization based on our data and results of geological and mineralogical study of the geologists from VSEGEI (LB Makariev, SK Voyakovsky) and Saint Petersburg Electrotechnical University (VV Knauf).

11.1 The Chitkanda Deposit

The Chitkanda deposit (Fig. 11.1) with the mean U grade of 0.043% was discovered in 1950 (Mineeva and Arkhangel'skaya 2007; Mashkovtsev et al. 2010). The deposit is composed of metasedimentary rocks pertaining to the Chitkanda and Aleksandrov Formations of the Chiney Group, deformed into the NE- and W-E-trending folds (Fig. 11.1). The Chitkanda Formation is subdivided into two subformations. The lower barren subformation (200–250 m) consists of gray fine-grained polymictic sandstone with interlayers of pink arkosic sandstone. The rocks of the upper subformation are represented by similar polymictic and arkosic sandstone. The number of arkosic interlayers and their thickness increase upward in the sequence. The magnetite and martite sandstone, bearing copper mineralization and orebodies of 0.2–0.5 to 20–45 m in thickness, are associated with widespread light gray quartzite-like sandstone interlayers. The total thickness of the Chitkanda Formation is 400–500 m. The area of the deposit is dissected by two large fault zones oriented in the latitudinal and northwestern directions. There are also gabbrodolerite dikes with augite phenocrysts.

The U-bearing rocks are represented by two groups. The first group comprises U-bearing magnetite sandstone of the Chitkanda Formation and partly of the Aleksandrov Formation. The U-bearing magnetite sandstone occurs as beds or series of en echelon lenticular beds, from tens to a few hundred meters in extent and 0.1–1.5 m in thickness. The magnetite-bearing sandstone associates with interlayers of quartzite-like sandstone. The sandstone was recrystallized in the course of regional metamorphism and was strongly altered by alkaline metasomatism. Biotitization, superimposed on magnetite sandstone, is especially widespread. Biotite, occasionally with sericite, develops in sandstone cement and replaces all rock-forming minerals up to magnetite. The amount of biotite reaches 60%. Albitization, silicification, and carbonation are also derivatives of the superimposed processes, as well as the formation of fine pyrite disseminations along with extremely small amount of chalcopyrite, marcasite, and galena.

The second group of the U-bearing rocks is related to albitite within tectonic zones, where they occur together with biotitized and slightly albitized U-bearing sandstone of the

© Springer Nature Switzerland AG 2019
B. Gongalsky, N. Krivolutskaya, *World-Class Mineral Deposits of Northeastern Transbaikalia, Siberia, Russia*,
Modern Approaches in Solid Earth Sciences 17, https://doi.org/10.1007/978-3-030-03559-4_11

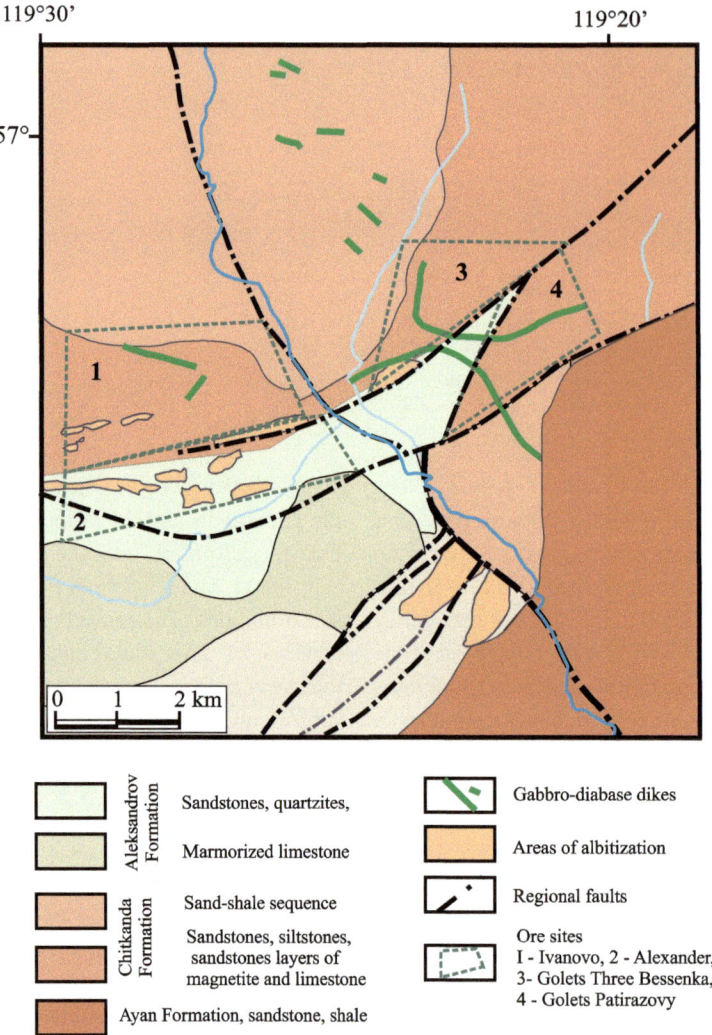

119°30'   119°20'

57°-

Fig. 11.1 Schematic geological map of the Chitkanda deposit. (After Mashkovtsev et al. 2007)

Chitkanda and Aleksandrov Formations. In structural terms, albitites are confined to the near-latitudinal fault zones, where they develop along bedding planes and spread off the faults. In other words, the albitites occur in wide halos of albitized rocks as en echelon lenticular bodies, a few tens of meters long and 1.5–3.0 m thick. Albitite consists of albite (75–80%), quartz (5–25%), rutile, relict K-feldspar grains (1–2%), chlorite, carbonate, hematite, zircon, apatite, and sulfides (Mashkovtsev et al. 2007).

Seventeen ore zones and three ore types are known at the deposit: (1) BIFs with two subtypes, (a) magnetite-mica and

albite-mica ore with brannerite, pitchblende, and uranium blacks and (b) pink to reddish lilac magnetite-free albitite ore, with finely dispersed hematite, and (3) mylonitized albite-mica rocks, with widespread malachite and uranium blacks. The uranium grade does not exceed 0.019 wt%.

The uranium mineralization is largely represented by brannerite, small amount of pitchblende and secondary uranium minerals, such as uranophane, autunite, and meta-torbernite. Chalcopyrite, bornite, covellite, and galena are associated with pitchblende in insignificant amounts. Molybdenite, powellite, and uranomolybdates (moluranite, iriginite) have been identified in one orebody. Brannerite is detected in crush zones dissecting albitites and commonly occurs in selvages of fractures often healed with gray quartz. Pitchblende is a rare mineral that occurs as outer rims around brannerite crystals and develops along thin fractures close to these crystals. Disseminations and small pockets of uranium minerals are distributed in orebodies nonuniformly. The ore is low grade with mean U grade of 0.043 wt% (occasionally up to 0.6 wt%). The Mo contents in particular samples reach 3–5 wt%.

11.2 Albitite Alteration in Country Rocks of the Chiney Pluton

New type of mineralization (U-REE with PGE), localized in the Chiney Group of the Udokan Supergroup, in exocontact of the Chiney pluton, was discovered by Knauf et al. (2000). The albitite bodies reveal vague contacts with protolith or gradual transitions to tectonized and graphitized silty sandstone of the Aleksandrov Formation with elevated uranium concentrations (Landa et al. 1999; Knauf et al. 2000). Micas (sericite and less frequent biotite) and chlorite constantly occur in these rocks. Chlorite forms veinlets and schlieren that contain chalcopyrite, pyrite, arsenopyrite, and molybdenite.

U-bearing albitites were discovered to the south of the Chiney pluton; the U content reaches n*0.1 wt%. The uranium minerals in the form of oxides and titanates (uraninite and brannerite) have been identified by Knauf et al. (2000) in U-bearing zones (Fig. 11.2). REE are concentrated as admixtures in all ore- and rock-forming minerals (titanite, perovskite), including radioactive varieties. The studied uraninite occurs as superimposed disseminations represented by several varieties (Fig. 11.2), Th-bearing cleveite and broggerite. The Th-free uraninite contains Ce_2O_3 (3.08–8.21 wt%) and PbO (5.15–15.56 wt%) as admixtures. The Th-bearing varieties consist of 15.16–22.64 up to 36.0 wt% ThO_2, 13.43–17.35 wt% PbO, 2.9–6.23 wt% Ce_2O_3, and 1.86 wt% Nd as admixtures. The U-Ti-Si mineral species, enriched in UO_2 (24.56–33.42 wt%) and containing 2.21–6.0 wt% ThO_2, have been identified at the uraninite-titanite boundary. Brannerite, which is associated with uraninite or makes up segregations in a solid mass of titanite, is Th-free with admixtures of 1.36–4.72 wt% Ce_2O_3 up to 1 wt% Nd_2O_3. In addition, quartz-carbonate polymetallic veins with large cleiophane crystals occur in the central part of the pluton.

Using combined fire assay and atomic absorption spectroscopy of heavy concentrates, it has been established that albitites contain 1.3 ppm Pd along with 0.2 ppm Au. The Pt content is below the detection limit. The palladium specialization has been proved using the ppm mineralogy technique for a 6 g sample residue. As a result, more than a dozen palladium phase grains have been identified as sobolevskite, froodite, merteieite, and paolovite. Electrum and hessite have been detected among precious metal minerals and chalcopyrite and uraninite as basic minerals of heavy concentrate.

The Pd-bearing phases detected under the microprobe are identified with confidence. However, the detailed study shows that the chemical composition within particular grains is nonuniform, and stoichiometric proportions in mineral formulas are distorted. In addition, there are no distinct phase boundaries between contacting minerals, and spotted impregnations of one mineral into another are characteristic. These features suggest that individual palladium phases are so small that the microprobe does not ensure sufficient locality and that identified minerals actually are aggregates of submicron phases. It is quite probable that these aggregates could have arisen as a product of initial amorphous mass crystallization or recrystallization of primary Pd-bearing phases accompanied by the destruction of crystal lattice under the effect of the late nonequilibrium processes. The latter suggestion matches the stadial formation of albitites and ore minerals contained therein, in particular, with the appearance of sinter varieties of uranium minerals related to the late hydrothermal activity.

The microprobe study of uraninites allowed to estimate the isotopic age of mineralization on the basis of measured Pb/U ratio. The reliability of such estimation is rather high, if two assumptions are fulfilled: (1) lead did not enter uraninite lattice as an initial isomorphic admixture, and (2) the lead produced by the radioactive decay of uranium was not removed in the course of subsequent processes. It is evident that these conditions are fulfilled in natural systems infrequently. In our case, the microprobe measurements of more than 20 uraninite grains from various samples have shown that Pb and U concentrations vary from 20 to 9 wt% and 68 to 79 wt%, respectively. It has been established that the chemical composition of uraninite differs not only between grains but also within one grain. This implies that lead was removed from uraninite by the superimposed processes. The porous structure of mineral grains corroborates this suggestion. The lowest Pb/U ratio (0.134) corresponds to 800 Ma; however, the highest Pb/U ratios in uraninites from different samples are close to one another (0.344–0.349) and correspond to 1870–1900 Ma. According to the conventional practice of using this method, the highest age value should be regarded as a good approximation of the true estimate.

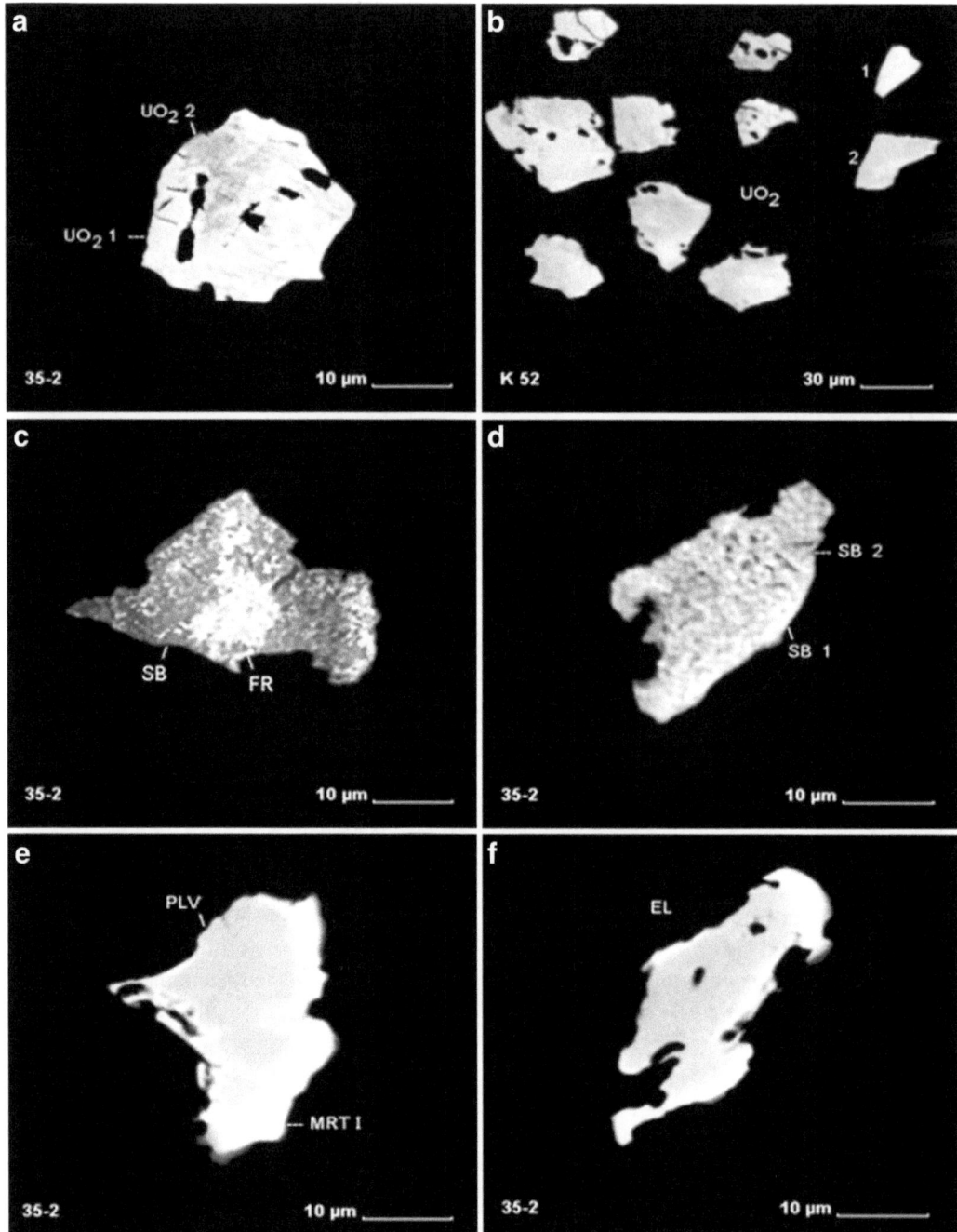

Fig. 11.2 Photomicrographs of precious metals and uranium minerals. (After Knauf et al. 2000). (**a**) Uraninite grains (UO_2) are heterogeneous in chemical composition: the amount of Pb in bright areas is higher than in the dark ones. The ratio of Pb/U in this grain varies from 0.134 to 0.344, corresponding to an absolute age of 0.8 and 1.87 Ga. (**b**) An uraninite grain in the heavy concentrate of the K-52 sample. Differences in brightness and "spotting" of grains are due to variations in composition with respect to Pb/U ratio: 1 – Pb/U = 0.349, 2 – Pb/U = 0.245. (**c**) A grain with chaotic microaggregates of sobolevskite (SB, PdBi) and frudite (FR, $PdBi_2$). The chemical composition within each phase varies and disturbs the stoichiometric relationships in the formula of minerals. (**d**) A grain of sobolevskite (SB 1, SB 2, PdBi), heterogeneous in chemical composition. The chemical composition within each phase varies. (**e**) Chemically heterogeneous grain of paolovite (PLV, Pd_2Sn) and mertieite (MRT, Pd_{-11} (Sb, As)$_{-4}$). There is no clear phase boundary between the minerals. Paolovite enriched in Sb and mertieite enriched in Sn (**f**) A grain of AuAg alloy (Let's pay attention to the possibilities of "ppm mineralogy": gold mineral was obtained from a 6 g sample with 0.2 ppm Au!)

Thus, the studied albitites could have been formed approximately 1.9 Ga ago, contemporaneously with sedimentary and plutonic complexes of the Kodar-Udokan basin, which date back to 2.1–1.8 Ga. It cannot be ruled out that albitites are coeval with the Kodar Granite Complex or the Chiney layered mafic intrusion dating back to 1.8–1.9 Ga.

The considered type of PGE mineralization is endogenic, and with the allowance for an abundance of U-bearing albitites, it may be suggested that the PGE mineralization is widespread in the Kodar-Udokan basin and probably also related to other types of metasomatic and metamorphic rocks. The given data confirm a conclusion on palladium specialization of products of hydrothermal processes. It is also noteworthy that the links of uranium and noble metal mineralization have been established in the metasomatic rocks that occur at a distance from black shale complexes, for which it is characteristic.

The uranium-bearing magnetite sandstone was initially regarded as sedimentary or metasedimentary rock. However, the hydrothermal origin of uranium mineralization is supported now by many lines of evidence. It is assumed that this mineralization is related to uranium redeposition under the effect of alkaline hydrothermal alteration, superimposed on U-enriched sandstone in regional long-lived fault zones and the adjacent areas. The gain of uranium by albitizing solutions is also possible. In this case, the ore formation could have been related to the magma source that produced multiphase granite and gabbro intrusions of the Kodar and Chiney Complexes.

11.3   Uranium Prospects in the Chiney Pluton

11.3.1   Geology

The uranium-bearing capacity of the Chiney pluton was discovered in the 1980s by the geologists of Chitageologia (Ditmar and Makariev 1987). The new uranium prospects discovered in 2004 were studied so far fragmentarily. The uranium mineralization is confined to stratiform propylitic zones in titanomagnetite-bearing rocks of the Chiney pluton (Fig. 11.3) in relation to the contrasting pyroxene-titanomagnetite orebodies at the Etyrko Fe-Ti-V deposit (Gongalsky et al. 2009; Makariev et al. 2009, 2010). The U-bearing propylites (Fig. 11.3) occasionally consist almost completely of newly formed titanite, perovskite, epidote, and zoisite and, to a lesser extent, of chlorite, calcite, and serpentine. Titanite and perovskite become main minerals in certain cases, and this is a characteristic and distinguishing attribute of the ore-bearing metasomatized rocks.

The economic uranium mineralization (Gudym prospect) has been established at the headwater of the Etyrko River after repeated logging of boreholes 92 and 95 in 2004. The intensity of anomalies is 1200–1300 microR/h. Radioactive anomalies with intensities higher than 700 microR/h were also detected at the surface. About 0.06–0.10 wt% U and 1.0–1.75 wt% REE are recorded in samples.

In 2007–2008, Makariev and Voyakovsky conducted the geological and radiometric studies, the results of which expanded the size of the uranium prospects in the western part of the Chiney layered mafic-ultramafic complex. In addition to the earlier studied occurrences, "Anomaly Magnitnaya" and "Bazaltovoe," with a number of new radioactive anomalies, have been identified. Five more uranium-bearing zones, sub-conformal with the stratification of the host rocks and having the form of layers (I, II, III, IV, VI), were first identified. Here and below, description of the mineralization is given after Makariev et al. (2010).

11.3.2   The Uranium-Bearing Strata

The uranium-bearing strata can be noticeably distinguished by the greenish color of metasomatized host rocks and radioactivity from 10 to 25 microR/h or more against the stable low background of 2–4 microR/h. Within the layer, radioactivity is variable but constantly anomalous, exceeding 100 microR/h in ore lenses and individual pods. The average (background) uranium content in the near-ore space is 8–10 ppm, with Th/U = 0.1–2.6 (on average 0.6). The ore lenses contain uranium mineralization. The numbering of the layers is given from the bottom to top of the pluton. The Layer 0 was assigned to the Bazaltovoe occurrence.

The Layer 0 (River Basalt area) is a uranium-bearing zone at the bottom of the Chiney pluton l, isolated near the deposit of titanomagnetite ores (Fig. 11.3). Its total length is 650–800 m, with a width of 100 m. It includes several ore lenses (up to 20 × 60 m) with a radioactivity of 100–1500 microR/h. Ore lenses are sphene-epidote metasomatites with brunnerite, uraninite, and thorite. In separate sections along the trenches, uranium content of 0.05% over a thickness of 1.5 m and 0.0966 wt% and 0.118 wt% over 0.5 m was recorded. In the core outcrops, 0.13 wt% of uranium is recorded over a thickness of 0.8 m. The ores are essentially uranium-rich, with thorium/uranium ratio Th/U = 0.22–2.9.

Layer I is also isolated in the marginal part of the pluton along the contact between the titanomagnetite rocks and the gabbronorite series. It is penetrated by boreholes 12 and 53. The ore lens is 3 × 7 m, with a radioactivity of 400–800 microR/h. It is composed of propylites among gabbro with ilmenite-titanomagnetite mineralization. The uranium content is 0.102 wt% (max 0.116 wt%) over a thickness of 0.4 m. Radioactive haloes were traced for 1 km.

Layer II has a thickness of 52 m and can be continuously traced by more than 1.2 km, with an average radioactivity of 80–100 microR/h. It is formed by numerous ore lenses, 1–5 m thick, with a radioactivity of up to 1200 microR/h.

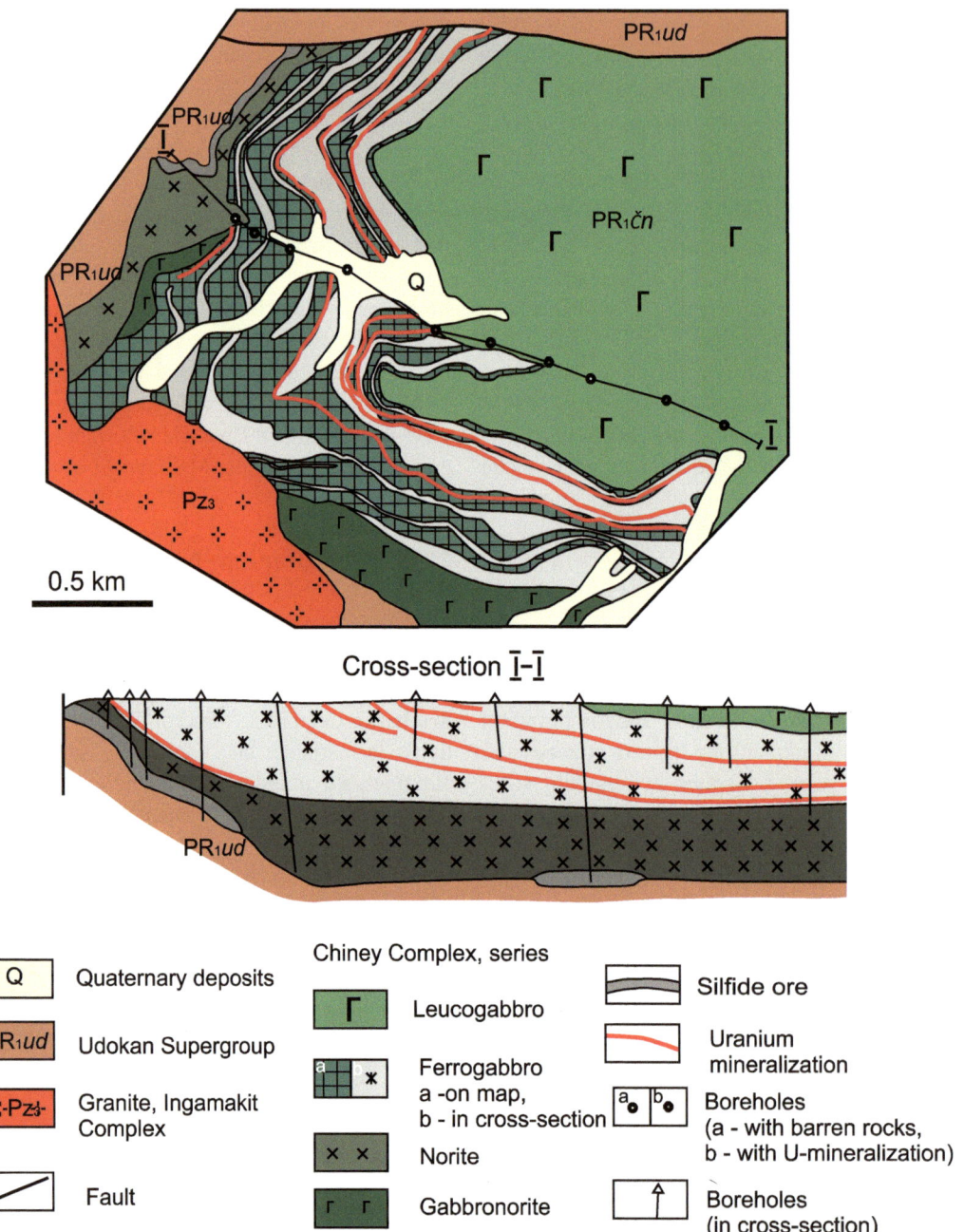

Cross-section Ī–Ī

	Chiney Complex, series		
Q Quaternary deposits	Г Leucogabbro		Silfide ore
PR₁ud Udokan Supergroup	a ⊠ ✳ Ferrogabbro a -on map, b - in cross-section		Uranium mineralization
∴Pz₃ Granite, Ingamakit Complex	× × Norite	a● b●	Boreholes (a - with barren rocks, b - with U-mineralization)
Fault	г г Gabbronorite	↑	Boreholes (in cross-section)

Fig. 11.3 Schematic geological map of the U-REE occurrences in the western part of the Chiney pluton. (After Makariev et al. 2010)

One of the lenses has a thickness of 15 m and was traced for more than 50 m. The host rocks are propylites among gabbro, with rich titanomagnetite mineralization. According to the sampling, the content of uranium is 0.0585% over a thickness of 7 m (max 0.0937 wt% over a thickness of 0.7 m). The radioactivity is essentially due to uranium.

Layer III includes one 750-m-long lens. The lens is composed of propylites in pyroxene-titanomagnetite rocks and has a size of 2 × 15 m. Radioactivity is 300–500 = 1500 microR/h. The content of uranium is 0.0777 wt% over a

thickness of 1.0 m at a maximum of 0.136 wt%. The radioactivity is due to uranium.

Layer IV has a thickness of 8–10 m and includes four closely spaced lenses with a radioactivity of 400–2000 microR/h in propylitized gabbronorite. The content of uranium varies from 0.0622 to 0.1 wt%. One of the lenses with a radioactivity of 2000 microR/h was traced over 5 × 2 m. The nature of radioactivity is essentially due to uranium.

Layer V includes such occurrences as "Anomaly Magnitnaya," continuously traced along the strike of 1 km with intensity of 100, 300, and 900 microR/h. The contents of uranium in the ore samples change from 0.154 to 0.175 wt%. The occurrence is exposed on the right side of the Etyrko Creek (near borehole 92), where it is represented by radioactive propylite (up to 1200 microR/h), located subconcordant with the horizon of ilmenite-titanomagnetite ores. The visible thickness is 35 m. Four orebodies, up to 60–70 m long, are distinguished, with uranium content of 0.0594–0.0924%. The nature of radioactivity is essentially due to uranium.

Layer VI combines three closely related ore lenses with radioactivity of 250–1300 microR/h. Their thickness is about 15 m. The length of the lenses changes from 5 to 32 m, and thickness varies between 1.2 and 1.7 m. The host rocks are relatively poorly propylitic gabbro, directly under the horizon of titanomagnetite ore. The maximum uranium content is 0.171 wt% at a thickness of 0.3 m. The nature of radioactivity is uranium.

There are following geological considerations: (1) the uranium mineralization mainly occurs in the rocks of the titanomagnetite gabbro series at the Chiney pluton; (2) it is related to propylitized rocks; and (3) uranium-bearing propylites contain sphene, perovskite, epidote, and zoisite, with less abundant chlorite, calcite, and serpentine.

11.3.3 Geochemistry

A positive correlation was established between TR and TiO_2, without correlation between U and Th. The anomalous TiO_2 (up to 21.2 wt%) and the increased total iron content are explained solely by the composition of the original host rocks. Concentration of rare earths is more epigenetic. The average content of rare earth elements is about 1.5 wt% (from 0.52 to 2.39 wt%), with the dominance of TRCe. Studies on noble metals showed that in samples with uranium, the content of Au and Ag does not exceed 0.01–0.052 g/t and 1–2 g/t, respectively. PGM anomalies were found at the Bazaltovoe occurrence (up to 0.13–0.36 ppm Pd) and in Layer II (up to 0.064–0.14 ppm Pt).

11.3.4 Mineralogy

For the first time, uranium minerals were discovered by Tatarinov et al. (1998) in sulfide ore at the Chiney pluton (at different types of sulfide mineralization). They were found as small inclusions in main ore-forming minerals, in pyrrhotite, and, especially, in chalcopyrite. Many rare minerals were identified. Besides uraninite UO_2, there are uranophane $Ca(UO_2)_2(SiO_3OH)_2 \cdot 5H_2O$, uranospinite $Ca(UO_2)$ $(AsO_4)_2 \cdot 10H_2O$, vanuralite $Al(UO_2)_2(VO_4)_2(OH) \cdot 11H_2O$, rutherfordine $(UO_2)CO_3$, and vandenbrandeite $Cu(UO_2)$ $(OH)_4$.

The mineralogical study of metasomatized rocks with elevated uranium concentrations was carried out by Makariev et al. (2007, 2010). They showed presence of uranium minerals in the form of oxides and titanates, uraninites and brunnerite. The formation of uranium titanates in the space of highly titanic host rocks is quite natural. Another significant feature of this mineralization is high concentrations of rare earths in all ore minerals (including radioactive ones) and in its own mineral species.

The uraninites form an impregnation in the rock and are represented by several species, including thorium-containing types of slandite and bergerite. The ferrous uraninites contain 3.08–8.21 wt% Ce_2O_3 and 5.15–15.56 wt% PbO, and thorium-bearing uraninites contain 15.16–22.64 up to 36 wt% ThO_2, 13.43–17.35 wt% PbO, 2.9–6.23 wt% Ce_2O_3, and 1.86 wt% Nd_2O_3. On the boundary of uraninite with titanite is the mineral of variable composition (U-Ti-Si), enriched in UO_2 (24.56, 33, and 42 wt%) and containing ThO_2 (2.21–6 wt%). Brannerite is associated with uraninite or forms grain aggregates with titanite. It is characterized by a theoretical composition with impurities 1.36–4.72 wt% Ce_2O_3 and up to 1 wt% Nd_2O_3.

Typical minerals of ore paragenesis, except perovskite and titanite, are leucoxene, ilmenite, and sulfides. Uranium-bearing rare earth minerals are less common. Perovskite contains up to 2 wt% Ce_2O_3. Niobium-bearing (0.37–1 wt% Nb_2O_5) and rare earth varieties were distinguished in titanites. The rare earth variety contains 10–13.63% Ge_2O_3 and up to 1.78 wt% UO_2. Sulfides are represented by Co-pyrite (up to 4.94% Co), galena, chalcopyrite, and sphalerite. Rare earth minerals are represented by monazite, orthite, cerite, as well as titanate with rare earths' Ti-Ce mineral with 1.57–11.18 wt% UO_2, up to 2.34 wt% ThO_2, 19.2–30.85 wt% Ce_2O_3, 3.44–8.38 wt% Nd_2O_3, and up to 0.5 wt% Nb_2O_5. Niobium minerals (columbite, pyrochlore, samarskite) are found only at the "Anomaly Magnitnaya" and "Bazaltovoe."

We studied metasomatized rocks from outcrops at the Etyrko deposit and collected some samples from the core of borehole 92 (Fig. 11.4). The minerals are shown in Fig. 11.5.

Our study revealed a number of uranium minerals, with a substantial content of PbO and, probably, H_2O, and rare earth carbonates. Uranium minerals are curite (?) $Pb_3(UO_2)_8O_8(OH)_6 \cdot 3H_2O$ (75.02 wt% UO_2, 20.99 wt% PbO, in 96.01% total; Fig. 11.6a) and fourmarierite (?) $Pb(UO_2)_4O_3(OH)_4 \cdot 4H_2O$ (most commonly, 79.63 wt% UO_2, 14.37 wt% PbO, 94.00 wt% total; Fig. 11.6b). Vandendriesscheite $(PbU_7O_{22} \cdot 12H_2O)$ is a very rare mineral

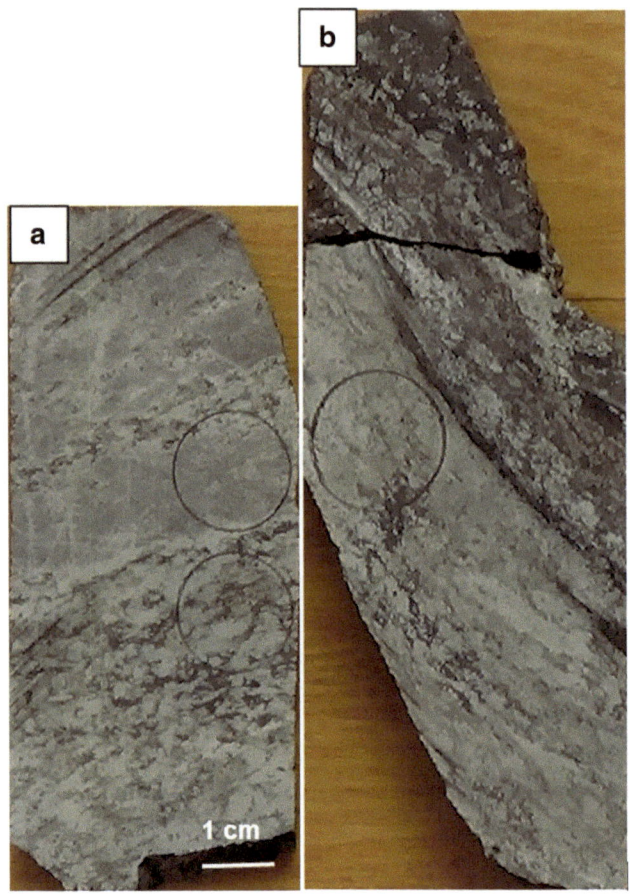

Fig. 11.4 Samples of metasomatized rocks from the western part of the Chiney pluton
Samples: (**a**) 8/72.6, (**b**) 92/76.7 (№ = № borehole/depth, m). Curcles demonstrate palces shown on Fig. 11.5

The high TiO_2 concentrations (15–20 wt% and more) are probably due to the presence of aqueous uranium hydroxides. Their development due to uraninite is characteristic of hydrothermal transformation (Brugger et al. 2004; Sidorenko and Doinikova 2009). This is more probable, given the Proterozoic age of the gabbroids of the Chiney pluton. According to Belova (2000), the qualitative composition of uranium hydroxides depends on the time of formation of uranium ores. In the Proterozoic, uranium deposit (Shinkolobwe) and complex cationic hydroxides (fourmarierite, curite, vandendrisscheite, woelsendorfite) are widely developed.

11.3.5 Genesis of Mineralization

The genesis of mineralization is quite complex. The presence of the multi-metal deposits suggests a long-term telescoping of ore-magmatic systems. Findings of various phases of rare earth and uranium minerals (in the form of micro-inclusions in titanomagnetite and chalcopyrite) may indicate to the appearance of U and TR even at the magmatic stages in formation of the layered intrusion. The formation of rare earth mineralization is most likely due to high-temperature epigenesis, manifested in the alkaline metasomatism in gabbroids (albitization, development of alkaline amphibole, etc.) and the associated skarn host rocks. The actual uranium (brannerite-uraninite) type of mineralization in propylites could presumably form both at the emplacement stage of the late Paleozoic granitoid intrusion (Ingamakit Complex) and in Mesozoic times.

Based on the ratio of uranium minerals, it can be concluded that they were formed as a result of a hydrothermal low-temperature process.

(82.27 wt% UO_2, 12.73 wt% PbO, 94.97 wt% total). A characteristic rare earth carbonate is synchysite-(Ce) (CaCe (CO_3) .2F).

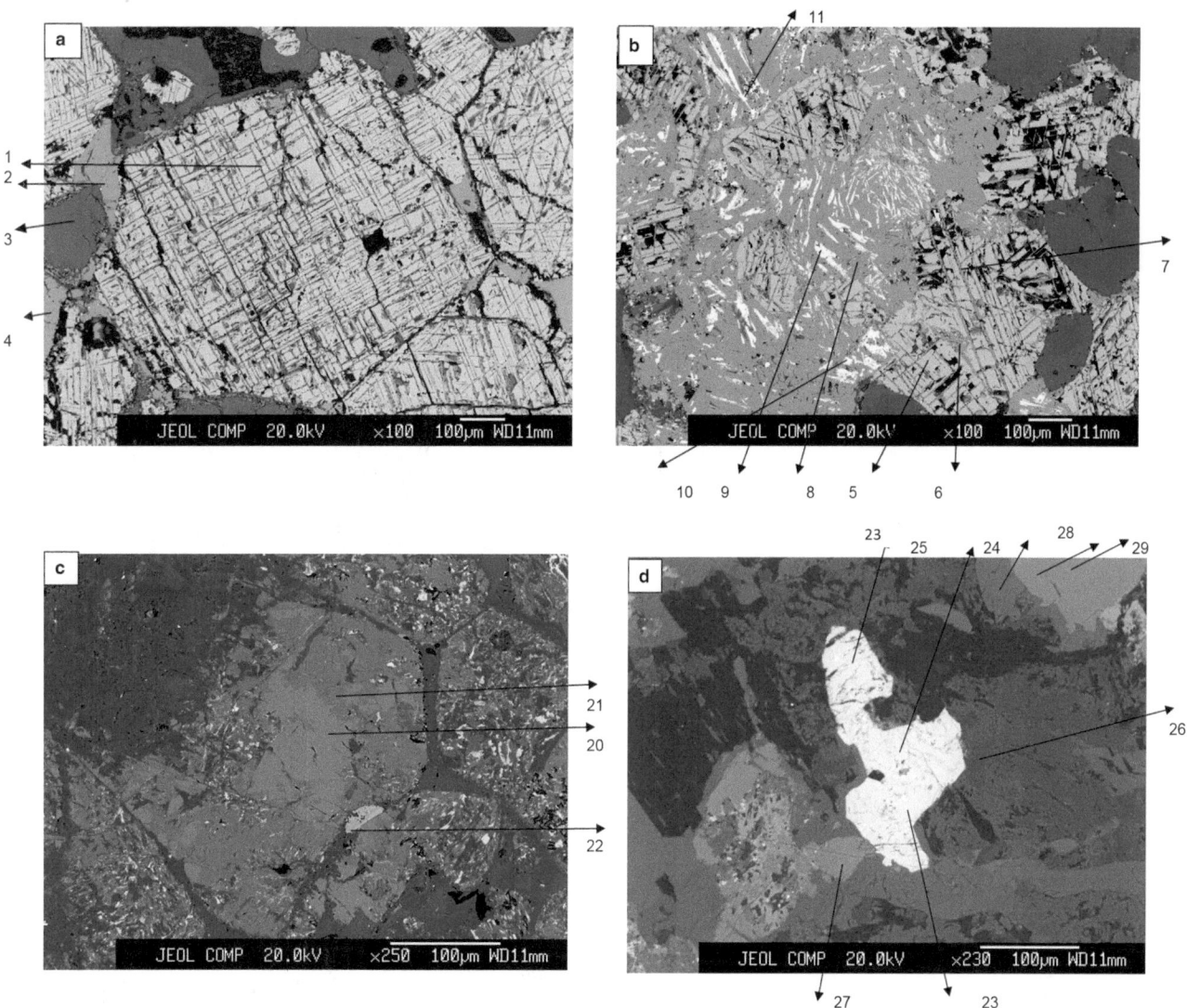

Fig. 11.5 Rare earth minerals in metasomatized rocks at the Etyrko deposit

BSE images were taken at IGEM (analyst EV Kovalchuk). № point in Figures = № analyses in Table 11.1. (**a**) Titanomagnetite grain, (**b**) altered titanomagnetite grain with REE phase (points 9–11, unknown), (**c**) perovskite (point 20) in magnetite, (**d**) brannerite (points 23, 25) and perovskite (points 27, 29), (**e**) perovskite (point 31) and titanite (points 32, 33), (**f**) brannerite (points 35, 37) and titanite (point 39), (**g**) perovskite (point 41) and titanite (points 42, 43) with magnetite (points 45–47), (**h**) titanite (point 50), uraninite (points 53, 54), and magnetite (point 52)

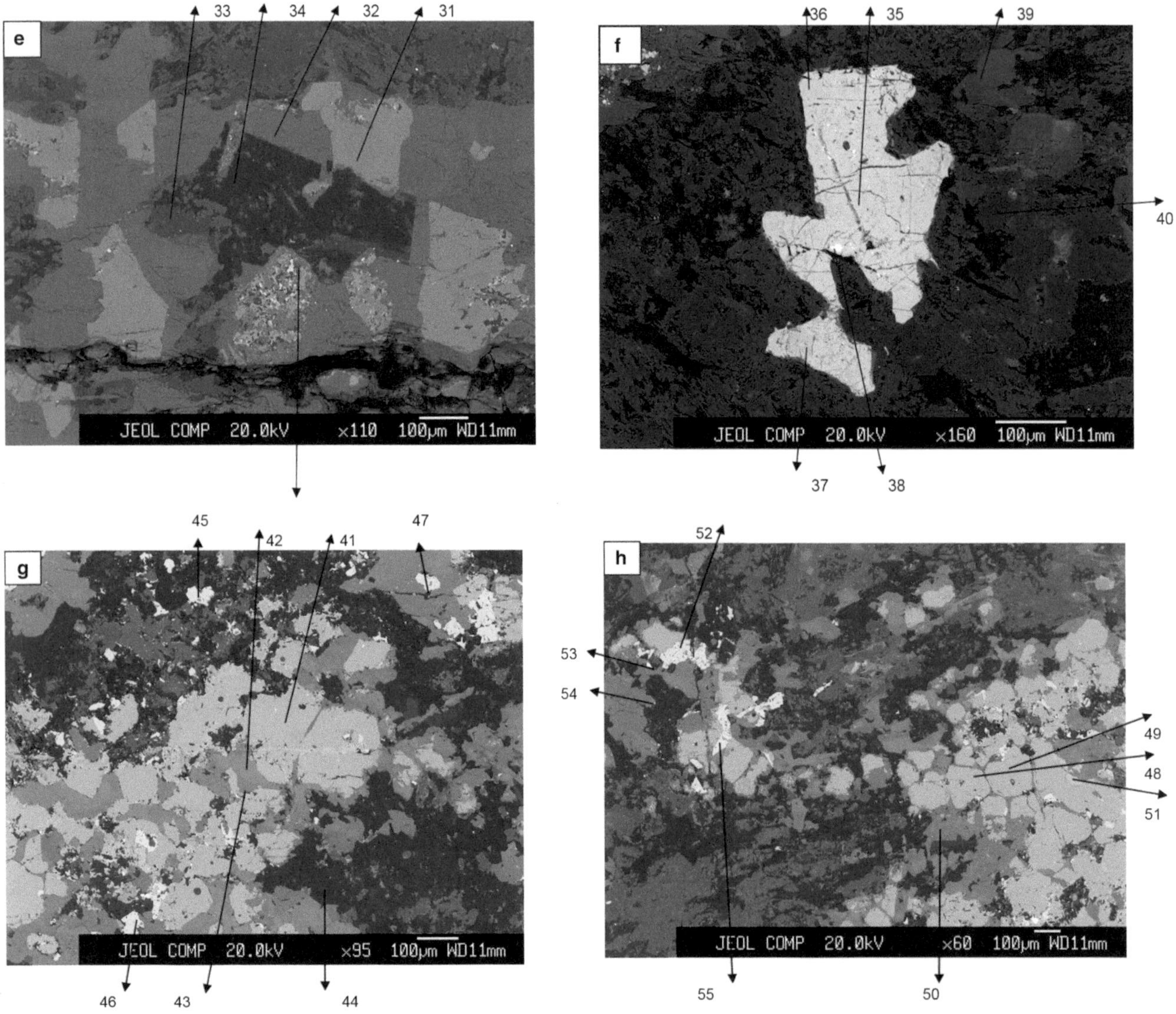

Fig. 11.5 (continued)

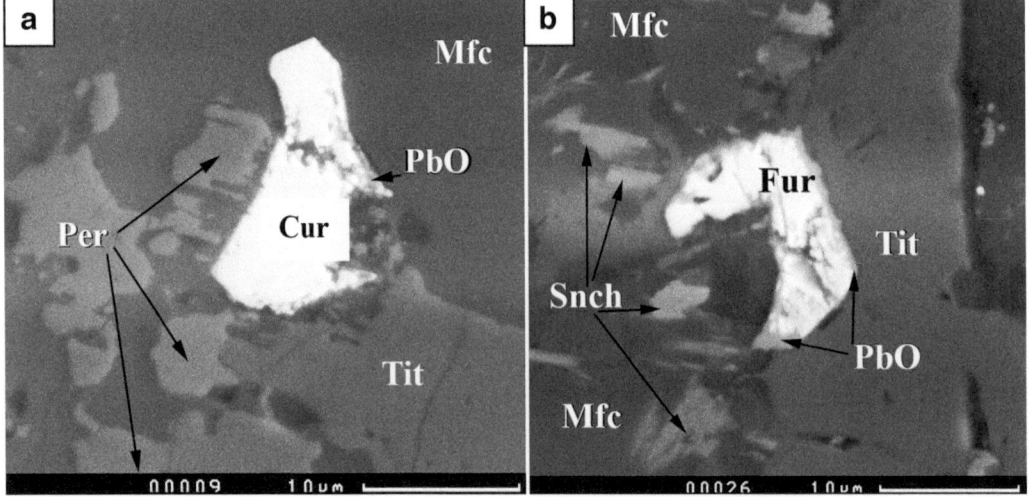

Fig. 11.6 Uranium minerals in the metasomatic rocks at the Chiney pluton (after Azarova et al. 2017)
(**a**) Curite in associations with perovskite and titanite, (b) curite with silicate minerals
Cur curite, *Tit* titanite, *Per* perovskite, *Mfc* mafic silicate minerals, *Fur* fourmarierite, *Snch* synchisite

Table 11.1 Composition of Ti minerals from hydrothermally altered rocks of the Chiney pluton, wt%

№	№ point	TiO₂	MgO	FeO	CaO	SiO₂	MnO	Al₂O₃	Nb₂O₅	V₂O₃	Cr₂O₃	Ce₂O₃	Total	Comment
1	2	51.36	1.57	45.36	0.03	0.00	1.23	0.03	0.00	0.28	0.00		99.87	Ilm
2	4	50.50	1.53	45.52	0.00	0.00	1.30	0.02	0.01	0.24	0.03		99.17	Ilm
3	7	50.04	1.39	46.40	0.06	0.03	1.31	0.04	0.01	0.49	0.00		99.77	Ilm
4	8	52.63	1.61	44.34	0.04	0.00	1.37	0.01	0.04	0.22	0.01		100.28	Ilm
5	13	52.41	1.59	40.53	0.53	0.01	1.47	0.00	0.19	0.11	0.06		96.96	Ilm
6	14	99.28	0.00	0.18	0.56	0.01	0.01	0.00	0.07	0.00	0.03		100.16	Rt
7	18	99.51	0.00	0.07	0.51	0.02	0.01	0.00	0.10	0.09	0.00		100.31	Rt
8	5	0.67	0.03	89.47	0.01	0.00	0.01	0.21	0.00	1.38	0.05		91.87	Mag
9	1	5.19	0.35	84.81	0.04	0.01	0.14	0.22	0.00	1.65	0.05		92.53	Mag
10	22	1.47	0.72	84.35	1.71	1.40	0.01	3.71	0.02	0.15	0.05		93.30	Mag
11	30	0.26	0.42	90.30	0.37	0.47	0.02	0.90	0.01	0.19	0.06		93.09	Mag
12	52	0.27	0.48	90.58	0.24	0.41	0.00	1.28	0.00	0.12	0.04		93.47	Mag
13	3	37.09	0.04	1.13	26.54	28.71	0.01	0.98	0.05		0.00	1.54	96.13	Tit
14	19	38.89	0.00	0.15	29.03	31.13	0.00	1.45	0.02		0.03	0.00	100.70	Tit
15	28	36.62	0.01	0.62	28.13	30.34	0.00	1.83	0.14		0.00	0.53	98.21	Tit
16	32	36.17	0.04	0.80	28.11	29.62	0.02	2.11	0.06		0.00	1.57	99.16	Tit
17	39	37.23	0.02	0.63	27.92	30.29	0.04	1.54	0.00		0.00	0.00	97.70	Tit
18	42	35.82	0.03	0.90	27.48	29.93	0.00	2.11	0.09		0.00	1.46	97.81	Tit
19	43	36.62	0.01	0.92	28.12	29.95	0.00	1.93	0.04		0.00	0.70	98.31	Tit
20	50	37.26	0.00	0.57	28.68	30.76	0.02	1.76	0.05		0.05	0.38	99.53	Tit

Note. N point corresponds to number in Fig. 11.5. Minerals: *Ilm* ilmenite, *Mag* magnetite, *Rt* rutile, *Tit* titanite
Here and in Tables 11.2 and 11.3, analyses were carried out at IGEM, analyst EV Kovalchuk

Table 11.2 Rare earth minerals from the Chiney pluton, wt%

№	No point	TiO₂	MgO	FeO	CaO	SiO₂	MnO	Nb₂O₅	Al₂O₃	Na₂O	Ce₂O₃	SrO	La₂O₃	Pr₂O₃	Total
1	9	45.86	0.01	1.85	0.71	0.04	0.00	0.13	0.02	0.00	39.78	0.00	14.94	0.00	103.34
2	10	46.15	0.01	0.87	0.23	0.01	0.00	0.10	0.02	0.00	22.73	0.00	9.47	1.54	81.14
3	11	47.99	0.55	12.57	2.44	0.18	0.37	0.12	0.21	0.00	15.14	0.00	3.04	1.10	83.72
4	12	47.20	0.00	0.70	2.83	0.10	0.00	0.26	0.05	0.06	20.69	0.00	6.50	1.32	79.72
5	20	55.29	0.05	1.56	39.76	0.00	0.01	0.10	0.51	0.13	2.43	0.02	1.21	0.00	101.06
6	21	63.58	0.04	1.26	22.07	0.07	0.00	0.07	0.66	0.00	2.68	0.03	1.34	0.00	91.80
7	27	54.99	0.00	1.40	38.82	0.00	0.08	0.14	0.63	0.30	2.97	0.03	1.61	0.00	100.97
8	29	55.38	0.00	1.37	39.02	0.00	0.01	0.10	0.60	0.30	2.77	0.02	1.47	0.00	101.04
9	31	55.01	0.00	1.32	38.41	0.24	0.00	0.11	0.49	0.24	2.23	0.00	0.00	0.00	98.10
10	41	55.69	0.01	1.31	39.85	0.01	0.01	0.17	0.49	0.22	2.40	0.06	1.13	0.00	101.35

Table 11.3 Analyses of pyrochlore and magnetite from the Chiney deposit, wt%

No.	TiO₂	FeO	CaO	SiO₂	Nb₂O₅	Al₂O₃	Ce₂O₃	La₂O₃	UO₂	PbO	Bi₂O₃	ThO₂	Y₂O₃	Total	Comment
16	46.44	0.61	6.94	0.26	0.43	0.17	14.72	10.03	1.59	0.00	0.00	0.00	0.21	81.46	
17	45.32	0.74	6.59	0.26	0.56	0.28	16.14	10.38	1.42	0.00	0.00	0.00	0.20	81.91	
23	33.09	0.89	16.97	0.70	5.34	0.06	1.86	0.04	33.04	0.06	0.00	0.34	0.05	92.46	
24	31.61	0.90	16.55	0.15	5.89	0.02	1.85	0.01	33.78	0.00	0.00	0.47	0.02	91.24	
25	31.41	0.91	17.47	0.08	5.14	0.00	1.58	0.03	34.00	0.06	0.11	0.48	0.00	91.33	
35	31.44	0.95	17.63	0.32	5.40	0.02	1.93	0.29	31.68	0.26	0.00	0.51	0.01	90.55	
36	33.25	0.18	14.10	0.23	6.10	0.01	2.05	0.17	32.50	0.00	0.10	0.60	0.03	89.35	
37	34.37	1.35	7.86	0.49	5.62	0.09	1.90	0.09	33.83	0.02	0.22	0.62	0.03	86.48	
45	0.09	90.27	0.05	0.01	0.00	0.27	0.04	0.02	0.00	0.00	0.00	0.08	0.00	90.84	Mag
46	0.35	90.55	0.18	0.01	0.04	0.25	0.05	0.00	0.00	0.00	0.27	0.00	0.00	91.72	Mag
47	0.36	89.41	0.24	0.15	0.00	0.64	0.03	0.00	0.00	0.02	0.21	0.00	0.00	91.14	Mag
53	0.24	0.29	6.15	0.02	0.00	0.00	2.98	0.59	70.73	13.05	0.45	0.07	0.02	94.70	
54	0.76	0.88	6.99	0.36	0.01	0.24	0.82	0.21	70.67	14.77	0.00	0.08	0.04	96.00	
55	0.39	91.39	0.19	0.06	0.00	0.35	0.05	0.00	0.00	0.00	0.25	0.01	0.01	92.69	Mag
56	0.32	0.16	4.04	0.08	0.00	0.02	2.41	0.41	71.94	14.31	0.13	0.57	0.02	94.49	

Note. *Mag* magnetite

11.4   Conclusions

1. There are some occurrences and prospects of U and REE in the Kodar-Udokan basin. The most important of them are Chitkanda and Etyrko (western part of the Chiney pluton) prospects.

2. The prospects are located in metasomatized rocks. Uranium mineralization at the Chitkanda prospect relates to two groups of rocks. The first group comprises U-bearing magnetite sandstone in the Chitkanda Formation and partly the Aleksandrov Formation. The second group of U-bearing rocks is related to albitite within tectonic zones. The uranium mineralization of the Chiney pluton is confined to stratiform propylitic zones and albitite in titanomagnetite-bearing rocks.

3. The Chitkanda uranium mineralization consists of brannerite, a small amount of pitchblende and secondary uranium minerals of uranophane, autunite, and metatorbernite. Chalcopyrite, bornite, covellite, and galena associate with pitchblende in insignificant amounts. The Chiney uranium-bearing rocks contain uraninite and brannerite.

4. The sources of metals and age of mineralization are unknown.

References

Azarova YuV, Krinov DI, Gongalsky BI (2017) About lead-contenting hydroxides of uranium and rare-earth minerals of u-ree-mineralization of the field of Etyrko (Chineysky complex). In: 200th Anniversary Meeting of the Russian Mineralogical Society, pp 157–159 (in Russian)

Belova LN (2000) Condition of formation for zone of oxidation at uranium deposits and aggregates of U-minerals in the supergene zone. Geol Ore Deposits 42(2):113–121 (in Russian)

Brugger J, Krivovichev SV, Berlepsch P (2004) Spriggite, $Pb_3[(UO_2)_6O_8(OH)_2](H_2O)_3$, a new mineral with β-U_3O_8–type sheets: description and crystal structure. Amer Miner 89:339–347

Ditmar GV, Makariev LB (1987) Thorium in early Proterozoic and Mesozoic hydrothermal- metasomatic processes (at the Chara uplift and western part of the Kodar-Udokan trough). In: Kuzmin VI (ed) Processes of Th concentration in the crust. VIMS, Moscow, pp 32–33 (in Russian)

Gongalsky BI, Makariev LB, Voyakovsky SK (2009) Mesozoic and Cenozoic magmatism of the Udokan–Chiney district and uranium mineralization. In: Gordeev EI (ed) Volcanism and geodynamics. Petropavlovsk-Kamchatsky, pp 321–323 (in Russian)

Knauf VV, Makariev LB, Landa EA (2000) A new type of PGE mineralization in the Kodar–Udokan trough. Dokl Earth Sci 371(3):423–425 (in Russian)

Makariev LB, Voyakovsky SK, Il'kevich IV (2009) Gold ore potential of uranium objects in the Kodar–Udokan trough. Rudy I Metally 6:56–64 (in Russian)

Makariev LB, Mironov YB, Voyakovsky SK (2010) The outlook for the discovery of new types of economic uranium deposits in the Kodar–Udokan zone of the Transbaikalian territory in Russia. Geology Ore Deposits 52(5):381–391 (in Russian)

Makariev LB, Bylinskaya LV, Pavlov MV, Pavshukov VV, Saltykova TE, Tolmacheva EV (2007) New type of gold and rare metal mineralization in ancient conglomerates of Eastern Siberia (Patom Highland). Region Geol Metallog 32:134–145 (in Russian).

Mashkovtsev GA, Konstantinov AK, Miguta AK, Shumilin MV, Shchetochkin VN (2010) Uranium in Russian subsurface. VIMS, Moscow, p 857 (in Russian)

Mineeva MG, Arkhangel'skaya VV (2007) A new line in methodology of revealing U and Au deposits in shields and Precambrian fold domains. Razvedka I Okhrana Nedr 11:18–25 (in Russian)

Sidorenko GA, Doinikova OA (2009) Uranium minerals as indicators of formation conditions for uranium deposits. Razvedka I Okhrana Nedr 2:14–22 (in Russian)

Tatarinov AV, Yalovik LI, Chechetkin VS (1998) A dynamometamorphic model for formation of layered basic massifs. Nauka, Novosibirsk, p 120 (in Russian)

Abstract

The Northeastern Transbaikalia is one of the largest metallogenic provinces in Russia. It comprises three world-class deposits: Fe-Ti-V Chiney, Zr-Ta-Nb Katugin, and Cu Udokan. Besides these giant deposits there are many small deposits and prospectings in this area. All these deposits belong to the Paleoproterozoic epoch. New data on the structure, geochemistry, and mineralogy of these deposits were obtained by authors for the last three decades. The main results demonstrate the huge volume of basic-ultrabasic magmas intruded in sedimentary rocks of the Udokan Supergroup. They formed intrusive bodies; part of them are exposed on the surface as the Chiney, Luktur, and Mylovskiy layered mafic-ultramafic plutons. The chemical and mineralogical similarities of massifs demonstrate the common source of their parental magmas. It was found that Cu deposits in sandstones are characterized by different chemical ore compositions (Cu, Ag, Fe, Bi, Ni, Co, PGE) that demonstrate the essential role of hydrothermal processes in their origin, especially in Pravoingamakitskoe deposit. It was suggested that not only magmatic fluids but metheoritic water took place in ore origin due to their involving in circulation under heating caused by a huge magmas volume intrusion. Thus, the modern look of the deposits is a result of many long-lived geological processes operating in this area.

12.1 Overview

Northeastern Transbaikalia is an area located to east of Lake Baikal in Eastern Siberia. Its resources are regarded as an important mineral source for sustainable development of Russia. The Kodar-Udokan mineral district occupies a significant part of the metallogenic province, comprising three world-class deposits: (1) Udokan copper deposit, containing 23 Mt. Cu in the deposit and additional 27 Mt. Cu in its satellites; (2) Fe-Ti-V deposits in the Chiney pluton (the largest vanadium deposit in Russia, with 30 Gt ore); and (3) Katugin Ta-Nb deposit with 2.7 Mt. Nb. Some of these deposits host PGE, Ag, Au and U as well.

All deposits are Paleoproterozoic in age. According to recent paleotectonic reconstructions, Siberian craton was part of the supercontinent Columbia (Rogers and Santosh 2009; Gladkochub et al. 2010; Ernst et al. 2015, 2016; Mekhonoshin et al. 2016). In the southern part of the Siberian craton, there are late Paleoproterozoic volcanic-sedimentary and sedimentary assemblages, interpreted to have accumulated in the Kodar-Udokan Zone, accompanied by multiple mafic-ultramafic plutons.

Katugin is the oldest deposit in the Kodar-Udokan basin (2066 ± 6 Ma; Larin et al. 2002). It is related to a zone of fault-controlled alkaline metasomatism at the contact of rapakivi granite pluton that cuts through the Paleoproterozoic schist and gneiss. The quartz-albite-microcline alteration in combination with alkaline granite occurs as a steeply dipping lode, exposed at surface over an area of 5 km^2 and traced to a depth of 800 m. Its mineralization was studied in several works (Bykov and Arkchangel'skaya 1995; Arkhangel'skaya et al. 2004, 2012; Mashkovtsev et al. 2010; see Chap. 10 for details).

Udokan Cu mineralization was most recently dated as 1896.2 ± 6.2 Ma (Perelló et al. 2017). It associates with the sedimentary rocks of the Paleoproterozoic Udokan Supergroup. But some researchers attributed the upper formation of the Kemen Group to Neoproterozoic (Burmistrov 1993; Vilmova 1990; Tombasov and Sinitsa 1990; Sinitsa 1996) based on transgressive relationships of sedimentary rocks and problematic finds of organic matter. The available data from different levels of stratigraphic sequence indicate a broad age interval of their formation: from 2102 Ma (Pokrovsky and Grigorev 1995) to 1832 Ma (Chechetkin and Kharitonov 2002).

The youngest mineralization occurs in the mafic-ultramafic rocks of the Chiney Complex (1858 ± 17 Ma,

© Springer Nature Switzerland AG 2019
B. Gongalsky, N. Krivolutskaya, *World-Class Mineral Deposits of Northeastern Transbaikalia, Siberia, Russia*,
Modern Approaches in Solid Earth Sciences 17, https://doi.org/10.1007/978-3-030-03559-4_12

1811 ± 11 Ma, according to Gongalsky et al. 2012; 1880 ± 16 Ma, according to Polyakov et al. 2006; 1867 ± 3 Ma, according to Popov et al. 2009).

The first uranium prospects (Chitkanda) were discovered here in the 1950s. But new occurrences of uranium mineralization were found in the rocks of the Chiney pluton and its country rocks only in the early 2000s. We noted occurrences of uranium mineralization in form of pitchblende and U-Th rims around chalcopyrite grains at the Unkur copper deposit, hosted in sedimentary rocks. The enrichment in U and Pb has been documented in crosscutting quartz-bornite veinlets at the Udokan deposit. Sporadic isotopic (U-Pb, Rb-Sr, Sm-Nd, Ar/Ar) datings, related to host rocks and without direct timing of ore minerals, leave room for debate concerning the age of mineralization.

Pd-U-REE mineralization was identified in albitized sandstones along the contact with the gabbro-diabase dike east of the Chiney massif (Knauf et al. 2000).

12.2 Ore Deposits: New Results

12.2.1 Sediment-Hosted Cu-Ag-Fe Deposits

The Udokan deposit and its satellites occur inside a huge sequence (10–12 km in thickness) of clastic rocks of the Udokan Supergroup. Mineralization is located at different levels spanning several kilometres of stratigraphy from the Chitkanda to the Sakukan Formations (Chaps. 3 and 4). Detailed geochemical-mineralogical study of the Udokan deposit and many other occurrences (Pravoingamakitskoe, Saku, Sulban, Unkur, Krasnoe, Burpala) demonstrate their essential diversity (Gongalsky 2015). Only the Udokan deposit contains chalcocite-bornite ore with subordinate chalcopyrite-pyrite, while mineralization from the other sedimentary rock-hosted deposits is more complex in composition (galena, sphalerite, pyrrhotite, pentlandite, ilmenite, native gold, silver, and tiny Pd phases). Chemically, the mineralization varies from Fe-Ag-Cu in the Udokan deposit to Au-Ag-Cu, Bi-Ag-Cu, with U and REE, in the other deposits.

Most contrast difference was recognized between the Pravoingamakitskoe and Udokan deposits that can be regarded as end members of deposit range with different percentage of hydrothermal ores. For the first time, quartz veins with chalcopyrite-pyrite mineralization were found at the Bazaltovoe site of the Pravoingamakitskoe deposit (Gongalsky et al. 2007) within sedimentary rocks. This type of mineralization was distinguished due to anomalous values of precious metals: 2.2 ppm Pt, 6.2 ppm Pd, and 0.4 ppm Au. It is also enriched in Ni (1.7 wt%) and Co (1.48 wt%). Such composition is not typical of sediment-hosted deposits.

Chalcopyrite, pyrrhotite, and pentlandite are major minerals, while the rare minerals are clausthalite, bravoite, bogdanovichite, intermetallic Pd compounds, and Au-Ag alloys. This is an example of unusual hydrothermal type of mineralization. High Ag and Au values (and elevated U-REE) in ore of the Krasnoe and Unkur deposits, in contrast to Udokan, point to their different genesis. They can be placed in the middle of the above-mentioned range of sedimentary-hydrothermal deposits in the Kodar-Udokan area. Diversity in geochemistry and mineralogy between the Udokan and satellites deposits is reflected in the mineral compositions of ore and their isotope composition of sulfur: $\delta^{34}S$ changes from $-24‰$ in the Udokan deposit to $+3.5‰$ in the Pravoingamakitskoe deposit (which is similar to magmatic ore from the Chiney pluton); the other deposits revealed intermediate values (Saku, Klyukvennoe, Unkur, Krasnoe, Burpala; see Chap. 3). These data suggest an essential contribution of magmatic fluids to the origin of sediment-hosted deposits (most pronounced in the Pravoingamakitskoe deposit). The geological facts support this suggestion as well. For instance, the stratiform lodes comprise crosscutting veins and orebodies concordant with bedding, and there are breccia zones with sulfide cement (Chap. 3).

Genesis of the Udokan ores is debated for a long time (Bakun et al. 1958; Reznikov 1965; Bogdanov et al. 1966; Narkelyun et al. 1968, 1983, 1987; Gablina 1983; Krendelev et al. 1983; Volodin et al. 1994; Chechetkin et al. 2000; Arkhangel'skaya et al. 2004). The ores were formed during several stages, continuing now even in the permafrost (Krendelev et al. 1983; Krendelev 1987; Ptitsyn et al. 2003; Yurgenson et al. 2017). It is typical of the other similar deposits. For example, the redeposition of sulfides in the Copperbelt of South Africa was demonstrated and discussed (Hitzman and Broughton 2017; Muchez et al. 2017; Sillitoe et al. 2017). Alderton et al. (2016) provided evidence for two-stage formation of copper ores in Poland.

The so-called cross-bedded stratification is an important argument in discussion on relationships between the rocks and ores of the Udokan deposit. It was employed to demonstrate the sedimentary origin of the Udokan ores. But sulfide ores are located in dilational fractures only of certain directions (Petrovsky 2003; Salikhov and Petrovsky 2004). Originally, the term "sedimentary" ore implied mechanical transportation of sulfide minerals by water streams and their settling in river deltas. However, sulfides sometimes occupy a discordant position in respect to sedimentary textures (Fig. 3.17a, b), interpreted as remobilization (Gongalsky et al. 2017).

The main problem in genesis of sedimentary-hosted deposits is a source of copper and sulfur. The existing hypotheses suggest (a) deep (hypogene) metal sources from mantle or crust (metals were brought by fluids, magma, and saline solutions) and (b) surface (supergene) sources from the areas of denudation together with clastic material. Our data indicate that sulfide minerals were formed in situ under specific water conditions in a reducing environment and enriched in SO_4 ion. It might have happened during sedimentation or immediately after it. Copper could be transported by deep fluids in the form of complex compounds rather than being transported in hard pieces. There is no evidence of Cu-bearing rocks in the Archean-Paleoproterozoic basement around the Kodar-Udokan district with $\delta^{34}S = -20-24‰$. The sulfur isotope data support a serious role of bacteria in ore genesis in water environment. Oxygen isotopes ($\delta^{18}O = 15-17‰$) for quartz from sandstone confirm its hydrothermal origin as well.

12.2.2 Fe-Ti-V and PGE-Cu Deposits in Magmatic Rocks

Mineralization in basic-ultrabasic plutons is the second (in importance) type of deposits in the Kodar-Udokan area. It comprises oxide and sulfide mineralization in the rocks of the Chiney Complex. First of all, this is a Chiney pluton with vanadium and PGE-Cu-Ni ores. Second, this is a Luktur massif with titanomagnetite and sulfide ore. And third, this is a Mylovskiy pluton, which was identified by us for the first time. Many dikes and small intrusive bodies were attributed to the Chiney Complex as well. Based on interpretation of geophysical data (Chap. 2), petrography and geochemical-mineralogical study of magmatic rocks (Chap. 5), and similarities in magma crystallization in different massifs (Chaps. 6 and 9), it was demonstrated that these intrusive bodies belong to the same magmatic system but represent its separately intruded fragments.

The 120 km²Chiney layered pluton (2500 m thick) was studied in more detail in comparison with the other intrusions due to its complex internal structure and related large deposits. It consists of four rock groups (Gongalsky and Krivolutskaya 1993; Gongalsky et al. 2008, 2015).The Chiney pluton shows rhythmic layering at several scales. Its microrhythms are characterized by compositional variation within specific layers of centimeter-to-decimeter size. The mesorhythms are layered at a meter scale, and there are macrorhythms, which are tens to hundreds of meters in thickness (Gongalsky 2015). The rocks of the Chiney pluton are enriched in Fe and Ti in gabbro, gabbronorite, and leucogabbro. Norite and pyroxenite are subordinate rocks. Inside rhythmic units, from the bottom to the top, SiO_2, Na_2O, and K_2O increase, while Fe_2O_3 decreases. The trace-element patterns of various rocks and the results of simulations using the COMAGMAT-3.5 computer program (Ariskin Barmina 2000) suggest that all four rock groups of the massif were

generated by the successive emplacements of several portions of initial magma, which was a complexly differentiated suspension of olivine, plagioclase, and magnetite crystals in ferrobasaltic melt at a temperature of approximately 1130 °C (Gongalsky et al. 2016). Thus, the exposed body of the Chiney pluton is composed of solidified evolved basaltic magma with elevated Fe/Mg ratio as compared with that of primary melt in equilibrium with a peridotitic mantle source. This implies that a complementary cumulate, enriched in Mg, partially present in the Luktur massif, should be expected at depth.

The following types of oxide ore are recognized in the Chiney pluton: (1) early and late magmatic Fe-Ti-V oxide, associated with high-Ti gabbro rocks, (2) syngenetic titanomagnetite disseminations in low-Ti gabbro rocks, (3) magmatic stockwork sulfide zones in high-Ti gabbro rocks, and (4) post-magmatic hydrothermal mineralization related to fluid-magmatic breccia (Gongalsky et al. 2012). Titanium concentration in titanomagnetite changes due to ilmenite lamellae content. According to Mössbauer spectroscopy data, magnetite matrix contains only 10% Ti determined by EPMA (Lyutoev et al. 2017). The T and fo_2 were measured for each type of mineralization based on Baddington-Lindsley thermometer, corrected by Poltavets.

The sulfide ores are divided into endocontact and exocontact types. Some mineralized zones were found in the internal part of the pluton. Their main feature is high Cu/Ni ratio, varying from 10 to 100. We studied mineralogy and geochemistry of sulfides (Krivolutskaya 1986; Krivolutskaya et al. 1997; Gongalsky and Krivolutskaya 2009). Systematic variations in ore composition from internal zones to exocontact part of pluton were recognized, with enrichment in Cu and PGE. The changes in ore-forming mineral composition were established in the same directions: high-Co pentlandite (12–18 wt% Co) in internal part of the pluton (and Co-minerals from linnaeite group) is replaced by medium (5–6 wt% Co, in endocontact zone) and then low-Co pentlandite (2 wt%, in exocontact zone). The list of rare minerals changes as well, including PGE minerals (Tolstykh 2008).

These data were used for determination of crystallization parameters for different ore types (T = 650–400 °C, and $lgf_{S2} = -11-9$). The shift of mineral assemblages from endo- to exocontact parts of the Chiney pluton confirms that a certain temperature gradient still existed at the time of sulfide formation. The copper sulfide prospects, situated at a distance from the contact of the pluton, are comparable to the exocontact ore in mineralogy and geochemistry despite explicable local deviations. At the same time, they drastically differ from the geological style of the Udokan deposit in all features, except for the appearance of sporadic late chalcocite-bornite mineralization. This is an evidence of long distance Cu-rich sulfide melt penetration.

12.3   Origin of Deposits

High copper concentration in different rocks (sedimentary, intrusive, effusive, including Cu prospects in basalts of the Udokan lava plateau (Stupak et al. 1987)) of the Kodar-Udokan area require understanding of time-space and genetic relationships between the deposits. Initially, it was suggested that Chiney magma assimilated sulfides from the sedimentary rocks of the Udokan Supergroup (partly Konnikov and Truneva 1982; Konnikov 1986; Krendelev 1987), explaining very high Cu/Ni ratio in the Chiney ore. The hypothesis of ore formation in sedimentary rocks from fluids related to basaltic melts attempted to explain its copper-rich composition. But none of these models are supported due to controversial sulfur isotope data in magmatic and sedimentary deposits. It is important to emphasize that sulfide ores at the Udokan (Figs. 3.18 and 3.19), Pravoingamakitskoe (Fig. 4.6), and Rudnoe (Fig. 8.14) deposits and in the Chiney massif are often cemented by magmatic and hydrothermal minerals, containing the fragments of country rocks. Similar magmatic minerals were described in deep horizons of the Dzhezkazgan deposit (Satpaeva 2007, 2008).

Based on the collected evidence, the long-standing discussion on magmatic versus sedimentary origin of sulfide ores in the Kodar-Udokan district is likely to develop a complex model, incorporating magmatic, hydrothermal, and sedimentary models with epigenetic-hydrogenic mineralization.

During the Late Paleoproterozoic times, rift basins of the Akitkan belt and Kodar-Udokan zone in the southern part of the Siberian craton were gradually filled with volcanic-sedimentary and sedimentary assemblages. Paleoproterozoic magmatic rocks of the Udokan-Chiney area are represented by large granitoid massifs of the Kodar Complex (Kodar, Kemen) and mafic-ultramafic massifs of the Chiney Complex (Chiney, Luktur, Mylovskiy, Verkhniy Sakukan). Mafic massifs were emplaced in a setting of post-collisional extension (Gladkochub et al. 2012).

Mafic-ultramafic and granite intrusions deformed terrigenous strata within anticlines and synclines (Fig. 12.1). Structural lows accommodated cupriferous sandstone deposits, such as Udokan and Unkur deposits within the Naminga and Unkur synclines. Anticlines correspond to the mafic-ultramafic massifs of the Chiney Complex. As obvious from 3D gravity and magnetic models, the layered massifs exposed at the surface represent merely the heads of magmatic columns that extend as deep as 20 km or deeper (Fig. 12.1).

Such magmatic columns could be emplaced during multiple magma intrusions.

It is quite possible that the source of copper was common for these deposits. The lower crust of the local segment of the lithosphere was enriched in copper, which was transported to the upper crust either by fluids or magmas (or both). In this case, the difference in the sulfur isotope composition between sediment- and magmatic-hosted deposits is not of fundamental importance, because sulfur fractionation occurred already at the paleo-surface under the action of sulfate-reducing bacteria, most likely in shallow basins, penetrated by deep fluids. Obviously, the question about source of these fluids remains open.

The problem of origin of uranium and rare metal mineralization is similar. The source of fluids is also unknown. In most cases, mineralization is not directly related to magmatic complexes. This question is actively debated for the Katugin ores, the oldest in the Kodar-Udokan basin (Larin et al. 2015; Sklyarov et al. 2016). However, the observed volume of granite near the Katugin deposit is unlikely to produce a great volume of rare minerals from magmatic rocks.

The reported results, therefore, just open a news stage in the study of unique metallogeny of Northeastern Transbaikalia. Many problems of mineral genesis and their relationships in the Kodar-Udokan region require further research.

12.4   Implication to Exploration

1. It has been recognized that hydrothermal processes participated in formation of copper mineralization in the Paleoproterozoic sedimentary rocks, especially in the satellite deposits of Udokan and unusual hydrothermal ores of the Pravoingamakitskoe deposit. These processes led to redistribution of the early mineralization and resulted in formation of new ore bodies.
2. The similarity in composition and crystallization sequences of the mafic rocks, attributed to the Chiney complex and belonging to a single deep chamber, was demonstrated. This opens possibilities for discovery of new Fe-Ti-V and PGE-Cu-Ni occurrences, both at the surface and at shallow depths in the southern part of the Kodar-Udokan basin.
3. The discovery of new occurrences of uranium and rare metal mineralization may open new directions for prospecting works in the Kodar-Udokan region.

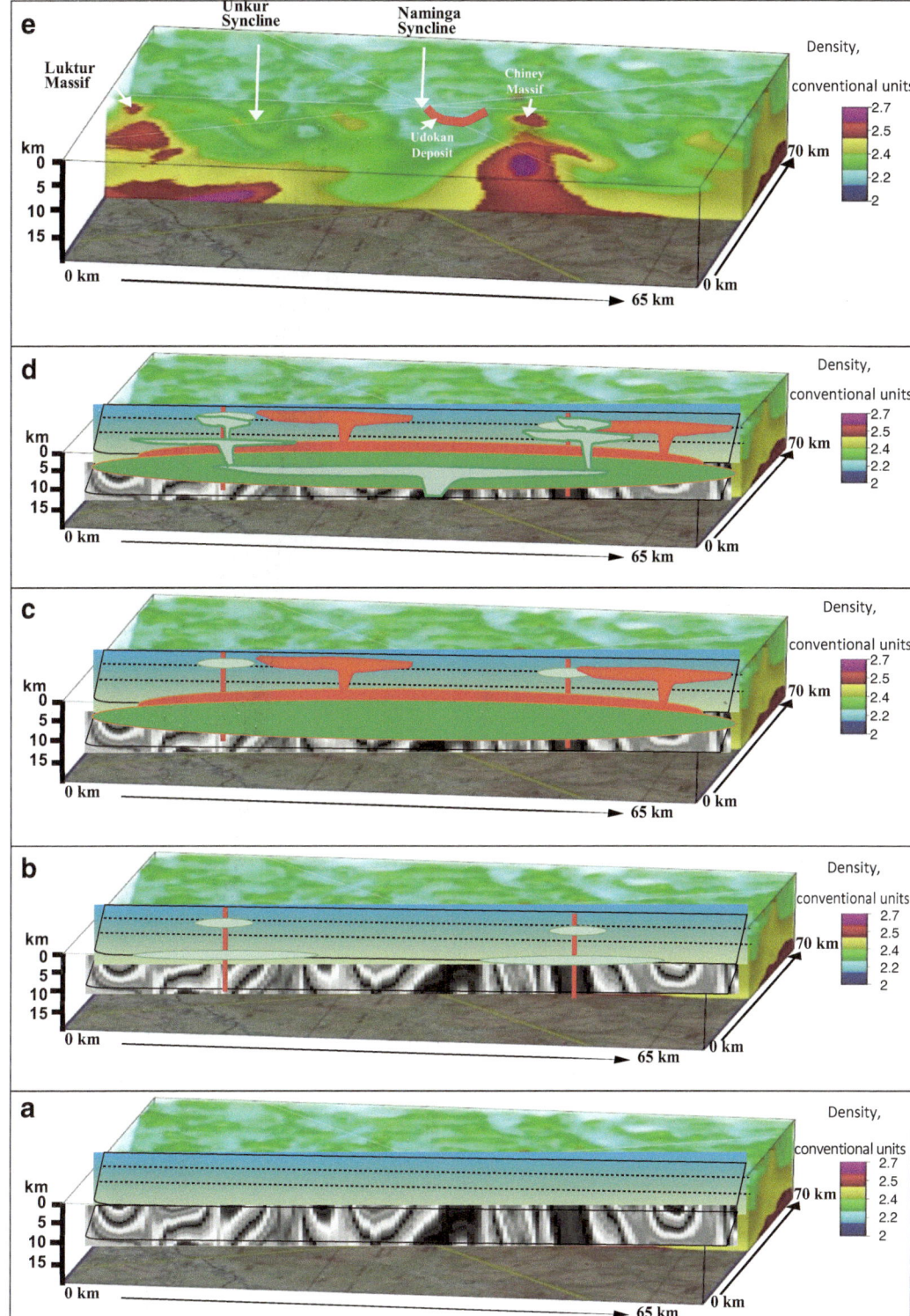

Fig. 12.1 Schematic model of evolution of the Udokan-Chiney area

Stages: (**a**) sedimentation, (**b**) faulting and intrusion of ultrabasic-basic magma at depth, (**c**) melting of the lower crust rocks with formation of the granite melts, (**d**) intrusion of basic magmas into individual chambers, (**e**) the resulting section of the deep structure based on the 3D modelling

References

Ariskin AA, Barmina GS (2000) Modeling of phase equilibria during the crystallization of basaltic magmas. Nauka, Moscow, p 363 (in Russian)

Alderton DHM, Selby D, Kucha H, Blundell DJ (2016) A multistage origin for Kupferschiefer mineralization. Ore Geol Rev 79:535–543

Arkhangel'skaya VV, Bykhov Yu, Volodin RN, Narkelyun LF, Skursky VC, Trubachev AI, Chechetkin VS (2004) Udokan copper and Katuginskoe rare metal deposits in the Chita region, Russia. Chita, Administration of Chita, p 519 (in Russian)

Arkhangel'skaya VV, Ryabtsev VV, Shuriga TN (2012) Geology and mineralogy of tantalum deposits in Russia. In: Mashkovtsev GA (ed) Mineralnoe Syr'e. VIMS, Moscow, p 318 (in Russian)

Bakun NN, Volodin RN, Krendelev FP (1958) Main features of geological structure of Udokan Cu sandstone deposit and direction of its prospecting. Geologiya i Razvedka 5:67–83 (in Russian)

Bogdanov YV, Kochin GG, Kutyrev EI, Travin LV, Feoktistov VP (1966) Geology, formation conditions and distribution of cupriferous sandstones in northeastern Olekma-Vitim mountain province. Int Geol Rev 8:1305–1315

Burmistrov VN (1993) Complex organisms from the Udokan complex of eastern Siberia with implications for its age. Palaeogeogr Palaeoclimatol Palaeoecol 104:3–12

Bykov YV, Arkhangel'skaya VV (1995) Katugin rare metals deposit. In: Laverov NP (ed) Deposits of Transbaikalia, vol 1, pp 76–85 (in Russian)

Chechetkin VS, Kharitonov YuF (2002) Geology and mineral deposits of the Chita segment of BAM. Chita, p 63 (in Russian)

Chechetkin VS, Yurgenson GA, Narkelyun LF, Trubachev AI, Salikhov VS (2000) Geology and ore of the Udokan copper deposit: a review. Russ Geol Geophys 41:710–722

Ernst RE, Söderlund U, Hamilton MA, Chamberlain KR, Bleeker W, Okrugin AV, LeCheminant AN, Kolotilina T, Mekhonoshin AS, Metelkin D, Buchan KL, Gladkochub DP, Didenko AN, Hanes JA (2015) Long-term neighbors: reconstruction of Southern Siberia and Northern Laurentia based on multiple LIP barcode matches over the interval 1.9—0.7 Ga. LIPs, mantle plumes and metallogeny in the Earth's history. Extended abstract of international conference. Irkutsk-Listvyanka, pp 29–31

Ernst RF, Hamilton MA, Söderlund U, Hanes JA, Gladkochub DP, Okrugin AV, Kolotilina TB, Mekhonoshin AS, Bleeker W, LeCheminant AN, Buchan KL, Chamberlain KR, Didenko AN (2016) Long-lived connection of Siberia and northern Laurentia in the Proterozoic. Nat Geosci 9:464–469

Gablina IF (1983) Copper accumulation conditions in continental red beds. Nauka, Moscow, p 112 (in Russian)

Gladkochub DP, Pisarevsky SA, Donskaya TV, Ernst RE, Wingate MT, Söderlund U, Mazukabzov AM, Sklyarov EV, Hamilton MA, Hanes JA (2010) Proterozoic mafic magmatism in Siberian craton: an overview and implications for paleocontinental reconstruction. Precambrian Res 183:660–668

Gladkochub DP, Donskaya TV, Ernst R, Mazukabzov AM, Sklyarov EV, Pisarevskiy SA, Wingate M, Sederlund W (2012) Mafic magmatism of the Siberian craton during the proterozoic: an overview of the main stages and their geodynamic interpretation. Geotektonika 4:28–41

Gongalsky BI (2015) Deposits of the unique metallogenic province of Northern Transbaikalia. VIMS, Moscow, p 248 (in Russian)

Gongalsky BI, Krivolutskaya NA (1993) Chiney layered pluton. Nauka, Novosibirsk, pp 184–(in Russian)

Gongalsky BI, Krivolutskaya NA (2009) Udokan-Chiney ore-magmatic system, Russia. Northwes Geol 42:180–184

Gongalsky BI, Safonov YG, Krivolutskaya NA, Prokof'ev VY, Yushin AA (2007) A new type of copper–noble metal mineralization in Northern Transbaikalia. Dokl Earth Sci 414(5):645–648

Gongalsky BI, Krivolutskaya NA, Ariskin AA, Nikolaev GS (2008) Internal structure, composition and genesis of the Chiney anorthosite–gabbronorite pluton, Northern Transbaikalia. Geochem Int 46:637–665

Gongalsky BI, Timashkov AN, Voyakovsky SL (2012) U-Pb dating results on Paleoproterozoic zircons from intrusions of the Udokan-Chiney ore district (Russia). In: Laverov NP (ed) Abstracts of materials in Russian conference on isotope geochemistry. IGEM, Moscow, pp 110–112 (in Russian)

Gongalsky BI, Galyamov AL, Pavlovich GD, Petrov AV, KYu M (2015) 3D model of the head of a paleoproterozoic plume in the Southern Siberian craton. Large igneous provinces, mantle plumes and metallogeny in the Earth's history. Publishing House of V.B. Sochava Institute of Geography SB RAS, Irkutsk, pp 45–46

Gongalsky BI, Krivolutskaya NA, Ariskin AA, Nikolaev GS (2016) The Chiney gabbronorite-anorthosite layered massif (Northern Transbaikalia, Russia): its structure, Fe-Ti-V and Cu-PGE deposits, and parental magma composition. Mineral Deposita 51(8):1013–1034

Gongalsky B, Belousova E, Petrov A, Pavlovich G, Murashov K, Krivolutskaya N, Timashkov A (2017) Magmatic deposits of the Kodar-Udokan area, Southern Siberia, Russia: geology, geochemistry, and modeling. Society of Economic Geologists, Inc. SEG 2017 Conference P208

Hitzman MW, Broughton D (2017) Discussion: age of the Zambian copper belt by Sillitoe et al. Mineral Deposita 52(8):1271

Knauf VV, Makariev LB, Landa EA (2000) A new type of PGE mineralization in the Kodar–Udokan Trough. Dokl Earth Sci 371(3):423–425 (in Russian)

Konnikov EG (1986) Precambrian differentiated mafic–ultramafic complexes in the Transbaikalia region. Nauka, Novosibirsk, p 224 (in Russian)

Konnikov EG, Truneva MF (1982) About the source of sulfide ores in the Chiney copper deposit (Northern Transbaikalia). Dokl Akad Nauk USSR 264:216–219

Krendelev FP (1987) About genesis of sulfide ore of the Udokan Cu sandstone deposit. Russ Geol I Geofiz 8:133–134 (in Russian)

Krendelev FP, Bakun NN, Volodin RN (1983) Udokan copper sandstone. Nauka, Moscow, p 248 (in Russian)

Krivolutskaya NA (1986) Sulfide mineralization of the Chiney pluton. Geol Ore Deposits 28(5):94–100 (in Russian)

Krivolutskaya NA, Gongalsky BI, Sergeeva NE (1997) Mineral composition of sulfide ore of the Chiney pluton. Gorny Zhurnal 7:187–201 (in Russian)

Larin AM, Kotov AB, Sal'nikova EB, Kovalenko VI, Kovach VP, Yakovleva SZ, Berezhnaya NG, Ivanov VE (2002) Age of the Katugin Ta–Nb deposit. Aldan–Stanovoi Shield: Evidence for the Identification of the Global Rare-Metal Metallogenic Epoch. Dokl Earth Sci 383:336–339

Larin AM, Kotov AB, Vladykin NV, Gladkochub DP, Kovach VP, Sklyarov EV, Donskaya TV, Veklikoslavinskiy SD, ZagornayaNYu SIA (2015) Rare metal granites of the Katugin complex (Aldan Shield): sources and geodynamic formation settings. Dokl Earth Sci 464:889–893

Lyutoev VP, Gongalsky BI, Makeev AB, Lysyuk AY, Magazina LO, Taskaev VI (2017) Titanomagnetite ores: mineral composition and Mössbauer spectroscopy. Fortschr Mineral 2:43–65 (in Russian)

Mashkovtsev AG, Konstantinov AK, Miguta AK, Shumilin MV, Shchetochkin VN (2010) Uranium in Russian subsurface. VIMS, Moscow, p 850 (in Russian)

Mekhonoshin AS, Ernst RE, Söderlund U, Hamilton MA, Kolotilina TB, Izokh AE, Polyakov GV, Tolstykh ND (2016) Relationship between platinun-bearing ultramafic-mafic intrusions and large igneous provinces (exemplified by the Siberian Craton). Russ Geol Geophys 57(5):1043–1057

Muchez P, André-Mayer AS, Dewaele S, Larg R (2017) Discussion: age of the Zambian Copperbelt. Mineral Deposita 52:1269

Narkelyun LF, Salikhov VS, Trubachev AI (1983) Copper sandstones and shales of the world. Nedra, Moscow, p 414 (in Russian)

Narkelyun LF, Bezrodnykh YuP, Trubachev AI, Yurgenson GA (1968) Geology and genesis of the Udokan copper sandstone deposit. In: Geology of some deposits in the Transbaikalian region. Chita, ZabNII, pp 70–90 (in Russian)

Narkelyun LF, Trubachev AI, Salikhov VS, Kunitsin VV, Chechetkin VS, Zinoviev YI, Krivolutskaya NA (1987) Oxidized ore of the Udokan deposit. Nauka, Novosibirsk, p 102 (in Russian)

Perelló J, Sillitoe RH, Yakubchuk AS, Valencia VA, Cornejo P (2017) Age and tectonic setting of the Udokan sediment-hosted copper-silver deposit, Transbaikalia, Russia. Ore Geol Rev 86:856–866

Petrovsky PP (2003) Boundary disturbances and their influence on the geological structure of the Udokan deposit. In: Mater 4th scientific and technical conference of the mining Institute of the Chita. State Technical University, Chita, pp 122–125 (in Russian)

Pokrovsky BG, Grigoriev SV (1995) New data on the age and geochemistry of isotopes of the Udokan series from the lower Proterozoic of eastern Siberia. Lithology and Mineral Dep 30(3):243–283

Polyakov GV, Isokh AE, Krivenko AP (2006) Platiniferous ultramafic–mafic formations of mobile belts of central and southeastern Asia. Russ Geol Geofiz 47(12):1227–1241

Popov NV, Kotov AB, Postnikov AA, Sal'nikova EB, Shaporina MN, Larin AM, Yakovleva SZ, Plotkina YV, Fedoseenko AM (2009) Age and tectonic position of the Chiney layered massif, Aldan shield. Dokl Earth Sci 424(1):64–67

Ptitsyn AB, Zamana LV, Yurgenson GA, Abramov BN, Bashurova NF, Vilmova EV, Eremin OV, Zheleznyak II, Malchikova IY, Petrovsky PP, Sinitsa SM, Trubachev AI, Turanova TK, Usmanov MT, Shesternev DM, Chechel AP (2003) Udokan: geology, mineralization, conditions of exploration. Nauka, Novosibirsk, p 160 (in Russian)

Reznikov IP (1965) Genesis of the Udokan deposit. Litol Miner Dep 2:85–94 (in Russian)

Rogers JJ, Santosh M (2009) Tectonics and surface effects of the supercontinent Columbia. Gondwana Res 15:373–380

Salikhov VS, Petrovsky PP (2004) Shift displacements in the structure of the Udokan copper deposit (Eastern Siberia). Dokl Acad Sci 394(6):787–790

Satpaeva MK (2007) Mercury-arsenic-silver mineralization on the lower horizons of Dzhezkazgan. Izv NAS RK Ser geol 5:17–36 (in Russian)

Satpaeva MK (2008) On the mantle plumes. Izv NAS RK Ser geol 1:15–24 (in Russian)

Sillitoe RH, Perelló J, Creaser RA, Wilton J, Wilson A, Dawborn T (2017) Age of the Zambian copper belt. Mineral Deposita 52/8:1245–1268

Sinitsa SM (1996) A problem of the Udokan biota in the Kodar–Udokan District, the Transbaikaliaian region. In: Problems of ore formation. Nedra, Novosibirsk, pp 177–181 (in Russian)

Sklyarov EV, Gladkochub DP, Kotov AB, Starikova AE, Sharygin VV, Velikoslavinskiy SD, Larin AM, Mazukabzov AM, Tolmacheva EV, Khromova EA (2016) Genesis of the Katugin rare metals deposit: magmatism contra metasomatism. Rus J Pac Geol 10:155–167

Stupak FM, Krendelev FP, Krivolutskaya NA, Stupak PM (1987) New type of copper mineralization in the Udokan ridge. Dokl Akad Nauk USSR 297(4):929–931 (in Russian)

Tolstykh ND (2008) PGE mineralization in marginal sulfide ores of the Chiney layered intrusion, Russia. Mineral Petrol 92:283–306

Tombasov IA, Sinitsa SM (1990) Stratigraphy of rocks from Udokan Compex in Ikabya-Chitkanda region. In: Stratigraphy of low Precambrian of Far East. Geol Inst, Vladivostok, pp 56–61 (in Russian)

Vil'mova ES (1990) A Possible reconstruction of Udokania colonies in Proterozoic sedimentary rocks of Southern Transbaikaliaian region: in topical problems of geosciences. Chita, pp 33–38 (in Russian)

Volodin RN, Chechetkin VS, Bogdanov Yu V (1994) The Udokan copper sandstone deposit, Eastern Siberia. Geol Ore Deposits 36(1):3–30

Yurgenson GA, Epova ES, Eremin OV (2017) Mineralogical and geochemical feature of the postglacial oxidation zone of the Udokan deposit. Izv Vuzov Geologiya i Razvedka 3:78–83

Appendices

Appendix 1

Here and in Fig. A1.2 and in Tables A1.1, A1.2, A1.3, A1.4, A1.5 and A1.6 analyses were carried out at IGEM RAS, analyst EV Kovalchuk.

Table A1.1 Chemical composition of sulfide in ore from the Udokan deposit, wt%

No point in Fig. A1.1	Fe	Zn	S	Cu	Total
1	0,09	0,12	20,85	77,82	98,89
2	10,98	0,02	25,49	62,04	98,53
12	11,32	0,06	25,25	62,44	99,06
13	0,90	0,03	21,23	77,70	99,85
17	11,05	0,04	25,80	61,84	98,72
18	0,02	0,06	22,15	77,01	99,24
19	11,20	0,01	25,59	62,25	99,05
26	0,36	0,05	21,36	77,98	99,74
27	11,34	0,06	25,81	61,44	98,64
28	1,17	0,06	21,37	77,14	99,74
29	11,41	0,11	26,13	61,57	99,22
31	1,84	0,04	21,88	76,26	100,02
36	0,04	0,08	21,19	78,29	99,60
37	11,44	0,04	25,67	62,28	99,43
41	0,36	0,03	21,24	77,59	99,21
42	12,28	0,05	25,63	61,14	99,10
47	11,22	0,06	25,65	62,20	99,13
49	1,39	0,05	21,10	77,08	99,61
52	0,17	0,03	20,90	78,54	99,64
53	10,97	0,00	25,46	62,64	99,07

© Springer Nature Switzerland AG 2019
B. Gongalsky, N. Krivolutskaya, *World-Class Mineral Deposits of Northeastern Transbaikalia, Siberia, Russia*,
Modern Approaches in Solid Earth Sciences 17, https://doi.org/10.1007/978-3-030-03559-4

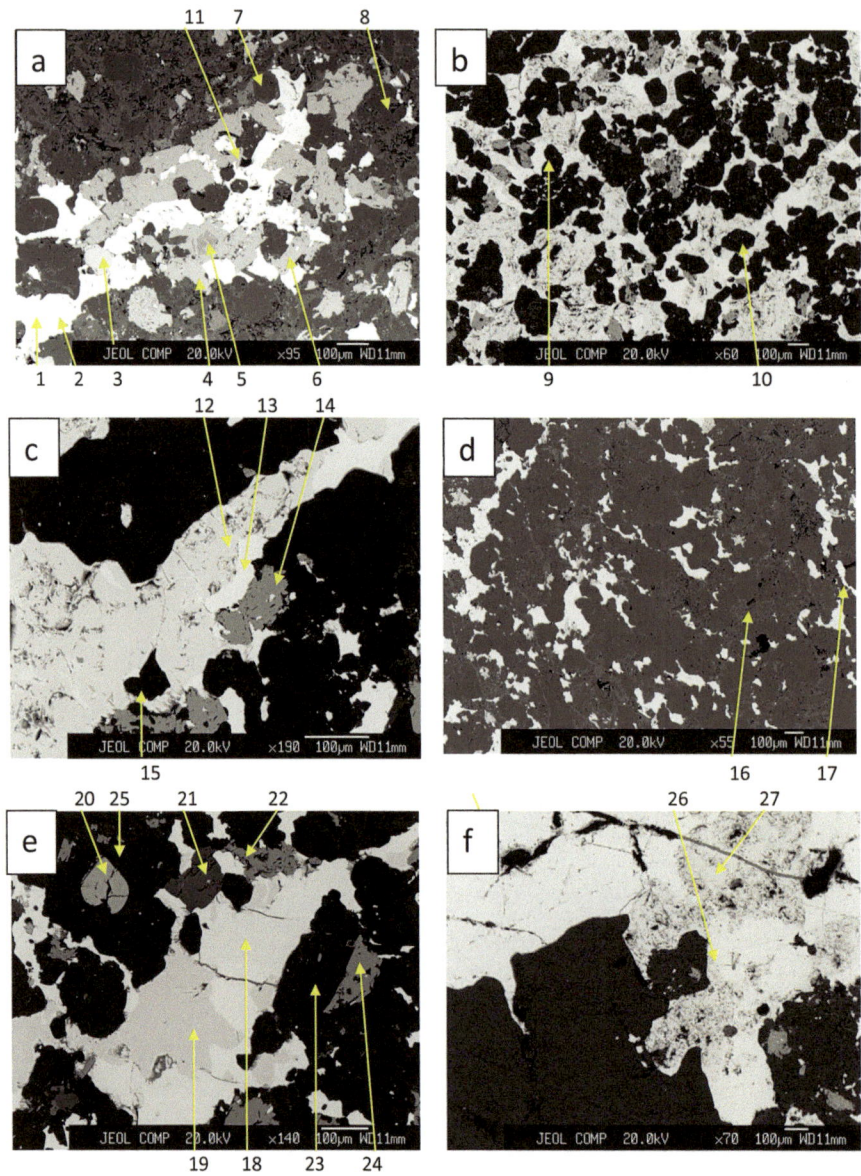

Fig. A1.1 BSE images of disseminated ore at the Udokan deposit, Levyi Bort Naminga site (**a**), (**b**) – Chalcocite-bornite association in sandstone; (**c**) – veinlet of chalcocite with border of bornite and magnetite inclusions; (**d**) – tiny sulfide dissemination in silicated matrix; (**e**) – bornite with rim of chalcocite and zircon grains between silicate minerals; (**f**) – chalcocite-bornite association; (**g**) – oval grains of magnetite in chalcocite-bornite aggregate; (**h**) – chalcocite-bornite assemblage in magnetite grain; (**i**) – aggregate of magnetite grains with chalcocite-bornite border; (**j**) – bornite and unknown phase among silicate minerals; (**k**) – irregular magnetite grains with bornite; (**l**) – chalcocite-bornated in silicate rock

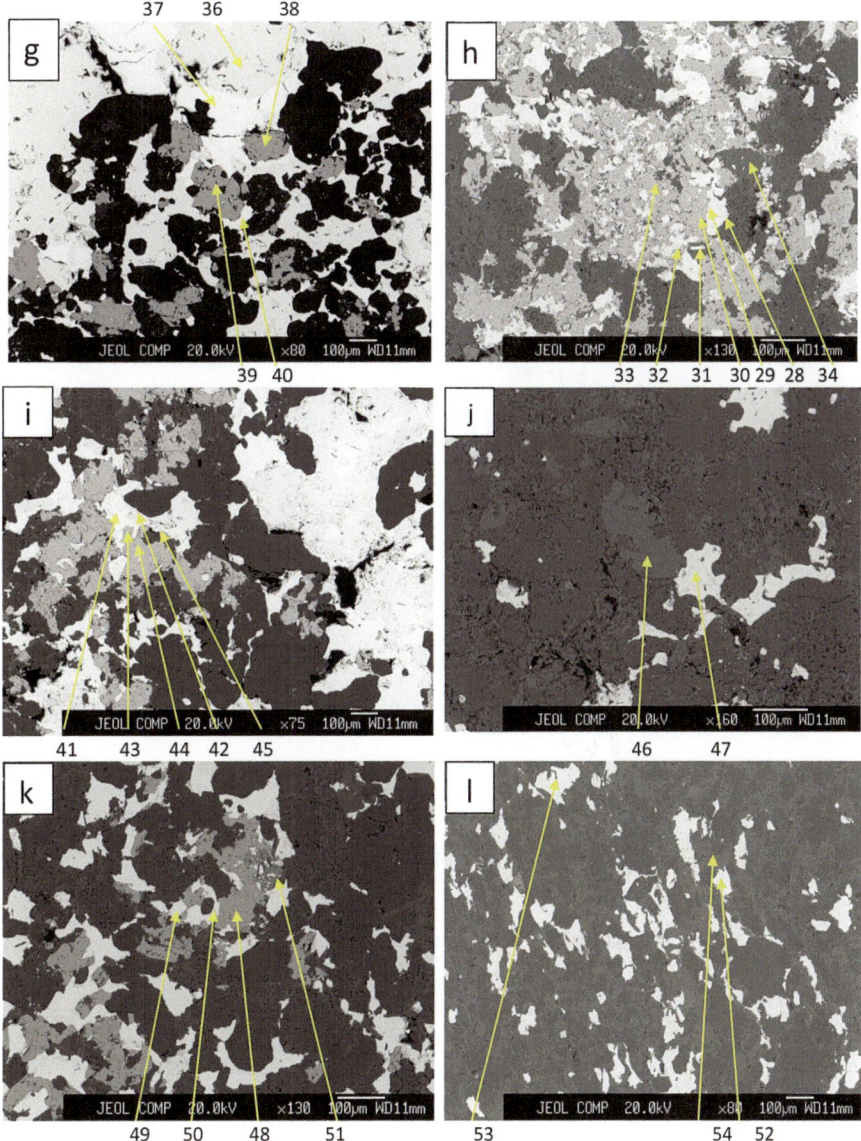

Fig. A1.1 (continued)

Table A1.2 Chemical composition of magnetite from the Udokan deposit, wt%

No point in Fig. A1.1	TiO$_2$	FeO	MnO	V$_2$O$_3$	Cr$_2$O$_3$	Total
4	4,01	84,68	0,00	0,02	0,08	88,90
6	3,72	85,81	0,02	0,15	0,07	89,86
14	3,24	85,41	0,00	0,00	0,00	88,76
22	3,11	85,07	0,01	0,08	0,01	88,38
23	4,36	85,23	0,03	0,04	0,06	89,82
30	3,97	84,46	0,06	0,03	0,07	88,65
32	3,31	83,84	0,03	0,02	0,11	87,38
33	2,15	87,41	0,04	0,00	0,00	89,67
38	3,76	84,88	0,00	0,06	0,06	88,95
39	3,83	85,05	0,02	0,13	0,09	89,23
43	3,27	85,24	0,05	0,00	0,04	88,66
44	3,99	85,36	0,00	0,18	0,05	89,65
48	4,31	84,79	0,06	0,10	0,07	89,38

Table A1.3 Chemical composition of rutile from the Udokan deposit, wt%

No point in Fig. A1.1	TiO$_2$	FeO	Nb$_2$O$_5$	V$_2$O$_3$	Al$_2$O$_3$	Cr$_2$O$_3$	Total
5	83,40	13,43	0,14	0,15	0,06	0,07	97,39
51	99,13	0,54	0,03	0,08	0,01	0,02	99,89

Table A1.4 Chemical composition of zircon from the Udokan deposit, wt%

No point in Fig. A1.1	FeO	SiO$_2$	UO$_2$	ZrO$_2$	HfO$_2$	ThO$_2$	SO$_3$	Total
3	0,13	34,37	0,02	66,17	0,98	0,03	0,07	101,77
20	0,15	34,33	0,05	65,68	0,90	0,00	0,09	101,22

Table A1.5 Chemical composition of zircon from the Udokan deposit, wt%

No point in Fig. A1.1	FeO	CaO	SiO$_2$	MnO	La$_2$O$_3$	SrO	SO$_3$	P$_2$O$_5$	F	Cl	Ce$_2$O$_3$	Total
21	0,04	54,35	0,65	0,07	0,29	0,07	0,16	41,19	3,73	0,07	0,61	99,75
35	0,22	55,11	0,25	0,03	0,09	0,06	0,15	42,06	3,83	0,04	0,28	100,54

Table A1.6 Chemical composition of albite, feldspar, chloride from the Udokan deposit, wt%

No point in Fig. A1.1	FeO	CaO	SiO	Al$_2$O$_3$	Na$_2$O	K$_2$O	Cr$_2$O$_3$	Total
8	0,16	0,03	68,51	18,67	11,39	0,08	0,00	98,88
9	0,08	0,04	68,06	18,70	11,88	0,08	0,02	98,91
10	0,08	0,05	68,33	18,72	11,48	0,08	0,00	98,75
11	0,15	0,04	67,61	18,44	11,41	0,11	0,02	97,80
15	0,21	0,06	68,48	18,81	11,45	0,10	0,02	99,14
16	0,08	0,08	68,34	18,87	11,55	0,13	0,00	99,08
24	0,26	0,05	68,27	18,67	11,33	0,11	0,00	98,70
25	0,10	0,02	68,79	19,01	11,52	0,08	0,00	99,54
40	0,10	0,02	68,59	18,86	11,44	0,06	0,02	99,10
50	0,57	0,05	69,52	19,15	11,80	0,09	0,00	101,24
34	0,32	0,00	64,32	17,76	0,35	16,17	0,03	99,06
45	0,93	0,05	67,79	18,67	0,01	11,44	0,08	99,01
46	0,01	0,00	65,17	18,01	0,05	0,28	16,43	100,02
54	0,12	0,00	65,00	17,99	0,00	0,34	16,29	99,77

Table A1.7 Chemical composition of feldspar from the Udokan deposit, wt%

No point in Fig. A1.2	FeO	SiO$_2$	Al$_2$O$_3$	Na$_2$O	K$_2$O	BaO	Total
5	0,05	64,46	18,33	0,22	16,11	0,54	99,75
6	0,11	64,25	18,14	0,21	15,91	0,42	99,10
8	0,02	64,34	18,30	0,25	16,00	0,42	99,41
9	0,02	64,65	18,15	0,18	16,23	0,34	99,65
12	0,03	64,54	18,60	0,24	15,90	0,47	99,85
13	0,04	64,39	18,13	0,25	16,22	0,39	99,47
14	0,04	64,63	18,12	0,30	15,95	0,17	99,26
20	0,02	64,44	18,49	0,25	16,10	0,60	99,96
24	0,20	65,04	18,46	0,19	16,22	0,23	100,41
29	0,12	64,02	18,35	0,25	15,58	0,36	98,73
31	0,06	64,67	18,41	0,29	16,04	0,62	100,13
36	0,05	64,09	18,10	0,25	15,98	0,38	98,91
38	0,28	64,65	18,24	0,22	16,20	0,35	99,98
40	0,22	64,43	18,39	0,40	15,66	0,62	99,75
43	0,02	64,79	18,48	0,45	15,78	0,58	100,18
45	0,11	64,57	18,35	0,39	15,11	0,53	99,11
46	0,08	64,78	18,60	0,39	15,95	0,97	100,83

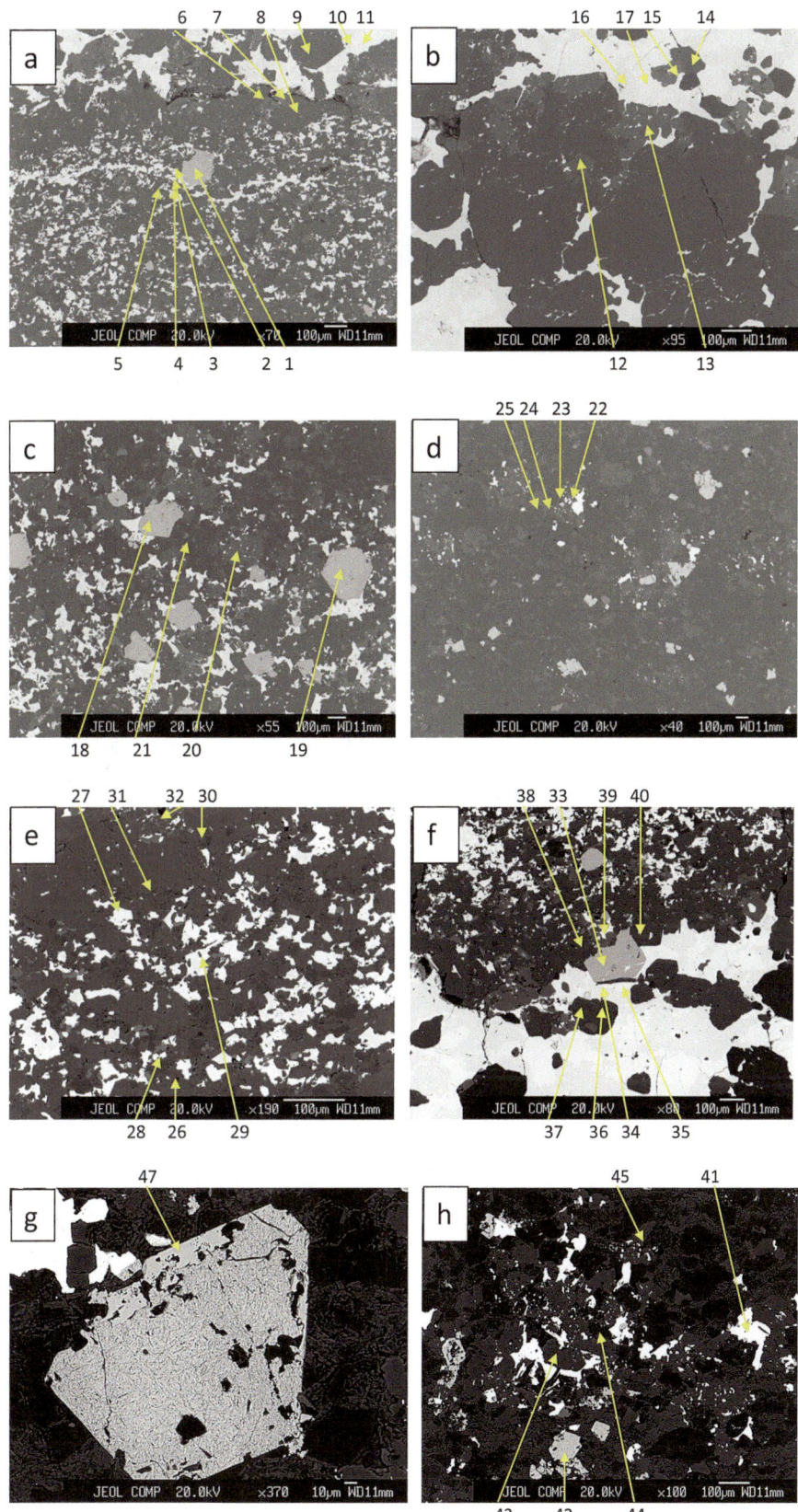

Fig. A1.2 BSE images of rock-forming minerals, titanomagnetite and sulfide from the Mednyi site, Udokan deposit (**a**) – Hematite and tiny sulfide dissemination in sandstone; (**b**) – sericite in rocks; (**c**) – magnetite grains in sandstone with small grains of bornite-chalcocite; (**d**), (**e**) – small interstitial sulfides in sandstone; (**f**) – magnetite crystal with chalcocite-bornite association; (**g**) – large magnetite crystal; (**h**) – magnetite and sulfides in silicate rock

Table A1.8 Chemical composition of amphibole from the Udokan deposit, wt%

No point in Fig. A1.2	FeO	CaO	SiO$_2$	MnO	Al$_2$O$_3$	Cr$_2$O$_3$	Total
7	14,29	22,68	37,44	0,20	20,75	0,04	95,47
15	13,34	22,52	37,92	0,27	21,42	0,07	95,56
32	14,96	22,53	37,50	0,12	20,31	0,03	95,58
37	13,36	22,94	37,39	0,20	21,17	0,24	95,44
39	14,07	22,94	37,54	0,21	20,70	0,38	95,99

Table A1.11 Chemical composition of albite from the Udokan deposit, wt%

No point in Fig. A1.2	FeO	CaO	SiO$_2$	Al$_2$O$_3$	Na$_2$O	K$_2$O	Total
21	0,03	0,28	67,66	19,20	11,44	0,07	98,72
25	0,08	0,04	68,29	19,26	11,19	0,07	99,02
28	0,07	0,07	68,09	19,21	11,53	0,07	99,14
30	0,14	0,17	68,33	19,55	11,45	0,18	99,86
44	0,08	1,62	66,23	20,45	10,40	0,09	98,88

Table A1.9 Chemical compositions of magnetite from the Udokan deposit, wt%

No point in Fig. A1.2	FeO	MnO	V$_2$O$_3$	Al$_2$O$_3$	Cr$_2$O$_3$	Total
1	92,10	0,11	0,26	0,00	0,08	92,59
18	91,98	0,11	0,07	0,03	0,12	92,40
19	91,73	0,14	0,18	0,02	0,12	92,23
23	91,46	0,05	0,08	0,02	0,08	91,75
33	92,47	0,09	0,01	0,05	0,04	92,67
42	88,42	0,13	0,17	0,11	0,05	88,96
47	88,43	0,09	0,10	0,00	0,01	88,71

Table A1.12 Chemical composition of chalcocite and bornite from the ores of the Udokan deposit, wt%

N° point in Fig. A1.3	Fe	S	Cu	Ag	Total
70	1,06	22,37	76,72	0,02	100,17
73	0,02	22,55	77,87	0,03	100,47
74	0,16	22,55	77,81	0,06	100,58
76	0,45	22,82	77,80	0,05	101,11
82	0,29	22,56	78,25	0,04	101,14
84	0,02	22,47	78,87	0,04	101,39
85	0,03	22,51	78,90	0,02	101,46
89	0,21	23,18	76,13	0,03	99,54
96	0,30	22,40	77,42	0,06	100,18
97	0,31	22,62	78,09	0,07	101,09
105	0,65	22,59	78,07	0,04	101,34
106	2,56	21,20	78,65	0,02	102,43
107	0,69	21,00	79,28	0,03	101,00
116	0,07	22,79	77,87	0,04	100,78
71	10,95	26,04	63,11	0,04	100,14
72	11,17	26,02	63,36	0,06	100,60
75	10,76	26,25	63,20	0,04	100,25
83	10,73	26,20	63,70	0,04	100,67
98	10,16	25,88	63,26	0,04	99,33
117	10,94	25,84	63,17	0,02	99,97

Table A1.10 Chemical composition of sulfide from the Udokan deposit, wt%

No Point in Fig. A1.2	Fe	Zn	S	Cu	Total
10	10,44	0,09	25,65	63,03	99,22
16	10,52	0,04	25,62	63,56	99,79
26	10,27	0,03	25,63	62,76	98,76
34	10,72	0,06	25,65	63,17	99,64
2	0,55	0,10	20,96	79,52	101,19
11	0,14	0,05	20,58	78,77	99,58
17	0,03	0,10	20,87	79,39	100,45
22	0,80	0,02	20,82	73,64	95,29
27	0,30	0,02	21,10	78,29	99,73
35	0,70	0,09	20,94	79,15	100,90
41	0,07	0,00	20,88	79,48	100,46

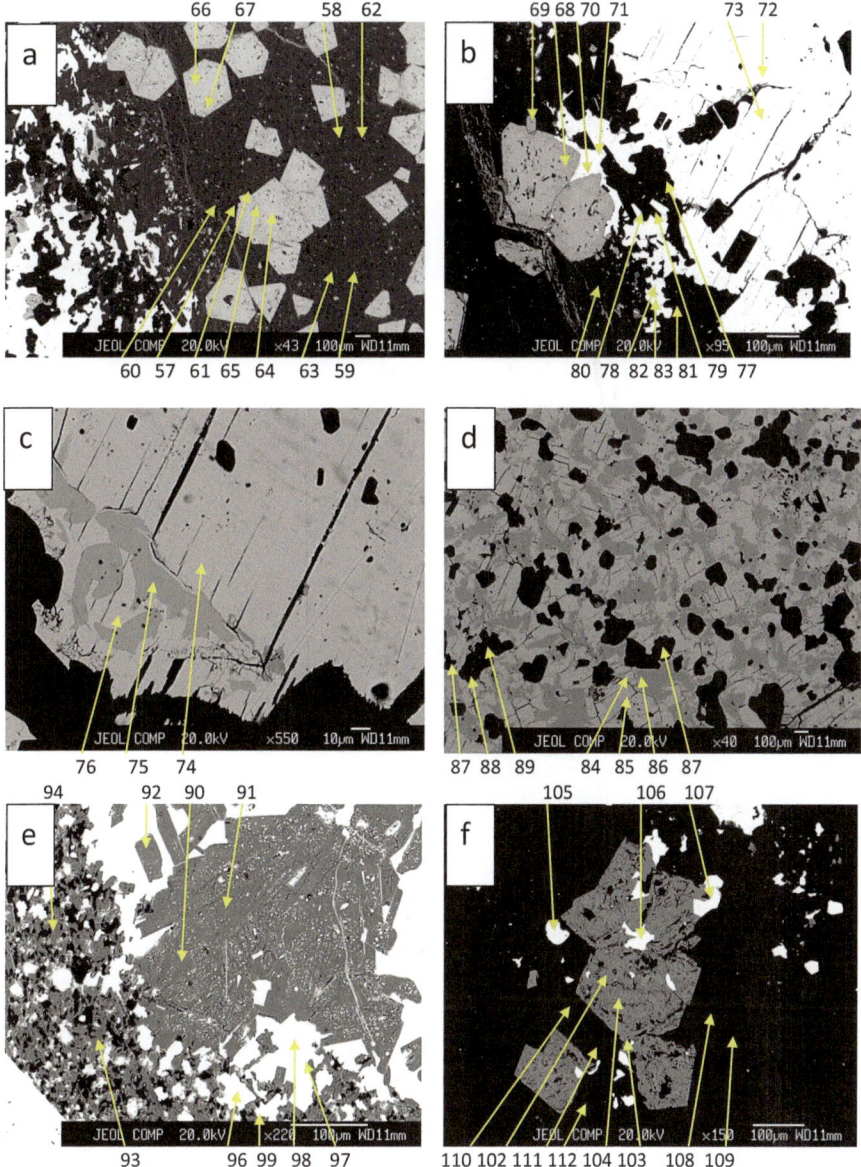

Fig. A1.3 BSE images of minerals at the Bluzhdayushchiy site, Udokan deposit (**a**), (**b**) – Eugedros magnetite crystals in disseminated sulfide ore; (**c**) – chalcocite grain (light) with bornite (dark); (**d**) – chalcocite-bornite aggregate; (**e**) – replacement of large magnetite grain by silicate and sulfide minerals; (**f**) – magnetite crystals in association with chalcocite; (**g**), (**i**) – pyrite crystals; (**h**) – chalcocite-bornite dissemination

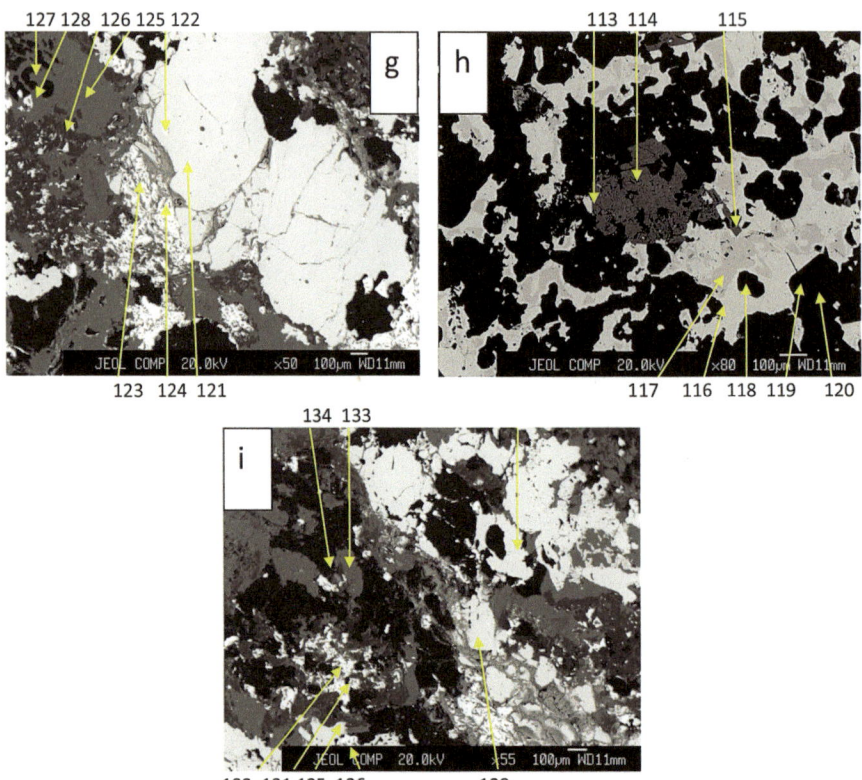

Fig. A1.3 (continued)

Table A1.13 Chemical composition of pyrite from the Udokan deposit, wt%

No point in Fig. A1.3	Fe	Ni	S	Co	Total
121	45,55	0,38	52,33	0,30	98,59
129	44,21	0,23	52,28	2,10	98,83
130	46,33	0,01	52,31	0,07	98,73
135	46,62	0,00	52,45	0,07	99,19

Table A1.14 Chemical composition of magnetite from the Udokan deposit, wt%

No point in Fig. A1.3	TiO_2	FeO	SiO_2	MnO	V_2O_3	Cr_2O_3	Total
65	0,04	89,15	0,01	0,05	0,00	0,04	89,32
66	0,00	91,16	0,00	0,08	0,03	0,08	91,42
67	0,08	90,42	0,05	0,05	0,03	0,08	90,83
68	0,05	89,19	0,13	0,01	0,03	0,04	89,57
69	0,07	87,02	0,22	0,04	0,00	0,12	87,59
99	1,08	85,42	0,02	0,00	0,00	0,10	86,69
102	0,00	87,81	0,01	0,00	0,10	0,12	88,10
103	0,03	87,66	0,08	0,01	0,27	0,47	88,60
104	0,14	87,79	0,05	0,06	0,02	0,36	88,55
113	0,00	87,90	0,01	0,08	0,04	0,19	88,33
114	0,02	91,86	0,00	0,04	0,04	0,07	92,14
115	0,01	88,55	0,00	0,02	0,00	0,03	88,67

Table A1.15 Chemical composition of mica from the Udokan deposit, wt%

No point in Fig. A1.3	MgO	FeO	SiO₂	Al₂O₃	Cr₂O₃	Na₂O	TiO₂	K₂O	Total
57	1,75	5,58	47,77	29,16	0,05	0,11	0,17	9,44	94,04
58	1,62	4,56	47,01	28,69	0,03	0,12	0,37	11,08	93,53
59	1,48	4,44	47,06	29,07	0,24	0,15	0,41	10,72	93,60
77	1,74	5,59	46,61	27,06	0,06	0,19	0,18	10,59	92,07
78	1,83	5,53	47,25	27,75	0,00	0,20	0,21	10,88	93,67
79	0,95	5,00	45,54	29,97	0,03	0,20	0,66	10,91	93,28
80	1,15	4,92	44,23	27,81	0,09	0,14	0,70	10,60	89,64
87	0,95	5,23	46,22	30,64	0,02	0,32	0,81	10,80	95,03
90	1,04	5,45	45,78	30,31	0,01	0,22	0,28	10,78	93,89
91	1,85	5,41	46,93	27,89	0,15	0,18	0,19	10,73	93,36
92	1,82	5,31	46,88	27,51	0,01	0,18	0,68	10,59	93,05
93	1,61	4,79	47,20	28,65	0,08	0,12	0,46	11,02	93,97
94	1,86	4,36	46,90	28,01	0,04	0,15	0,36	10,91	92,63
95	1,87	5,46	46,96	27,54	0,08	0,18	0,27	10,84	93,23
100	1,64	4,62	47,47	28,54	0,12	0,15	0,33	10,82	93,70
109	0,00	0,07	63,46	17,94	0,00	0,25	0,03	16,14	97,90
111	1,58	4,72	46,43	28,48	0,05	0,17	0,33	10,63	92,44
126	1,02	6,84	46,98	28,84	0,03	0,06	0,20	10,71	94,76

Table A1.16 Chemical composition of albite from the Udokan deposit, wt%

No point in Fig. A1.3	FeO	CaO	SiO₂	Al₂O₃	Na₂O	K₂O	Total
60	0,16	0,05	67,20	18,73	10,87	0,13	97,21
62	0,14	0,41	69,24	19,88	11,34	0,15	101,20
63	0,19	0,05	68,02	18,98	11,12	0,09	98,51
88	0,08	0,08	68,71	19,16	11,11	0,06	99,26
101	0,17	0,39	67,17	19,19	10,70	0,13	97,80
110	0,36	0,38	67,99	19,14	11,10	0,08	99,14
118	0,01	0,04	67,88	19,00	11,24	0,11	98,34
119	0,05	0,10	68,16	19,12	11,32	0,08	98,84
120	0,03	0,27	67,67	19,03	10,97	0,08	98,06
101	0,16	0,37	67,50	19,28	10,88	0,09	98,34
120	0,09	0,33	67,62	19,32	10,82	0,14	98,39

Table A1.17 Chemical composition of minerals from the Udokan deposit, wt%

No point in Fig. A1.4	SiO₂	TiO₂	Al₂O₃	FeO	MnO	MgO	CaO	Na₂O	K₂O	F	BaO	
18	37,49	0,03	20,66	14,31	0,62	0,00	21,78	0,00	0,01	0,02	0,10	95,01
13	37,21	0,08	21,59	12,98	0,25	0,00	22,45	0,00	0,02	0,02	0,00	94,60
10	37,59	0,04	22,02	12,19	0,31	0,00	22,59	0,00	0,03	0,00	0,00	94,77
14	46,16	0,39	28,67	4,12	0,00	1,64	0,02	0,12	9,81	0,00	0,12	91,06
5	44,85	0,67	28,04	4,84	0,03	2,00	0,01	0,11	9,98	0,00	0,34	90,86
11	47,59	0,44	26,17	4,44	0,05	2,54	0,03	0,11	9,98	0,00	0,30	91,65
8	45,21	0,30	29,02	4,47	0,04	1,71	0,01	0,13	10,05	0,00	0,07	91,01
10	64,93	0,03	18,54	0,00	0,00	0,01	0,00	0,60	15,55	0,84	0,00	100,50

Fig. A1.4 BSE images of minerals at the Ozernyi site, Udokan deposit (**a**) – Large magnetite grain surrounded by chalcocite, (**b**) – small magnetite grains among silicate minerals; (**c**), (**d**) – interstitial position of sulfides in rock; (**e**), (**f**) – euhedral magnetite grains in disseminated sulfide ore

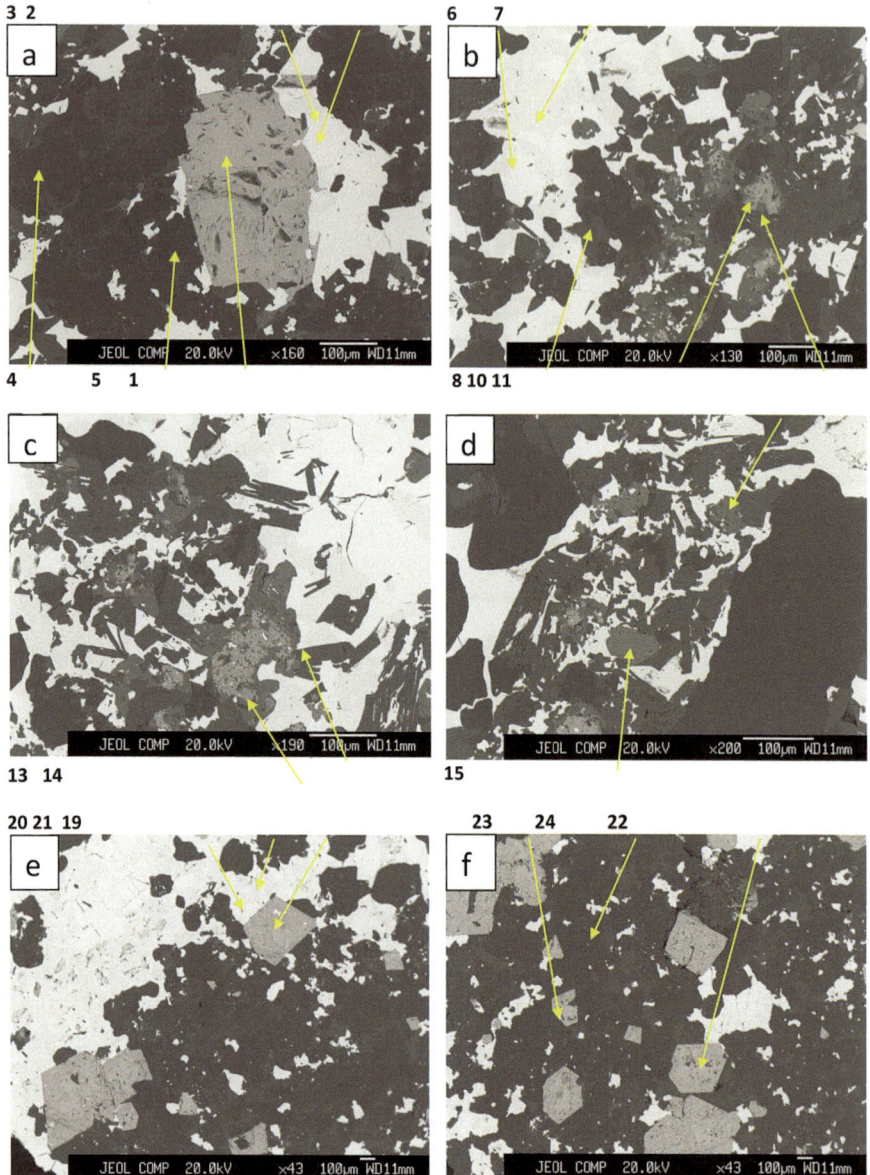

Table A1.18 Chemical composition of albite from the Udokan deposit, wt%

No point in Fig. A1.4	SiO$_2$	Al$_2$O$_3$	CaO	Na$_2$O	K$_2$O	Total
4	67,77	19,21	0,22	11,33	0,16	98,68

Table A1.19 Chemical composition of magnetite from the Udokan deposit, wt%

No point in Fig. A1.4	FeO	MnO	V$_2$O$_3$	Total
1	93,17	0,18	0,02	93,37
9	93,58	0,00	0,00	93,58
22	93,52	0,06	0,10	93,68
23	94,09	0,01	0,10	94,20

Table A1.20 Chemical composition of sulfide from the Udokan deposit, wt%

No point in Fig. A1.4	FeO	ZnO	S	Cu	Total
3	11,15	0,03	25,81	63,30	100,29
6	11,19	0,03	25,79	63,05	100,06
21	11,18	0,01	25,91	63,13	100,23
2	0,41	0,08	21,31	79,64	101,44
7	0,08	0,01	21,30	79,70	101,08
20	0,23	0,07	21,51	79,52	101,34

Table A1.21 Chemical composition of apatite from the Udokan deposit, wt%

No point in Fig. A1.4	FeO	CaO	MnO	Ce$_2$O$_3$	P$_2$O$_5$	F	Cl	Total
15	0,17	54,64	0,39	0,19	42,85	3,86	0,20	102,30

Appendix 2

Table A2.1 Chemical compositions of rock of the Chiney massif, wt%

N Analysis	No borehole	Depth, cm	SiO₂	TiO₂	Al₂O₃	Fe₂O₃	FeO	MnO	MgO	Cao	Na₂O	K₂O	P₂O₅	LOI	Total
	1	2	3	4	5	6	7	8	9	10	11	12	13	14	15
1	11	58.30	51.76	0.60	23.91	6.15		0.10	1.27	10.55	3.89	0.62	0.04	0.52	99.41
2	11	68.64	38.35	3.42	13.66	26.05		0.17	5.04	10.44	1.89	0.25		0.12	99.39
3	11	90.20	37.92	3.89	15.99	25.05		0.16	3.35	7.64	2.77	0.99		1.14	98.90
4	11	91.10	52.75	0.63	23.27	5.70		0.10	1.64	10.86	3.90	0.70		1.24	100.79
5	11	94.00	52.83	0.53	24.25	5.60		0.09	1.34	11.06	4.48	0.66		1.30	102.14
6	11	94.60	51.80	0.61	23.71	6.42		0.09	1.42	10.96	4.84	0.59		0.92	101.36
7	11	96.70	52.31	0.43	22.63	4.36		0.09	1.16	11.47	3.10	0.88		5.31	101.74
8	11	97.20	51.42	0.46	23.69	6.45		0.09	1.35	11.37	4.13	1.05		1.90	101.91
9	11	97.80	51.60	0.68	22.12	6.39		0.11	2.02	9.98	4.04	0.73		2.05	99.72
10	11	99.10	44.10	2.36	13.30	21.30		0.18	5.12	10.20	2.64	0.53	0.07	0.25	100.05
11	11	99.1	44.11	2.36	13.3	21.36		0.19	5.12	10.28	2.64	0.54	0.08	0.25	100.23
12	11	106.60	52.69	0.78	15.29	12.02		0.14	7.92	8.15	2.43	0.99	0.07	0.76	101.24
13	11	108.80	50.25	1.67	13.97	15.49		0.17	4.99	8.31	2.77	1.09	0.10	0.53	99.34
14	11	109.10	46.47	2.07	13.43	17.79		0.17	5.19	11.34	2.65	0.69	0.07	1.08	100.95
15	11	109.20	48.34	1.91	13.76	17.72		0.17	5.60	9.76	2.26	0.85	0.08	0.38	100.83
16	11	109.80	47.92	1.88	13.62	4.74	11.57	0.18	5.84	9.37	2.16	0.63	0.07	0.21	98.19
17	11	111.10	47.94	1.97	13.70	0.53	15.45	0.18	5.68	9.62	2.62	0.59	0.07	0.31	98.66
18	11	118.70	49.44	1.30	13.05	16.69		0.18	8.54	7.19	2.06	0.84	0.07	0.65	100.01
19	11	121.20	50.82	1.09	14.14	14.06		0.17	7.52	7.74	2.09	1.06	0.08	1.23	100.00
20	11	124.20	51.42	1.19	15.95	12.76		0.16	5.39	8.19	2.34	1.27	0.10	1.28	100.05
21	11	125.40	49.09	2.36	15.40	18.15		0.15	3.45	5.41	3.26	1.20	0.80	1.73	101.00
22	11	126.40	51.16	1.19	14.25	14.00		0.16	6.86	7.35	2.11	1.48	0.09	1.38	100.03
23	11	137.80	46.58	1.24	16.38	14.95		0.16	6.08	10.57	2.87	0.26	0.02	0.85	99.96
24	11	138.00	45.89	1.02	17.87	4.85	9.23	0.02	6.77	10.77	2.36	0.23	0.17	0.00	99.18
25	11	138.10	44.34	1.40	17.19	7.06	9.34	0.02	6.20	11.14	2.19	0.22	0.17	0.00	99.27
26	11	139.70	47.42	1.28	17.24	14.60		0.15	5.28	11.57	2.21	0.25	0.05	0.00	100.05
27	11	145.60	52.50	0.19	18.34	9.27		0.17	7.30	9.32	2.65	0.27	0.03	0.05	100.09
28	11	148.30	45.70	1.52	16.20	16.88		0.17	5.93	11.19	1.95	0.23	0.02	0.25	100.04
29	11	158.50	54.39	0.19	18.57	4.24		0.19	7.54	9.35	2.03	0.27	0.01	0.36	97.14
30	11	165.00	50.30	1.01	22.74	8.89		0.11	2.05	10.76	2.34	0.44	0.06	0.42	99.12
31	11	165.70	52.38	0.45	18.51	8.70		0.14	6.07	11.28	1.94	0.29	0.03	0.17	99.96
32	11	184.70	49.79	1.14	20.58	10.02		0.12	2.85	11.29	3.22	0.51	0.12	0.63	100.27
33	11	189.30	47.65	1.46	16.15	14.42		0.15	4.54	11.28	2.94	0.66	0.14	1.36	100.75
34	11	198.20	43.58	2.16	14.56	19.20		0.16	5.33	12.00	2.25	0.28	0.13	0.96	100.61
35	11	200.00	43.59	2.06	13.02	20.06		0.18	6.12	11.28	1.59	0.31	0.03	0.59	98.83
36	11	204.00	42.40	2.12	13.20	10.13	10.34	0.17	6.29	11.88	2.70	0.38	0.03	0.13	99.77
37	11	246.90	39.49	3.22	14.51	6.72	17.20	0.16	4.98	9.72	2.03	0.25	0.10	0.34	98.72
38	11	246.90	40.15	2.92	14.87	12.31	12.28	0.17	5.16	9.19	2.25	0.32	0.06	0.13	99.81
39	11	249.60	38.41	3.52	8.09	30.25		0.24	8.71	9.21	1.92	0.24	0.03	−0.64	99.98
40	11	250.00	50.28	0.75	23.66	2.53	4.09	0.06	1.77	10.98	4.00	0.60	0.05	1.10	99.87
41	11	251.86	15.56	8.31	3.05	62.36		0.23	5.39	2.63	0.73	0	0.03	1.5	99.79
42	11	251.9	25.6	6.00	3.27	49.27		0.23	7.95	4.81	1.08	0.01	0.04	1.5	99.76
43	11	251.96	47.25	1.23	18.86	14.87		0.15	3.72	9.78	1.87	0.33	0.06	1.5	99.62
44	11	252.00	53.48	0.65	24.11	3.94	1.65	0.10	1.13	8.40	5.07	1.55	0.00	2.01	102.09
45	11	254.20	53.27	0.71	22.58	5.49		0.09	2.80	10.70	3.35	0.67	0.03	0.28	99.97
46	11	260.2	43.68	2.11	7.16	22.47		0.27	10.29	12.14	1.06	0.22	0.03	0.72	100.15
47	11	262.50	50.20	1.55	14.90	3.07	12.18	0.19	6.34	6.95	2.25	0.77	0.08	1.55	100.03
48	11	275.40	40.12	3.18	16.34	24.71		0.14	3.59	8.85	2.55	0.47	0.04	0.00	99.99
49	11	277.50	42.78	3.54	14.11	3.40	16.85	0.17	6.05	8.99	2.16	0.42	0.07	0.21	98.75
50	11	279.20	49.06	1.42	22.42	11.39		0.07	1.41	10.00	3.37	0.67	0.07	0.15	100.03

(continued)

Table A2.1 (continued)

N	No borehole	Depth, cm	SiO₂	TiO₂	Al₂O₃	Fe₂O₃	FeO	MnO	MgO	Cao	Na₂O	K₂O	P₂O₅	LOI	Total
Analysis	1	2	3	4	5	6	7	8	9	10	11	12	13	14	15
51	11	281.50	49.05	1.38	22.33	10.80		0.07	1.70	10.30	3.55	0.68	0.14	0.00	100.00
52	11	283.40	46.97	1.74	21.85	14.27		0.09	1.88	9.37	3.25	0.56	0.08	−0.08	99.98
53	11	285.00	53.13	0.21	24.61	3.31		0.09	0.92	9.37	4.87	1.47	0.10	1.44	99.52
54	11	292.20	37.44	3.73	14.99	28.58		0.16	4.34	8.71	2.41	0.33	0.03	−0.71	100.01
55	11	295.00	48.84	1.25	22.48	2.08	7.95	0.11	1.45	10.16	3.42	0.56	0.14	0.39	98.83
56	11	300.40	49.56	1.36	22.34	10.87		0.08	1.60	10.51	3.02	0.62	0.05	0.07	100.08
57	11	302.5	31.76	5.04	9.24	39.05		0.2	5.77	7.47	2.2	0.24	0.02	−0.94	100.05
58	11	308.80	46.96	1.85	21.75	15.06		0.08	1.70	9.67	2.27	0.53	0.04	0.08	99.99
59	11	311.36	45.41	2.07	20.94	16.52		0.09	2.19	9.44	2.90	0.40	0.05	,	100.01
60	11	315.40	45.49	2.20	20.95	16.83		0.09	1.87	9.53	2.30	0.57	0.05	0.09	99.97
61	11	324.00	42.20	2.98	18.91	21.31		0.12	2.74	9.37	2.00	0.43	0.03	−0.08	100.01
62	11	336.00	38.29	3.58	16.07	27.01		0.15	3.78	8.29	2.29	0.28	0.03	0.21	99.98
63	11	339.20	42.98	2.40	19.40	18.90		0.10	2.25	8.98	4.31	0.42	0.04	0.22	100.00
64	11	344.70	54.05	0.51	24.57	4.26		0.04	1.23	8.59	3.12	1.46	0.04	2.11	99.98
65	11	356.50	53.31	0.65	24.00	5.25		0.05	1.31	10.52	2.60	0.71	0.07	1.50	99.97
66	11	363.00	40.62	2.93	15.82	23.91		0.15	4.19	10.01	1.89	0.27	0.03	0.17	99.99
67	11	364	51.45	0.31	6.8	7.11		0.2	11.68	20.26	1.21	0.21	0.32	0.49	100.04
68	11	365.70	36.74	3.67	12.36	28.91		0.18	5.72	9.87	2.07	0.26	0.02	0.20	100.00
69	11	367.00	36.02	3.73	12.06	28.75		0.18	5.54	9.98	3.48	0.27	0.02	0.00	100.03
70	11	369.80	35.04	4.01	12.02	31.51		0.17	5.01	9.70	2.51	0.35	0.05	−0.38	99.99
71	11	374.3	29.47	7.18	8.23	37.86		0.23	5.69	7.95	2.29	0.26	0.02	0.78	99.96
72	11	379.80	48.65	1.05	16.77	4.84	7.68	0.16	6.77	10.54	2.60	0.26	0.01	0.40	99.73
73	11	383	34.09	6.21	10.19	33.1		0.2	5.58	8.16	1.88	0.21	0.02	−0.09	99.55
74	11	385.20	30.90	5.75	10.31	39.31		0.23	6.52	5.22	1.76	0.28	0.02	−0.32	99.98
75	11	385.60	33.44	4.95	11.18	35.54		0.22	6.18	6.64	2.16	0.22	0.02	−0.35	100.20
76	11	389.90	45.17	2.48	18.84	16.41		0.14	3.66	7.89	3.72	0.76	0.00	1.80	100.87
77	11	393	35.86	3.99	7.8	32.97		0.25	9.49	6.84	1.63	0.34	0.033	0.65	99.85
78	11	455.9	44.61	2.1	16.91	19.14		0.15	4.85	8.03	2.4	0.35	0.1		98.64
79	11	458.60	37.17	5.89	18.83	5.29	19.45	0.14	0.91	7.12	2.70	0.37	0.07	0.32	98.26
80	11	459.4	51.8	1.27	16.55	14.45		0.14	5.89	4.33	3.11	1.21	0.15	0.65	99.55
81	11	463.30	34.25	4.18	13.09	32.73		0.16	3.41	6.62	2.73	0.52	0.05	0.31	98.05
82	11	463.70	34.56	7.04	17.85	7.13	21.50	0.15	1.05	6.51	2.99	0.32	0.08	0.37	99.55
83	11	467.40	47.57	1.15	15.40	2.37	9.35	0.14	6.54	7.68	1.44	2.57	0.23	3.70	98.14
84	11	481.50	30.50	5.23	11.55	39.62		0.20	4.53	6.16	1.76	0.24	0.04	−0.09	99.74
85	11	481.70	30.75	5.23	11.55	39.62		0.20	4.53	6.16	1.76	0.24	0.04	−0.09	99.99
86	11	484.00	49.14	1.75	21.57	11.03		0.09	1.92	10.34	3.12	0.60	0.09	0.33	99.98
87	11	484.30	48.97	1.34	21.47	5.59	5.82	0.12	2.21	9.34	2.88	0.58	0.16	0.58	99.06
88	11	489.00	51.60	1.13	19.80	10.47		0.09	2.54	9.91	2.42	0.81	0.07	0.36	99.20
89	11	491.30	37.92	3.48	15.42	12.86	14.37	0.16	4.40	6.99	2.08	0.43	0.09	0.24	98.44
90	11	503.80	40.55	3.16	16.84	25.51		0.13	2.64	7.82	2.31	0.55	0.04	−0.03	99.52
91	11	507.30	43.55	2.43	20.63	9.07	9.84	0.12	1.51	8.35	2.94	0.52	0.14	0.39	99.49
92	11	518.80	47.22	1.71	19.15	13.97		0.09	1.80	8.82	2.95	1.37	0.08	0.85	98.01
93	11	522.40	33.76	4.95	13.93	34.60		0.16	3.06	6.91	2.06	0.62	0.04	−0.03	100.06
94	11	528.40	44.49	2.38	13.08	7.19	12.50	0.18	5.77	9.48	1.70	0.61	0.12	0.63	98.13
95	11	529.50	45.01	2.50	12.83	8.09	12.71	0.21	5.89	9.19	2.25	0.56	0.09	0.40	99.73
96	11	530.00	44.97	2.53	14.16	8.32	11.78	0.02	5.53	9.65	2.14	0.39	0.11	0.18	99.78
97	11	530.60	25.27	7.22	10.10	46.57		0.22	4.39	5.33	1.39	0.30	0.02	−0.30	100.51
98	11	532.40	39.52	3.45	17.83	12.03	12.50	0.15	2.51	7.65	2.28	0.51	0.11	0.04	98.58
99	11	532.60	44.79	2.28	20.34	6.74	8.84	0.12	1.95	8.83	3.12	0.46	0.10	0.58	98.15
100	11	560.50	40.30	3.50	11.80	10.34	17.42	0.22	6.66	6.28	1.73	0.52	0.06	1.52	100.35
101	11	560.70	30.60	5.33	9.30	18.58	21.52	0.23	6.45	4.26	1.13	0.46	0.03	1.95	99.84
102	11	560.80	45.20	2.55	12.65	7.08	14.77	0.20	6.81	5.61	2.08	0.66	0.08	1.52	99.21
103	11	560.90	36.70	4.34	10.85	13.36	18.93	0.23	6.95	4.71	1.60	0.46	0.04	1.56	99.73

(continued)

Table A2.1 (continued)

N Analysis	No borehole	Depth, cm	SiO$_2$	TiO$_2$	Al$_2$O$_3$	Fe$_2$O$_3$	FeO	MnO	MgO	Cao	Na$_2$O	K$_2$O	P$_2$O$_5$	LOI	Total
	1	2	3	4	5	6	7	8	9	10	11	12	13	14	15
104	11	561.10	45.30	2.33	13.10	6.05	14.98	0.21	6.83	6.51	2.01	0.61	0.07	1.41	99.41
105	11	561.30	44.21	2.34	12.32	23.14		0.19	7.89	6.89	1.72	0.54	0.08	0.17	99.49
106	11	561.60	17.50	8.08	6.10	31.25	26.23	0.26	5.23	2.24	0.50	0.29		2.08	99.76
107	11	561.80	49.50	1.46	15.10	2.94	12.29	0.18	5.85	7.18	2.50	0.64	0.08	1.33	99.05
108	11	561.90	46.90	2.10	12.50	4.66	15.56	0.22	7.34	6.73	2.01	0.50	0.05	1.31	99.88
109	11	562.00	30.10	5.47	8.50	20.34	21.02	0.26	6.63	4.94	0.96	0.31	0.03	1.52	100.08
110	11	582.50	44.47	2.59	12.95	8.60	11.39	0.18	5.71	9.61	2.04	0.49	0.15	0.20	98.38
111	11	603.30	28.14	6.29	6.41	46.33		0.04	9.21	3.96	0.41	0.22	0.28	0.00	101.29
112	11	603.42	45.23	2.09	14.37	18.23		0.10	6.32	10.12	2.93	0.40	0.17	−0.23	99.73
113	11	603.43	43.99	2.25	14.12	19.90		0.07	6.35	9.87	3.23	0.41	0.18	0.00	100.37
114	11	603.44	44.62	2.27	13.23	19.17		0.12	6.78	10.72	2.80	0.35	0.17	−0.38	99.85
115	11	603.45	42.66	2.98	10.13	23.68		0.10	7.79	11.01	1.55	0.28	0.20	−0.32	100.06
116	11	619.78	26.46	5.85	4.92	47.28		0.30	10.22	3.56	0.21	0.14		1.66	100.60
117	11	619.80	24.17	6.24	5.33	49.85		0.27	8.57	3.28	0.80	0.13		0.47	99.11
118	11	619.82	22.66	6.56	5.22	50.93		0.27	8.51	3.38	0.89	0.12		0.69	99.23
119	11	619.85	44.57	1.98	10.01	24.04		0.24	1.25	5.79	0.95	0.85		1.21	90.89
120	11	619.87	40.89	2.74	11.09	27.61		0.23	8.63	5.78	1.12	0.47		1.17	99.73
121	11	619.88	28.20	5.48	6.32	43.83		0.27	10.01	3.83	0.40	0.15		1.44	99.93
122	11	619.89	30.43	5.23	6.56	42.25		0.28	9.37	3.85	0.43	0.20		1.88	100.48
123	11	619.91	22.74	6.65	5.23	50.71		0.27	8.68	3.39	0.88	0.14		1.57	100.26
124	11	619.95	15.02	8.74	4.79	61.45		0.27	5.28	2.08	0.65	0.08		1.65	100.01
125	11	619.97	17.99	7.97	4.52	58.11		0.25	6.04	2.78	0.74	0.10		1.68	100.18
126	11	619.99	16.54	7.67	5.71	57.15		0.27	6.97	2.87	0.99	0.10		1.65	99.92
127	11	622.68	15.83	7.85	3.62	24.22	26.22	0.26	5.98	3.08	1.59	0.18	0.01	−1.00	87.84
128	11	622.69	17.59	8.23	3.40	26.77	26.44	0.29	6.20	3.71	1.39	0.18	0.01	−0.85	93.36
129	11	622.71	41.65	3.38	7.25	10.18	16.99	0.26	10.02	9.40	0.99	0.37	0.08	0.21	100.78
130	11	622.73	36.10	4.75	6.93	15.37	20.18	0.29	8.96	7.89	0.96	0.26	0.03	−0.08	101.64
131	11	622.74	30.85	5.75	6.04	19.43	21.98	0.30	9.13	5.31	1.01	0.30	0.03	−0.62	99.51
132	11	622.75	20.60	7.82	4.17	25.47	26.72	0.31	6.78	3.15	0.95	0.23	0.03	−0.89	95.34
133	11	622.76	16.14	8.58	3.22	28.37	28.09	0.30	6.55	2.03	1.36	0.18	0.01	−1.18	93.65
134	11	622.78	32.93	4.95	5.27	16.39	22.05	0.31	10.12	4.06	0.90	0.28	0.05	−1.06	96.25
135	11	622.79	28.59	6.73	5.74	21.17	23.85	0.31	9.20	3.70	0.94	0.25	0.02	−0.89	99.61
136	11	622.80	23.50	7.48	4.74	26.63	25.08	0.32	7.27	3.19	0.94	0.23	0.02	−1.30	98.10
137	11	622.81	19.23	8.18	4.26	26.45	26.87	0.29	7.15	2.75	0.83	0.20	0.01	−0.90	95.32
138	11	622.83	48.69	1.54	18.02	1.04	12.43	0.13	4.05	8.80	2.30	0.48	0.06	−0.24	97.30
139	11	622.84	46.64	2.59	8.79	8.09	17.60	0.29	11.21	6.06	1.31	0.43	0.04	−1.01	102.04
140	11	622.85	45.28	3.04	8.10	10.79	18.02	0.31	11.46	5.93	1.14	0.35	0.06	−1.52	102.96
141	11	622.86	36.79	4.41	5.33	13.68	21.77	0.32	11.54	4.11	0.98	0.24	0.05	−1.20	98.01
142	11	622.87	29.36	5.95	3.99	18.59	24.04	0.31	10.18	3.48	0.95	0.19	0.02	−1.44	95.62
143	11	622.89	48.82	1.97	14.39	8.43	9.70	0.19	6.60	8.53	2.15	0.49	0.08	0.42	101.77
144	11	627.77	11.66	10.08	3.42	65.29		0.26	5.5	1.41	2.8	0.14	0.03	−0.78	99.81
145	11	627.77	11.66	10.08	3.42	65.29		0.26	5.5	1.41	2.8	0.14	0.03	−0.78	99.81
146	11	631.10	48.91	1.43	13.60	2.15	11.78	0.16	6.19	9.55	2.47	0.41	0.10	0.83	97.58
147	11	632.5	46.69	1.96	13.83	18.72		0.16	5.27	10.49	2.39	0.5	0.09		100.10
148	11	633.60	49.44	1.16	14.44	2.16	11.06	0.16	6.55	10.20	1.76	0.67	0.06	0.78	98.44
149	11	634.50	42.35	2.05	15.01	4.51	11.78	0.16	5.04	9.24	2.01	0.56	0.05	0.31	93.07
150	11	637.99	14.82	9.15	4.34	31.74	28.88	0.25	6.31	1.62	0.51	0.14	0.01		97.77
160	11	638.00	14.53	8.58	3.42	30.74	28.45	0.23	7.04	1.38	0.64	0.12	0.02	0.00	95.15
161	11	638.01	47.14	0.91	14.90	1.18	14.37	0.17	5.79	10.40	2.47	0.35	0.05	0.80	98.53
162	11	638.02	44.96	0.47	14.54	2.67	16.81	0.15	0.42	11.21	2.52	0.32	0.04	1.88	95.99
163	11	638.05	36.64	4.06	5.16	14.01	21.16	0.25	10.74	6.48	0.51	0.18	0.06		99.25
164	11	638.07	25.49	6.18	3.92	20.97	25.08	0.24	8.43	4.26	0.79	0.13	0.02	0.00	95.51
165	11	638.08	21.65	7.23	4.16	23.70	27.12	0.25	7.37	4.10	0.18	0.11	0.02	0.00	95.89

(continued)

Table A2.1 (continued)

N Analysis	No borehole	Depth, cm	SiO$_2$	TiO$_2$	Al$_2$O$_3$	Fe$_2$O$_3$	FeO	MnO	MgO	Cao	Na$_2$O	K$_2$O	P$_2$O$_5$	LOI	Total
1	2	3	4	5	6	7	8	9	10	11	12	13	14	15	
166	11	638.10	13.68	8.88	3.45	30.79	29.46	0.24	6.47	1.48	0.52	0.08	0.01		95.06
167	11	638.12	40.61	2.93	6.50	9.82	20.51	0.25	11.68	6.02	0.63	0.21	0.03		99.19
168	11	641.89	17.93	8.16	4.37	59.30		0.23	5.72	1.91	0.73	0.06	0.03		98.44
169	11	641.90	45.33	1.70	12.94	20.17		0.19	6.00	10.30	1.23	0.22	0.06		98.14
170	11	641.91	44.70	1.97	9.15	26.00		0.22	8.38	6.01	1.28	0.26	0.06		98.03
171	11	641.93	36.90	3.82	6.51	35.89		0.23	8.40	5.06	1.22	0.17	0.08		98.28
172	11	641.94	30.18	5.18	4.88	44.70		0.25	8.49	3.73	0.75	0.10	0.06		98.32
173	11	641.96	17.93	8.21	3.31	59.83		0.24	6.21	1.71	0.73	0.03	0.04		98.24
174	11	641.97	12.93	9.23	3.30	65.53		0.24	5.06	1.26	0.75	0.01	0.04		98.35
175	11	642.00	38.90	3.24	5.57	31.51		0.22	8.18	9.67	0.64	0.17	0.18		98.28
176	11	651.80	49.50	1.46	15.10	2.94	12.29	0.18	5.85	7.18	2.50	0.64	0.08	1.33	99.05
177	11	661.00	30.46	5.02	7.76	39.64		0.25	7.05	7.44	2.48	0.28	0.05	−0.13	100.30
178	11	667.00	26.83	5.79	5.20	42.59		0.28	7.20	5.51	2.27	0.23	0.03	1.29	97.22
179	11	671.00	10.71	9.11	4.15	67.09		0.24	3.84	2.81	2.22	0.13	0.03		100.33
180	11	680.7	24.76	6.51	5.87	46.99		0.22	7.77	4.39	2.45	0.51	0.04	0.41	99.92
181	11	680.7	24.76	6.51	5.87	46.99		0.22	7.77	4.39	2.45	0.51	0.04	0.41	99.92
182	11	682.00	8.99	9.21	4.46	68.42		0.23	6.30	0.70	1.40	0.06	0.03	0.24	100.04
183	11	712.00	17.03	1.40	9.79	56.31		0.21	2.44	3.83	2.15	0.21	0.01	0.76	94.14
184	11	724.30	46.40	1.50	12.11	6.02	12.28	0.20	9.19	8.07	2.00	0.60	0.09	1.50	99.96
185	11	726.40	42.75	2.55	11.00	3.38	18.20	0.25	10.80	7.96	0.88	0.26	0.07	1.63	99.73
186	11	727.20	53.10	0.95	17.09	3.00	6.86	0.12	4.27	9.19	3.25	0.92	0.17	0.88	99.80
187	11	727.80	48.90	1.47	12.10	3.28	14.95	0.21	10.15	5.61	1.73	0.57	0.05	1.58	100.60
188	11	733.00	49.05	1.23	14.18	4.22	10.49	0.18	7.95	7.34	1.71	0.51	0.06	0.36	97.28
189	11	745.80	51.79	0.75	12.47	2.57	10.71	0.19	10.14	6.35	1.44	0.57	0.06	0.95	97.99
190	11	753.00	50.80	0.89	15.03	3.15	9.41	0.17	7.31	7.54	1.66	0.61	0.06	0.89	97.52
191	11	762.50	51.90	0.82	17.30	2.12	7.98	0.15	6.01	8.32	2.31	0.76	0.05	1.25	98.97
192	11	769.5	50.87	0.86	14.98	13.11		0.17	7.27	7.58	2.35	0.68	0.05	0.11	98.03
193	11	776.3	51.74	0.88	16.63	11.81		0.16	5.81	8.27	2.65	0.67	0.05	0.11	98.78
194	11	783.00	53.96	0.70	15.24	12.57		0.19	8.73	7.43	1.62	0.73	0.05	1.24	102.46
195	11	788.70	54.08	0.68	13.73	13.08		0.19	9.15	7.55	3.22	0.76	0.06	0.00	102.50
196	11	803.30	50.08	0.97	11.21	16.64		0.22	11.79	6.01	1.25	0.58	0.05	1.14	99.94
197	11	821.30	51.93	1.00	18.60	12.27		0.12	5.30	7.20	2.39	1.21	0.08	2.31	102.41
198	11	833.30	50.90	1.01	11.10	1.87	14.51	0.23	12.30	4.94	1.60	0.55	0.05	1.43	100.49
199	11	837.20	49.90	0.96	11.95	1.38	14.15	0.21	11.14	5.16	1.67	0.55	0.05	1.99	99.11
200	11	844.00	51.74	0.71	16.16	11.39		0.17	7.88	8.27	2.05	0.68	0.07	0.94	100.06
201	11	870.20	52.16	1.00	16.42	2.37	9.05	0.16	7.09	7.96	2.60	0.74	0.07		99.62
202	11	875.50	51.60	0.86	16.06	3.12	9.41	0.17	8.06	7.62	2.34	0.60	0.07		99.91
203	11	876.40	52.80	0.78	16.80	2.03	8.69	0.14	6.27	8.20	2.62	0.70	0.06	0.20	99.29
204	11	895.00	50.96	0.95	14.98	3.14	10.02	0.17	8.80	7.44	2.19	0.67	0.07		99.39
205	11	907.50	51.90	0.72	13.64	2.35	10.98	0.20	10.56	6.28	2.28	0.60	0.07		99.58
206	11	912.20	51.80	0.75	16.84	3.12	7.90	0.15	7.18	7.51	2.86	0.60	0.07	0.80	99.58
207	11	918.00	51.20	0.89	14.75	1.01	11.82	0.18	9.00	6.51	2.17	0.61	0.06	1.05	99.25
208	11	920.50	46.51	0.96	19.61	4.63	6.79	0.13	5.63	11.24	3.13	0.71	0.11	0.42	99.87
209	11	920.50	48.71	1.45	9.69	4.88	9.27	0.18	9.61	13.37	1.76	0.57	0.09	0.37	99.95
210	11	920.60	49.69	1.16	15.45	4.62	6.90	0.15	5.90	12.02	2.46	0.59	0.15	0.82	99.91
211	11	924.30	51.26	0.75	12.33	2.37	11.53	0.19	11.38	6.54	1.93	0.52	0.06	1.68	100.54
212	11	938.80	52.10	0.72	15.00	4.24	8.04	0.18	9.27	7.06	2.45	0.60	0.06		99.72
213	11	944.60	43.14	2.23	11.40	11.68	11.03	0.14	9.59	5.47	1.92	0.46	0.05	0.61	97.72
214	11	970.70	50.32	1.02	13.00	4.39	11.27	0.20	10.32	5.83	2.82	0.56	0.07	0.17	99.97
215	11	984.50	50.36	1.14	11.75	15.55		0.18	10.00	7.35	1.98	0.61	0.08	1.54	100.54
216	11	988.30	50.71	0.72	9.83	15.32		0.24	12.81	5.64	1.25	0.81	0.06	1.48	98.87
217	11	988.60	52.32	0.74	11.66	14.82		0.23	11.83	6.37	1.54	0.56	0.06	1.15	101.28

(continued)

Table A2.1 (continued)

N Analysis	No borehole 1	Depth, cm 2	SiO_2 3	TiO_2 4	Al_2O_3 5	Fe_2O_3 6	FeO 7	MnO 8	MgO 9	Cao 10	Na_2O 11	K_2O 12	P_2O_5 13	LOI 14	Total 15
218	11	989.20	49.87	0.83	10.64	15.16		0.21	11.74	5.91	1.43	0.52	0.05	1.28	97.64
219	11	989.30	51.98	0.92	9.46	16.60		0.24	12.36	6.07	1.51	0.76	0.05	1.33	101.28
220	11	989.40	52.34	1.02	10.04	16.93		0.24	11.97	6.57	1.63	0.79	0.06	1.36	102.95
221	11	989.50	48.05	1.31	8.88	15.89		0.20	10.07	9.61	1.78	0.53	0.04	0.66	97.02
222	11	989.60	49.42	1.11	12.12	15.90		0.18	10.90	6.95	1.58	0.73	0.06	1.27	100.22
223	11	990	46.07	2.21	16.12	20.01		0.14	4.68	10.71	2.78	0.67	0.07	0.15	103.61
224	11	990.1	40.78	3.11	18.19	25.6		0.12	2.44	8.22	3.42	0.8	0.07	−0.33	102.42
225	11	990.4	46.43	2.18	15.58	20.83		0.15	5.1	11.02	2.61	0.67	0.08		104.65
226	11	991.3	41.95	2.52	14.36	23.49		0.14	5.06	9.67	2.28	0.56	0.08	−0.11	100.00
227	11	991.4	40.22	3.24	17.1	27.93		0.13	3.14	8.45	2.83	0.66	0.06	−0.2	103.56
228	11	991.6	42.92	2.49	16.93	21.84		0.13	3.47	9.09	2.36	0.72	0.06		100.01
229	11	991.7	40.6	2.81	16.91	25.1		0.13	3.11	8.48	2.33	0.65	0.06	−0.17	100.01
230	11	991.9	43.71	1.96	17.11	18.32		0.11	3.55	9.16	2.82	0.65	0.06	0.16	97.61
231	11	992	42.53	2.35	16.72	22.63		0.13	4.1	8.73	2.02	0.83	0.06		100.10
232	11	992	42.53	2.35	16.72	22.63		0.13	4.1	8.73	2.02	0.83	0.06		100.10
233	11	992.2	39.23	3.04	17.49	26.77		0.13	2.55	8.00	2.22	0.69	0.06	−0.19	99.99
234	11	994.90	42.90	2.69	9.29	23.17		0.21	11.13	8.62	0.75	0.26	0.08	0.14	99.24
235	11	997.30	41.24	2.40	16.96	22.59		0.15	3.35	8.55	2.24	0.90	0.08	1.26	99.72
236	11	1040.70	52.00	0.70	14.81	2.84	9.48	0.18	9.27	6.84	2.45	0.74	0.07	0.13	99.51
237	11	1041.00	52.00	0.70	14.81	2.84	9.48	0.18	9.27	6.84	2.45	0.74	0.07	0.13	99.51
238	11	1058.10	50.07	0.73	5.34	22.52		0.37	13.49	4.72	0.88	0.74	0.09	1.13	100.08
239	11	1079.80	68.20	0.72	13.57	1.06	3.30	0.05	1.21	3.47	4.83	1.15	0.28	0.17	98.01
240	11	1080.80	51.70	1.17	10.49	3.10	13.40	0.25	9.78	5.70	1.50	1.85	0.13	0.73	99.80
241	11	1088.5	50.25	1.67	13.98	15.49		0.17	4.99	8.32	2.77	1.09	0.1		98.83
242	11	1089	49.51	1.57	13.65	15.76		0.18	5.45	8.99	2.86	1.06	0.08		99.11
243	11	1090.3	48.67	1.54	13.78	16.18		0.18	6.15	9.7	2.99	0.78	0.07		100.04
244	11	1090.6	49.00	1.83	14.31	16.65		0.17	5.24	8.8	2.66	0.92	0.09		99.67
245	11	1091.2	46.47	2.07	13.43	17.8		0.17	5.2	11.34	2.65	0.69	0.07		99.89
246	11	1092.3	48.34	1.91	13.76	17.72		0.18	5.6	9.77	2.26	0.85	0.08		100.47
247	11	1098	47.93	1.89	13.63	17.62		0.18	5.84	9.37	2.16	0.64	0.08		99.34
248	11	1111	47.95	1.97	13.71	17.71		0.18	5.69	9.63	2.62	0.59	0.07		100.12
249	11	1126.4	48.65	1.83	13.04	16.55		0.18	5.08	8.45	2.55	1.01	0.12		97.46
250	11	1129.70	51.94	1.45	14.40	4.97	9.51	0.18	5.00	7.29	2.90	1.43	0.19	0.25	99.51
251	11	1129.7	51.21	1.57	14.2	15.72		0.18	4.42	7.58	2.69	1.37	0.17	0.38	99.49
252	11	1142.70	52.00	0.95	14.84	3.44	9.12	0.15	7.18	6.73	3.00	1.18	0.11	1.00	99.70
253	11	1149.40	51.60	0.96	15.46	2.81	9.33	0.17	7.68	6.85	2.76	1.08	0.11	0.70	99.51
254	11	1157.90	40.46	3.09	13.95	27.97		0.18	5.37	8.62	2.29	0.41	0.07	0.00	102.41
255	11	1161.40	51.79	0.88	14.44	12.72		0.17	7.47	7.73	2.26	1.06	0.10	1.37	99.99
256	11	1205.10	50.65	0.62	11.81	15.16		0.19	10.11	6.66	1.30	1.66	0.09	1.83	100.08
257	11	1212.00	45.09	2.02	13.11	20.95		0.18	7.19	7.82	1.75	0.67	0.08	1.13	99.99
258	11	1218.80	52.29	1.02	15.90	12.64		0.16	6.47	8.20	1.98	0.88	0.11	0.43	100.08
259	11	1231.10	52.50	1.05	15.44	12.52		0.15	6.30	7.86	1.83	1.16	0.12	0.97	99.90
260	11	1233.00	54.62	0.77	19.86	8.24		0.08	2.89	8.58	2.78	0.99	0.13	1.51	100.45
261	11	1234.0	56.66	0.75	16.03	8.84		0.11	4.07	2.52	4.48	3.21	0.10	1.15	97.92
262	11	1257	48.41	2.01	13.03	16.56		0.19	5.00	7.92	3.21	0.84	0.12		97.29
263	11	1276.00	50.99	1.07	16.17	11.39		0.12	4.52	7.75	2.23	1.22	0.12	0.95	96.53
264	11	1281.20	48.08	1.29	13.66	13.35		0.18	6.63	8.94	1.99	0.44	0.09	1.77	96.42
265	11	1284.50	53.81	0.81	18.10	9.76		0.12	4.42	8.58	2.45	0.95	0.12	0.94	100.06
266	11	1289.20	48.52	1.76	13.86	18.03		0.19	6.04	8.34	1.93	0.87	0.08	0.32	99.94
267	11	1301.50	42.05	2.33	12.92	25.66		0.17	5.71	7.97	1.98	1.17	0.09	0.00	100.05
268	11	1316.00	45.48	2.12	14.91	19.39		0.19	5.02	8.10	1.96	1.48	0.07	1.24	99.96
269	12	328.0	49.69	1.27	13.27	15.60		0.17	7.41	7.74	2.79	0.93	0.09	0.88	99.84

(continued)

Table A2.1 (continued)

N Analysis	No borehole	Depth, cm	SiO$_2$	TiO$_2$	Al$_2$O$_3$	Fe$_2$O$_3$	FeO	MnO	MgO	Cao	Na$_2$O	K$_2$O	P$_2$O$_5$	LOI	Total
	1	2	3	4	5	6	7	8	9	10	11	12	13	14	15
270	12	459.0	56.21	0.18	15.34	6.65		0.14	7.68	6.21	3.28	1.42	0.04	0.00	97.15
271	14	400.0	61.01	0.93	13.67	14.8		0.18	9.29	6.89	2.49	0.71	0.07		110.04
272	14	650.0	49.78	12.69	17.09	12.62		0.16	4.3	9.67	2.8	0.85	0.07		110.03
273	16	500.0	50.81	1.60	14.19	15.98		0.18	5.43	7.42	3.03	1.17	0.12	0.05	99.98
274	19	26.7	42.07	3.03	11.72	23.07		0.19	6.04	9.98	2.39	0.49	0.11		99.09
275	19	67.0	44.9	2.43	12.72	19.29		0.18	5.4	9.64	2.2	0.81	0.16		97.73
276	19	479.7	49.01	1.88	15.71	14.78		0.18	3.45	8.09	3.39	1.51	0.17		98.17
277	19	6.7	44.90	2.44	12.72	19.29		0.18	5.40	9.64	2.20	0.81	0.16	0.26	98.00
278	19	26.7	42.07	3.03	11.72	23.07		0.19	6.04	9.98	2.39	0.49	0.11	0.10	99.19
279	19	436.0	50.46	1.74	13.96	15.59		0.18	4.80	7.94	2.80	1.23	0.16	1.13	99.99
280	19	479.7	49.53	1.88	15.83	14.72		0.17	3.48	8.16	3.30	1.52	0.17	1.20	99.96
281	19	26.7	42.07	3.03	11.72	23.07		0.19	6.04	9.98	2.39	0.49	0.11		99.09
282	19	67	44.9	2.43	12.72	19.29		0.18	5.4	9.64	2.2	0.81	0.16		97.73
283	19	479.7	49.01	1.88	15.71	14.78		0.18	3.45	8.09	3.39	1.51	0.17		98.17
284	20	152.5	70.93	0.43	11.72	4.14		0.12	1.68	2.74	3.88	1.37	0.11	1.3	98.42
285	20	13.60	54.85	0.82	14.94	11.88		0.15	3.85	5.04	3.83	2.57	0.19	1.50	99.62
286	20	38.50	53.40	1.19	14.71	12.84		0.17	3.95	7.58	3.31	1.25	0.12	1.30	99.82
287	20	48.00	53.97	1.21	16.59	10.84		0.15	2.69	7.63	3.47	1.68	0.13	0.63	98.99
288	20	84.50	54.67	1.29	15.36	11.91		0.16	2.31	6.91	3.66	2.04	0.14	1.20	99.65
289	20	113.50	56.25	1.07	17.15	9.72		0.15	1.88	7.96	3.34	1.78	0.13	1.06	100.49
290	20	123.80	54.85	1.18	13.76	13.05		0.17	4.27	7.49	3.24	1.71	0.14	0.50	100.36
291	20	139.80	53.50	1.14	16.64	10.54		0.14	2.40	8.04	3.63	1.58	0.12	0.85	98.58
292	20	152.5	70.93	0.43	11.72	4.14		0.12	1.68	2.74	3.88	1.37	0.11	1.3	98.42
293	36	116	55.16	1.39	15.66	12.03		0.16	2.3	7.71	3.6	1.74	0.12		99.87
294	36	153	66.68	0.67	17.58	0.91		0.08	0.13	0.77	9.01	0.29	0.09		96.21
295	36	200	66.72	0.6	16.18	2.57		0.09	1.69	1.94	5.87	1.38	0.09		97.13
296	36	0.6	52.74	1.94	12.9	15.84		0.18	3.69	7.57	2.92	1.65	0.12		99.55
297	36	39.5	54.52	1.08	15	1.24		0.17	4.16	7.74	2.87	1.47	0.09		88.34
298	36	19.3	49.61	1.64	13.39	16.19		0.18	4.8	8.66	2.41	1.02	0.08		97.98
299	36	46.5	54.94	1.11	14.89	12.79		0.17	4.47	7.74	3.08	1.43	0.1		100.72
300	36	90.5	54.55	1.24	16.09	11.64		0.15	2.4	7.62	3.38	1.82	0.1		98.99
301	36	96.5	55.07	1.25	16.6	11.35		0.15	2.35	7.87	3.35	1.67	0.11		99.77
302	36	115.6	55.35	1.33	15.4	11.91		0.16	2.31	7.49	3.34	1.79	0.12		99.20
303	39	38.5	53.5	1.14	16.64	10.54		0.14	2.4	8.04	3.63	1.57	0.12		97.72
304	39	6.00	52.38	1.96	12.71	16.25		0.18	3.62	7.52	3.07	1.61	0.14	0.18	99.62
305	39	13.80	55.64	1.02	17.27	9.62		0.14	2.19	7.90	3.47	1.62	0.11	0.96	99.94
306	39	34.00	54.24	1.25	16.22	11.44		0.15	2.38	7.93	3.25	1.56	0.14	0.60	99.16
307	39	44.50	54.36	1.53	16.10	12.37		0.16	2.41	7.67	3.33	1.76	0.13	0.84	100.66
308	39	62.40	54.07	1.27	15.27	12.40		0.16	2.71	6.88	3.54	1.87	0.15	1.90	100.22
309	39	87.60	55.47	1.57	12.61	13.71		0.16	3.79	7.21	2.97	1.59	0.17	0.24	99.49
310	39	89.00	53.80	1.75	13.06	15.59		0.18	3.84	7.55	2.95	1.69	0.14	0.18	100.73
311	39	92.50	51.53	1.59	13.43	15.74		0.18	4.60	8.78	2.74	1.19	0.10	0.52	100.40
312	39	100.10	52.60	1.61	13.30	14.77		0.18	3.60	7.67	3.02	1.27	0.19	0.84	99.05
313	39	101.60	54.88	1.62	12.80	14.49		0.17	3.30	6.91	3.12	1.60	0.16	0.54	99.59
314	39	103.00	54.20	1.14	16.64	10.54		0.14	2.40	8.04	3.63	1.58	0.12	0.85	99.28
315	57	33.00	44.95	2.42	13.28	20.20		0.17	5.37	9.60	3.81	0.58	0.04	−0.42	100.00
316	57	38.00	48.96	1.40	16.07	15.94		0.17	6.02	8.43	2.17	0.65	0.05	0.12	99.98
317	57	46.40	49.92	1.12	18.54	12.89		0.14	4.79	8.85	2.59	0.75	0.06	0.34	99.99
318	57	46.60	41.59	3.16	14.39	25.31		0.16	5.04	8.76	1.85	0.38	0.05	−0.68	100.01
319	57	53.50	50.27	1.23	18.16	13.04		0.15	4.63	9.10	2.44	0.78	0.06	0.14	100.00
320	57	77.50	47.20	1.76	15.64	15.10		0.15	5.02	10.91	2.15	0.76	0.04	1.34	100.07
321	57	81.00	52.78	0.69	17.20	11.20		0.16	6.35	8.70	2.25	0.74	0.09	−0.17	99.99

(continued)

Table A2.1 (continued)

N Analysis	No borehole	Depth, cm	SiO$_2$	TiO$_2$	Al$_2$O$_3$	Fe$_2$O$_3$	FeO	MnO	MgO	Cao	Na$_2$O	K$_2$O	P$_2$O$_5$	LOI	Total
	1	2	3	4	5	6	7	8	9	10	11	12	13	14	15
322	57	126.40	49.69	1.32	15.90	13.67		0.15	5.44	10.60	2.37	0.78	0.06	0.00	99.98
323	57	145.20	53.77	0.98	18.22	8.78		0.10	3.39	10.18	3.13	1.04	0.13	0.22	99.94
324	57	176.00	52.87	0.77	17.26	10.52		0.14	6.39	8.44	2.61	0.88	0.09	0.00	99.97
325	57	190.40	46.37	2.02	15.33	17.72		0.15	4.60	9.93	2.78	0.90	0.08	0.13	100.01
326	57	191.80	43.29	2.94	19.58	20.24		0.10	1.95	8.66	2.99	0.64	0.06	−0.50	99.95
327	57	192.80	51.72	0.79	16.00	12.50		0.18	7.11	8.17	2.46	0.83	0.07	0.14	99.97
328	57	196.30	46.35	1.10	20.31	14.54		0.10	2.76	10.58	2.70	0.59	0.04	0.14	99.21
329	57	197.20	52.39	0.68	16.30	12.09		0.19	7.11	8.03	2.17	0.80	0.08	0.12	99.96
330	57	210.00	52.28	0.95	16.10	12.52		0.16	7.02	8.03	2.10	0.81	0.11	−0.10	99.98
331	57	255.60	54.16	0.93	21.22	7.59		0.08	2.44	9.67	2.68	1.04	0.09	0.10	100.00
332	57	259.80	52.56	0.86	15.76	12.39		0.17	7.76	7.90	1.84	0.80	0.07	−0.12	99.99
333	57	265.40	52.98	0.81	16.61	11.74		0.15	7.16	7.82	1.80	0.79	0.08	0.00	99.94
334	57	266.40	51.22	1.00	13.24	15.72		0.20	9.44	6.66	1.91	0.78	0.07	−0.25	99.99
335	57	266.90	47.83	1.28	13.20	16.70		0.19	9.03	6.95	3.83	0.78	0.07	0.15	100.01
336	57	267.10	50.22	1.09	12.06	17.70		0.22	10.83	6.30	1.52	0.63	0.06	−0.62	100.01
337	57	267.30	50.50	0.65	16.09	11.58		0.15	7.13	7.67	4.94	0.98	0.08	0.21	99.98
338	57	267.50	52.08	0.82	16.22	12.22		0.16	7.08	7.87	2.47	0.81	0.09	0.21	100.03
339	57	267.80	52.33	0.86	15.90	12.52		0.16	7.58	7.83	1.82	0.76	0.08	0.13	99.97
340	57	269.00	51.69	0.75	15.63	11.74		0.17	7.40	7.81	3.67	0.90	0.10	0.17	100.03
341	57	293.00	53.87	0.81	16.76	11.68		0.16	7.34	8.42	1.97	0.82	0.08	0.12	102.03
342	57	303.30	52.73	0.76	17.54	10.61		0.14	6.31	8.25	2.58	0.89	0.09	0.14	100.04
343	57	317.60	51.96	0.80	13.89	13.76		0.20	9.15	7.36	2.07	0.74	0.08	0.00	100.01
344	57	324.00	50.04	0.87	14.79	13.62		0.19	9.08	8.00	2.53	0.73	0.07	0.12	100.04
345	57	412.80	52.58	0.86	16.75	11.07		0.15	6.29	8.07	1.91	0.90	0.10	1.30	99.98
346	57	427.20	50.68	1.16	15.94	13.39		0.17	6.72	8.72	2.32	0.78	0.10	0.00	99.98
347	57	440.30	52.43	1.05	18.19	11.15		0.14	4.66	9.69	2.50	1.02	0.10	−0.93	100.00
348	57	442.00	52.51	0.79	16.63	10.92		0.15	6.48	7.96	2.14	1.06	0.10	1.26	100.00
349	57	453.30	52.02	0.90	13.96	13.81		0.20	8.34	7.50	2.35	0.94	0.09	−0.08	100.03
350	57	459.00	52.35	0.84	15.81	12.39		0.17	7.39	8.01	2.02	0.92	0.09	0.00	99.99
351	57	468.80	52.33	0.22	12.43	14.66		0.22	10.43	6.79	1.62	0.86	0.09	−0.17	99.48
352	57	493.70	50.71	0.82	11.48	14.85		0.21	10.89	6.29	1.73	0.68	0.09	0.14	97.89
353	58	105.50	45.81	1.64	7.11	10.02		0.15	11.54	21.69	0.64	0.17	0.04	1.22	100.03
354	58	213.77	51.50	0.67	18.35	10.62		0.15	6.56	8.12	3.25	0.71	0.07	0.00	100.00
355	58	213.80	48.67	1.08	15.09	15.36		0.18	9.22	7.24	2.28	0.55	0.04	0.29	100.00
356	58	213.83	47.14	1.38	13.88	17.59		0.19	9.17	6.90	2.61	0.51	0.04	0.60	100.01
357	58	213.86	45.28	1.68	12.99	20.00		0.00	9.35	6.24	3.09	0.46	0.05	0.89	100.03
358	58	213.90	46.49	1.57	12.85	19.16		0.20	9.75	6.59	2.30	0.48	0.06	0.61	100.06
359	58	213.94	51.22	0.70	16.87	11.32		0.16	7.55	8.70	2.66	0.67	0.06	0.11	100.02
360	59	30.70	49.33	1.11	16.59	11.64		0.13	4.62	10.98	1.82	0.49	0.15	−0.13	96.73
361	59	326.30	45.38	2.02	15.08	17.07		0.15	4.56	9.43	1.88	0.59	0.07	0.33	96.56
362	59	326.40	46.86	2.03	12.51	18.74		0.20	6.40	10.68	1.77	0.47	0.06	0.12	99.84
363	59	326.50	46.57	2.41	12.08	21.54		0.21	6.48	10.89	1.52	0.45	0.07	0.00	102.22
364	59	326.70	47.39	1.60	18.59	12.33		0.09	2.31	8.97	2.22	0.94	0.10	0.88	95.42
365	59	422.00	49.42	1.44	10.69	19.98		0.24	10.74	6.88	1.38	0.46	0.05	−1.63	99.65
366	59	486.00	40.13	2.61	12.69	23.47		0.18	4.93	8.73	1.62	0.50	0.10	1.51	96.47
367	59	707.00	48.53	1.38	21.01	9.93		0.07	1.17	8.99	2.62	1.19	0.05	−0.16	94.78
368	59	742.80	11.70	9.34	3.85	62.90		0.28	5.59	1.64	1.54	0.08	0.01	−0.84	96.09
369	59	749.00	50.10	0.80	11.58	12.78		0.18	8.74	9.24	1.46	0.55	0.05	−0.15	95.33
370	59	900.80	45.48	3.24	15.00	20.34		0.18	5.98	8.79	2.24	0.50	0.05	−0.50	101.30
371	59	901.00	51.65	1.32	14.90	13.60		0.18	7.54	9.15	1.77	0.52	0.09	−0.39	100.33
372	63	25.1	48.08	1.13	16.02	11.73		0.12	4.83	13.77	2.59	0.85	0.06	0.89	100.07
373	63	42.25	40.27	2.60	14.72	22.24		0.17	5.71	11.45	1.62	0.71	0.03	0.51	100.02

(continued)

Table A2.1 (continued)

N Analysis	No borehole	Depth, cm	SiO$_2$	TiO$_2$	Al$_2$O$_3$	Fe$_2$O$_3$	FeO	MnO	MgO	Cao	Na$_2$O	K$_2$O	P$_2$O$_5$	LOI	Total
1	2	3	4	5	6	7	8	9	10	11	12	13	14	15	
374	63	42.25	40.27	2.60	14.72	22.24		0.17	5.71	11.45	1.62	0.71	0.03	0.51	100.03
375	63	46.6	50.15	0.82	16.27	11.59		0.14	5.40	11.85	2.46	0.72	0.09	0.49	99.98
376	63	47.4	51.08	0.87	16.26	11.20		0.16	5.77	11.34	2.90	0.69	0.09	−0.35	100.01
377	63	50	45.81	1.89	13.13	18.16		0.18	7.09	11.03	2.20	0.33	0.04	0.08	99.94
378	63	53.9	45.56	1.82	16.14	16.21		0.14	4.82	10.78	2.39	0.61	0.05	1.45	99.97
379	63	60.1	51.58	0.33	15.69	9.55		0.15	8.33	10.91	2.23	0.28	0.02	0.93	100.00
380	63	62.4	44.60	2.31	12.16	21.38		0.20	6.60	10.69	2.05	0.47	0.03	−0.47	100.02
381	63	66.7	51.34	0.93	22.04	8.27		0.08	1.88	10.93	3.27	0.58	0.12	0.44	99.88
382	63	67.4	52.93	0.47	22.76	5.56		0.07	1.97	10.88	3.25	0.78	0.09	1.20	99.96
383	63	94.3	50.51	1.16	16.14	13.77		0.20	6.29	9.37	1.97	0.55	0.03	0.00	99.99
384	63	98.8	47.31	2.08	17.78	16.18		0.15	4.62	9.07	2.17	0.53	0.03	0.10	100.02
385		99.8	4.22	8.80	6.09	72.42		0.22	4.91	0.96	1.63	0.11	0.04	−1.78	97.62
386	63	100	47.56	1.87	16.13	16.45		0.17	6.08	8.98	2.12	0.50	0.04	0.09	99.99
387	63	102.8	50.09	1.24	15.41	14.42		0.17	7.92	8.43	1.96	0.56	0.04	−0.17	100.07
388	63	110	49.79	1.62	17.10	15.28		0.16	5.46	10.19	2.26	0.62	0.06	0.22	102.76
389	63	124.4	50.47	0.79	11.64	17.03		0.22	11.39	6.86	1.72	0.39	0.06	−0.62	99.95
390		128.8	49.33	1.07	4.89	22.07		0.30	16.91	4.82	0.89	0.36	0.06	−0.68	100.02
391	63	147	50.91	1.06	19.02	10.98		0.12	5.05	9.73	2.41	0.66	0.03	0.00	99.97
392	63	148	47.45	1.79	18.01	15.16		0.15	5.66	9.35	2.26	0.44	0.02	−0.29	100.00
393	63	210.5	32.30	4.80	15.67	35.49		0.14	2.52	7.26	2.15	0.36	0.02	−0.70	100.01
394	63	216.6	43.45	2.60	13.31	22.11		0.18	5.65	10.17	2.13	0.52	0.06	−0.19	99.99
395	63	220.7	46.91	1.72	12.78	15.06		0.16	6.38	13.96	1.79	0.70	0.08	0.45	99.99
396	63	227	41.31	3.94	14.64	24.73		0.18	4.52	8.62	2.06	0.61	0.04	−0.64	100.01
397	63	230	46.64	2.40	13.01	18.94		0.19	6.48	10.18	2.04	0.60	0.05	−0.52	100.01
398	63	290.7	49.23	1.33	6.88	16.15		0.26	10.05	14.36	1.23	0.46	0.05	0.00	100.00
399	63	296.5	52.02	0.76	22.73	6.76		0.07	1.85	10.83	3.41	0.74	0.05	0.81	100.03
400	63	300.7	55.07	0.54	23.84	5.20		0.05	1.10	10.38	2.49	0.73	0.08	0.49	99.97
401	63	309	52.59	0.80	17.11	11.59		0.16	6.41	8.86	1.75	0.80	0.09	−0.18	99.98
402	63	309.2	52.03	0.77	14.55	13.63		0.19	8.54	8.01	1.69	0.67	0.06	−0.15	99.99
403	63	312.8	52.41	1.13	15.66	12.43		0.16	6.55	9.11	1.70	0.81	0.07	0.00	100.03
404	63	316	51.70	0.85	14.90	13.52		0.18	8.41	7.88	1.90	0.71	0.07	−0.16	99.96
405	63	319	50.02	0.54	13.64	12.65		0.16	7.37	13.87	1.56	0.51	0.04	−0.31	100.05
406	63	319.4	48.06	0.52	13.04	12.98		0.16	7.05	13.41	2.37	0.62	0.04	1.77	100.02
407	63	321.2	46.06	2.24	15.68	17.68		0.14	5.16	10.76	2.07	0.56	0.05	−0.40	100.00
408	63	324	45.50	2.55	14.64	19.27		0.17	5.45	10.50	1.73	0.53	0.05	−0.38	100.01
409	63	324.4	51.93	0.83	15.27	13.73		0.21	7.67	8.35	1.77	0.67	0.05	−0.50	99.98
410	63	325.3	45.85	2.32	14.84	18.62		0.16	5.17	10.68	2.10	0.58	0.05	−0.38	99.99
411	63	342.8	45.66	2.11	13.60	19.66		0.18	6.24	10.20	1.84	0.69	0.06	−0.24	100.00
412	63	346.5	38.27	3.72	17.10	27.95		0.14	2.34	8.15	2.09	0.71	0.03	−0.51	99.99
413	63	350.5	44.17	2.51	13.45	22.44		0.20	5.68	10.03	1.57	0.52	0.04	−0.60	100.01
414	63	355.5	38.14	3.73	14.08	28.87		0.17	4.23	9.12	1.70	0.57	0.03	−0.64	100.00
415	63	361.5	42.28	2.89	13.58	23.97		0.19	5.37	9.86	1.83	0.63	0.04	−0.58	100.06
416	63	366.5	42.90	2.83	14.76	23.89		0.17	4.64	9.24	1.59	0.57	0.04	−0.62	100.01
417	63	370	41.28	3.20	14.90	25.02		0.16	4.24	9.31	1.77	0.54	0.04	−0.43	100.03
418	63	380	44.18	2.56	14.53	21.92		0.17	4.93	9.85	1.85	0.62	0.04	−0.62	100.03
419	63	385	42.25	2.87	12.09	25.08		0.20	6.05	9.84	1.66	0.54	0.04	−0.64	99.98
420	63	390	44.98	2.32	14.16	20.92		0.18	5.58	10.04	1.77	0.57	0.05	−0.57	100.00
421	63	394	43.26	2.59	13.71	23.20		0.18	5.65	9.62	1.78	0.57	0.04	−0.60	100.00
422	63	399.5	42.53	2.75	13.75	23.68		0.18	5.17	9.47	2.16	0.53	0.04	−0.26	100.00
423	63	401.9	44.35	2.43	12.70	22.46		0.22	6.71	9.16	1.38	0.56	0.05		100.02
424	63	403.5	52.45	0.93	16.59	11.84		0.16	6.78	8.15	2.11	0.90	0.10		100.01
425	63	412.5	50.31	0.85	11.13	15.36		0.22	10.28	6.47	4.58	0.92	0.09	−0.23	99.98

(continued)

Table A2.1 (continued)

N Analysis	No borehole 1	Depth, cm 2	SiO$_2$ 3	TiO$_2$ 4	Al$_2$O$_3$ 5	Fe$_2$O$_3$ 6	FeO 7	MnO 8	MgO 9	Cao 10	Na$_2$O 11	K$_2$O 12	P$_2$O$_5$ 13	LOI 14	Total 15
426	63	415.5	52.20	0.87	9.46	17.30		0.26	12.93	5.69	1.01	0.79	0.07	−0.55	100.03
427	63	418	43.47	2.63	15.13	22.62		0.15	4.58	9.51	1.82	0.59	0.06	−0.55	100.01
428	63	419	52.36	0.90	11.89	14.93		0.21	10.76	6.56	1.60	0.76	0.08	−0.10	99.95
429	63	419.5	55.23	1.04	18.69	6.94		0.14	5.79	9.10	2.12	0.86	0.06		99.97
430	63	422.4	50.67	0.93	10.11	17.58		0.25	11.44	5.80	2.28	0.88	0.09		100.03
431	63	422.5	51.06	0.90	14.31	14.74		0.18	7.65	7.17	2.90	0.96	0.10	0.00	99.97
432	63	425	55.91	0.93	13.77	8.34		0.21	10.26	7.16	2.43	0.93	0.04	0.00	99.98
433	63	427	51.47	0.95	11.70	15.64		0.22	10.12	6.56	2.41	0.86	0.09	0.00	100.02
434	63	427.2	53.74	0.95	9.72	15.25		0.23	9.27	6.04	3.81	1.08	0.09	0.00	100.18
435	63	427.3	52.37	0.90	11.80	15.47		0.22	10.82	6.49	1.35	0.80	0.08	−0.30	100.00
436	63	428	52.52	1.09	11.88	15.63		0.22	9.99	6.61	1.37	0.94	0.09	−0.30	100.04
437	63	429	52.75	0.94	13.41	13.70		0.20	8.68	6.94	2.35	0.97	0.11	0.00	100.05
438	63	433.9	35.56	3.92	13.54	33.84		0.17	5.07	6.49	1.69	0.49	0.04	−0.84	99.97
439	63	436.5	51.36	1.03	21.98	8.76		0.08	1.86	9.96	3.07	0.86	0.10	0.99	100.05
440	63	437.2	51.99	0.72	18.35	10.18		0.13	6.20	8.98	1.98	0.78	0.08	0.57	99.96
441	63	438.1	49.95	0.85	10.15	12.27		0.19	9.99	14.45	1.13	0.45	0.04	0.52	99.99
442	63	438.4	31.91	4.69	17.30	34.16		0.13	2.46	7.84	1.59	0.43	0.03	−0.55	99.99
443	63	438.7	29.70	8.19	17.16	33.19		0.15	2.58	7.89	1.40	0.37	0.03	−0.64	100.02
444	63	439.3	48.92	1.11	14.01	15.20		0.18	9.20	8.68	1.52	0.78	0.07	0.29	99.96
445	63	442.1	41.62	2.86	15.27	24.21		0.14	4.09	8.77	2.41	0.62	0.05	0.00	100.04
446	63	443.5	40.08	2.88	15.51	26.01		0.15	4.47	9.38	1.64	0.50	0.04	−0.69	99.97
447	63	446	37.43	3.34	13.58	30.94		0.16	4.73	8.41	1.78	0.49	0.03	−0.93	99.96
448	63	448.3	41.41	3.11	13.51	24.28		0.16	5.50	10.32	1.50	0.52	0.05	−0.37	99.99
449	63	453.5	37.04	3.98	15.52	29.30		0.15	3.43	8.49	1.81	0.85	0.05	−0.60	100.02
450	63	457.5	34.02	4.51	16.41	32.61		0.15	2.11	7.04	1.83	0.95	0.04	0.37	100.04
451	63	462.7	36.34	4.05	15.16	29.73		0.18	3.02	6.84	2.71	0.93	0.03	0.83	99.82
452	63	468.7	36.49	3.81	15.56	29.33		0.17	2.84	8.00	2.10	0.76	0.03	0.87	99.96
453	63	471.9	51.98	1.33	13.65	15.02		0.21	7.04	7.76	1.75	1.11	0.06	0.14	100.05
454	63	478.8	51.77	1.14	17.04	11.04		0.12	4.75	8.30	2.75	1.11	0.10	1.88	100.00
455	63	493.3	46.56	1.87	13.03	20.24		0.20	6.43	9.23	1.65	0.86	0.08	−0.12	100.03
456	63	496.2	47.17	2.05	12.82	18.45		0.20	6.18	9.64	2.51	0.75	0.07	0.17	100.01
457	63	500.6	51.82	1.34	14.73	14.75		0.19	6.18	8.41	1.66	0.90	0.09	−0.09	99.98
458	63	511.3	41.40	3.09	14.21	23.97		0.17	4.27	9.20	2.91	0.64	0.05	0.09	100.00
459	63	512.3	44.52	2.37	13.84	19.63		0.18	6.21	10.57	2.34	0.57	0.05	−0.28	100.00
460	63	521.2	50.93	1.06	16.65	12.75		0.15	6.22	9.87	1.71	0.61	0.06	0.00	100.01
461	63	525.5	43.54	2.44	12.80	21.06		0.17	6.68	10.67	2.32	0.49	0.04	−0.21	100.00
462	63	527.4	57.48	1.06	13.93	10.65		0.14	4.13	5.01	2.64	3.28	0.11	1.43	99.86
463	63	528.2	42.04	3.15	12.24	23.79		0.21	6.59	9.94	2.20	0.48	0.04	−0.67	100.01
464	66	37.00	50.89	0.84	17.45	11.90		0.15	6.78	8.79	1.80	0.63	0.04		99.27
465	66	41.50	48.62	0.72	15.33	14.09		0.18	8.67	8.85	1.86.	0.63	0.04		98.99
466	66	47.60	51.81	0.56	18.19	8.86		0.12	5.30	10.87	2.40	0.90	0.06		99.07
467	66	92.00	49.93	0.72	8.71	19.25		0.27	13.93	6.35	1.19	0.39	0.03		100.77
468	66	105.30	51.40	0.72	22.68	7.36		0.07	2.28	11.47	2.93	0.71	0.05		99.67
469	66	113.70	45.85	1.74	15.47	17.75		0.15	5.80	10.76	2.21	0.52	0.05		100.30
470	66	119.60	51.41	0.81	15.12	13.73		0.19	9.04	7.61	1.88	0.55	0.05		100.39
471	66	130.60	44.24	2.10	14.47	19.82		0.15	6.11	11.32	1.79	0.37	0.03		100.40
472	66	318.70	52.94	0.69	16.10	11.44		0.16	7.50	8.01	2.69	0.76	0.07		100.36
473	66	422.50	49.92	1.62	14.34	16.82		0.19	4.13	8.72	2.78	1.25	0.14		99.91
474	70	113.60	46.67	1.92	15.80	16.82	10.35	0.17	6.13	9.75	2.59	0.74	0.11	0.18	111.23
475	70	119.90	48.84	1.17	16.90	13.96	8.77	0.15	6.72	8.87	2.06	0.70	0.07	0.08	108.29
476	70	120.10	51.11	1.16	18.07	11.95	7.11	0.14	4.83	8.85	2.64	1.12	0.12	0.12	107.22
477	70	121.00	48.88	1.33	13.73	18.17	10.13	0.20	9.05	8.14	1.89	0.63	0.06	0.42	112.63

(continued)

Table A2.1 (continued)

N Analysis	No borehole	Depth, cm	SiO$_2$	TiO$_2$	Al$_2$O$_3$	Fe$_2$O$_3$	FeO	MnO	MgO	Cao	Na$_2$O	K$_2$O	P$_2$O$_5$	LOI	Total
	1	2	3	4	5	6	7	8	9	10	11	12	13	14	15
478	70	128.20	49.34	1.10	16.08	13.75		0.18	7.57	8.43	2.58	0.68	0.05	0.11	99.87
479	70	137.50	49.27	1.26	13.57	15.12		0.19	8.31	8.36	2.66	0.67	0.11	0.23	99.75
480	70	142.20	37.39	4.33	4.47	38.33	10.71	0.34	14.13	2.48	1.31	0.34	0.14	1.29	115.26
481	70	142.30	20.87	7.12	4.45	55.92	22.27	0.28	9.58	1.87	1.16	0.16	0.02	1.95	125.65
482	70	147.50	47.75	1.85	11.49	20.72		0.24	10.40	7.01	1.53	0.59	0.07	0.40	102.05
483	70	156.00	49.00	0.99	13.26	16.74		0.21	10.91	6.90	2.14	0.51	0.08	0.09	100.83
484	70	150.00	48.80	1.18	15.17	15.77		0.19	8.17	7.88	2.46	0.68	0.12	0.20	100.62
485	70	177.70	51.34	0.84	15.38	12.85		0.20	8.59	8.64	2.55	0.76	0.09		101.24
486	70	213.40	50.86	0.82	17.48	11.95	6.75	0.16	7.29	8.75	2.32	0.79	0.07		107.24
487	70	215.50	49.00	1.27	10.36	19.73		0.25	12.58	6.30	1.75	0.61	0.06	0.39	102.30
488	70	435.00	51.43	1.05	20.54	9.64	3.95	0.10	2.94	9.80	2.89	0.99	0.14	0.56	104.03
489	70	438.00	47.94	1.96	19.59	13.97		0.11	2.13	9.39	2.82	0.96	0.08	0.56	99.51
490	70	438.60	48.23	2.25	13.53	17.08	7.83	0.20	5.73	10.58	2.44	1.06	0.12	0.07	109.12
491	70	440.00	47.34	2.63	12.73	18.26		0.23	5.51	11.09	2.33	0.86	0.10	0.29	101.37
492	70	446.60	48.63	2.18	13.88	15.81		0.21	5.50	10.50	2.52	1.31	0.12	0.39	101.05
493	70	452.30	45.97	2.65	14.67	17.58		0.16	4.85	12.40	2.28	0.81	0.05	0.18	101.60
494	70	455.30	46.67	2.26	12.04	17.03	11.28	0.14	4.07	13.65	3.14	0.83	0.12	0.96	112.19
495	70	462.50	45.31	2.35	14.03	18.24		0.14	3.80	14.09	2.10	0.81	0.12	0.24	101.23
496	70	462.70	53.61	0.93	17.53	10.76	6.54	0.15	5.51	9.27	2.51	1.06	0.09	0.14	108.10
497	70	463.80	52.65	0.85	13.40	13.40	9.27	0.21	9.56	7.58	2.29	0.96	0.09	0.21	110.47
498	70	464.20	51.95	1.04	13.72	13.05	14.66	0.19	8.03	7.36	2.72	1.06	0.21	0.13	114.12
499	70	501.00	52.98	0.85	18.86	9.13	4.74	0.12	3.41	9.32	2.78	1.24	0.13	0.60	104.16
500	70	508.00	52.48	0.91	8.19	17.48	9.84	0.28	13.70	5.46	2.18	0.88	0.18	0.08	111.66
501	70	530.00	43.46	2.51	19.36	20.85		0.11	2.21	8.48	3.00	0.74	0.05	0.55	101.32
502	70	545.90	51.45	1.18	12.79	15.12	8.26	0.22	9.14	7.78	1.87	0.99	0.09	0.35	109.24
503	70	550.20	52.38	0.75	12.22	14.04		0.22	11.13	6.72	1.93	0.72	0.14	0.09	100.34
504	70	558.30	52.91	0.98	14.49	12.49		0.18	7.48	7.49	2.23	1.16	0.22		99.63
505	70	561.00	46.94	2.91	13.64	18.51		0.21	5.39	10.61	2.58	0.84	0.17	0.21	102.01
506	70	730.00	53.35	0.56	24.45	5.83		0.06	1.31	10.38	3.38	1.11	0.02	0.64	101.09
506	70	834.00	52.12	0.72	21.93	7.34		0.12	2.53	12.14	3.20	0.88	0.20	0.25	101.43
507	70	925.00	34.99	4.27	9.12	37.53	18.18	0.25	8.55	6.34	1.49	0.35	0.04	1.02	122.13
508	83	20.10	49.55	0.78	22.50	8.55		0.08	2.47	10.51	3.22	0.61	0.14	1.11	99.52
509	83	20.20	51.50	0.83	24.88	5.88		0.06	0.65	11.31	3.51	1.09	0.21	1.38	101.30
510	83	38.80	46.54	1.43	16.20	15.97		0.16	7.65	8.04	2.49	0.53	0.09	0.51	99.61
511	83	60.00	43.03	2.82	18.83	19.35		0.13	3.00	8.89	3.05	0.49	0.08	-0.13	99.54
512	83	74.20	44.77	2.83	18.62	16.98		0.13	2.82	8.62	3.55	0.47	0.06	0.00	98.85
513	83	76.40	42.43	4.17	15.07	26.08		0.20	5.44	7.74	3.19	0.40	0.13	-0.55	104.30
514	83	80.00	33.36	4.69	16.49	32.19		0.12	1.70	7.47	3.68	0.39	0.03	0.52	100.64
515	83	84.00	41.26	3.28	13.92	21.25		0.17	4.94	10.42	3.61	0.44	0.06	0.22	99.57
516	83	86.50	24.75	7.98	13.24	43.11		0.19	1.66	5.59	3.48	0.43	0.03	0.16	100.62
517	83	87.50	33.40	6.20	16.44	33.42	19.97	0.14	2.07	6.69	4.16	0.51	0.05	-0.16	122.89
518	83	87.60	47.90	1.66	17.38	12.93	8.33	0.08	3.54	9.07	4.61	0.77	0.15	1.40	107.82
519	83	88.60	12.26	11.52	8.26	60.41	29.53	0.26	2.94	2.68	3.31	0.21	0.03	-0.75	130.66
520	83	102.00	26.95	6.10	14.01	35.61	18.32	0.13	1.93	5.60	5.06	0.49	0.04	-0.14	114.10
521	83	104.20	34.49	3.54	16.03	22.50	11.99	0.10	1.35	7.86	4.14	0.48	0.03	0.00	102.51
522	83	113.00	36.41	3.29	17.26	23.32	11.92	0.10	1.56	8.31	3.21	0.37	0.04	0.85	106.64
523	83	115.40	40.96	2.55	18.89	18.63	9.12	0.08	1.13	9.25	3.97	0.47	0.04	0.53	105.62
524	83	117.50	34.90	3.86	16.59	27.75	13.58	0.11	1.60	7.70	3.74	0.42	0.04	0.00	110.29
525	83	119.80	21.04	7.75	11.57	46.84	24.07	0.17	2.45	4.30	2.41	0.24	0.03	0.25	121.12
526	83	121.10	0.49	11.40	3.22	71.98	34.34	0.24	2.58	0.61	1.73	0.05	0.02	-2.20	124.46
527	83	121.60	7.42	11.42	7.45	65.50	31.64	0.24	2.52	1.56	2.00	0.10	0.03	-0.62	129.26
528	83	140.00	32.32	4.62	17.91	32.95		0.12	1.87	6.44	3.16	0.46	0.02	0.30	100.17

(continued)

Table A2.1 (continued)

N Analysis	No borehole	Depth, cm	SiO_2	TiO_2	Al_2O_3	Fe_2O_3	FeO	MnO	MgO	Cao	Na_2O	K_2O	P_2O_5	LOI	Total
	1	2	3	4	5	6	7	8	9	10	11	12	13	14	15
529	83	160.00	23.34	7.19	14.34	45.65		0.16	2.07	5.45	2.98	0.33	0.02	−0.34	101.19
530	83	178.60	42.96	2.76	15.91	19.91		0.14	4.48	10.82	2.64	0.39	0.03	−0.09	99.95
531	83	200.00	45.28	2.17	11.96	19.81		0.19	6.68	12.10	2.25	0.49	0.03	0.29	101.25
532	83	220.00	44.93	1.96	13.86	17.69		0.17	5.98	11.26	2.58	1.09	0.02	1.85	101.39
533	83	240.00	43.31	2.80	16.56	22.04		0.15	4.80	8.23	2.74	0.53	0.03	0.23	101.42
534	83	260.00	52.03	0.64	17.21	10.36		0.16	5.76	9.88	2.83	0.65	0.04	0.68	100.24
535	83	20.00	51.68	0.64	17.05	11.49		0.14	6.14	8.96	2.69	0.73	0.06	1.11	100.69
536	83	300.00	45.94	1.36	16.38	15.90		0.14	4.38	10.10	4.75	0.69	0.08	1.55	101.27
537	83	320.20	45.85	1.83	13.92	17.28		0.16	6.91	10.14	2.48	0.49	0.03	2.80	101.89
538	83	336.60	43.12	2.92	13.27	20.22		0.18	6.46	10.33	2.24	0.52	0.03	1.15	100.44
539	83	365.50	44.65	1.84	12.36	17.16		0.17	7.96	11.75	2.37	0.49	0.03	0.37	99.15
540	83	380.50	46.16	1.69	15.08	16.47		0.17	6.12	12.02	2.79	0.61	0.02	0.31	101.44
541	83	400.00	45.00	1.77	13.11	18.88		0.20	7.41	11.34	2.43	0.32	0.02	−0.31	100.17
542	83	420.00	46.46	1.91	13.70	16.85		0.18	6.42	12.07	2.27	0.40	0.03	0.71	101.00
543	83	440.10	49.83	0.87	15.11	12.06		0.14	6.52	13.15	2.18	0.32	0.02	0.42	100.62
544	83	460.20	40.26	2.73	11.87	24.15		0.19	6.92	10.87	2.41	0.34	0.03	0.25	100.02
545	83	480.00	44.63	2.26	15.42	18.51		0.17	5.76	10.83	2.71	0.48	0.03	0.33	101.13
546	83	520.20	35.68	4.29	13.21	32.34		0.16	4.42	6.61	3.21	0.43	0.02	0.51	100.88
547	83	540.00	43.44	2.77	13.18	22.28		0.20	6.30	10.95	2.33	0.42	0.04	−0.36	101.55
548	83	560.00	37.99	3.49	14.30	26.55		0.16	4.57	9.51	2.93	0.45	0.03	−0.39	99.59
549	83	561.70	45.25	3.01	13.82	19.99		0.19	5.69	2.90	2.53	0.45	0.03	0.43	94.29
550	83	580.00	44.23	3.03	12.90	20.69		0.20	6.30	10.31	2.32	0.48	0.03	−0.08	100.41
551	83	600.00	34.07	4.20	14.35	31.88		0.16	4.21	8.59	2.42	0.38	0.02	0.00	100.28
552	83	620.00	37.71	3.63	14.08	27.67		0.16	3.74	9.35	3.50	0.66	0.04	0.28	100.82
553	83	640.00	34.32	4.54	12.37	32.12		0.14	4.77	8.28	3.31	0.34	0.02	0.84	101.05
554	83	660.00	47.79	1.50	15.20	15.35		0.17	5.85	12.14	2.97	0.59	0.06	0.00	101.62
555	83	680.30	43.64	2.50	13.60	20.75		0.19	6.18	10.06	2.78	0.68	0.09	0.22	100.69
556	83	700.00	45.33	1.60	13.27	19.22		0.22	7.89	10.74	2.50	0.39	0.03	0.42	101.61
557	83	719.70	44.97	1.68	14.96	15.46		0.16	5.85	9.55	2.62	0.87	0.07	4.43	100.62
558	83	738.00	40.29	3.29	12.28	24.38		0.21	6.78	10.25	2.65	0.45	0.05	−0.46	100.17
559	83	760.00	40.44	2.70	14.86	23.77		0.15	4.80	9.57	2.48	0.48	0.04	−0.23	99.06
560	83	781.50	35.00	5.48	16.97	27.81		0.14	1.77	7.63	3.36	0.56	0.03	0.31	99.06
561	83	800.00	45.95	1.97	12.37	18.39		0.19	6.92	9.71	2.39	1.05	0.07	0.37	99.38
562	83	820.00	44.62	2.23	13.58	20.43		0.20	7.89	7.39	2.50	0.54	0.04	0.23	99.65
563	83	840.00	48.93	1.22	14.97	14.56		0.20	8.43	7.68	2.66	0.81	0.08		99.54
564	83	862.80	50.99	0.78	15.14	12.61		0.19	8.26	7.26	2.58	0.78	0.07	0.61	99.27
565	83	887.50	50.57	0.91	18.96	10.45		0.13	5.33	8.95	2.58	1.02	0.10	1.96	100.96
566	83	903.20	51.25	0.90	15.54	12.46		0.17	7.68	8.08	2.28	1.06	0.09	1.84	101.35
567	83	927.00	49.04	1.33	13.98	15.92		0.19	9.32	6.97	2.21	0.99	0.09	1.34	101.38
568	83	928.60	47.41	1.76	9.20	21.69		0.27	12.40	5.67	1.79	0.71	0.08	0.07	101.05
569	83	940.60	48.29	1.42	12.28	18.80		0.20	9.18	6.37	2.56	0.77	0.06	−0.10	99.83
570	83	961.00	49.21	0.93	12.45	16.04		0.20	10.59	6.84	2.87	0.92	0.09	0.68	100.82
571	83	981.20	49.13	0.91	15.47	13.16		0.16	7.64	8.15	2.44	0.95	0.08	0.95	99.04
572	83	1000.00	49.72	1.20	11.61	16.73		0.23	10.41	6.88	3.64	0.92	0.09	0.29	101.72
573	83	1020.80	42.45	2.10	9.93	26.72		0.23	10.93	5.48	2.39	0.57	0.06	−0.61	100.25
574	83	1036.50	49.94	1.22	10.88	16.55		0.22	11.18	6.67	2.73	0.85	0.07	0.43	100.74
575	83	1041.00	46.58	0.87	12.92	17.55		0.18	8.96	7.12	2.30	0.73	0.06	1.85	99.12
576	83	1062.00	56.26	1.01	19.02	9.62		0.07	3.46	1.21	3.45	4.28	0.11	0.82	99.31
577	83	1080.00	50.63	1.36	11.73	16.64		0.21	9.19	7.55	3.68	0.86	0.08	0.00	101.93
578	83	1101.20	53.51	1.28	18.24	8.44		0.18	7.23	9.18	2.48	0.82	0.03	0.29	101.68
579	83	1122.60	51.03	1.16	12.37	13.60		0.23	12.17	6.80	1.80	0.66	0.04	−0.17	99.69
580	83	1143.50	46.88	2.13	14.31	18.76		0.16	4.75	9.50	2.42	0.91	0.09	0.00	99.91

(continued)

Table A2.1 (continued)

N Analysis	No borehole	Depth, cm	SiO₂	TiO₂	Al₂O₃	Fe₂O₃	FeO	MnO	MgO	Cao	Na₂O	K₂O	P₂O₅	LOI	Total
	1	2	3	4	5	6	7	8	9	10	11	12	13	14	15
581	83	1160.00	41.29	3.17	12.18	26.56		0.24	6.38	5.95	2.35	1.41	0.12	0.36	100.01
58	83	1180.00	50.18	0.67	9.07	10.09		0.16	10.91	17.88	1.76	0.37	0.07	0.65	101.81
583	83	1200.40	40.67	3.94	16.83	25.36		0.16	3.34	7.83	2.91	0.99	0.09	−0.40	101.72
584	83	1220.00	46.65	2.44	14.79	18.19		0.17	4.94	9.14	3.18	1.15	0.12	0.34	101.11
585	83	1240.00	43.52	2.67	15.44	21.89		0.14	4.16	8.03	2.57	0.98	0.09	0.00	99.49
586	83	1261.00	37.80	3.66	13.99	28.33		0.15	4.26	6.85	3.29	1.68	0.08	0.34	100.43
587	92	66.00	51.89	0.66	22.22	7.12		0.07	2.39	11.37	2.55	0.70	0.06	0.96	99.99
588	92	80.00	48.54	1.34	15.82	14.54		0.17	6.67	8.30	2.10	0.55	0.10	−0.23	97.90
589	92	108.00	44.12	2.10	13.22	19.32		0.19	7.13	11.18	2.00	0.32	0.02	0.42	100.02
590	92	111.00	44.28	2.47	15.89	20.91		0.15	5.42	10.58	2.00	0.31	0.02	0.38	102.41
591	92	151.00	44.75	1.75	18.58	14.00		0.13	4.98	11.89	1.85	0.57	0.02	1.38	99.90
592	92	189.60	37.37	3.50	15.47	28.25		0.15	4.36	8.98	1.90	0.27	0.01	−0.23	100.03
592	92	192.00	41.54	2.56	14.08	23.44		0.16	6.01	10.31	1.86	0.30	0.02	−0.28	100.00
594	92	305.30	39.71	3.03	12.68	25.65		0.18	7.18	11.59	1.57	0.46	0.01	0.43	102.49
595	92	408.00	48.12	1.13	15.60	12.91		0.14	6.50	12.30	1.97	0.27	0.02	0.97	99.93
596	93	31.00	52.14	0.97	16.30	12.13		0.16	5.86	10.84	2.47	0.59	0.08	0.69	102.23
597	93	38.00	43.99	2.79	12.72	23.75		0.21	7.32	10.28	1.82	0.35	0.06	−0.45	102.84
598	93	51.20	48.55	1.92	13.95	17.88		0.18	6.18	10.18	2.29	0.52	0.07	0.00	101.72
599	93	67.20	49.51	1.86	13.94	17.58		0.20	6.65	10.66	2.12	0.55	0.09	−0.34	102.82
600	93	83.80	50.05	1.43	14.16	16.52		0.21	7.00	8.57	2.09	0.64	0.09	0.69	101.45
601	93	97.40	51.12	1.39	16.14	14.39		0.17	5.62	8.82	2.19	0.58	0.08	0.18	100.68
602	93	99.20	51.53	1.28	15.92	14.32		0.18	5.72	9.00	2.52	0.61	0.09	0.57	101.74
603	93	103.00	23.15	7.46	11.32	46.53		0.19	2.72	4.22	2.03	0.22	0.03	0.00	97.87
604	93	104.90	46.15	2.23	20.50	15.94		0.09	1.89	8.95	2.58	0.49	0.10	1.40	100.32
605	93	105.30	23.87	7.17	11.83	45.04		0.15	2.59	3.59	2.44	0.29	0.03	1.11	98.11
606	93	113.20	55.21	0.33	21.86	5.62		0.06	1.97	9.94	2.81	0.64	0.14	1.49	100.07
607	93	114.40	46.82	1.96	13.52	18.26		0.17	5.88	11.53	2.22	0.36	0.03	0.00	100.75
608	93	115.10	49.67	1.38	14.25	16.14		0.21	6.87	8.42	2.09	0.56	0.12	0.00	99.71
609	93	115.10	47.67	1.39	15.03	16.12		0.21	6.73	8.52	2.25	0.57	0.11	0.00	98.60
610	93	115.50	49.85	1.27	15.22	15.48		0.20	7.26	8.36	1.99	0.53	0.10	−0.28	99.98
611	93	119.30	48.68	1.74	19.51	12.95		0.09	2.67	10.85	2.91	0.61	0.11	0.39	100.51
612	93	119.30	49.17	1.77	19.79	13.23		0.10	2.68	11.28	3.11	0.62	0.12	0.39	102.26
613	93	123.30	43.06	2.80	18.84	20.89		0.13	2.58	10.28	2.31	0.52	0.02	0.99	102.42
614	93	129.20	52.90	0.76	21.78	6.93		0.07	2.10	11.24	2.76	0.82	0.03	1.37	100.76
615	93	138.50	41.64	2.83	18.35	20.88		0.14	2.66	10.72	2.79	0.59	0.04	0.00	100.64
616	93	141.00	38.49	3.30	8.09	29.98		0.22	7.70	12.44	2.34	0.26	0.04	1.90	104.76
617	93	141.90	44.64	2.57	18.66	20.54		0.11	2.63	9.91	2.37	0.55	0.08	−0.22	101.84
618	93	147.30	40.58	3.02	18.27	24.66		0.12	1.67	8.41	2.35	0.49	0.06	0.41	100.04
619	93	151.80	45.08	2.53	18.27	19.93		0.11	2.66	9.64	2.54	0.56	0.07	0.42	101.81
620	93	157.35	49.29	2.46	22.25	6.72		0.09	0.86	9.77	2.57	0.76	0.02	1.09	95.88
621	93	159.80	46.61	1.16	10.58	17.25		0.18	7.15	12.46	1.89	0.38	0.11	1.38	99.15
622	93	165.20	50.64	0.73	14.41	11.08		0.14	6.39	13.26	1.82	0.40	0.03	0.25	99.15
623	93	174.30	47.83	1.87	14.11	17.66		0.17	5.35	11.53	2.30	0.39	0.04	0.32	101.57
624	93	182.20	43.10	2.85	11.59	25.22		0.20	6.14	10.44	2.01	0.35	0.02	0.16	102.08
625	93	184.90	42.41	2.69	13.37	22.12		0.17	4.67	10.18	2.27	0.39	0.03	0.00	98.30
626	93	191.80	40.83	3.95	12.54	25.91		0.19	5.66	10.21	1.96	0.30	0.03	−0.31	101.27
6227	93	198.30	45.47	2.86	16.80	21.31		0.16	4.23	8.55	2.32	0.39	0.02	0.77	102.88
628	93	201.30	52.87	0.37	23.59	6.97		0.05	1.26	9.21	2.66	1.76	0.10	0.26	99.10
629	93	201.60	23.51	5.97	10.53	45.59		0.21	2.30	5.64	2.95	0.31	0.02	−0.55	96.48
630	93	202.00	33.39	3.61	15.00	30.97		0.12	1.59	6.60	2.46	0.37	0.04	0.90	95.05
631	93	205.10	40.65	2.06	12.61	24.81		0.15	5.29	9.75	1.95	0.31	0.02	1.07	98.67
632	93	207.00	50.87	0.73	8.80	14.93		0.19	9.09	13.62	1.48	0.42	0.05	1.87	102.05

(continued)

Table A2.1 (continued)

N Analysis	No borehole	Depth, cm	SiO$_2$	TiO$_2$	Al$_2$O$_3$	Fe$_2$O$_3$	FeO	MnO	MgO	Cao	Na$_2$O	K$_2$O	P$_2$O$_5$	LOI	Total
	1	2	3	4	5	6	7	8	9	10	11	12	13	14	15
633	93	209.30	43.77	2.53	11.67	24.98		0.19	5.89	10.48	2.27	0.49	0.05	1.99	104.31
634	93	210.00	46.31	1.28	14.63	16.97		0.15	5.60	10.84	2.17	0.57	0.02	1.45	99.99
635	93	212.70	40.08	2.12	13.23	25.45		0.16	4.64	9.46	2.15	0.33	0.02	2.38	100.02
636	93	216.80	44.26	2.96	14.66	22.32		0.17	4.95	10.25	2.17	0.36	0.04	0.30	102.44
637	93	221.30	39.27	4.19	11.91	28.40		0.18	5.33	10.10	2.05	0.28	0.02	0.66	102.39
638	93	224.40	43.82	3.80	13.63	21.88		0.17	5.48	10.91	2.19	0.32	0.02	−0.39	101.83
639	93	228.40	47.79	2.28	13.78	12.84		0.15	5.32	10.81	2.29	0.53	0.37	0.97	97.13
640	93	233.60	36.41	2.86	15.22	25.49		0.12	2.30	7.01	2.31	0.35	0.02	0.96	93.05
641	93	240.30	37.49	4.92	17.12	25.60		0.11	2.00	9.64	2.05	0.31	0.05	1.67	100.96
642	93	241.10	51.43	1.16	21.44	8.42		0.06	1.31	9.66	2.79	0.67	0.13	1.74	98.81
643	93	245.80	45.04	2.54	18.37	15.44		0.12	2.47	9.97	2.36	0.76	0.07	1.73	98.87
644	93	257.80	50.83	1.63	20.17	11.86		0.07	1.52	9.25	3.34	0.94	0.09	0.93	100.63
645	93	263.00	54.71	0.45	17.59	7.64		0.08	5.97	8.48	2.54	1.05	0.05	2.31	100.87
646	93	268.30	51.06	1.39	16.55	15.37		0.18	5.52	10.45	2.48	0.57	0.07	0.21	103.85
647	93	269.70	47.43	1.52	17.14	13.05		0.10	4.34	11.60	2.18	0.51	0.07	1.86	99.80
648	93	273.00	49.25	1.62	21.82	9.33		0.07	1.46	10.10	2.98	0.60	0.03	0.22	97.48
649	93	285.70	30.38	6.05	15.32	34.93		0.14	1.74	6.56	2.25	0.33	0.03	−0.29	97.44
650	93	290.20	33.16	5.72	17.51	32.41		0.12	2.03	6.50	2.23	0.29	0.03	0.00	100.00
651	93	292.60	50.22	1.22	15.89	12.66		0.12	5.07	11.45	1.95	0.50	0.05	1.56	100.69
652	93	295.20	36.98	5.22	17.06	29.73		0.13	1.64	7.54	2.74	0.36	0.03	0.26	101.69
653	93	303.40	34.65	5.14	16.17	30.11		0.13	1.63	7.47	2.13	0.32	0.03	−0.10	97.68
654	93	305.10	48.34	1.63	18.64	11.23		0.09	1.95	9.54	2.79	0.64	0.07	0.95	95.87
655	93	315.80	45.58	2.69	12.82	20.48		0.18	6.19	11.93	1.96	0.34	0.03	0.13	102.33
656	93	323.60	49.81	2.04	18.27	14.52		0.12	2.81	10.83	2.47	0.59	0.05	0.25	101.76
657	93	325.60	54.02	1.56	19.10	3.82		0.12	3.89	11.56	2.60	0.64	0.01	0.33	97.65
658	93	339.00	39.69	3.94	12.12	27.76		0.19	5.86	11.85	1.69	0.45	0.03	0.00	103.58
659	93	363.00	44.82	3.85	17.91	20.90		0.13	2.69	8.79	2.50	0.54	0.03	0.19	102.35
660	93	370.60	55.76	0.76	21.44	6.72		0.08	1.97	11.76	2.54	0.63	0.04	0.57	102.27
661	93	376.10	49.67	1.97	16.11	15.34		0.16	4.64	11.81	2.31	0.48	0.03	0.24	102.76
662	93	378.10	45.34	2.51	15.37	18.16		0.15	4.24	10.15	2.20	0.42	0.03	0.68	99.25
663	93	383.30	53.56	0.91	21.64	6.13		0.07	1.31	11.16	2.45	0.79	0.08	1.89	99.99
664	93	386.50	47.09	2.21	13.95	18.03		0.18	5.50	11.99	1.92	0.34	0.03	0.45	101.69
665	93	409.90	46.11	2.21	12.37	18.36		0.18	6.39	11.26	2.07	0.47	0.03	0.64	100.09
666	93	411.00	48.14	1.79	13.74	15.53		0.15	5.58	11.04	2.47	0.57	0.04	0.23	99.28
667	93	418.40	37.25	5.22	17.24	29.26		0.13	1.66	7.58	2.46	0.35	0.03	0.08	101.26
668	93	429.80	42.98	3.05	11.32	23.40		0.22	6.97	10.51	2.18	0.36	0.02	0.00	101.01
669	93	430.00	46.27	2.16	10.92	19.53		0.21	7.76	11.56	1.53	0.50	0.02	0.33	100.79
670	93	434.60	40.80	3.55	12.87	27.37		0.18	5.21	9.87	2.04	0.43	0.02	−0.08	102.26
671	93	444.10	38.51	4.13	14.61	26.65		0.19	3.37	8.29	2.11	0.46	0.02	0.70	99.04
672	93	451.70	49.53	1.55	21.03	10.46		0.07	1.09	10.13	2.76	0.76	0.09	2.80	100.27
673	93	454.00	42.86	2.03	14.87	19.71		0.16	4.98	8.89	2.51	0.64	0.06	0.48	97.19
674	93	464.20	42.68	3.54	14.27	23.11		0.17	5.00	8.74	2.14	0.53	0.02	1.78	101.98
675	93	532.60	48.05	1.84	15.39	16.62		0.17	6.14	11.79	2.05	0.41	0.03	−0.40	102.09
676	94	120.00	40.51	3.27	10.78	26.16		0.22	7.38	10.38	1.66	0.27	0.06	−0.65	100.04
677	94	121.60	46.08	2.02	14.35	19.06		0.20	6.92	8.98	1.92	0.48	0.07	−0.09	99.99
678	94	187.50	51.94	1.02	17.93	12.17		0.16	5.48	10.19	2.24	0.70	0.09	0.15	102.07
679	94	214.00	52.18	0.93	20.85	7.87		0.09	2.74	10.30	2.40	0.75	0.14	1.74	99.99
680	94	403.20	29.50	5.68	15.86	38.36		0.13	2.18	5.86	2.23	0.37	0.02	−0.20	99.99
681	94	454.70	44.80	2.22	12.66	19.02		0.19	7.04	11.58	1.70	0.38	0.03	0.40	100.02
682	94	478.00	44.34	2.64	16.35	19.81		0.16	4.80	9.65	2.05	0.41	0.03	−0.22	100.02
683	94	516.00	42.54	2.97	14.69	22.09		0.16	5.57	10.18	2.03	0.29	0.02	−0.54	100.00
684	94	542.20	47.46	1.81	15.08	16.20		0.17	5.63	11.52	2.11	0.47	0.05	−0.44	100.06

(continued)

Table A2.1 (continued)

N Analysis	No borehole	Depth, cm	SiO$_2$	TiO$_2$	Al$_2$O$_3$	Fe$_2$O$_3$	FeO	MnO	MgO	Cao	Na$_2$O	K$_2$O	P$_2$O$_5$	LOI	Total
	1	2	3	4	5	6	7	8	9	10	11	12	13	14	15
685	94	543.00	44.24	2.53	13.07	20.79		0.20	6.97	10.66	1.72	0.31	0.05	−0.52	100.02
686	94	571.40	49.58	1.50	15.85	15.29		0.18	6.57	11.62	2.14	0.41	0.03	−0.23	102.94
687	95	74.20	51.42	1.38	23.24	9.27		0.06	0.87	9.47	2.91	0.45	0.09	0.85	100.01
688	95	97.50	53.03	0.26	23.59	3.43		0.04	0.93	9.37	3.12	0.75	0.07	1.52	96.11
689	95	125.20	31.20	5.34	16.28	36.99		0.15	2.34	6.73	2.24	0.32	0.02	−1.60	100.01
690	95	143.50	47.33	1.64	15.76	15.69		0.17	6.12	11.00	2.05	0.40	0.05	−0.24	99.97
691	95	159.50	46.01	2.09	10.84	21.04		0.23	9.37	10.84	1.56	0.30	0.03	−0.22	102.09
692	95	172.50	38.36	3.80	19.49	25.91		0.11	1.66	8.08	2.42	0.38	0.02	−0.18	100.05
693	95	192.50	47.18	1.51	12.05	15.60		0.21	8.60	11.28	1.76	0.29	0.02	−0.56	97.94
694	95	202.50	46.53	1.97	21.84	15.32		0.09	1.61	9.71	2.39	0.42	0.05	0.00	99.93
695	95	214.00	36.79	3.35	13.66	29.70		0.17	5.39	6.47	2.06	0.88	0.03	1.51	100.01
696	95	250.70	41.68	2.83	12.79	23.73		0.20	6.67	10.00	1.72	0.35	0.02	0.00	99.99
697		4138	47.61	1.6	19.59	13.42		0.14	1.95	9.29	3.67	0.85	0.18		98.30
698		4151	5.1	11.16	5.29	72.35		0.22	1.7	0.94	1.35	0.04	0.08	1.5	99.73
699		8430	48.3	2.1	12.9	18.15		0.18	4.47	8.82	2.96	0.09	0.08		98.05
700		522	42.44	2.93	11.37	21.68		0.22	6.63	10.74	2.38	0.62	0.10	1.16	100.27
701		521	43.61	2.49	13.07	20.10		0.18	5.79	10.20	2.79	0.92	0.09	0.06	99.30
702		520	44.98	2.48	13.62	19.87		0.18	5.80	11.00	2.96	0.67	0.11	0.18	101.85
703		519	43.09	2.69	12.42	20.05		0.21	6.91	12.11	2.88	0.42	0.07	0.00	100.85
704		518	42.94	2.68	13.21	19.23		0.20	6.68	11.52	2.87	0.39	0.11	0.38	100.21
705		517	45.58	2.29	13.87	17.84		0.19	6.44	11.96	2.79	0.47	0.07	0.34	101.84
706		516	44.83	2.31	12.85	17.70		0.19	6.52	12.21	2.64	0.52	0.06	0.14	99.97
707		515	39.77	3.34	9.88	23.49		0.22	7.77	11.93	2.28	0.35	0.08	0.29	99.40
708		514	51.42	0.61	13.91	9.99		0.21	11.42	8.93	2.40	0.35	0.06	0.44	99.74
709		513	47.97	1.59	23.69	9.77		0.06	1.27	10.34	4.07	0.73	0.08	1.35	100.92
710		512	43.40	2.15	18.87	17.30		0.11	3.03	9.67	4.35	0.68	0.09	0.20	99.85
711		511	46.16	1.64	14.54	15.88		0.16	6.16	12.26	3.71	0.39	0.07	0.38	101.35
712		510	52.61	0.31	24.14	4.93		0.05	1.10	9.90	4.14	1.25	0.28	1.62	100.33
713		509	46.15	2.17	14.86	16.16		0.02	5.65	12.20	2.96	0.44	0.07	0.16	100.84
714		508	47.98	1.66	12.06	15.85		0.21	7.47	12.08	3.39	0.60	0.12	0.00	101.42
715		507	46.11	1.89	14.26	15.35		0.16	6.07	11.31	3.03	0.62	0.10	0.20	99.10
716		506	48.67	1.49	14.05	14.15		0.20	7.12	11.41	3.00	0.49	0.07	0.17	100.82
717		505	45.77	2.02	22.29	13.69		0.07	1.28	9.00	4.13	1.00	0.09	1.32	100.66
718		504	46.82	1.79	14.94	15.97		0.18	6.13	10.79	3.93	0.61	0.08	0.18	101.42
719		503	38.09	3.77	8.07	26.59		0.29	9.28	11.74	2.46	0.28	0.05	−0.47	100.15
720		502	43.45	2.43	13.76	18.46		0.20	6.17	11.34	3.42	0.53	0.11	0.00	99.87
721		501	45.98	2.01	14.10	16.29		0.19	6.30	11.38	2.89	0.51	0.08	0.11	99.84

Note. Empty cell – not analysed. Analeses were carried out at CHIPR, analysts N Baluev and V Subbotina

Appendix 3

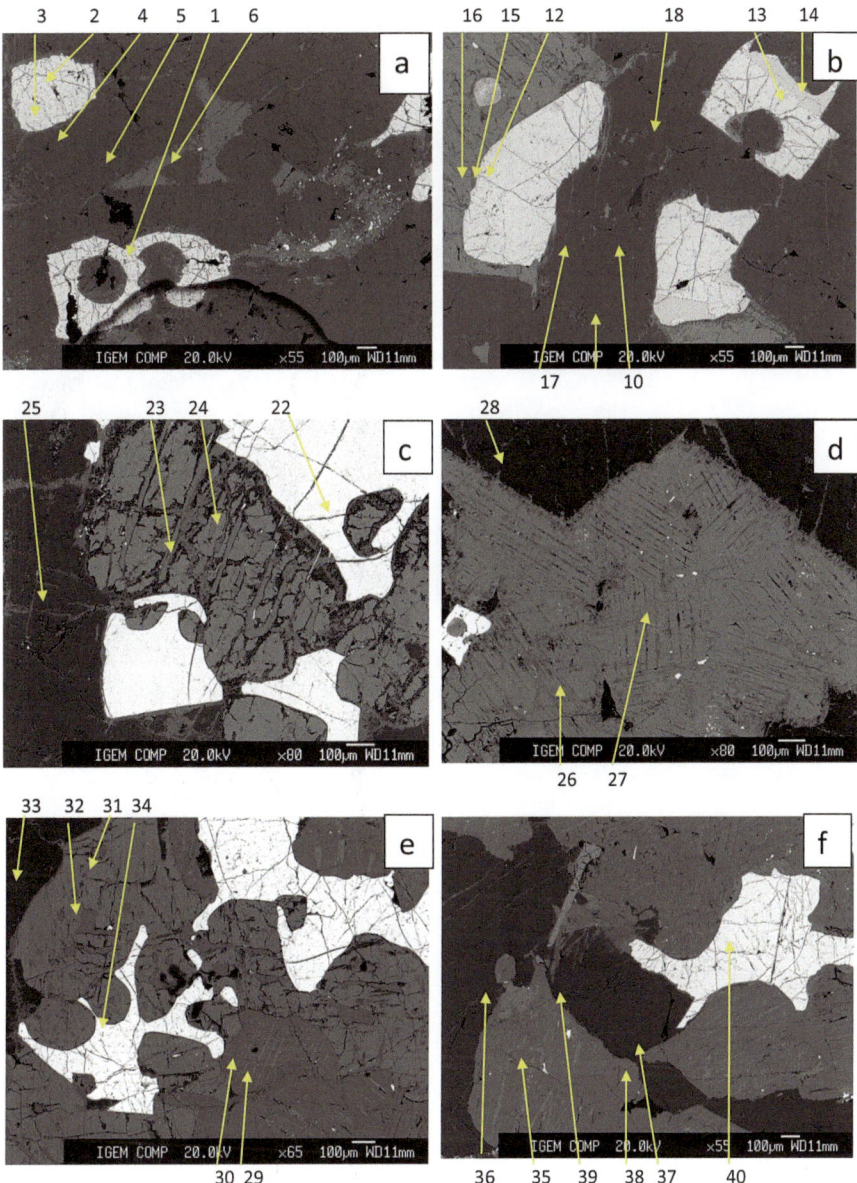

Fig. A3.1 BSE images of titanomagnetite grains in rocks from the Chiney pluton penetrated by borehole 83 Samples, depth in borehole 83, m: a-121, b-174, c-240, d-320, e-400, f-500, h-600, g-760 I-780, j-848, k-1036, l-1100, m-1101, n-1285. Number points correspond numbers in Table A3.1-A3.4

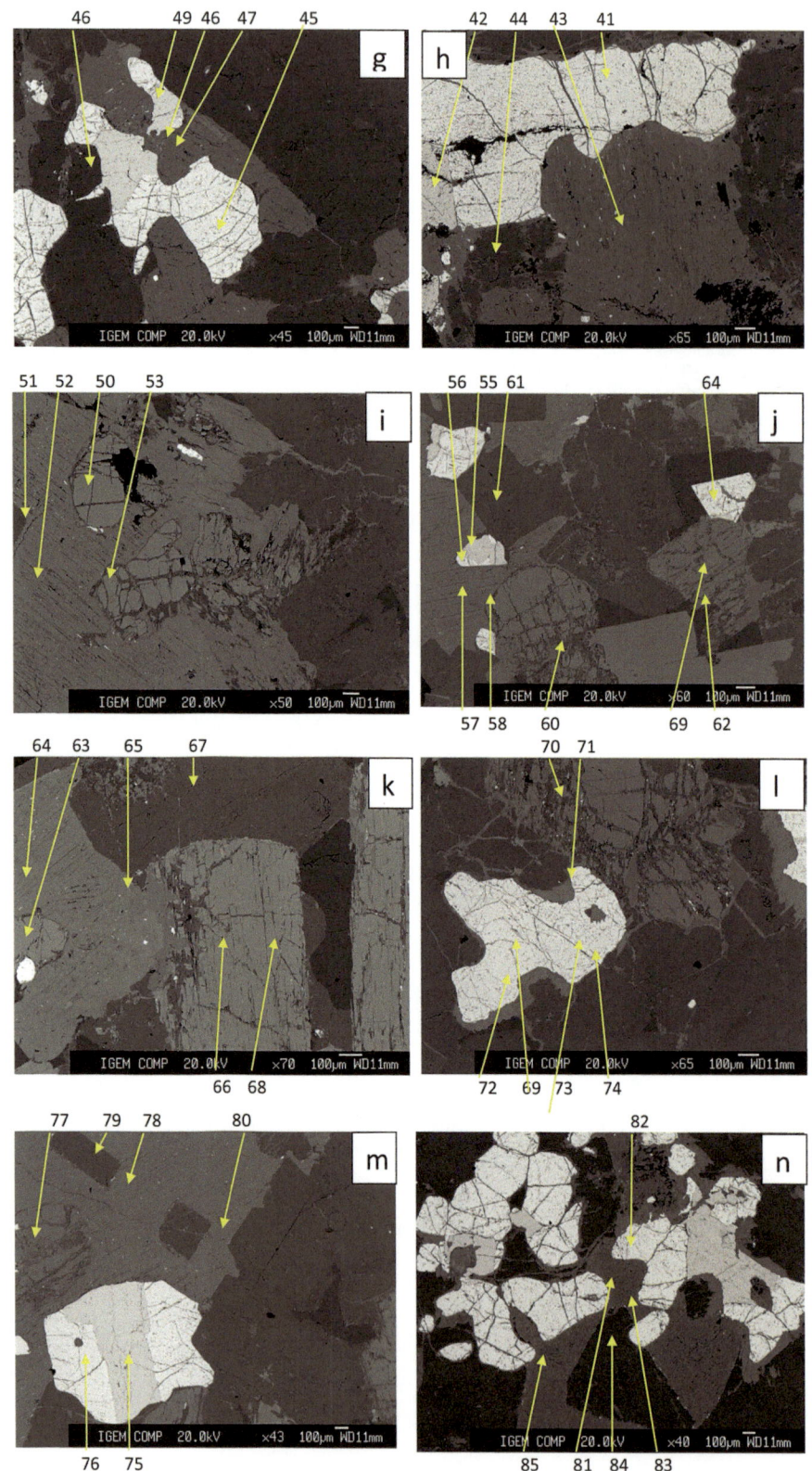

Fig. A3.1 (continued)

Table A3.1 Chemical composition of plagioclase from the Etyrko deposit (Borehole 83), wt%

No point in Fig. A3.1	SiO_2	Al_2O_3	FeO	CaO	Na_2O	K_2O	Total	Depth, m
4	53,71	28,83	0,41	11,57	4,80	0,47	99,79	121
6	53,32	28,51	0,36	11,25	5,02	0,54	99,00	121
17	54,28	28,13	0,34	10,74	5,61	0,46	99,56	174
18	53,98	28,24	0,36	10,86	5,29	0,42	99,15	174
20	54,25	28,09	0,44	10,73	5,55	0,41	99,46	174
25	52,83	28,89	0,42	11,58	4,95	0,41	99,08	240
28	54,32	28,65	0,26	10,92	5,31	0,53	99,99	320
33	53,44	29,11	0,29	11,62	4,92	0,45	99,84	400
39	53,73	28,40	0,31	11,05	5,22	0,43	99,14	500
44	54,22	28,55	0,30	10,82	5,26	0,38	99,51	600
49	53,40	28,61	0,32	11,33	5,02	0,48	99,15	760
52	51,80	28,71	0,31	12,09	4,57	0,37	97,85	780
61	54,93	28,38	0,28	10,70	5,73	0,35	100,38	848
62	54,77	29,10	0,27	11,13	5,57	0,29	101,13	848
67	55,64	29,07	0,28	10,90	5,67	0,35	101,90	1036
68	56,02	28,91	0,25	10,42	6,02	0,27	101,88	1036
73	53,16	29,44	0,20	11,80	5,09	0,20	99,90	1100
74	56,80	26,99	0,16	8,95	6,69	0,27	99,86	1100
79	53,15	29,67	0,27	11,98	4,83	0,34	100,22	1101
80	53,12	29,46	0,28	11,92	4,87	0,38	100,04	1101
85	53,25	29,49	0,24	11,89	4,92	0,34	100,13	1285

Table A3.2 Chemical composition of clinopyroxene from the Etyrko deposit (Borehole 83), wt%

No point in Fig. A3.1	SiO_2	TiO_2	Al_2O_3	FeO	MnO	MgO	CaO	Na_2O	K_2O	Total	Depth, m
5	50,96	0,52	1,58	12,91	0,34	13,50	19,30	0,22	0,01	99,32	121
16	50,63	0,50	1,45	11,33	0,38	12,93	21,45	0,22	0,00	98,87	174
21	49,78	0,57	2,43	11,06	0,34	12,75	21,29	0,37	0,00	98,59	200
24	51,28	0,55	1,54	10,34	0,25	13,63	21,85	0,25	0,01	99,69	240
26	50,77	0,58	1,85	11,24	0,27	12,71	21,86	0,27	0,00	99,56	320
27	51,05	0,37	1,54	11,04	0,31	12,89	21,85	0,30	0,00	99,35	320
29	50,99	0,44	1,52	10,37	0,27	13,38	21,73	0,25	0,01	98,94	400
32	50,98	0,52	1,68	11,63	0,25	13,54	20,55	0,24	0,02	99,40	400
35	50,78	0,54	1,82	11,81	0,31	12,24	21,65	0,33	0,00	99,49	500
37	50,02	0,55	2,04	12,72	0,29	12,89	20,11	0,25	0,00	98,87	500
43	50,65	0,57	2,33	9,36	0,26	13,47	21,68	0,35	0,00	98,66	600
47	50,55	0,47	1,72	10,58	0,32	12,99	21,45	0,29	0,00	98,36	760
51	51,03	0,51	1,99	11,36	0,31	13,02	20,70	0,39	0,00	99,31	780
57	50,82	0,60	2,51	9,68	0,17	13,10	21,92	0,46	0,00	99,26	848
64	51,75	0,25	1,34	9,88	0,22	13,89	22,40	0,34	0,00	100,06	1036
65	49,64	0,34	4,94	13,06	0,17	15,34	11,77	0,71	0,34	96,32	1036
71	51,23	0,52	3,12	10,23	0,21	17,42	11,73	0,60	0,28	95,34	1100
78	52,15	0,32	1,41	19,08	0,40	17,85	9,44	0,13	0,00	100,77	1101
84	50,31	0,40	2,03	8,49	0,35	13,64	22,18	0,42	0,00	97,84	1285

Table A3.3 Chemical composition of orthopyroxene from the Etyrko deposit (Borehole 83), wt%

Nopoint in Fig. A3.1	SiO_2	TiO_2	Al_2O_3	FeO	MnO	MgO	CaO	Total	Depth, m
15	50,45	0,30	0,77	25,44	0,57	19,27	1,16	97,98	174
23	51,21	0,26	0,80	22,91	0,56	21,48	1,11	98,32	240
30	51,02	0,22	0,95	24,56	0,58	20,03	1,77	99,14	400
31	51,61	0,26	0,83	24,17	0,57	20,59	1,04	99,07	400
36	51,08	0,27	0,72	25,04	0,58	17,71	3,86	99,26	500
38	50,91	0,15	0,86	27,28	0,63	18,53	0,87	99,23	500
50	50,52	0,56	0,64	24,64	0,52	19,02	2,57	98,46	780
53	51,91	0,32	0,72	25,78	0,53	19,90	0,96	100,12	780
58	51,70	0,29	0,89	22,87	0,60	21,70	1,15	99,19	848
59	52,10	0,22	0,95	22,49	0,52	21,41	1,80	99,48	848
63	51,85	0,25	0,94	24,03	0,48	20,85	1,10	99,49	1036
66	52,29	0,26	1,13	22,47	0,50	22,16	1,41	100,22	1036
70	51,93	0,32	1,25	19,60	0,40	22,78	2,04	98,31	1100
77	51,51	0,15	1,26	20,31	0,43	22,16	2,85	98,67	1101

Table A3.4 Chemical compositions of iron-titanium oxides from the Etyrko deposit (Borehole 83), wt%

No point in Fig. A3.1	SiO$_2$	TiO$_2$	Al$_2$O$_3$	FeO	MnO	MgO	ZnO	V$_2$O$_3$	Cr$_2$O$_3$	Total	Depth, m
1	0,01	5,87	0,13	87,65	0,10	0,02	0,00	1,50	0,06	95,33	121
2	0,13	12,98	3,23	79,25	0,32	0,18	0,36	1,22	0,04	97,71	121
3	0,12	7,12	12,43	77,78	0,19	0,03	0,03	1,77	0,24	99,70	121
7	0,09	23,29	0,14	69,82	0,54	0,13	0,02	1,17	0,18	95,37	121
8	0,17	48,14	0,13	45,38	0,86	0,26	0,04	0,80	3,13	98,89	121
9	0,18	11,93	0,18	81,01	0,23	0,03	0,00	1,19	0,27	95,01	121
10	0,04	11,85	0,69	82,85	0,23	0,03	0,04	0,90	0,13	96,75	174
11	0,08	51,27	0,00	49,37	1,10	0,11	0,03	0,06	0,07	102,08	174
12	0,04	12,70	1,83	80,20	0,39	0,05	0,01	0,91	0,08	96,23	174
13	0,08	4,76	0,99	89,76	0,15	0,05	0,07	0,81	0,04	96,71	174
14	0,01	51,48	0,00	48,40	1,10	0,09	0,00	0,05	0,00	101,12	174
19	0,08	8,95	0,20	84,20	0,50	0,00	0,00	0,96	0,03	94,92	200
22	0,14	11,18	3,97	79,28	0,48	0,06	0,06	1,08	0,01	96,27	240
34	0,04	2,49	0,88	90,14	0,09	0,01	0,02	1,38	0,05	95,09	400
40	0,01	3,99	1,28	90,45	0,11	0,05	0,06	1,47	0,04	97,46	500
41	0,06	9,13	0,76	84,65	0,24	0,02	0,10	1,42	0,01	96,39	600
42	0,01	47,96	0,01	50,41	1,25	0,04	0,04	0,24	0,02	99,98	600
45	0,05	4,40	0,49	88,85	0,13	0,03	0,03	1,53	0,10	95,60	760
46	0,04	50,67	0,02	48,73	1,47	0,08	0,00	0,21	0,02	101,24	760
54	0,04	0,24	0,39	92,17	0,02	0,00	0,05	1,23	0,19	94,32	848
55	0,05	46,87	0,04	52,84	1,80	0,13	0,00	0,24	0,05	102,02	848
56	0,03	0,23	0,26	92,67	0,03	0,02	0,00	1,10	0,29	94,62	848
69	0,09	0,54	0,22	93,01	0,03	0,01	0,00	1,05	0,13	95,07	1100
75	0,04	1,42	0,34	92,01	0,06	0,01	0,04	1,22	0,17	95,31	1101
76	0,02	47,77	0,07	51,96	1,39	0,22	0,04	0,19	0,00	101,65	1101
81	0,03	0,52	0,25	92,17	0,04	0,04	0,03	1,38	0,27	94,72	1285
82	0,00	43,06	0,06	56,27	1,58	0,09	0,00	0,13	0,04	101,21	1285